T0146051

Snakes of Central and Western Africa

Snakes of Central and Western Africa

Jean-Philippe Chippaux and Kate Jackson

 Johns Hopkins University Press · Baltimore

© 2019 Johns Hopkins University Press
All rights reserved. Published 2019
Printed in Canada on acid-free paper
9 8 7 6 5 4 3 2 1

Johns Hopkins University Press
2715 North Charles Street
Baltimore, Maryland 21218-4363
www.press.jhu.edu

Library of Congress Cataloging-in-Publication Data

Names: Chippaux, Jean-Philippe, author. | Jackson, Kate, 1972– author.
Title: Snakes of Central and Western Africa / Jean-Philippe Chippaux and Kate
    Jackson.
Description: Baltimore : Johns Hopkins University Press, 2019. | Includes
    bibliographical references and index.
Identifiers: LCCN 2018023756 | ISBN 9781421427195 (hardcover : alk. paper) |
    ISBN 1421427192 (hardcover : alk. paper) | ISBN 9781421427201 (electronic)
    | ISBN 1421427206 (electronic)
Subjects: LCSH: Snakes—Africa, Central—Identification. | Snakes—Africa,
    West—Identification.
Classification: LCC QL666.O6 C5352 2019 | DDC 597.960967—dc23
LC record available at https://lccn.loc.gov/2018023756

A catalog record for this book is available from the British Library.

Frontispiece: *Dipsadoboa viridis* defensive posture, Republic of Congo.
By K. Jackson

*Special discounts are available for bulk purchases of this book. For more information,
please contact Special Sales at 410-516-6936 or specialsales@press.jhu.edu.*
Johns Hopkins University Press uses environmentally friendly book materials,
including recycled text paper that is composed of at least 30 percent post-
consumer waste, whenever possible.

For Monique Bourgeois, in honor of her paradigm-shifting recognition of the true phylogenetic affiliations of *Atractaspis*, an independently derived lineage of front-fanged venomous snakes previously assumed to be vipers. Bourgeois made this discovery using graph paper and a dissecting microscope, at the University of Lubumbashi in the early 1960s, decades before technological advances made it possible for molecular systematics to prove her right.

Contents

Acknowledgments

We acknowledge the following people and institutions for their help in making this book possible. We thank the National Science Foundation (grant 1145437 to KJ) and Whitman College for funding. We thank the museums whose collections we referred to in the course of research for this book, and the curators and collections managers who facilitated our efforts, particularly: Garin Cael and Danny Meirte at the Royal Museum of Central Africa, Tervuren; Jeremy Jacobs, Roy McDiarmid, James Poindexter, Rob Wilson, Addison Wynn, and George Zug at the Smithsonian Institution, Washington, DC; Alan Resatar at the Field Museum of Natural History, Chicago; Patrick Campbell and Colin McCarthy at the Natural History Museum, London; David Kizirian and Chris Raxworthy at the American Museum of Natural History, New York; Andreas Schlüter and Günter Stephan at the Staatliches Museum für Naturkunde, Stuttgart; Ivan Ineich, Laure Paul, and Nicolas Vidal at the Museum National d'Histoire Naturelle, Paris. We thank the students from Whitman College (United States) and from Université Marien Ngouabi (Republic of Congo) who worked with KJ on various aspects of this project: Jordan Benjamin, Sylvestre Boudzoumou, Nat Clarke, Evan Conner, Andrew Hill, Eric Hsu, Willie Kunkel, Lise-Bethy Mavoungou, Ange Mboungou-Louiki, Kevin Moore, Katie Moyer, Anna Ripley, Claire Snyder, Ange-Ghislain Zassi-Boulou. We thank colleagues and coworkers who lent their expertise in geography and geographic information system (GIS) technology to aspects of this project: Austun Ables, Nick Bader, Bob Carson, Amy Molitor. We thank friends and colleagues who gave feedback and suggestions as we developed this book, and we thank fellow herpetologists and photographers of African snakes who allowed us to use their photographs in our book. We thank Tuhin Giri for applying his artistic talent and eye for scientific accuracy to the head scale drawings used throughout this book. We thank the peer reviewer for constructive comments and suggestions provided. Finally, we thank our editor, Tiffany Gasbarrini, and the staff of Johns Hopkins University Press for bringing it all together.

Jean-Philippe Chippaux and Kate Jackson

The decade since I began work on this book has seen both the death of my father and the birth of my son. Some may wonder that I did not choose to dedicate this book to the memory of my father, J. R. de J. Jackson, a professor of English literature specializing in poetry of the romantic period. Those who knew him, however, and of his scholarly efforts to bring to light the overlooked work of female poets (1770–1835), will see that my co-author and I have instead dedicated this book in a manner he would have wholeheartedly supported. Finally, I thank Andrea and Robin for keeping me away from extended field expeditions for the past several years, thereby making possible the completion of this book.

Kate Jackson

Snakes of Central and Western Africa

Identification of African Snakes

This book is intended to enable the reader to correctly identify snakes from central and western Africa. In the chapters that follow, instructions for identification, in the form of dichotomous keys and genus and species descriptions, are written in a specialized and precise vocabulary used for the description of snakes. It is a vocabulary with a grand history, with many of its terms having been in use for centuries. Although these terms are much used, they are seldom explained. We devote a chapter to defining and explaining the descriptive characters used throughout the book. Most of these are scale characters, of which the majority refers to the scales of the head. Our goal is to provide precise and accurate definitions of traditional characters, and especially to present enough examples of variations in these characters to prepare the reader for the diversity of form displayed by almost half a continent of snakes. Our hope is that this information may be of use and interest to both the novice and the specialist.

What Is and What Is Not a Snake?

It is worth keeping in mind that several other types of elongate, limbless animals besides snakes are found in central and western Africa (Fig. 1.1).

Terminology Basics

Anatomical directions help to explain the scale characters used to identify snakes. "Anterior" means in the direction of the snout; "posterior" toward the tip of the tail; "ventral" toward the underside; "dorsal" toward the back, or upper side of a snake lying on its belly; "lateral" toward the sides; and "medial" toward the snake's midline. Snakes are bilaterally symmetrical animals, and descriptive characters (such as scales or markings) are "median" if the scale or marking is on the snake's midline so that there is just one of it, or "paired" when an identical scale or marking appears on both sides of the snake. Note that paired structures are generally written about as if there were only one. For example, a snake that has 8 scales along either side of its upper lips is said to have 8 upper labials, not 16. The statement "There are 2 postoculars" (postoculars are scales immediately posterior to the eye) means that there are 4 postoculars in total—2 on each side of the head.

The Body

Ventral Scales

Ventral scales ("ventrals") are the broad scales that cover the underside of the snake's body. Ventrals are counted starting from the first ventral on the throat. Dowling (1951) defined the first ventral scale as the

FIGURE 1.1. Snake-like animals of central and western Africa. *A* shows a caecilian, *Herpele squalostoma*, an elongate, limbless amphibian. Though superficially similar to a snake in shape, its skin is moist and slimy, not dry and scaly like that of a reptile. Caecilians are burrowing creatures that prefer wet places and are rarely encountered above ground except after heavy rain. *B* shows an amphisbaenian, *Monopeltis guentheri*, an elongate, limbless burrowing lizard seldom found above ground. Amphisbaenians have dry, scaly skin, but unlike snakes their scales are arranged in distinctive rings, encircling the whole body. *C* shows a legless lizard, *Feylinia currori*, another burrowing reptile. It is easily distinguishable from nearly all snakes by the fact that the scales are the same size all around its body rather than having broad scales covering its underside and small ones on its back and sides. It is distinguishable from absolutely all snakes by the presence of eyes with move-able eyelids and external auditory openings. *D* shows a snake. This is the Blind Snake, *Afrotyphlops lineolatus*, a small burrowing creature like the others shown here. Blind Snakes belong to the Scolecophidia, a group distantly related to other snakes and not covered in detail in this book. Blind Snakes differ from other snakes in many ways, for example, in having scales of equal size all around their bodies rather than enlarged ventral scales along their underside. They may superficially resemble limbless burrowing lizards such as *F. currori*, but they differ in that they lack eyelids of any kind. The eyes of the Blind Snake are greatly reduced and covered by the scales of the head. *A*, by A. Kupfer; *B*, by S. Boudzoumou; *C*, by K. Jackson; *D*, by S. Spawls

first to line up with the first dorsal scales. This method of counting ventrals has not been used as consistently by researchers working with African snakes as it has in North America, so we also mention here the simpler traditional definition (Thompson 1914) in which the first ventral is said to be the first scale on the underside of the snake that is broader than it is long. Continuing from that first ventral, the ventrals are counted posteriorly. When counting ven-trals, it is a good idea to stick a pin in a scale at regular intervals (e.g., every 10 or every 50 scales) so that if a distraction causes you to lose count, you won't have to start counting all over again.

Cloacal Scale

The last ventral scale is the one immediately before the cloacal scale (also called the anal plate), the scale that covers the cloaca (the cloaca is the single, combined opening of the digestive, urinary, and reproductive tracts). The cloacal scale may be single ("entire") or divided. If it is divided, the division is asymmetrical, with the scale on one side overlapping the other. This division is often hard to see, but it can be found by running a dissecting needle along the edge of the cloacal scale while viewing the cloacal region with a microscope, and by shifting the snake slightly in one's grip so that the light catches the line of contact between the two parts of the cloacal scale.

Subcaudal Scales

Just as the ventrals cover the underside of the body, the subcaudal scales (subcaudals) cover the underside of the tail. Like the cloacal scale, the subcaudals may be single or divided, or in rare cases single for part of the tail and divided for another part of the tail. When counting divided subcaudals, each pair counts as one, so you shouldn't end up with twice as many subcaudals if they are divided as you would if they were single. When counting subcaudals, it is a good idea to start counting at the tip, working your way toward the body. This is because you are more likely to lose count among the small scales at the tip of the tail than among the larger scales closer to the cloacal scale, and it is less frustrating to lose count in the first dozen scales than in the last few, after you have counted a hundred or more subcaudals. The scale at the very

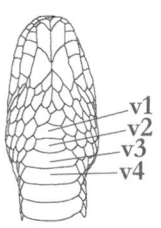

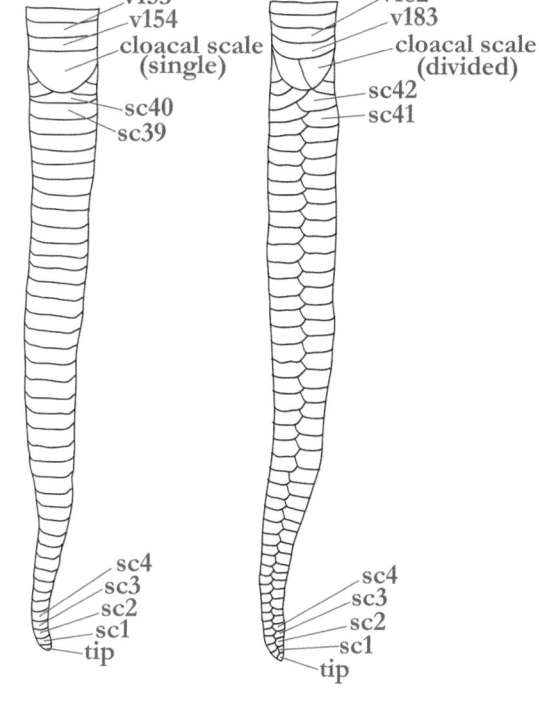

FIGURE 1.2. Ventral view of the head and tail, showing the ventrals (v) and subcaudals (sc). Single subcaudals (*left*) and divided subcaudals (*right*). By K. Jackson from a drawing by W. Kunkel

tip of the tail does not count as a subcaudal. When counting subcaudals, be sure to examine the tip of the tail for evidence that the end of the tail has been broken off or damaged. If it has, the number of subcaudals you count will be fewer than for an intact snake.

Dorsal Scales

Dorsal scales are the small (relative to the ventrals) scales that cover the back and sides of the snake from the neck to the end of the tail.

Number of Dorsal Scale Rows

The dorsal scales are arranged in longitudinal rows along the snake. The number of rows is counted around the body, starting with the first dorsal in contact with a ventral on one side and proceeding over the back and down the other side, so that the last scale counted is the dorsal scale in contact with a ventral on the other side of the snake. In some snakes, the number of dorsal scale rows varies along the length of the snake. Unless otherwise specified, the number of dorsal scale rows refers to the number at midbody (roughly halfway between the neck and the cloaca). Sometimes numbers of dorsal scale rows are given in the format "17-19-19," indicating the number of dorsal scale rows at three places (in order): the neck, midbody, and a bit before the cloaca. Figure 1.3 shows dorsal scale rows being counted while viewed from the side.

Instructions for counting dorsal scale rows are usually depicted as though the snake's skin had been cut down the ventral midline and spread out, so that all the dorsals on both sides can be seen, as in Figure 1.4.

Straight versus Oblique Dorsal Scale Rows

Dorsal scales rows are described as "straight" when they are arranged so that the scales are generally non-overlapping and appear symmetrical. Dorsal scales may also be arranged in closely overlapping rows, giving the individual scales an asymmetrical appearance, a condition described as "oblique." The distinction between straight and oblique is not always clear-cut. The arrangement of scales may appear to lie somewhere between the two extreme examples shown below. A further complication is presented by snakes whose dorsal scales are oblique on part of their body but straight in other places. As a taxonomic character, its use is best avoided in such ambiguous cases.

Smooth versus Keeled Dorsal Scales

An individual dorsal scale may have a ridge extending lengthwise along its midline. Such scales are described as "keeled," and the overall effect is to give the snake's skin a rough appearance. When no such ridge is present, the scales are described as "smooth." A snake with smooth scales looks

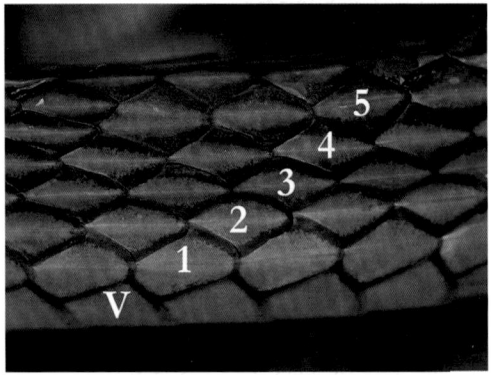

FIGURE 1.3. Dorsal scale rows of the Emerald Snake, *Hapsidophrys lineatus*, in lateral view (v = ventral). By K. Jackson from a photo by K. Moore

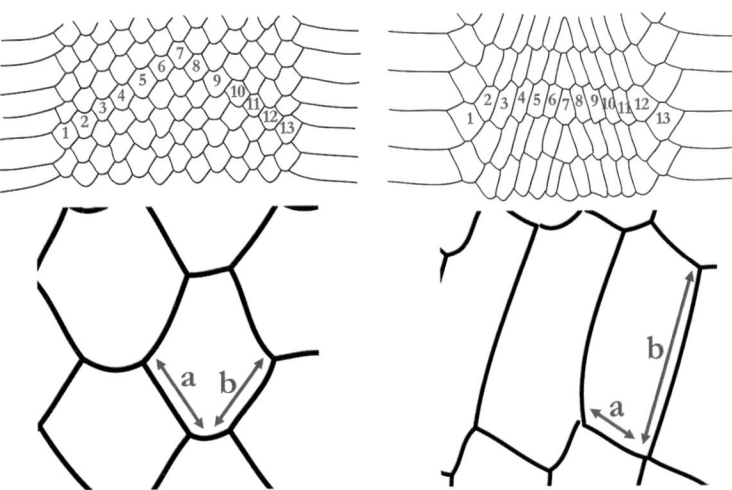

FIGURE 1.4. Straight (*left*) versus oblique (*right*) dorsal scale rows. Magnified view (*lower left*) shows straight scales with posterior edges a and b roughly equal in length. Magnified view (*lower right*) shows oblique scales with difference in length between posterior edges (a < b). By K. Jackson from a drawing by W. Kunkel

generally smooth and glossy. Figure 1.5 shows close-ups of one snake with smooth scales and one with keeled scales.

Apical Pits

Under a microscope, it is sometimes possible to see one (or more) small dots near the tip of each dorsal scale. The function of these pits is not certain. One suggestion is that they serve a structural function, helping to anchor the internal and external epidermal layers and thus preventing their separating prematurely during ecdysis (Chiasson 1981). Another possibility is that they serve a sensory function, since the apical pits are particularly rich in nerve endings, suggesting that they might be sensors for touch, light, or possibly heat, similar to the labial pits seen in some Boidae (Gray 2011). Though a useful taxonomic character, apical pits may be difficult to see, even under a microscope. Turning the snake under the microscope, so as to let the skin catch light from different angles, sometimes helps to reveal them (Fig. 1.6*E*).

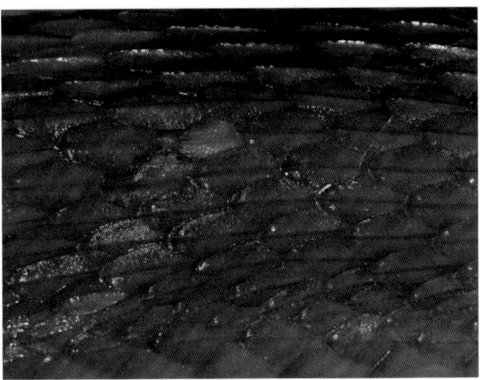

FIGURE 1.5. *Top*, smooth dorsal scales in the House Snake, *Boaedon olivaceus*. *Bottom*, keeled dorsal scales in the Egg-eating Snake, *Dasypeltis atra*. By K. Moore

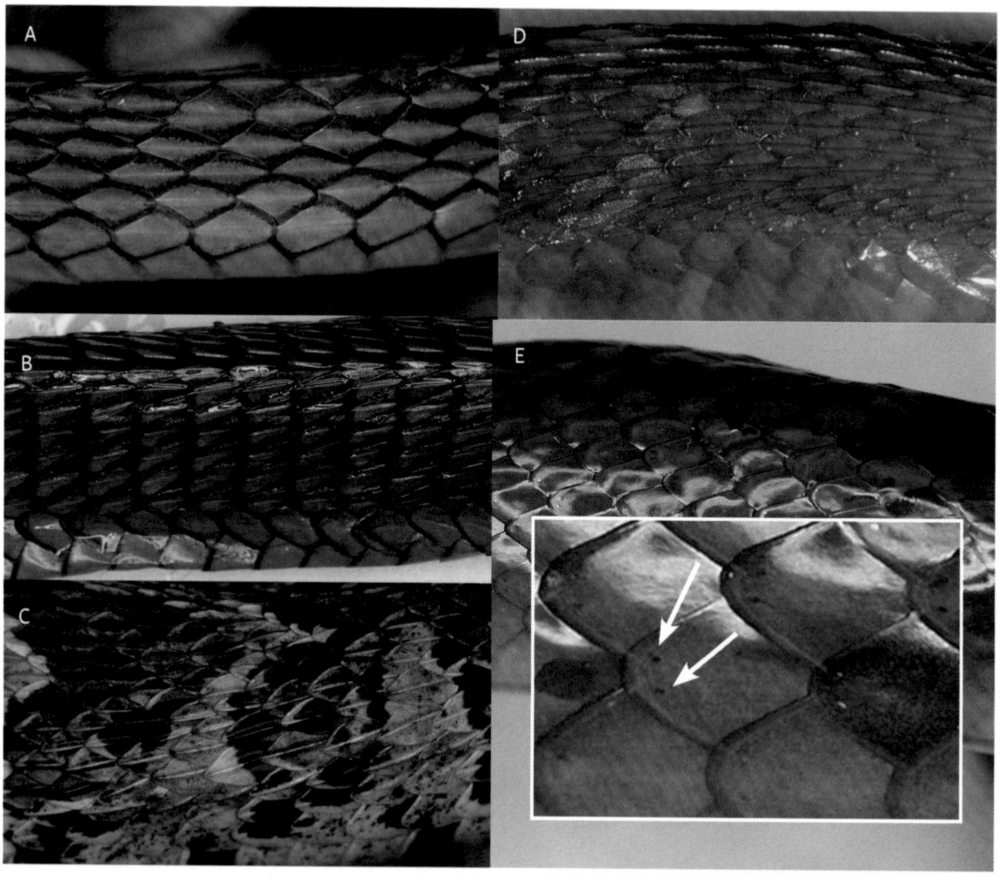

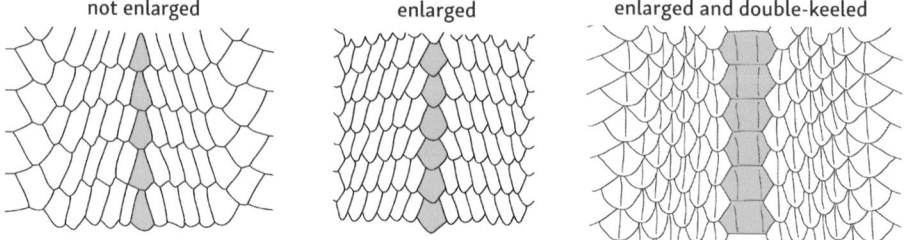

FIGURE 1.6. Dorsal scale examples of variation:
(*A*) the Emerald Snake, *Hapsidophrys lineatus*, has smooth scales arranged in straight rows; (*B*) the Boom-slang, *Dispholidus typus*, has smooth scales that are unambiguously oblique; (*C*) the Puff Adder, *Bitis arietans*, has keeled scales that are technically slightly oblique, though they could easily be misinterpreted as straight; (*D*) the Egg-eating Snake, *Dasypeltis atra*, has keeled scales that are, like the Puff Adder, so slightly oblique that they resemble the straight rows of the Emerald Snake more than the oblique rows of the Boomslang; and (*E*) the House Snake, *Boaedon fuliginosus*, has smooth scales arranged in straight rows. The arrows in the inset indicate the two apical pits on each scale. By K. Jackson using photos by K. Moore

not enlarged   enlarged   enlarged and double-keeled

FIGURE 1.7. Variation in the vertebral row. By K. Jackson from a drawing by W. Kunkel

Enlarged Vertebral Row

The vertebral row is the row of dorsal scales along the midline of the snake's back. Sometimes the scales in this row are enlarged relative to the other dorsal scales. In snakes with oblique dorsal scales, the scales of the vertebral row may appear larger than the surrounding scales because there is less overlap over them than the scales on the snake's sides (Fig. 1.7, *left*). In such a case the vertebral row is not considered enlarged. By comparison (Fig. 1.7, *middle*), the vertebral row is considered enlarged because the scales of the dorsal midline are larger than the other dorsal scales. More extreme examples exist (Fig. 1.7, *right*) where the scales of the vertebral row are not only greatly enlarged relative to the other dorsal scales, but also distinctive in being "double keeled," meaning there are two parallel ridges on each scale.

Keeled Ventrals and Subcaudals

Dorsals are not the only scales that can be keeled. Characters that occasionally come up, particularly in arboreal snakes, are keeled ventrals and keeled subcaudals. Ventrals are considered to be keeled when, instead of being rounded in cross section, a crease extends along each side of the underside of the snake, formed by a bend or fold in each side of the ventrals. The term "angulate" is sometimes used to describe weakly keeled ventrals. Figure 1.8 shows a cross section of a snake with normal, rounded ventrals (*left*) and keeled ventrals (*right*). The same principle can be applied to subcaudals.

Color Patterns

Coloration in snakes is highly variable (especially in preserved snakes in museum collections). Scale characters are much more stable and useful as taxonomic characters, which is why this book, like many others, places so much emphasis on them. Because color and (more importantly) pattern are described where possible, however, a few terms are briefly explained here.

Stripes

Stripes are lines that run lengthwise along the snake. Snakes are sometimes simply described generally as "striped," but sometimes stripes are specific to particular places on the snake's back and sides. The photo of the Striped Skaapsteker (*Psammophylax tritaeniatus*) in Figure 1.9 is labeled to illustrate the terminology for stripes in different positions. A mid-dorsal stripe runs along the vertebral row, and if it is broad, it may include adjacent scale rows. It differs from other stripes in being on the midline and therefore median and single rather than paired, as other stripes are. Paravertebral stripes are paired, one on each side of the dorsal midline. Lateral stripes are paired and extend along the snake's sides. There may be paired stripes on the paraventral

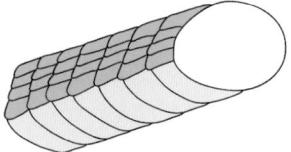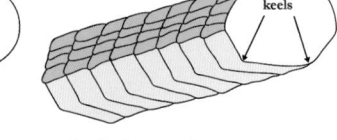

FIGURE 1.8. Smooth versus keeled ventrals (ventrals shown in yellow, dorsals shown in green). By K. Jackson from a drawing by W. Kunkel

rounded ventrals   keeled ventrals

rows, meaning the dorsal scale rows immediately adjacent to the ventrals.

Bands

While stripes run lengthwise along the snake, bands extend transversely across the snake's dorsal scales. (If they go all the way around the snake, including both dorsals and ventrals, they are considered "rings.") Bands are a pattern usually repeated along most or all of the snake's length, as seen in the Ornate Water Snake, *Grayia ornata* (Fig. 1.10). A single band across the back of the neck (or "nape") is a "nuchal band," seen in the Collared Snake-eater, *Polemon collaris* (Fig. 1.11).

The Head

Lateral View

Rostral

The rostral (dark blue "R") is a median scale, and the anteriormost scale on the upper lip. Seen from the front, there is a notch in its lower edge to allow the tongue to flick in and out without the mouth being opened.

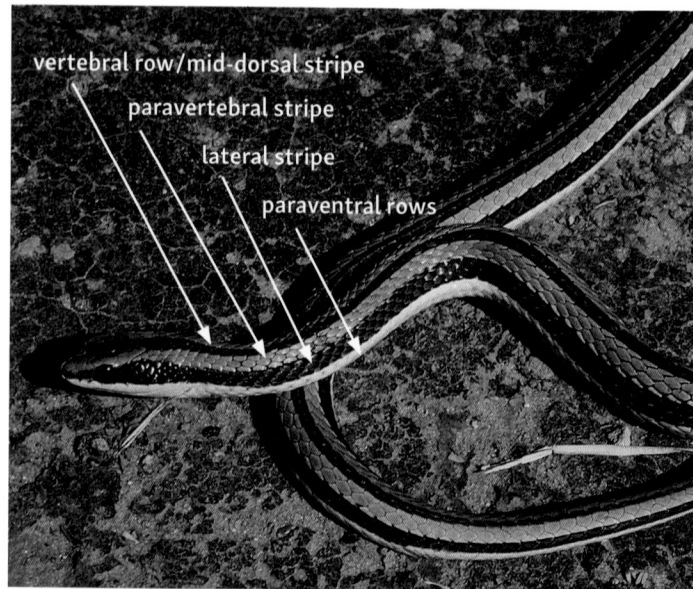

FIGURE 1.9. Stripes on the Striped Skaapsteker, *Psammophylax tritaeniatus*. By K. Jackson from a photo by S. Spawls

FIGURE 1.10. Banded: Ornate Water Snake, *Grayia ornata* (juvenile), Democratic Republic of Congo. By E. Greenbaum

FIGURE 1.11. Nuchal band: Collared Snake-eater, *Polemon collaris*, Congo. By M. Burger

FIGURE 1.12. Lateral view of the head of the African Brown Water Snake, *Afronatrix anoscopus*. By K. Jackson from a line drawing by T. Giri

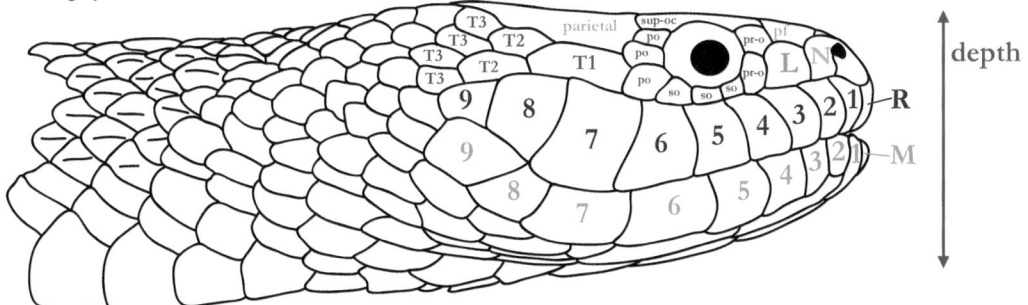

Upper Labials

The upper labials, or "supralabials" (dark blue numbers), are the scales that make up the upper lip. They are counted from the first, the upper labial in contact with the rostral, to the last (in this example, the ninth), at the corner of the mouth. Upper labials are often described by a number followed by a second number in parentheses. The first number represents the number of upper labials, while the number(s) in parentheses indicate which of the upper labials are in contact with the eye. In this example, there are no upper labials in contact with the eye, so it would be expressed as 9(0).

Mental

The mental, or "symphysial" (light blue "M"), is a median scale, and the equivalent of the rostral for the lower lip.

Lower Labials

The lower labials, or "infralabials" (light blue numbers), form the border of the lower lip much as the upper labials do for the upper lip. They are counted from the first, the upper labial in contact with the mental, to the last (in this example, the ninth) at the corner of the mouth. Lower labials are often described by a number followed by a second number in parentheses. This will be explained below, in the section on the ventral view of the head.

Supraocular

The supraocular (red "sup-oc") is the scale above the eye, in contact with the eye and the frontal (to be explained below, in the section on dorsal view of the head).

Preoculars

The preoculars (red "pr-o") are the scales in contact with and immediately anterior to the eye. In this example, there are two preoculars.

Postoculars

The postoculars (red "po") are the scales in contact with and immediately posterior to the eye. In this example, there are three postoculars.

Suboculars

The suboculars (red "so") are the scales that separate the eye from the upper labials and are in contact only with upper labials, preoculars, postoculars, or other suboculars. In this example, there are three suboculars.

Temporal Formulae

Temporal scales (the pink T1, T2s, and T3s) are the scales that fill the space between the upper labials below and the parietal above (to be explained below, in the section on dorsal view of the head). For efficiency, the arrangement of the temporals is described using a formula. In this example, the temporal formula would be 1+2+4, with the numbers corresponding to the number of scales in the first, second, and third rows of temporals. The first row of temporals is in contact with, and immediately posterior to, the postocular(s), and in contact with the upper labials below and the parietal above. If there is more than 1 scale in the first row, the first row includes all the scales in contact with the postoculars, and the upper labials and parietal or each other. The scales in the second row of temporals are in contact with the first row, and with the

upper labials below and parietal above and each other if there are 2 or more. The scales making up the third row of temporals are in contact with the second row of temporals, the upper labials below, the parietal above, and each other in between. There are usually 2 or 3 rows of temporals (very rarely 4). Often, in a species description, only the first two rows of the temporal formula are given if the numbers in subsequent rows vary a lot among individuals.

Loreal Scale

The loreal (orange "L"), sometimes called the "frenal" in the older literature, is a scale of great taxonomic importance. It is absent in all elapids and atractaspidids, as well as in a few lamprophiines and colubrids (e.g., *Dasypeltis*). The loreal is in contact with the preocular posteriorly and with the nasal anteriorly, unless the prefrontal (to be explained below, in dorsal view section) is in contact with the upper labials, in which case the loreal is in contact anteriorly with the prefrontal and upper labials. It is impossible for it to be in contact with the eye because any scale in contact with the anterior edge of the eye is, by definition, a preocular.

Nasal Scale

The nasal (green "N") is the scale perforated by the nostril. If the nasal scale completely surrounds the nostril, it is considered single (or "entire"). If the nostril is in contact with another scale in one place, the nasal is semi-divided (as in Fig. 1.10). If the nostril divides the nasal into two separate scales, the nasal is considered divided, and the two parts are referred to as the anterior nasal and the posterior nasal. In some snakes lacking a loreal, it is easy to mistake the posterior nasal for a loreal, so it is important to be aware of the position of the nostril.

Depth

"Depth" is the term traditionally used in the taxonomic literature for what would be commonly referred to as "height" in the head of a snake lying with its chin on the ground. Depth is important in describing proportions (e.g., "There is one preocular equal in depth to the diameter of the eye").

Dorsal View
Internasals

Internasals (pink "in") are (almost always) the paired scales between the nasals.

Prefrontals

Prefrontals (light blue "pf") are (almost always) the paired scales immediately posterior to the internasals and in contact with the frontal.

Frontal

The frontal (dark blue "frontal") is a large median scale on top of the head. Its proportions are important taxonomically, both its own ratio of length to breadth and its breadth relative to that of the supraoculars. The breadth of the frontal and supraoculars is measured at the scale's midpoint.

Supraocular

The supraocular (red "sup-oc") is the scale above and in contact with the eye, and in contact medially with the frontal.

Parietals

The parietals (light blue "parietal") are a pair of large scales covering most of the top of the head posterior to the eyes. They are in contact with the posterior edge of the frontal.

Occipitals

Occipitals (dark blue "o") are the name for the rather vaguely defined small scales posterior to the parietals that are still head scales, but not dorsals yet.

Breadth

"Breadth" is equivalent to "width," but it is the term traditionally used in snake descriptions, particularly in the older literature.

Ventral View
Chin Shields

Chin shields (red "cs"), also called "sub-linguals," are one or two pairs of scales on the underside of the lower jaw. They are defined as being in contact with the lower labials and in contact with each other at the midline for at least part of their length. Gular scales (see below) may sometimes resemble chin shields. Such gulars are known as "false chin shields" because they superficially resemble chin shields but do not meet the definition (Fig. 1.17).

Lower Labials

Lower labials, or "infralabials" (light blue numbers), are the scales that make up the edge of the lower lip. Lower labials are described by a number followed by a second number in parentheses. For the example above, the lower labials would be 9(5), meaning that there are a total of 9 lower labials, of which the first 5 are in contact with the anterior chin shields. In species that have only one pair of chin shields, the number in parentheses indicates which of the first lower labials are in contact with the one pair of chin shields.

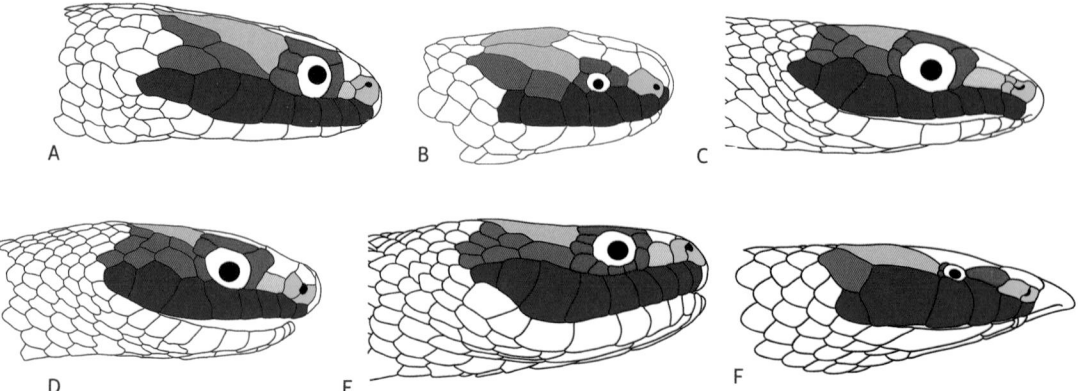

FIGURE 1.13. Lateral view of variation in head scales. Blue = upper labials; red = preoculars, postoculars, su-praoculars, and suboculars; orange (anterior to eye) = loreal; orange (posterior to eye) = parietal; green = nasal; pink = temporals. (*A*) The Swamp Snake, *Limnophis bicolor*, has a semi-divided nasal, the loreal is present, the upper labials are 8(3,4), meaning 8 upper labials total, with the third and fourth in contact with the eye, and the temporal formula is 1+2. (*B*) The Snake-eater, *Polemon neuwiedi*, has a single nasal, the loreal is absent, the upper labials are 7(3,4), and the temporal formula is 1+1. (*C*) The Sand Snake, *Psammophis leopardinus*, has a nasal divided into three parts (posterior nasal divided into upper and lower sections), the loreal is present, the upper labials 8(4,5), and the temporal formula 1+1+3. (*D*) The Skaapsteker, *Psammophylax rhombeatus*, has a divided nasal, the loreal is present, the upper labials 8(4,5), and the temporal formula 2+2+3. (*E*) The Brown Water Snake, *Afronatrix anoscopus*, has a semi-divided nasal, the loreal is present, the upper labials 9(0), and the temporal formula 1+2+4. (*F*) The Quill-snouted Snake, *Xenocalamus mechowii*, has a divided nasal, the loreal is absent, the upper labials 6(3,4), and the temporal formula 0+1. (Because there is no temporal in contact with the postoculars, there is by definition no first row of temporals.) By K. Jackson using line drawings by T. Giri

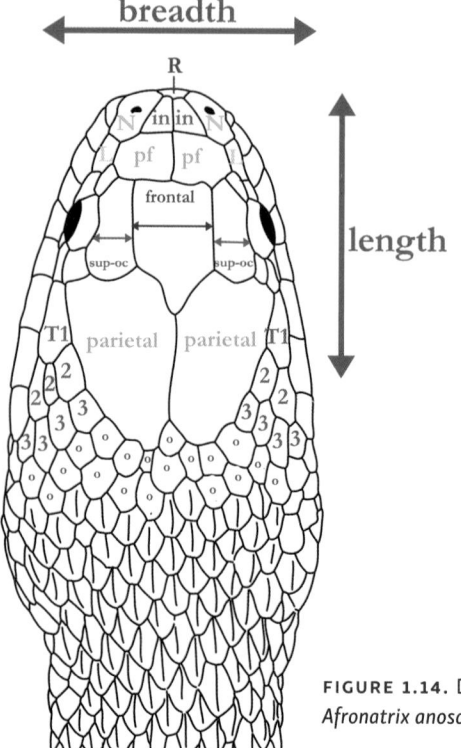

FIGURE 1.14. Dorsal view of the head of the African Brown Water Snake, *Afronatrix anoscopus*. By K. Jackson from a line drawing by T. Giri

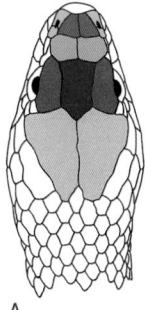 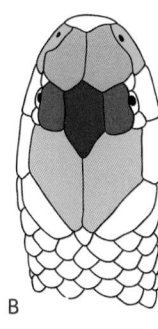 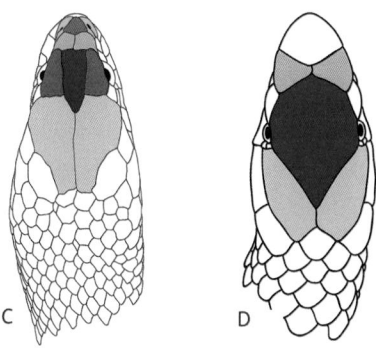

A   B   C   D

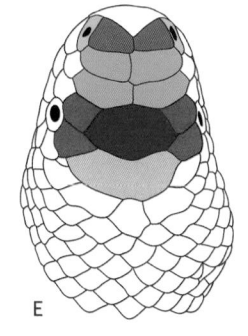 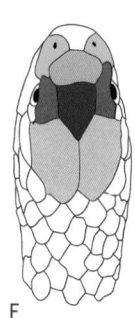

E   F

FIGURE 1.15. Dorsal view of variation in head scales. Dark blue = frontal; red = supraoculars; orange = parietals; light blue = prefrontals; pink = internasals; green = nasal; gray = scales fused to one another. (*A*) The Marsh Snake, *Natriciteres olivacea*, shows the typical pattern, with internasals, prefrontals, and parietals paired, with frontal and supraoculars present. Others deviate from this pattern as follows. (*B*) The Two-headed Snake, *Chilorhinophis gerardi*, has prefrontals fused to internasals but still paired. (*C*) The Swamp Snake, *Limnophis bicolor*, is unusual in having internasals fused into a single median scale. (*D*) The Quill-snouted Snake, *Xenocalamus mechowii*, has internasals fused to prefrontals but still paired. The supraocular is absent (or one could say "supraocular fused to frontal"). (*E*) The Ground Boa, *Calabaria reinhardtii*, has, in addition to paired internasals, 2 pairs of prefrontals, 2 supraoculars (defined as such by being in contact with both eye and frontal), and parietals fused into a single median scale. (*F*) The Cameroon Racer, *Poecilopholis cameronensis*, is distinctive in having internasals and prefrontals all fused into a single median scale. By K. Jackson using line drawings by T. Giri

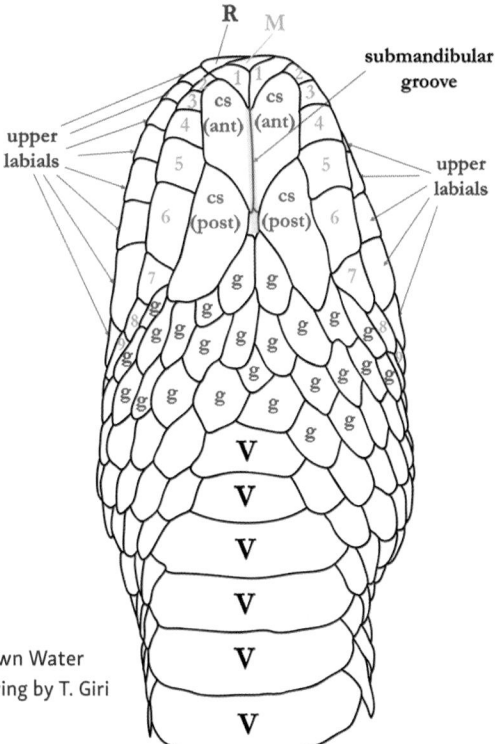

FIGURE 1.16. Ventral view of the head of the African Brown Water Snake, *Afronatrix anoscopus*. By K. Jackson from a line drawing by T. Giri

Gulars

Gulars (pink "g"), or "sublabials," are the scales that cover the underside of the lower jaw that are not lower labials, ventrals, dorsals, or chin shields. They often resemble chin shields but are considered "false" chin shields because they fail to conform to the definition of chin shields as described in Figure 1.17.

Submandibular Groove

The submandibular groove (translucent gray line) is an area of stretchy skin along the midline of the underside of the lower jaw, between the chin shields. It allows expansion of the mouth during swallowing of large prey items. Its taxonomic relevance is that it is pronounced in some species and less so in others.

The Eye

Eye Size

Eye size is a potentially useful taxonomic character for snakes. But because eye size (small, medium, or large) is necessarily relative to something (so as to be descriptive of proportion rather than simply size of snake), it is important to be clear about which definition of eye size is being used. The definition we follow here is the diameter of the eye relative to the distance between the anterior edge of the eye and the nostril (Fig. 1.18). If the diameter of the eye is less than the distance between the eye and the nostril, the eye is small (Fig. 1.18A). If the diameter of the eye is equal to the distance between the eye and the nostril, the eye is medium (Fig. 1.18B). If the diameter of the eye is greater than the distance between the eye and the nostril, the eye is large (Fig. 1.18C). Another definition used by some authors is the diameter of the eye relative to the distance between the lower edge of the eye and the upper lip.

Pupil Shape

Pupil shape is important in the identification of snakes but may be difficult to assess in preserved specimens. Figure 1.19 displays some of the variation in pupil shape (*from left to right*): the Irregular Green Snake, *Philothamnus irregularis*, has a round pupil; the Bush Viper, *Atheris chlorechis*, has a vertical pupil; this preserved museum specimen of the Egg-eating Snake, *Dasypeltis atra*, had a vertically elliptical pupil in life, but it could easily be mistaken for round now; finally, the Twig Snake, *Thelotornis capensis*, has a horizontally elliptical pupil.

Characters of Specific Groups of Snakes

Psammophiinae

Canthus Rostralis

A canthus rostralis is a ridge extending from above the eye to the snout, giving the area in front of the eye a sunken appearance. It is seen especially in psammophiines such as this Olive Sand Snake, *Psammophis phillipsii* (Fig. 1.20).

Elapidae

Specific to the cobras, genus *Naja*, such as the Black-necked Spitting Cobra, *Naja nigricollis* (Fig. 1.21).

Nuchals

Nuchals (green) are all the temporal and occipital scales in contact with the parietals. Postoculars, if they contact the parietals, are not counted. The number of nuchals

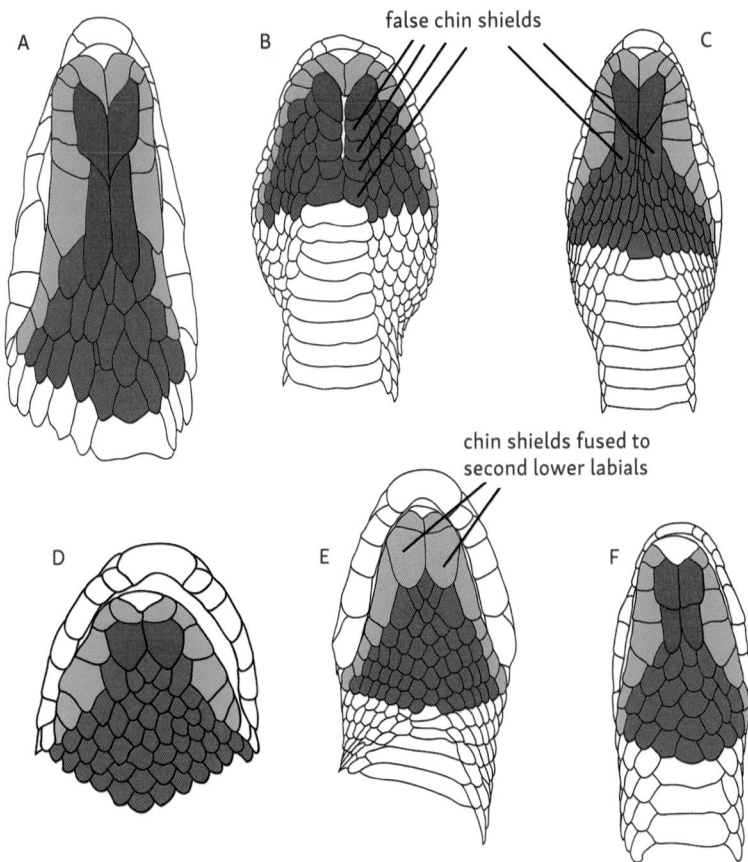

FIGURE 1.17. Ventral view of variation in head scales. Red = chin shields; light blue = lower labials; pink = gulars; gray = scales fused to each other. (A) The Sand Snake, *Psammophis namibensis*, has 2 pairs of chin shields, the lower labials are 10(5) meaning that of a total of 10 lower labials, 5 of which are in contact with the anterior chin shields. (B) The Carpet Viper, *Echis pyramidum*, has 1 pair of true chin shields. Gulars resembling chin shields are "false" because they are not in contact with the lower labials, lower labials 10(4). (C) The Diadem Snake, *Spalerosophis diadema*, has 1 pair of true chin shields. The gulars resembling a posterior pair of chin shields are "false" because there are gulars separating them from one another so that there is no point of contact between them, lower labials 12(5). (D) The Small-scaled Stiletto Snake, *Atractaspis microlepidota*, has 1 pair of chin shields, lower labials 6(3). (E) The Fat Stiletto Snake, *Atractaspis corpulenta*, has chin shields fused to second pair of lower labials, lower labials 6(0). (F) The Centipede-eater, *Aparallactus moeruensis*, has 2 pairs of chin shields, lower labials 6(3). This species is unusual in that the chin shields are in contact with the mental whereas in most snakes the first pair of lower labials separates the chin shields from the mental. By K. Jackson using line drawings by T. Giri

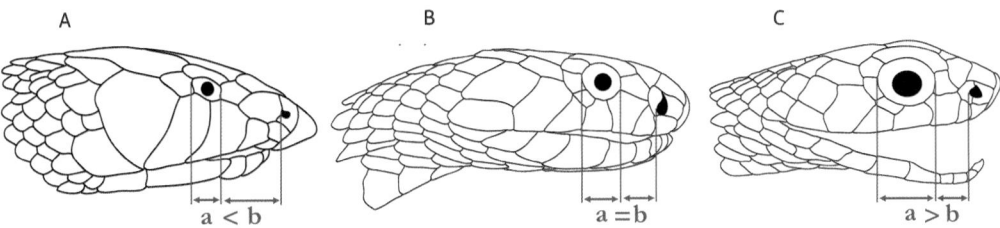

FIGURE 1.18. Eye size variation, where a = diameter of eye and b = distance between eye and nostril. (A) Small eye (a < b): *Atractaspis dahomeyensis*. (B) Medium eye (a = b): *Naja melanoleuca*. (C) Large eye (a > b): *Pseudohaje goldii*. By K. Jackson using line drawings by T. Giri

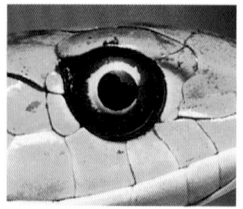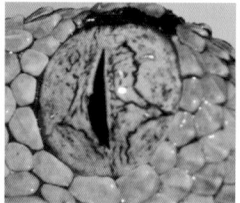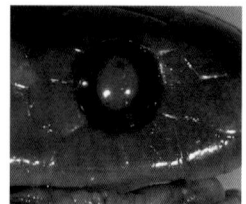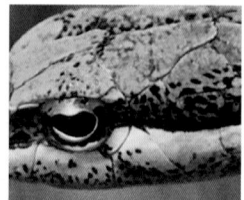

FIGURE 1.19. Pupil shape variation. *Left to right: Philothamnus irregularis, Atheris chlorechis, Dasypeltis atra, Thelotornis capensis.* By S. Spawls, P. Naskrecki, K. Moore, and S. Spawls, respectively

FIGURE 1.20. Canthus rostralis of the Olive Sand Snake, *Psammophis phillipsii.* By K. Jackson from a photo by C. de Haan

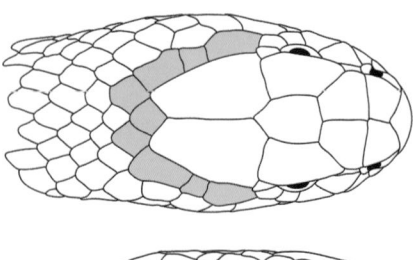
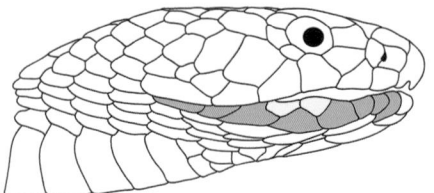

FIGURE 1.21. Dorsal and lateral views of the head of the Black-necked Spitting Cobra, *Naja nigricollis,* showing the nuchals (green) and cuneates (yellow). By K. Jackson using line drawings by T. Giri

(9 in Fig. 1.21) is a useful taxonomic character.

Cuneates

Cuneates (yellow) are scales along the edge of the lower lip that partially separate the lower labials (light blue) from one another. They do not count as lower labials because they do not completely separate adjacent lower labials from one another. The presence and precise position of cuneates are subject to considerable individual variation, making them not useful as a taxonomic character. But they are explained here so as to avoid confusion when counting lower labials in cobras. In Figure 1.21, there are cuneates between the fourth and fifth and the fifth and sixth lower labials.

Viperidae

African Viperidae (with the important exception of adders of the genus *Causus*) have head scales so different from those of other snakes that different characters are used to identify them, and these characters have a different set of terms to describe them. We begin with the Egyptian Saw-scaled Viper, *Echis pyramidum* (Fig. 1.22).

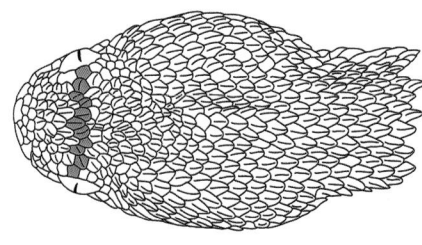

FIGURE 1.22. Dorsal and lateral views of the head of the Egyptian Saw-scaled Viper, *Echis pyramidum*, showing the interorbitals (light blue), perioculars (red), and subocular scale rows (yellow). By K. Jackson using line drawings by T. Giri

Interorbitals

Interorbitals (light blue) are the scales across the top of the head that separate the eyes. The number of interorbitals is a useful taxonomic character. In this example, there are 11 interorbitals.

Perioculars

Perioculars (red) are all the scales in contact with the eye. This example has 17 perioculars.

Subocular Scale Rows

Subocular scale rows (yellow) are rows of scales between the perioculars and the upper labials (dark blue). The number of subocular scale rows is a useful character. There is just one row in this example.

Keeled Gulars

In some viperids, the gulars are keeled, a character useful in the identification of species. Of the Bush Vipers pictured in Fig-ure 1.23, (A) *Atheris squamigera* has keeled gulars, while (B) *A. nitschei* has smooth ones.

Scales of the Rostral Region

Diagrams of the scales of the snout of a viperid viewed head-on are sometimes used in taxonomic works. We explain here how to interpret those diagrams. In Figure 1.24, a diagram of the scales of the snout has been superimposed on a Bush Viper, *Atheris chlorechis*.

Suprarostrals

Suprarostrals (yellow) are all the scales in contact with the rostral (light blue) that are not upper labials (dark blue) or nasals (green).

Internasals

Internasals (pink) are the scales in between the nasals (other than suprarostrals).

Typhlopidae and Leptotyphlopidae

The Scolecophidia ("worm-like" snakes), represented in our zone by the families Typhlopidae and Leptotyphlopidae (Blind Snakes and Thread Snakes), diverged from the true snakes (Alethinophidia) 130 million years ago. They are consequently so drastically different in appearance from any other groups that they require a separate system of terminology for the characters used to identify them. With some notable exceptions, these characters are so small that they require a microscope to see them, and expertise at identifying them takes practice and patience.

Midbody Scale Rows

Midbody scale rows are the number of scales *around* the body, about halfway along the snake. Counting scale rows can be tricky

FIGURE 1.23. (A) Keeled gulars of *Atheris squamigera* and (B) smooth gulars of *A. nitschei*. Red arrows point to some of the keels. By K. Jackson using photos by K. Moore

FIGURE 1.24. Rostral scale diagram superimposed on the Bush Viper, *Atheris chlorechis*. By K. Jackson from a photo by R. Carmichael

because typhlopids and leptotyphlopids lack ventral scales as a helpful marker of where to start and stop counting. Stick a pin to mark where you started, and count all the way around the body. Typhlopids have far more midbody scale rows than leptotyphlopids, so this scale count is used to tell members of the two families apart.

Mid-dorsal Scale Rows

Mid-dorsal scale rows are the number of scales *along* the full length of the snake, counted along the dorsal midline, starting with the first scale posterior to the rostral and ending with the last scale before the final scale at the tip of the tail, which is pointed.

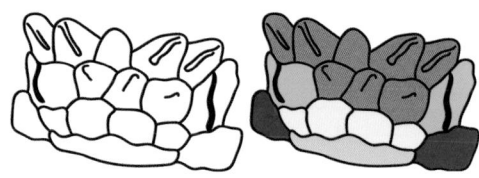

FIGURE 1.25. Rostral scale diagram of the Bush Viper, *Atheris chlorechis*. Figure by K. Jackson from a line drawing by T. Giri

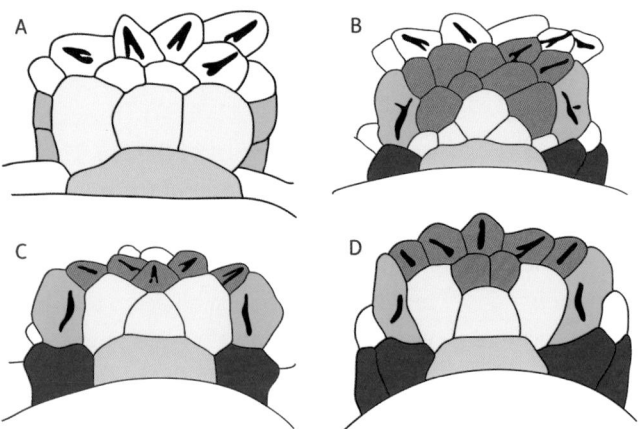

FIGURE 1.26. Examples of scale variation in the viperid rostrum: (*A*) *Atheris hispida*: 3 large suprarostrals and no internasals. (*B*) *A. anisolepis*: 5 suprarostrals and numerous small internasals. (*C*) *A. squamigera*: 3 large suprarostrals and a line of 5 internasals. (*D*) Another *A. squamigera*. This pattern is identical to that of the other *A. squamigera* except for 2 extra scales (gray) that do not fit the definitions of either suprarostrals or internasals. This probably represents an aberrant individual, in which the middle suprarostral split up, forming the 2 extra scales. Adapted by K. Jackson from Broadley (1998)

FIGURE 1.27. *Leptotyphlops scutifrons*. By S. Spawls

Head Scales

The heads of typhlopids and leptotyphlopids differ from those of other snakes, though some of the same terms are applied to the scales. The eye, for example, is underneath the head scales (an adaptation for burrowing) and is often difficult to see. Three typhlopids are presented here as examples in Figure 1.28, with their head scales color-coded to allow easy comparison: (A) *Afrotyphlops elegans*, (B) *A. schlegelii*, and (C) *Indotyphlops braminus*. Each species is shown in lateral, dorsal, and ventral view. The *rostral* (light blue) is the prominent midline scale at the front of the upper jaw and may be rounded and blunt or wedge shaped with a sharp edge for digging. The *nasal* (orange) is immediately posterior and lateral to the rostral. It is pierced by the nostril. The examples in Figures 1.28A and B have semi-divided nasals, with the line

extending from the nostril to the edge of the scale referred to as the internasal suture. The example in Figure 1.28C has a completely divided nasal so that there is a superior nasal suture extending upward from the nostril as well as an inferior nasal suture extending beneath. The *preocular* (green) is the scale immediately posterior and lateral to the nasal. The *ocular* (yellow) is immediately posterior and lateral to the preocular and covers all or part of the eye. The *upper labials* form the border of the upper lip and are indicated in medium blue and are visible in lateral and ventral views. The *frontal*, also indicated in medium blue, is a midline scale visible only in dorsal view, immediately posterior to the rostral on the top of the head. The *supraocular*, indicated in red, is in contact with the ocular and the frontal. The *postoculars*, indicated in pink, are scales that are in contact with the posterior side of the ocular and are not upper labials or supraoculars.

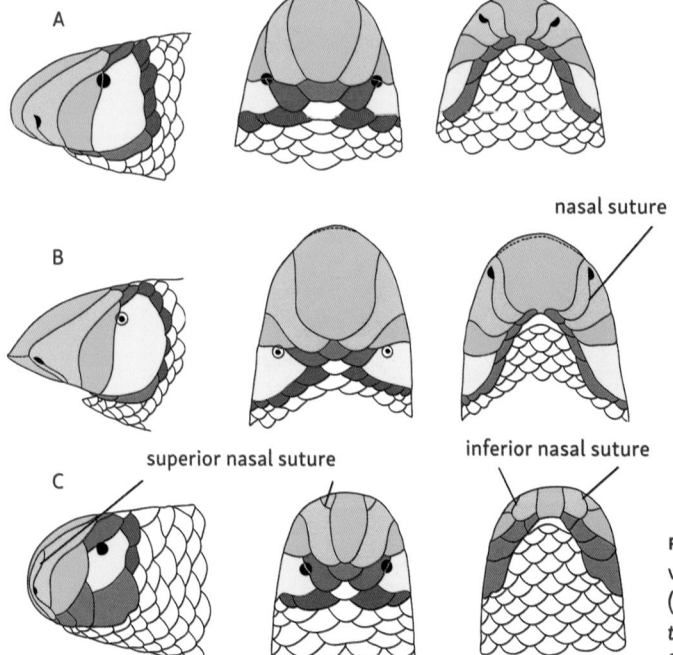

FIGURE 1.28. Lateral, dorsal, and ventral views of (A) *Afroyphlops elegans*, (B) *Afrotyphlops schlegelii*, and (C) *Indotyphlops braminus*. Adapted by K. Jackson from drawings by Roux-Estève (1975)

Key to Genera of Central and Western African Snakes

A multi-access version of this genus key is also available (Hsu et al. 2017).

1	Body scales undifferentiated, same all around	2 (Scolecophidia)
1'	Body scales differentiated into broad ventral scales below and smaller dorsal scales above	3 (Alethinophidia)

2(1)	Fewer than 16 scale rows around body..... Leptotyphlopidae (see genus key in chap. 5)	
2'	More than 16 scale rows around body .... Typhlopidae (see genus key in chap. 5)	

3(1)	Ventral scales distinctly narrower than ventral surface of body	4
3'	Ventral scales as broad as ventral surface of body	6 (Colubroidea)

4(3)	Subcaudals divided; 15 or more lower labials	Python
4'	Subcaudals single; 14 or fewer lower labials	5

5(4)	Eye in contact with upper labials	Calabaria
5'	Eye separated from upper labials by suboculars	Eryx

6(3)	Head scales small and undifferentiated	7
6'	Head scales large and platelike	10

7(6)	Subcaudals single	8
7'	Subcaudals divided	9

8(7)	Fewer than 45 subcaudals; sahel or savanna habitat	Echis
8'	More than 35 subcaudals; forest habitat	Atheris

9(7)	Ventrals rounded	Bitis
9'	Ventrals keeled	Cerastes

10(6)	Internasals absent	11
10'	1 or 2 internasals	12

11(10)	1 prefrontal	Poecilopholis
11'	2 prefrontals (paired prefrontals)	Chilorhinophis

12(10)	1 internasal	13
12'	2 internasals (paired internasals)	15

13(12)	1 prefrontal	Prosymna
13'	2 prefrontals (paired prefrontals)	14

14(13)	19 dorsal scale rows at midbody	Limnophis
14'	21 or more dorsal scale rows at midbody	Hydraethiops

15(12)	Preocular absent; upper labials in contact with the internasal and separating the eye from the nasal .... *Amblyodipsas*	
15'	1 or more preoculars separating the upper labials from the internasals or the prefrontal 16	

16(15)	Prefrontal absent; rostral pointed .... *Xenocalamus*	
16'	At least 1 prefrontal 17	

17(16)	Loreal absent, potentially with a preocular extending behind a lower preocular to resemble a loreal 18	
17'	Loreal present 39	

18(17)	Cloacal scale divided 19	
18'	Cloacal scale single 21	

19(18)	More than 60 subcaudals .... *Dendroaspis*	
19'	Fewer than 40 subcaudals 20	

20(19)	15 dorsal scale rows at midbody .... *Polemon*	
20'	17 or more dorsal scale rows at midbody .... *Atractaspis*	

21(18)	Subcaudals single 22	
21'	Subcaudals divided 25	

22(21)	17 or more dorsal scale rows at midbody .... *Atractaspis*	
22'	15 dorsal scale rows at midbody 23	

23(22)	Fewer than 120 ventrals .... *Hypoptophis*	
23'	More than 125 ventrals 24	

24(23)	29 or fewer subcaudals .... *Polemon*	
24'	30 or more subcaudals .... *Aparallactus*	

25(21)	Dorsal scales keeled 26	
25'	Dorsal scales smooth 27	

26(25)	Vertebral scale row not enlarged relative to other dorsal scale rows .... *Dasypeltis*	
26'	Vertebral scale row enlarged and double-keeled .... *Gonionotophis*	

27(25)	Pupil vertically elliptical 28	
27'	Pupil round 29	

28(27)	Fewer than 200 ventrals and fewer than 60 subcaudals .... *Chamaelycus*	
28'	More than 220 ventrals and more than 65 subcaudals .... *Dendrolycus*	

29(27)	15 or fewer dorsal scale rows 30	

29'	17 or more dorsal scale rows	35
30(29)	More than 70 subcaudals	*Pseudohaje*
30'	Fewer than 60 subcaudals	31
31(30)	More than 150 ventrals	32
31'	Fewer than 150 ventrals	34
32(31)	Eye small; not front fanged	*Polemon*
32'	Eye medium or large; proteroglyphous	33
33(32)	30 or more subcaudals	*Naja*
33'	Fewer than 30 subcaudals	*Elapsoidea*
34(31)	Eye small; nasal scale single	*Duberria*
34'	Eye medium; nasal scale divided	*Elapsoidea*
35(29)	Nasal undivided	*Chamaelycus*
35'	Nasal divided	36
36(35)	Rostral scale enlarged and shield-like	*Aspidelaps*
36'	Rostral unspecialized, rounded	37
37(36)	Opisthoglyph maxillary dentition; 152 or fewer ventrals	*Bothrolycus*
37'	Proteroglyph or solenoglyph maxillary dentition; 153 or more ventrals	38
38(37)	Solenoglyph maxillary dentition; 6 or fewer upper labials	*Atractaspis*
38'	Proteroglyph maxillary dentition; 6 or more upper labials	*Naja*
39(17)	No upper labials in contact with the eye	40
39'	At least 1 upper labial in contact with the eye	43
40(39)	Solenoglyph maxillary dentition; fewer than 40 subcaudals	*Causus*
40'	Aglyph maxillary dentition; more than 45 subcaudals	41
41(40)	Cloacal scale single	*Spalerosophis*
41'	Cloacal scale divided	42
42(41)	Dorsal scales smooth; more than 166 ventrals	*Scaphiophis*
42'	Dorsal scales keeled; fewer than 164 ventrals	*Afronatrix*
43(39)	Vertebral row enlarged relative to other dorsal scale rows	44
43'	Vertebral row not enlarged relative to other dorsal scale rows	50
44(43)	Vertebral row double-keeled	*Gonionotophis*
44'	Vertebral row smooth or bearing a single keel like other dorsal scales	45

60(50)	17 or fewer dorsal scale rows	61
60'	19 or more dorsal scale rows	80
61(60)	Cloacal scale divided	62
61'	Cloacal scale single	70
62(61)	Canthus rostralis present	63
62'	No canthus rostralis	66
63(62)	Rostral scale pointed and beak shaped	*Rhamphiophis*
63'	Rostral scale unspecialized, rounded	64
64(63)	Oblique scale rows	*Psammophis*
64'	Straight dorsal scale rows	65
65(64)	Anterior chin shields longer than the posterior pair	*Psammophylax*
65'	Anterior chin shields shorter than the posterior pair	*Malpolon*
66(62)	Oblique dorsal scale rows on the anterior third of the body	67
66'	Straight dorsal scale rows	68
67(66)	15 or fewer dorsal scale rows	*Philothamnus*
67'	17 or more dorsal scale rows	*Telescopus*
68(66)	Nasal undivided or semi-divided	*Hemirhagerrhis*
68'	Nasal divided	69
69(68)	1 anterior temporal	*Natriciteres*
69'	2 anterior temporals	*Grayia*
70(61)	Pupil vertically elliptical	71
70'	Pupil round	73
71(70)	Nasal undivided	*Chamaelycus*
71'	Nasal divided	72
72(71)	Fewer than 60 subcaudals	*Lycophidion*
72'	More than 65 subcaudals	*Dendrolycus*
73(70)	Oblique dorsal scale rows	74
73'	Straight dorsal scale rows	76
74(73)	15 or fewer dorsal rows	75
74'	17 dorsal rows	*Psammophis*
75(74)	Nasal undivided; proteroglyph maxillary dentition	*Pseudohaje*

75'	Nasal divided; aglyph or opistodont maxillary dentition	*Philothamnus*

76(73)   Large eye; proteroglyph maxillary dentition .... *Pseudohaje*
76'   Small or medium eye; aglyph or opisthoglyph maxillary dentition 77

77(76)   2 anterior temporals .... *Grayia*
77'   1 anterior temporal 78

78(77)   Nasal divided .... *Natriciteres*
78'   Nasal undivided 79

79(78)   Fewer than 150 ventrals .... *Duberria*
79'   More than 160 ventrals .... *Chamaelycus*

80(60)   Canthus rostralis present 81
80'   No canthus rostralis 85

81(80)   Rostral scale enlarged and leaf-shaped .... *Lytorhynchus*
81'   Rostral scale unspecialized and rounded 82

82(81)   Only 1 upper labial in contact with the eye; more than 23 dorsal scale rows ....
.... *Pseudaspis*
82'   More than 1 upper labial in contact with the eye; fewer than 23 dorsal scale rows ....
.... 83

83(82)   1 pair of chin shields; slight canthus rostralis .... *Meizodon*
83'   2 pairs of chin shields; distinct canthus rostralis 84

84(83)   Oblique dorsal scale rows .... *Psammophis*
84'   Straight dorsal scale rows .... *Malpolon*

85(80)   Cloacal scale single 86
85'   Cloacal scale divided 95

86(85)   2 or more anterior temporals 87
86'   1 anterior temporal 90

87(86)   Fewer than 21 dorsal scale rows 88
87'   More than 21 dorsal scale rows 89

88(87)   Fewer than 170 ventrals .... *Grayia*
88'   More than 190 ventrals .... *Telescopus finkeldeyi*

89(87)   Only 1 upper labial in contact with the eye .... *Hemorrhois*
89'   At least 2 upper labials in contact with the eye .... *Boaedon*

2

Evolution of African Snakes

The 3,500 or so species of living snakes in the world belong to two lineages, the Scolecophidia ("worm-like snakes") and the Alethinophidia ("true snakes"). These two lineages are thought to have diverged from one another about 130 million years ago (mya). The Scolecophidia include about 385 species worldwide. Within our zone, they are represented by 49 species belonging to the families Typhlopidae (Blind Snakes) and Leptotyphlopidae (Thread Snakes). The remaining 3,136 or so snakes worldwide belong to the Alethinophidia (true snakes). Most of these (2,926 or so) are the Caenophidia ("recent snakes"), a clade sharing a common ancestor about 70 mya that has undergone massive diversification and speciation over the past 60 million years. The remaining 210 or so alethinophidians are a paraphyletic assemblage sometimes called "Henophidia," often with common names "boa" and "python." These are represented in our zone by 7 species: 4 species of python (family Pythonidae, genus *Python*) and 3 species of boa: the fossorial Ground Boa, *Calabaria rheinhardtii* (family Boidae, subfamily Calabariinae), and two species of Sand Boa, genus *Eryx* (family Boidae, subfamily Erycinae).

The Caenophidia is a large and diverse clade containing approximately 2,926 species worldwide. Within our zone, the Caenophidia are represented by 240 spe-

cies, belonging to four families: Viperidae, Elapidae, Lamprophiidae, and Colubridae.

The family Viperidae, the vipers and adders, is a family of 305 species with an almost worldwide distribution. Within our zone, it is represented by 27 species in five genera. These include the Bush Vipers (genus *Atheris*), Night Adders (genus *Causus*), Sand Vipers (genus *Cerastes*), Saw-scaled Vipers (genus *Echis*), and Giant Vipers (genus *Bitis*).

The family Elapidae, the cobra family, is a family of 347 species worldwide. The Elapidae are represented within our zone by 23 species in five genera. These include Cobras (genus *Naja*), Tree Cobras (genus *Pseudohaje*), Mambas (genus *Dendroaspis*), and African Garter Snakes (genus *Elapsoidea*).

The family Lamprophiidae is an ecologically diverse, mainly African clade of 292 species. Within our zone, it is represented by 115 species in 26 genera, including House Snakes (genus *Boaedon*), Wolf Snakes (genus *Lycophidion*), Sand Snakes (genus *Psammophis*), Shovel-snouts (genus *Prosymna*), Centipede-eaters (genus *Aparallactus*), the venomous front-fanged Stiletto Snakes (genus *Atractaspis*), and many others.

The family Colubridae is a large family with 1,732 species worldwide. It is represented within our zone by 75 species in 23 genera, including Bush Snakes (genus *Philothamnus*), Egg-eating Snakes (genus *Dasypel-*

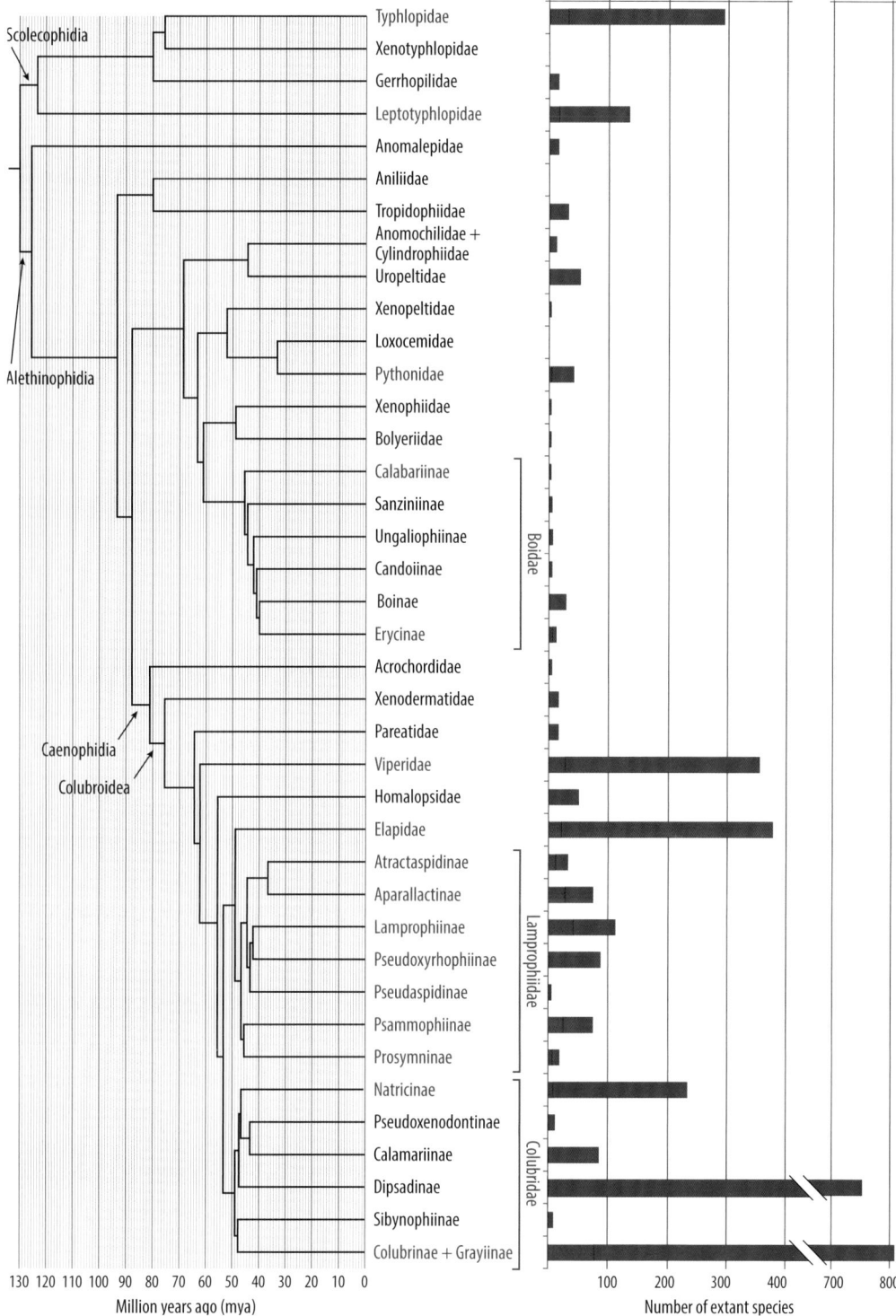

FIGURE 2.1. Higher-level phylogeny of snake families and subfamilies occurring worldwide, with timescale indicating dates of divergence between lineages (phylogeny adapted from Zheng and Wiens 2016). Families and subfamilies with representatives occurring within our zone are highlighted in red. Bar chart to the right indicates numbers of species in each terminal taxon, with the entire bar (blue + red) indicating total number of species worldwide, and the red portion of the bar indicating the number of species occurring within our zone. By K. Jackson

tis), Vine Snakes (genus *Thelotornis*), several genera of arboreal snakes (e.g., *Dipsadoboa*, *Toxicodryas*, *Thrasops*), and semiaquatic snakes belonging to the subfamilies Graylinae and Natricinae.

Reconstructing the Evolutionary History of African Caenophidia

The Caenophidia, represented in Africa by the families Viperidae, Elapidae, Lamprophiidae, and Colubridae, comprise most of the species and genera of African snakes. Of these, the Colubridae together with the group now recognized as the Lamprophiidae have proven the most difficult to make sense of in a phylogenetic context. Until late in the twentieth century, the Caenophidia were understood as comprising three groups. Two of these were the two front-fanged lineages, the viperids (solenoglyph) and the elapids (proteroglyph). The third group was by far the largest, the "colubrids" (quotation marks are used here to indicate the name used in its historical sense rather than restricted to the family Colubridae as currently understood), encompassing all aglyph and opisthoglyph caenophidians. Evolutionary relationships among "colubrid" genera were not well understood, nor was it clear where the more clearly defined viperid and elapid lineages fit among this underbrush of "colubrid" lineages. The diversification of the Caenophidia ("recent snakes") into the large number of genera found in Africa today took place during the Tertiary, and Africa has little in the way of fossil-bearing deposits from this period. Owing to the lack of Tertiary fossils for reconstructing the history of caenophidian diversification in Africa, knowledge of

the evolutionary relationships among African snake genera must be based on what can be learned from extant taxa.

Bogert (1940) was the first to take on the problem of relationships among African "colubrids." His arrangement of African "colubrid" genera on the basis of internal morphological characters formed the foundation for subsequent studies. Though his system was not explicitly evolutionary, and though he recognized the artificialness of some of the groups he arrived at, several of Bogert's groups have stood the test of time, and his system remains relevant today as a point of departure for subsequent studies of phylogenetic relationships among African snakes.

Bogert's "tentative arrangement of African colubrid genera" was based on a similar system for North American colubrids (Dunn 1928), using internal morphological characters such as vertebral hypapophyses and hemipenal morphology pioneered by in the nineteenth century by herpetologists such as Duméril (1853), Boulenger (1893, 1894, 1896), and Cope (1900). Bogert's system used just three dichotomous characters, in hierarchical order as follows: (1) presence or absence of hypapophyses on vertebrae from the posterior part of the body, (2) sulcus spermaticus forked or not forked, and (3) aglyph or opisthoglyph maxillary dentition. Some of Bogert's groups correspond to monophyletic groups recognized today. Group 4 is the subfamily Natricinae. Group 16 is the subfamily Psammophiinae. Groups 1 and 2 make up the subfamily Lamprophiinae, and Groups 8, 9, 10, 11, 13, 14, 15, and 18 correspond to the subfamily Colubrinae. Bogert's Groups 7 and 17 together correspond to the Aparallactinae. Note that Bogert did not include *Atractaspis* in his arrangement because at the time it was assumed to belong to the Viperidae.

Some of Bogert's groups that were problematic reflect genera that have proven in subsequent studies to be the most difficult to determine the taxonomic affiliations of (e.g., Group 5: *Grayia* [now Grayinae; Colubridae], *Pseudaspis* [now Pseudaspidinae; Lamprophiidae], and *Duberria* [now tentatively Pseudoxyrhophiinae; Lamprophiidae]) as well as genera with special adaptations that may have affected the morphological characters his system relied on (e.g., *Dasypeltis* by itself in Group 18, with dentition and vertebral hypapophyses specialized for egg-eating, and *Prosymna* by itself in Group 12, with hemipenal morphology affected by shortening of the tail associated with fossoriality).   The next important advance in understanding the evolutionary relationships among African snakes was made by Bourgeois (1961, 1965). Bourgeois examined the cranial morphology of Congo snakes with the goal of elucidating relationships among the African "colubrids." In contrast to Bogert's system of three simple dichotomous characters, Bourgeois's approach involved detailed, qualitative descriptions of the entire skull in several categories (nasal region, braincase, orbital region, palato-maxillary arcade, mandible, and mandibular suspension), including the shapes of the component bones and their positions relative to one another. Bourgeois assigned subfamily names to the groups of genera that her observations indicated were closely related, and several of these were consistent with Bogert's groups (e.g., Psammophiinae for Bogert's Group 16, Aparallactinae for Bogert's Groups 7 and 17). The

greater depth of Bourgeois's observations yielded important new insights, most notably her recognition of *Atractaspis* as a close relative of the Aparallactinae, and not a viperid as previously thought.

The late twentieth century saw the introduction of molecular tools for phylogenetic studies. Cadle (1994) was the first to apply these to the problem of African "colubrids," using immunological distance methods. Cadle's results lent support to some of the groups identified by Bogert (1940) and by Bourgeois (1965). Cadle (1984a, 1984b, 1984c, 1985) had previously used immunological distance methods to uncover the phylogenetic relationships among Neotropical colubrids, with clearer results. He recognized that the crux of the problem with the African "colubrids" was that in Africa, the extant "colubrid" genera originated from a large number of clades that had diverged from one another long ago, each leaving few living representatives, compared with the Neotropical colubrid fauna, which all originated from the diversification of three main lineages. Cadle (1994) recognized the value in identifying monophyletic groups within the African Caenophidia, even if the relationships of these groups relative to one another were not yet clear.

The start of the twenty-first century has seen the application of molecular phylogenetic methods using DNA and RNA sequence data to the problem of phylogenetic relationships among African snakes. Improvements in sequencing technology and computing as well as increased availability of sequence data for many taxa have made it possible to generate trees based on more sequences and more taxa, leading to increased consensus among researchers about the diversification

TABLE 2.1. Classification of African "colubrid" genera according to Bogert (1940)

A Tentative Arrangement of African Colubridae

 A. Hypapophyses present posteriorly

  B. Sulcus spermaticus forked

   C. No grooved teeth

     Group 1:

       Lycodonomorphus

       Lamprophis*

       Boaedon

       Pseudoboodon*

       Bothrolycus

     Group 2:

       Hormonotus

       Gonionotophis

       Lycophidion

       Chamaelycus

   CC. With grooved posterior maxillary teeth

     Group 3:

       Buhoma

  BB. Sulcus spermaticus not forked

   C. No grooved teeth

     Group 4:

       Afronatrix

       Natriciteres

       Hydraethiops

       Limnophis

   CC. With grooved posterior maxillary teeth

 AA. Hypapophyses absent posteriorly

  B. Sulcus spermaticus forked

   C. No grooved teeth

     Group 5:

       Duberria

       Grayia

       Pseudaspis

   CC. With grooved posterior maxillary teeth

     Group 6:

       Amplorhinus*

     Group 7:

       Aparallactus

       Polemon

       Amblyodipsas

  BB. Sulcus spermaticus not forked

   C. No grooved teeth

     Group 8:

       Hemorrhois

       Meizodon

    *Aeluroglena**

    Spalerosophis

    Lytorhynchus

   Group 9:

    Philothamnus

    Hapsidophrys

   Group 10:

    Thrasops

   Group 11:

    Scaphiophis

   Group 12:

    Prosymna

  CC. With grooved posterior maxillary teeth

   Group 13:

    Toxicodryas

    Crotaphopeltis

    Dipsadoboa

   Group 14:

    Telescopus

    *Macroprotodon**

   Group 15:

    Dispholidus

    Thelotornis

   Group 16:

    Hemirhagerrhis

    Psammophylax

    Malpolon

    Psammophis

    Rhamphiophis

   Group 17:

    Xenocalamus

    Chilorhinophis

    *Macrelaps**

    *Micrelaps**

AAA. Hypapophyses absent posteriorly, strongly enlarged in the region of the esophagus

 B. Sulcus spermaticus not forked

  C. Teeth vestigial

   Group 18:

    Dasypeltis

Note: Generic names and synonymies have been updated. Asterisks indicate genera occurring outside of the zone covered in this book.

of caenophidian snakes in Africa (Nagy et al. 2003, 2005; Vidal et al. 2007, 2008; Kelly et al. 2008, 2009; Pyron et al. 2011, 2013; Streicher and Wiens 2016; Zheng and Wiens 2016). Although minor disagreements about branching order and dates of divergences of lineages remain, resulting from differences in mathematical and statistical models used by different researchers, an increasingly clear and consistent picture of the higher-level relationships among African snakes has emerged from these studies.

Morphological Characters Useful for the Systematic Study of Snakes

The earliest systematic studies of snakes made use of external morphological features, such as the scale characters used for snake identification explained in chapter 1. In the second half of the nineteenth century, herpetologists (e.g., Duméril 1853; Boulenger 1893, 1894, 1895a, 1895b, 1896; Cope 1900) pioneered the use of internal morphological characters. Internal morphological characters—such as vertebral hypapophyses, maxillary dentition, and hemipenes—necessitated dissection or special preparation, but offered the possibility of new insights into the taxonomic affiliations of snakes.

Maxillary Dentition

The skulls of alethinophidian snakes are highly kinetic and have five different bones that may bear teeth. The upper jaw bears two rows of teeth: an outer (lateral) row of teeth on the maxilla (and premaxilla in some henophidians), and an inner (medial) row of teeth on the palatine and ptery-

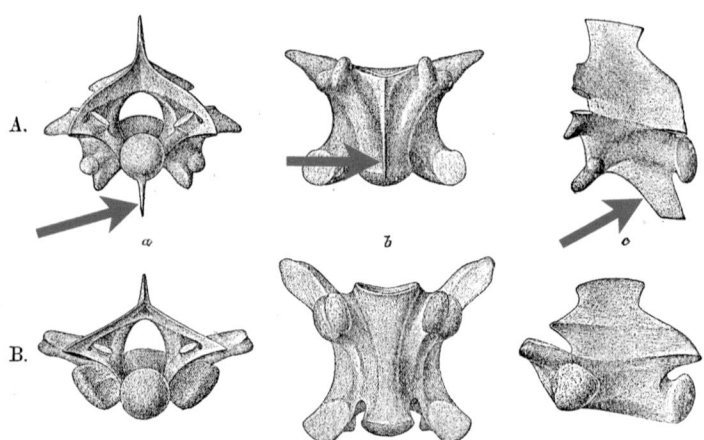

Posterior dorsal vertebræ of :—
A. *Lioheterodon madagascariensis.* B. *Heterodon nasicus.*
a. Back view. b. Lower view. c. Side view.

FIGURE 2.2. The presence or absence of vertebral hypapophyses was one of the three morphological characters used by Bogert (1940) in his "tentative arrangement of African colubrid genera." Shown here are vertebrae from the posterior part of the body in the Madagascan pseudoxyrhophiine, *Lioheterodon madagascariensis* (*top*), and the North American colubrid, *Heterodon nasicus* (*bottom*), showing the presence of hypapophyses (indicated by red arrows) in *L. madagascariensis* and their absence in *H. nasicus*. Vertebrae are shown in posterior view (*left*), ventral view (*center*), and dorsal view (*right*). Modified by K. Jackson from Boulenger (1893)

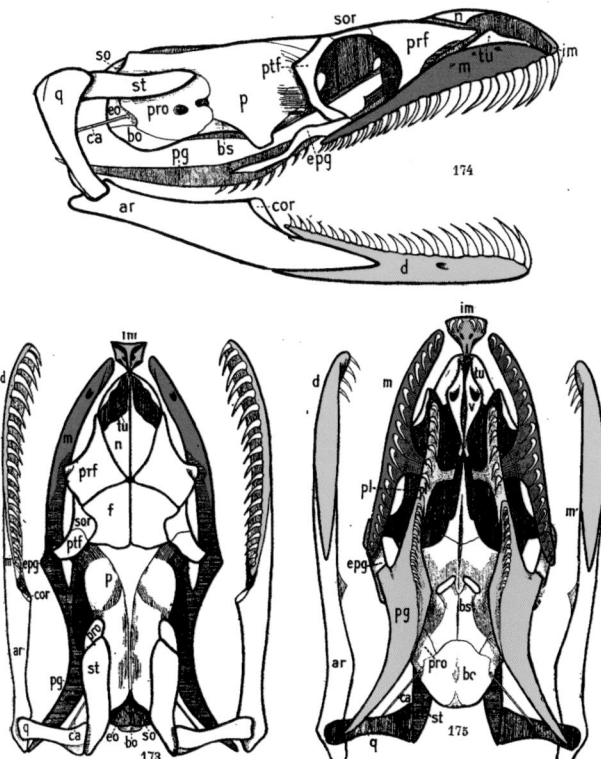

FIGURE 2.3. Skull of *Python regius* in lateral view (*top*), dorsal view (*bottom left*), and ventral view (*bottom right*), with tooth-bearing bones indicated as follows: maxilla (red), premaxilla (orange), dentary (green), palatine (yellow), pterygoid (blue). Modified by K. Jackson from Phisalix (1922)

goid bones. Finally, the dentary bone bears teeth in the lower jaw (Phisalix 1922). The (paired) dentary bones are not fused at the anterior midline, so that the two sides of the lower jaw are connected only by soft tissues that can stretch to engulf large prey items whole. The palatine-pterygoid teeth are used for prey transport, and the two sides can move independently of one another, ratcheting the prey item into the snake's throat and "walking" the cranial skeleton over the prey. This prey-transport role of the palatine-pterygoid row has led the palatine and pterygoid dentition to remain fairly conserved across taxa, freeing up the maxillary dentition to evolve and to take on specialized feeding roles in the Caenophidia (especially venom delivery) (Savitzky 1983; Cundall and Greene 2000; Cundall 2001).

For this reason, the maxillary dentition has long been of interest to snake taxonomists and to those looking for clues to reconstructing the evolutionary relationships among caenophidian snakes (e.g., Boulenger 1893; Anthony 1955).

A special vocabulary consequently exists to describe patterns of maxillary dentition seen in Caenophidian snakes. Two types of maxillary dentition are recognized to describe the dentition of venomous snakes with tubular fangs at the front of the mouth: proteroglyph and solenoglyph. Two other terms, aglyph and opisthoglyph, cover all other caenophidian maxillary dentitions. Proteroglyph maxillary dentition is seen in elapid snakes, in which the tubular fang is positioned at the anterior end of the maxilla. There may be additional (grooved

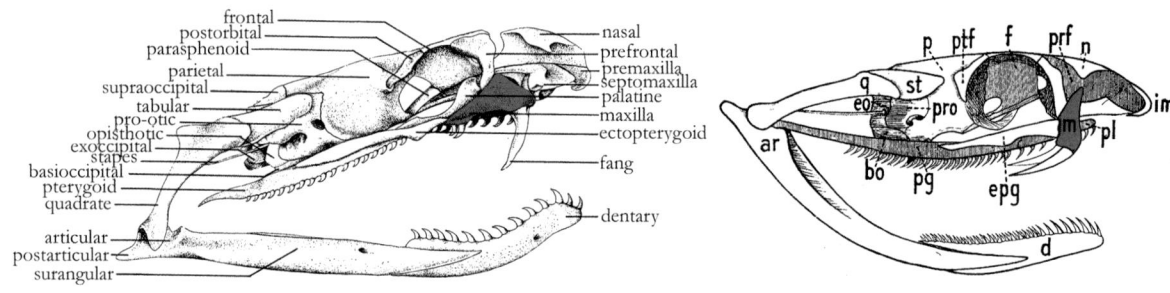

FIGURE 2.4. Lateral view of skull of the elapid *Naja nigricollis* (*left*) illustrating proteroglyph maxillary dentition, and the viperid *Causus rhombeatus* (*right*) illustrating solenoglyph maxillary dentition. Maxillae are indicated in red. Modified by K. Jackson from Bogert 1943 (*left*) and Phisalix 1922 (*right*)

but not tubular) maxillary teeth posterior to the tubular fang, or the tubular fang may be the only maxillary tooth (Bogert 1943). Solenoglyph maxillary dentition is seen in viperids and in the lamprophiid, *Atractaspis*. The solenoglyph dentition differs from the proteroglyph dentition in that the tubular fang is long and the maxilla reduced to a short stub. This shortening of the maxilla associated with elongation of the fang allows the fang to lie horizontally when the mouth is closed and to rotate into a vertical position for striking by rotational movement of the stub-like maxillary bone relative to adjacent skull bones (Klauber 1956; Kardong 1980). Tubular front fangs have evolved independently in three lineages of caenophidian snake: once in elapids (proteroglyph), once in viperids (solenoglyph), and once in *Atractaspis* (solenoglyph) (Jackson 2003; Vonk et al. 2008). In caenophidian snakes other than those with tubular front fangs, the maxillary dentition typically consists of unspecialized teeth for most of the length of the maxilla, and often some specialization of the posteriormost maxillary tooth or teeth: aglyph or opisthoglyph maxillary dentition. Specialization of the posterior maxillary teeth can consist of enlargement of the posteriormost teeth,

sometimes associated with a groove along the lateral surface of the tooth (Jackson and Fritts 1995). When the posterior maxillary tooth is grooved, the maxillary dentition is described as opisthoglyph. When the posterior maxillary tooth is not grooved (but may be enlarged, bladelike, etc.), the maxillary dentition is described as aglyph. The term "opisthodont" is sometimes encountered, generally (but not always) intended to indicate a posterior maxillary tooth that is enlarged but not grooved. The posterior maxillary teeth of aglyph and opisthoglyph snakes line up with the Duvernoy's gland (homologue of the venom gland) located in the upper jaw posterior to the eye. In opisthoglyph snakes, a duct helps to transfer secretions from Duvernoy's gland to the base of the grooved and enlarged posterior teeth, allowing secretions to be introduced into a bite wound (Taub 1966; Savitzky 1980). The venom gland of proteroglyph and solenoglyph snakes is also located posterior to the eye, but in the case of these front-fanged taxa, a duct carries venom from the venom gland to the base of the tubular fang at the front of the mouth (Cundall 2001).

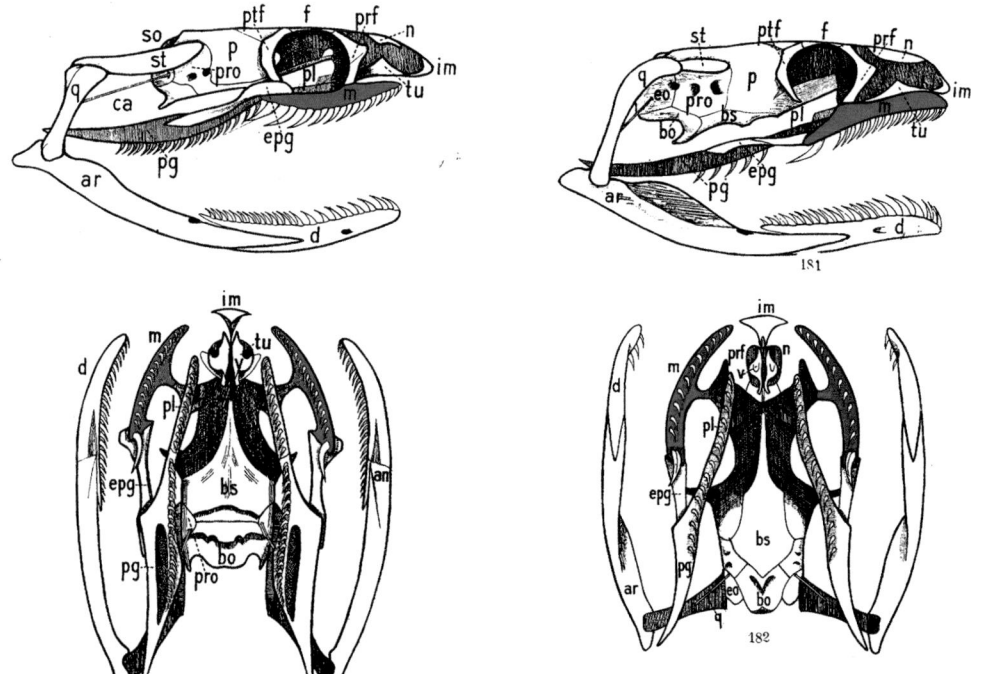

FIGURE 2.5. Lateral (*top*) and ventral (*bottom*) view of skull of *Hemorrhois hippocrepis* (*left*) illustrating aglyph maxillary dentition and *Malpolon monspessulanus* (*right*) illustrating opisthoglyph maxillary dentition. Maxillae are indicated in red. Modified by K. Jackson from Phisalix (1922)

Hemipenes

Hemipenes are the paired copulatory organs of male snakes. When not in use, the hemipenes are a pair of sacs extending posteriorly from the cloacal opening inside the base of the tail. The tails of male snakes tend to be longer than those of females in order to accommodate the hemipenes, and for this reason the number of subcaudal scales is usually higher in males than in females. Live or freshly dead snakes can be sexed in the field using a blunt probe. In males, a probe pushed posteriorly in the cloaca will slide a distance of several subcaudal scales into the base of the tail where the hemipenes are stored. At rest, the hemipenes lie retracted, in the base of the tail. During mating, one hemipenis fills with blood and lymphatic fluid and is everted from the cloaca of the male, inside out, and into the cloaca of the female.

The everted hemipenis has a specific and often elaborate shape and surface ornamentation that is stable at the species level and often at the level of genera or higher-level groupings. A whole vocabulary exists to describe the morphology of hemipenes. The overall shape of each hemipenis may be simple, bilobed, or divided. The sulcus spermaticus, a groove that serves to channel spermatozoa from the male into the oviducts of the female, runs along one side of the hemipenis. The sulcus spermaticus may be bifurcate (forked) if it splits into two on its way toward the tip of the hemipenis, or it may be simple (not forked) if it does not split into two. Note that the shape

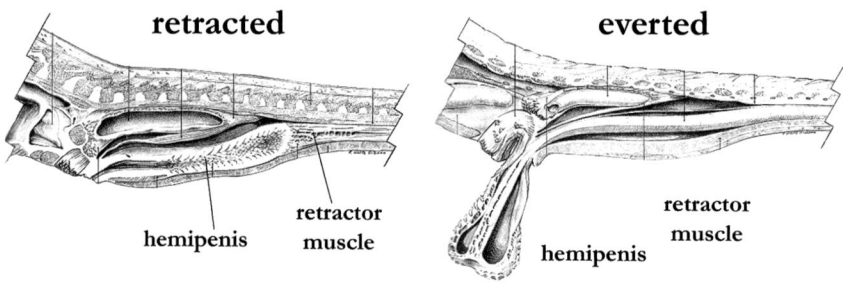

FIGURE 2.6. Hemipenis shown retracted (*left*) and everted (*right*). Modified by K. Jackson from Dowling and Savage (1960)

of the sulcus is not dependent on the shape of the hemipenis, so that the sulcus can be forked or not forked on a simple hemipenis, and forked or unforked on a bilobed or divided hemipenis. For example, a forked sulcus may extend up each side of a bilobed hemipenis or split into two partway toward the tip of a simple hemipenis, or a simple sulcus may extend up just one side of a bilobed hemipenis. Beyond the shape of the hemipenis and of the sulcus spermaticus, additional terminology exists to describe the different types of surface ornamentation possible on different parts of the hemipenis. The surface of the hemipenis may be ornamented with flounces, calyces, papillae, or spines; the base may be ornamented with hooks or pockets; and the apex may be ornamented with awns or discs. Dowling and Savage (1960) provided a detailed explanation and definitions of terminology used for describing hemipenes, in an effort to standardize their usage by different authors.

Descriptions of hemipenes are usually based on everted hemipenes, since ideally this is the best way to view them. In order to preserve everted hemipenes, however, the hemipenes must be everted when the snake is freshly dead but not yet fixed. This is done by injecting fluid into the base of the tail to evert a hemipenis, and then filling

one hemipenis with material such as wax, latex, or petroleum jelly, and tying off the base of the hemipenis with thread, before fixing the snake in formalin for long-term storage. This procedure requires considerable skill and practice to do well, and, consequently, most snakes in museum collections have been fixed without first having a hemipenis prepared in this way. Yet it is also possible to describe the morphology of hemipenes in situ, by dissection of the retracted hemipenis in the base of the tail in a preserved snake. Some authors, such as Bogert (1940), use the in situ method almost exclusively because it greatly increases the number and variety of specimens available to them. The in situ method does not give as accurate a view of the shape and ornamentation of the hemipenis, but the length of the in situ hemipenis as well as the length of the sulcus spermaticus and the distance along the hemipenis at which the sulcus forks can be described relative to number of subcaudals. For example, Bogert (1940) found the length of the hemipenis to range from the smallest, *Psammophis*, in which the retracted hemipenis extended only to the third subcaudal (counting posteriorly from the cloaca), to the largest in *Pseudaspis*, in which the retracted hemipenis extended to the thirty-second subcaudal.

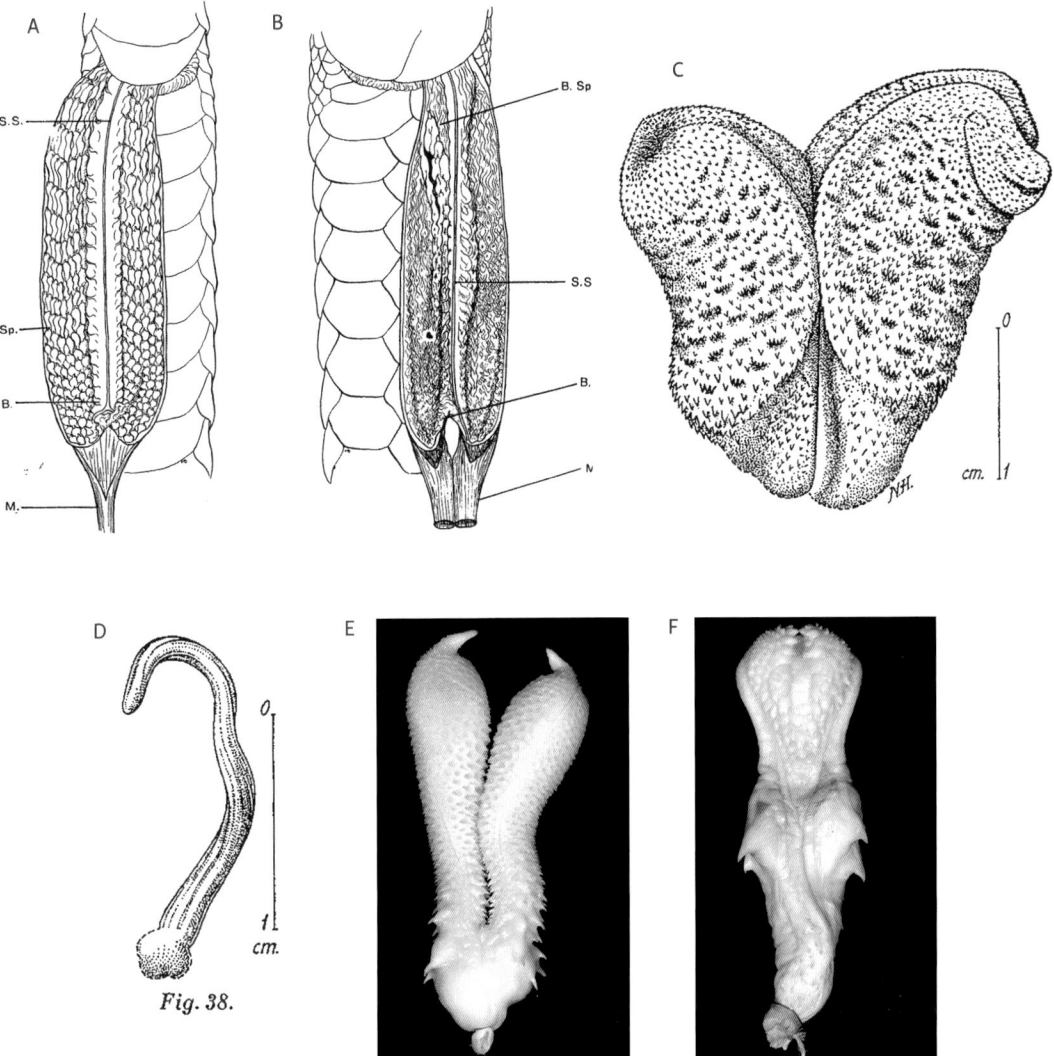

FIGURE 2.7. Examples of illustrations of hemipenes. (*A*) Right hemipenis of *Aparallactus modestus*, illustrated in situ in the retracted position, slit down the dorsal aspect and laid open, showing sulcus spermaticus (S. S.), spines (Sp.), and vestigial bifurcation of sulcus (B.). (*B*) Left hemipenis of *Hydraethiops melanogaster* illustrated in situ in the retracted position, slit down the dorsal aspect and laid open, showing enlarged basal spines (B. sp.), sulcus spermaticus (S. S.), bifurcation at distal end (B.), and retractor muscle (M.). (*C*) Hemipenis (everted) of *Naja nigricollis*. (*D*) Hemipenis (everted) of *Psammophis phillipsii*. (*E*) Right hemipenis (everted) of *Bitis arietans*. (*F*) Right hemipenis (everted) of *Grayia smithii*. Adapted by K. Jackson from (*A* and *B*) Bogert 1940; (*C* and *D*) Doucet 1963; (*E* and *F*) photo by W. Branch

3

Biogeography of African Snakes

The region covered in this book is completely contained between the Tropics of Cancer and Capricorn, and includes much of the Afrotropical zoogeographic region. All sub-Saharan biomes are represented, from desert to primary forest and from coasts to mountains. It is bounded in the north by the Sahara Desert, in the south by the Kalahari Desert, and in the east by the Albertine Rift Valley Lakes. It includes several mountain ranges (Fouta Djallon, a highland region in Guinea in the west; the mountainous Adamawa Region of Cameroon in the north; the Mitumba Mountains in the east; and the Central Plateau of Angola in the south). Along the West African coast there are two heavily degraded blocks of forest, the Guinean Forest to the west and the central Congolian Forest. These forested areas are bordered by large stretches of savanna, extending north to the desert. Each of these zones is characterized by a distinctive fauna. Human alteration of the landscape, including urbanization and agricultural development, results in a modified fauna composed of surviving species from the original fauna as well as colonizing species from bordering habitats (Chippaux and Bressy 1981; Akani et al. 2007). As natural habitats are degraded by humans, populations of snakes and other organisms that previously occurred there are affected. A new equilibrium is eventually reached and

a new population structure arrived at, dictated by forces such as the limitations of available resources and the pressures exerted by predators (Akani et al. 2002a, 2007; Piquet et al. 2012; Akaffou et al. 2017). The new ecosystem that establishes itself in this degraded habitat lacks the species diversity of the original one, as a consequence of the loss of the more vulnerable species from the original ecosystem (Akani et al. 2007; Akani and Luiselli 2009; Masterson et al. 2009). The species that appear to be most advantaged are the least specialized species, some of which are highly venomous (Luiselli 2006a, 2006b). Additionally, some species are apparently ubiquitous—found in a variety of habitat types—though in some cases, more detailed taxonomic study, especially with the use of molecular tools, may reveal instances of cryptic speciation, in which snakes from separate locations appear to be of the same species but turn out on closer examination to represent distinct species.

More than two-thirds of African snake species are restricted to a single habitat type. Twenty percent cross between two habitat types, and barely 5% are found in more than three habitat types, the latter demonstrating a high capacity for adaptation. Few species are truly invasive. The Guinean and Congolian Forests have lost much of their dense primary forest, characterized by large endemic trees and abun-

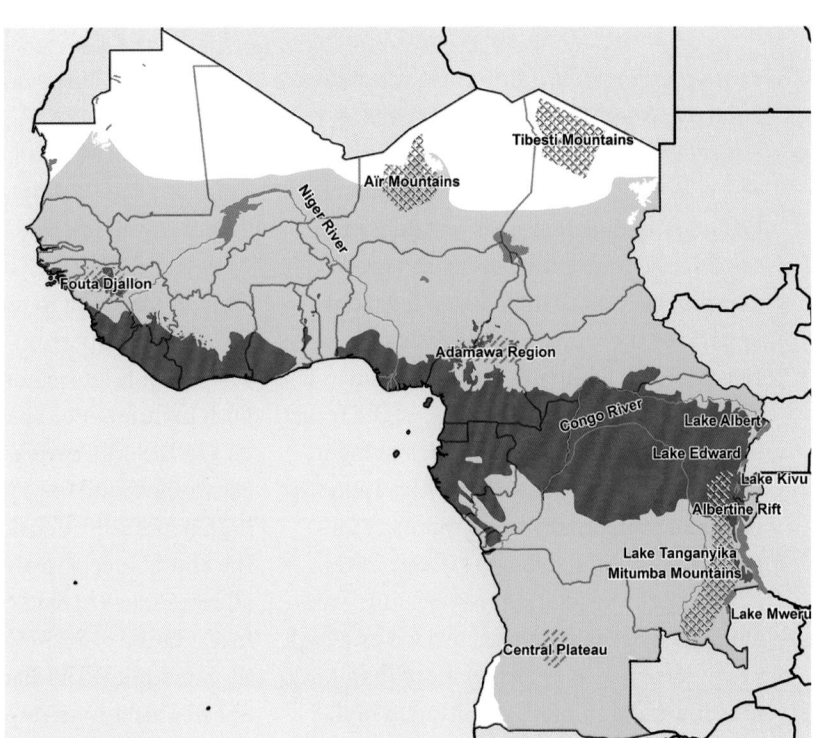

FIGURE 3.1. Ecoregions in the zone covered by this book: tropical forests (dark green), tropical savanna (light green), desert (yellow). By Eric Hsu

FIGURE 3.2. Political map of Africa with our zone highlighted. By Eric Hsu

dant vegetation providing cover for many animal species. These areas of primary forest gradually give way to degraded or secondary forest as a result of uncontrolled logging. Degraded forest is less dense than the original forest and favors fast-growing plant species, often introduced from other parts of the world such as Asia and the Neotropics, leading to a loss of animal and plant biodiversity. In some places, the forest has been completely destroyed, leaving in its place patches of savanna with a reduced flora and fauna comprising remnant forest species and introduced savanna ones. Although the vegetation may be quite variable, these areas are characterized by a high precipitation rate, typically more than 1,500 mm of annual rainfall, resulting in high humidity year-round. Each year is divided into a long dry season and a long rainy season, with the long rainy season interrupted by a short dry period.

The two forest blocks are separated by the Dahomey Gap, a strip of woodland savanna extending all the way to the coast between Ghana and Nigeria. The two blocks are quite similar, although in terms of snakes, each has a high proportion of endemic species. The forest fauna is notable for its richness, both in number of species (66% of the species in the area covered by this book are forest species) and in population density. This high population density can be attributed to the abundance of prey in all seasons, which allows snakes to feed frequently. The rich forest fauna also allows snakes to be easily camouflaged by the dense vegetation. Species occupy an impressive variety of forest microhabitats, including arboreal, terrestrial, underground, and aquatic ones.

FIGURE 3.3. Primary forest, Gabon, 2005. By M. Gérard © IRD

FIGURE 3.4. Flooded forest, Republic of Congo, 2005. By K. Jackson

FIGURE 3.5. Secondary forest, Republic of Congo, 2010. By K. Jackson

FIGURE 3.6. Forest-savanna mosaic, Republic of Congo, 2010. By K. Jackson

FIGURE 3.7. Woodland savanna, Republic of Congo, 2010. By K. Jackson

The Congolian Block appears more diverse than the Guinean Block (Hughes 1983), but this may be simply because the former has been more intensively studied, particularly in the Democratic Republic of Congo, and also because the Congolian Block occupies a larger area. Many genera and species are characteristic of the forest fauna: *Polemon sp.*, *Philothamnus sp.*, *Hapsidophrys sp.*, *Toxicodryas sp.*, *Atheris sp.*, *Dipsadoboa sp.*, *Thrasops sp.* Many of these species are arboreal. Most aquatic species are essentially forest species (*Afronatrix anoscopus*, *Hydraethiops sp.*, *Limnophis bicolor*, *Grayia sp.*, *Naja annulata*), although some of these may venture temporarily into gallery forest or even woodland savanna (*Natriciteres olivacea*, possibly *A. anoscopus*). Only a few terrestrial species from the forest snake fauna, such as *Bitis nasicornis*, are restricted to forest habitats. Most terrestrial species found in the forest are also found in other habitat types, such as woodland savanna or brush-grass savanna.

The area of transition from forest to savanna includes a variety of habitat types such as secondary or degraded forest, gallery forest along rivers in savanna, woodland savanna, and areas inhabited by humans. Where people live there exists an artificial flora, most often subsistence crops occupying small plots scattered across the landscape. These habitats fall into the category of woodland savanna. The trees are smaller than in primary forest and grow farther apart. The ground is covered by a layer of vegetation whose height varies depending on the season; inaccessible because of the height and density of the grasses (up to 2.5 m) during the rainy season, the savanna opens up during the dry season as the grasses disappear. In addi-

tion, these areas are regularly burned in the course of slash-and-burn cultivation, a traditional agricultural practice still in use. Although these fires do not change the species composition, or the natural seasonal changes in species, they reduce the density of snake populations by two-thirds (Barbault 1970, 1971, 1974). There is a single rainy season each year, marked by thunderstorms and heavy rainfall, and a single dry season during which there is little precipitation and the air is dry. This area of forest-to-savanna transition is characterized by an annual rainfall of between 1,000 and 1,500 mm. These patches of savanna are inhabited by a transitional fauna made up of several abundant species, ones well able to adapt to a habitat modified by human activity, some from the forest (*Hapsidophrys smaragdina*, *Philothamnus irregularis*, *Dipsadoboa unicolor*), and others from the savanna (*Psammophis phillipsi*, *Naja nigricollis*). Snake species that do well around human habitations make up approximately 20% of the species in this book, and the savanna is home to 40% of the species included in this book. Species found in both forest and savanna make up approximately 20%, including *Boaedon fuliginosus*, *Natriciteres olivacea*, *Philothamnus semivariegatus*, *Crotaphopeltis hotamboeia*, *Causus maculatus*.

In the northern part of the area covered by this book is the Sahel, an arid belt along the southern edge of the Sahara Desert. The seasons become increasingly distinct approaching the Sahel, and with this transition there is an accompanying gradual change in species composition. Woodland savanna gradually gives way to brush grass savanna, with smaller trees, fewer in number and consequently farther apart. The grasses that cover the ground are shorter than in woodland savanna, making this habitat easily accessible to humans in all seasons. One characteristic of brush grass savanna is landscapes dominated by a single tree species, resulting in a specialization of this habitat type. Tree species frequently found in this context include the African baobab, the Ronier palm, the Shea tree, and some species of *Acacia*. During the rainy season, which is fairly short (approximately four months), heavy rainfall promotes intensive growth of vegetation. The average annual precipitation ranges from 800 to 1,200 mm. Snake species found in tropical forest (*Natriciteres variegata*, *A. anoscopus*, *Dendroaspis viridis*, *D. jamesoni*, *B. nasicornis*) are completely absent, and some Sahel and desert species are found, such as *Eryx muelleri*, *Spalerosophis diadema*, *Cerastes cerastes*. Brush grass savanna is characterized by steppe, a ground cover of short grasses (an average height of 50 cm, sometimes less), providing adequate concealment for many terrestrial species (e.g., *Python regius*, *Echis ocellatus*, *Bitis arietans*). There may be occasional trees of 2–3 m in height that have little foliage but nonetheless provide a home for the few arboreal species found in this habitat type (e.g., *Telescopus tripolitanus*). Annual rainfall averages 500 to 1,000 mm and is concentrated into a period of two to three months of the year. This habitat type represents a zone of transition between wet and dry regions. The species composition changes dramatically, for example, with the absence of *E. ocellatus*, sometimes an abundant species in savanna, and the presence of *Echis leucogaster*, a brush grass savanna and Sahel species. Some desert species are also found in the steppe, including *Cerastes vipera* and *Hemorrhois algirus*.

In the Sahel, the annual precipitation ranges from 300 to 500 mm. Outside of the short rainy season, the humidity is low and there is little vegetation. The ground cover is made up of very short grasses (10–20 cm, sometimes less) and dries up quickly. Trees are short (rarely more than 2 m) and widely scattered, often with no trees at all for several square kilometers. The high temperature of the ground (up to 60°C) forces snakes to hide in burrows or to bury themselves in the sand during the day. The process of desertification leads to a relentless retreat of Sahel into what was once savanna. The boundaries between brush grass savanna, Sahel, and desert are not well defined, and some 10% of species included in this book circulate freely among them. There appear to be only two true desert species (*C. cerastes*, *C. vipera*) that can also be found in the Sahel, but many Sahel species are sometimes found in the desert, especially close to oases: *Eryx colubrinus*, *Hemorrhois algirus*, *Lytorhynchus diadema*, *Spalerosophis diadema*, *Malpolon moilensis*, *Telescopus tripolitanus*, *T. obtusus*, *Naja nubiae*, *Echis leucogaster*, *E. pyramidum*.

Mountain habitats are rich in snake species considering the lack of ground cover and often unfavorable conditions. Within the region covered by this book, 58 species (close to 25%) are found in mountain habitats, and 32 of these species are endemic. These species are often highly specialized both morphologically and ecologically, and among these are many species known only from a few specimens, in some cases only from the type specimen (*Buhoma marlieri*, *Duberria lutrix*, *Atheris hispida*, *A. nitschei*). This phenomenon is attributable to the isolation of mountain populations from those of surrounding areas.

Finally, some species are ubiquitous:

FIGURE 3.8. Sahel, Senegal, 2003. By M. Dukham © IRD

FIGURE 3.9. Desert dunes, Niger, 2003. By P. Blanchon © IRD

Natriciteres olivacea, Crotaphopeltis hotam-
boeia, Philothamnus semivariegatus, P. het-
erodermus, Boaedon fuliginosus, B. lineatus,
Grayia smithii, Psammophis sibilans, P. phil-
lipsii, Naja nigricollis, N. melanoleuca, Causus
maculatus, E. ocellatus. These species manage to adapt to many forested or savanna
areas, but they seem to thrive especially
well in semi-urban areas and other areas
developed by humans. African cities often
expand rapidly and with little planning,
leaving areas without construction, which

FIGURE 3.10. Mt. Cameroon, Cameroon, 1995. By M. Gérard © IRD

are used for small-scale agriculture, habitats to which bolder snake species are able to adapt. In addition, the presence of household garbage and sewage may attract rodents, which can result in relatively dense populations of snakes in heavily human-modified habitats. This is how several species, including some highly venomous ones, come to be found in areas of human habitation: *Naja melanoleuca* and *Dendroaspis jamesoni*, found in residential areas of Yaoundé, Cameroon; *Bitis gabonica* and *B. arietans*, often found on the outskirts of big cities in forest and savanna; and *E. ocellatus*, found in downtown Ouagadougou, Burkina Faso (Revault 1994). This explains the relatively high incidence of snakebite in urban areas, where one might expect the dense human population to limit the number of snakes (Chippaux 2011). Plantations,

agro-industrial complexes, dams, and other human developments may offer attractive habitats where snakes may find a desirable microclimate, abundant prey, and good hiding places.

Sub-Saharan Africa harbors a great diversity of snake species. But although some species may remain to be discovered in remote and isolated areas, such areas are becoming increasingly rare. As habitats are destroyed, snake populations are declining and species are being lost. A few hardy ubiquitous species, including some venomous ones, persist in human-altered habitats to which they seem to adapt well. Even so, urban sprawl and increasingly dense human populations can only result eventually in a reduction in snake populations and, quite possibly, over the long term, in the extinction of many species.

4

Snakebite in Sub-Saharan Africa

Snakebite is a significant public health hazard in sub-Saharan Africa, with a conservative estimate of 500,000 venomous bites to humans occurring every year, more than 25,000 of which prove fatal (Chippaux 1998a, 2011).

How and Why Do Snakebites Occur?

The incidence and severity of snakebites in a given area are determined primarily by two factors: (1) the overall abundance of snakes and composition of species present, meaning which species (venomous or not) are encountered in the habitat type in question, and (2) the fact that cultivated areas and industrial plantations often represent an attractive habitat for particular species because of an abundance of prey and microhabitats (e.g., places for hiding and breeding) that favor population growth (Chippaux and Bressy 1981; Oyaberu and Shokpeka 1984; Malukisa et al. 2005).

For reasons that are not well understood, venomous snakes do not always inject venom when they bite. The proportion of bites by venomous snakes that are "dry" varies depending on the species, but generally it is in the range of 20% to 60% (Silveira and Nishioka 1995).

Most snakebites in Africa occur in the course of agricultural or pastoral activities (Chippaux 1999, 2000). Another 20% or so occur during hunting, gathering firewood, or carrying water, which women often must transport several kilometers. Human habitations often harbor snakes, which are attracted by warmth and prey (e.g., rodents also attracted to human habitations). Snakes easily penetrate garden fences and walls of houses, resulting in snakebites around and even inside houses. Many bites inside houses occur at night while victims are sleeping. The Spitting Cobras, *Naja nigricollis* and *N. mossambica*, have been particularly noted for this behavior by Chippaux et al. (1978) and Greenham (1978).

In all the countries within our zone, most snakebites occur during the rainy season. This is the result of an increase in abundance and activity of snakes coinciding with an increase in agricultural work at this time of the year. In most places, active men between the ages of 15 and 50 are the most frequent bite victims, while women in the same age bracket are significantly less affected.

Many factors are involved in determining how severe the result of an envenomation will be. The primary variables determining the ultimate severity of an envenomation can be classified into two general groups: snake factors and human factors (Chippaux 2009; Hayes and Mackessy 2010). Snake factors include the venom's composition and amount of venom injected, and these can vary significantly depending on the species responsible, size of the snake, biogeographical variation, and varying degrees of intraspecific variation in venom composition among snakes within a single population of a single species. Human factors include body size, conditions of the bite, location of the bite, presence or absence of clothing over the bite site, preexisting medical conditions (asthma, anemia, allergies, prior history of envenomation, and/or treatment with antivenom, which leads to higher risk factors of anaphylactic reactions, etc.), and the quality and speed of medical treatment rendered across the spectrum of care from first aid to definitive treatment. Delay in receiving treatment is without question the leading cause of treatment failure.

Composition and Effect of Snake Venom

The venoms of African snakes are primarily made up of proteins belonging to two groups: toxins and enzymes (Chippaux 2006b).

Toxins generally have a low molecular mass. They work by binding to a cell surface receptor, thereby disrupting or completely preventing its functioning. Venoms composed primarily of toxins are most frequently associated with neurotoxic envenomations, which lead to descending paralysis and respiratory arrest.

Enzymes are molecules whose presence is needed to catalyze and allow a reaction between other molecules to take place, but that are not themselves consumed in the reaction because their role is purely as a facilitating substrate. Venoms rich in enzymes tend to have digestive cytotoxic

properties that can lead to tissue destruction or hemorrhagic complications. Clinical signs, whether caused by toxins or by enzymes, are immediate, because the venom diffuses quickly through the body (Blaylock 1983; Chippaux 1999).

Characteristics of Envenomations Caused by Different Snake Families

Viperid Envenomations

Viperid bites always cause inflammation and often bleeding and tissue necrosis. The venom of viperids is made up of enzymes that disrupt the process of blood coagulation or lead to the destruction of tissues, including skin, muscle, and bone. The processes set in motion during these hemorrhagic syndromes are complex owing to frequent, often contradictory interactions between venom molecules.

Sequelae are common. These result from necrosis caused by (1) traditional local treatment, (2) venom proteases, or (3) secondary infection, which may ultimately make an amputation necessary, or from a post-thrombotic syndrome that can lead to a visceral infarction far from the site of the bite wound. Renal lesions are the most common of these and result in an anuria that lasts for weeks following the bite, even though the patient appears to be recovering well.

Elapid Envenomations

The venom of elapids is rich in neurotoxins that bind to the cholinergic receptors or ion channels of nerve cells, thereby blocking nerve signals. Envenomations by African elapids cause death by respiratory arrest but cause survivors no permanent aftereffects. In contrast to viperid envenomations, lasting neurological, cardiovascular, or renal damage has never been reported from elapid bites that were correctly treated. Such consequences as do occur are generally the result of inappropriate medical intervention, with the one exception of localized tissue necrosis sometimes caused by Spitting Cobra bites.

Atractaspis Envenomations

Envenomations by solenoglyph lamprophiid snakes of the genus *Atractaspis* are distinctive in their cardiotoxicity (disrupting the function of the heart) and represent a fairly constant 1% to 2% of venomous bites throughout the zone covered by this book. Bites from these snakes are rarely fatal but can cause significant pain and local tissue destruction. For further detail, see chapter 9.

Colubrid Envenomations

Duvernoy's gland secretions have become better understood since the development of technical analysis of proteins ("omics") (Hill and Mackessy 2000; Junqueira-de-Azevedo et al. 2016). Enzymes—mainly metalloproteinases, serine proteases, phospholipases A2, acetylcholinesterases, hyaluronidases, and activators of blood coagulation—are abundant. They explain the local inflammatory symptoms and hemostasis disorders observed in patients. Three-finger toxins, which are similar to those from elapids, have been described in several non-African rear-fanged snakes but could be found in African genera such as *Toxicodryas* or *Telescopus*. These toxins are responsible for paralysis of the striated muscles. Larger glycoproteins (cysteine-rich secretory protein) cause paralysis of the smooth muscles. Finally, proteins of variable size—or genomic sequences suggestive of their

presence—have been identified, including enzymatic inhibitors (Kunitz domain proteins, phospholipase A2 inhibitors), cell-cell adhesion proteins (C-type lectins), immune proteins (defensins, waprin-like proteins, cobra-venom factor), and proteins that promote growth of vascular endothelial (vascular endothelial growth) or neurons (nerve growth factor), the role of which in Duvernoy's gland secretions remains unknown. African opisthoglyph colubrids from two genera are known to have caused envenomations fatal to humans: the Boomslang (*Dispholidus typus*) and the Vine Snake or Twig Snake (genus *Thelotornis*).

Treatment of Snakebite

The use of traditional medicine to treat snakebite is universal throughout Africa. Too often, first aid measures are extreme and do more harm than good, as seen with mutilation that results in infection and reduced circulation to the injured region. Examples of such practices include scarification, excision of flesh around the bite wound, burning the site of the bite, application of foreign bodies and substances to the wound, and the like. On no account should a tourniquet be used, however, as it all but ensures the loss of a limb due to occlusion of blood flow as well as additional complicating factors such as blood acidosis, which can prove fatal to the patient upon release of the tourniquet. The use of "black stone," a piece of charred bone, is almost universal. Its complete ineffectiveness has been demonstrated experimentally (Chippaux et al. 2007a).

First aid for snakebite should be carried out quickly so as not to delay evacuation to a health center, which is the top priority.

The victim should be calmed so as to slow circulation of the venom and to prevent manifestation of psychosomatic symptoms that can complicate assessment and treatment. Constricting bands such as rings and watches should be removed from the bitten limb, and the site of the bite should be washed with soap and water or any available disinfectant. The bitten limb should be immobilized to slow the spread of venom through the body. In the case of elapid bites (neurotoxic envenomations), the entire limb should be wrapped in a pressure bandage (an elastic bandage—or several—of the kind used to treat a sprain) from below the site of the bite to the major joint of the bitten limb, and then back down again (Sutherland et al. 1979; Bush et al. 2004). Note that the pressure bandage technique is not recommended for viperid bites with tissue-destroying venoms (Seifert et al. 2011). If a non-aspirin pain reliever (e.g., acetaminophen = paracetamol) is available, it can be given to the patient. Aspirin and other nonsteroidal anti-inflammatory drugs should be avoided, however, as they can complicate the hemorrhagic effects of hemotoxic and cytotoxic venoms.

Antivenom remains the only specific treatment for venomous snakebites (Chippaux and Goyffon 1998). The latest-generation antivenoms are composed of highly purified antibodies, which have dramatically reduced both the incidence and the severity of adverse events (Chippaux et al. 2015a). Currently, however, antivenom is rarely available in health centers (Chippaux 1998b, 2002b). It must be administered intravenously to maximize its speed of action and its effectiveness.

Respiratory paralysis (resulting from severe elapid envenomation) requires the use of artificial respiration. This should be maintained until the victim starts to breathe again, which may take several days or even weeks (Visser and Chapman 1982).

Patients presenting with hemorrhagic syndromes (resulting from systemic viperid envenomation) can be difficult to resuscitate, since this depends heavily on the clinical course and etiology of symptoms. Surgical intervention performed too soon after the bite has the potential to result in bleeding and infection (Chippaux 1982).

The disappearance of antivenoms from the market over the past 30 years is a source of major concern (Chippaux 1998b, 2002b; Theakston and Warrell 2000). The primary reason is its cost, which has increased on average by more than 200%, while the economic crisis that has hit Africa harder than the rest of the world has further reduced the affordability of antivenom to most demographics in Africa. Reduced availability of antivenom—for reasons other than its cost and inappropriate application of the treatment—has undermined confidence in this vitally important treatment.

Families Typhlophidae and Leptotyphlopidae

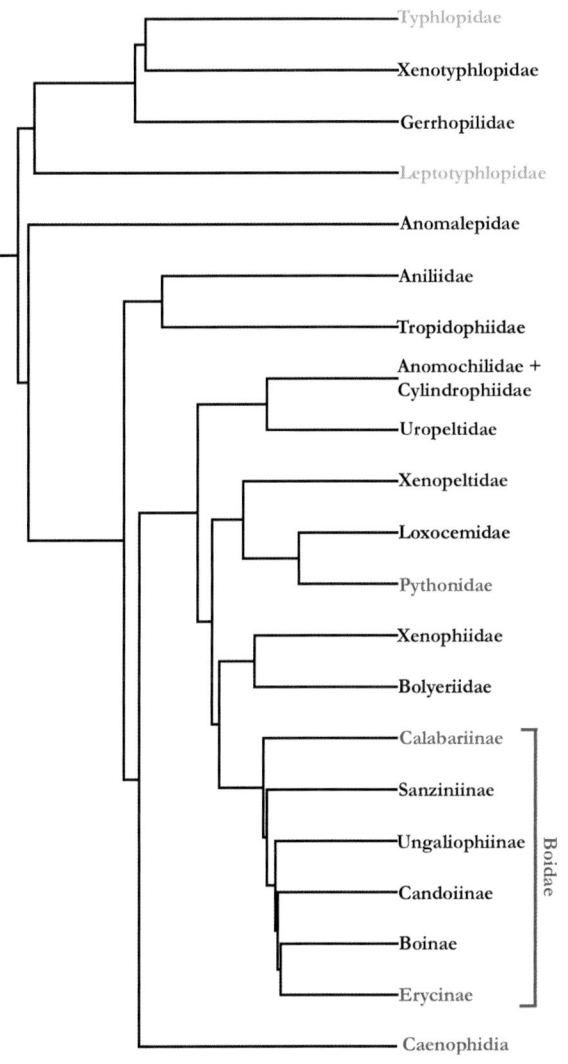

Typhlopidae

Xenotyphlopidae

Gerrhopilidae

Leptotyphlopidae

Anomalepidae

Aniliidae

Tropidophiidae

Anomochilidae + Cylindrophiidae

Uropeltidae

Xenopeltidae

Loxocemidae

Pythonidae

Xenophiidae

Bolyeriidae

Calabariinae

Sanziniinae

Ungaliophiinae

Candoiinae

Boinae

Erycinae

Boidae

Caenophidia

FIGURE 5.1. Higher-level phylogeny of snake families and subfamilies, with families and subfamilies occurring within our zone highlighted (red or orange). Scolecophidians occurring within our zone (Typhlopidae and Leptotyphlopidae) are highlighted in orange. Modified by K. Jackson from Zheng and Wiens (2016)

The Scolecophidia ("worm-like snakes") include 385 or so species worldwide. Within our zone, they are represented by 49 species belonging to the families Typhlopidae (Blind Snakes) and Leptotyphlopidae (Thread Snakes). Although the Scolecophidia are the nearest relatives of the Alethinophidia (true snakes), they diverged from them about 130 mya (Zheng and Wiens 2016), and the Typhlopidae and Leptotyphlopidae diverged from each other only shortly (in geological time) after that (Adalsteinsson et al. 2009; Vidal et al. 2009; Zheng and Wiens 2016), and they bear little resemblance to other snakes. We will deal with them in less detail than we devote to the alethinophidian snakes. The taxonomy followed here comes primarily from Hedges et al. (2014), from Pyron and Wallach (2014) for Typhlopidae, and from Adalsteinsson et al. (2009) for Leptotyphlopidae. A section near the end of chapter 1 provides a brief overview of taxonomic characters used for the study of typhlopids and leptotyphlopids, as these differ from those used for other snakes.

Both typhlopids and leptotyphlopids are fossorial egg-layers, adapted for life digging through soil. The body is cylindrical, and the head is often difficult to distinguish from the tail. Sometimes the rostral has a sharp edge for digging. The eyes are vestigial and often faintly visible underneath the large scales that cover the eye and part of the head. Members of the family Typhlopidae have a few teeth only on the maxilla, while members of the family Leptotyphlopidae have a few teeth only on the mandible. They feed primarily on insect larvae—mostly of social, burrowing insects such as ants and termites—which they ratchet into their tiny mouths using these small teeth (Kley and Brainerd 1997). In parts of central Africa, Blind Snakes and Thread Snakes are great-

ly feared and known as the "Five-minute Snake," referring to the belief that a person has only five minutes to live after being bitten by one. Some people believe that these snakes can sting with their tail, which sometimes ends in a tiny, pointed scale. All typhlopids and leptotyphlopids are completely harmless to humans.

African Giant Blind Snakes: Genus *Afrotyphlops* Broadley and Wallach (2009)

Afrotyphlops is a genus of 29 species distributed throughout most of sub-Saharan Africa. Thirteen species occur within our zone: *Afrotyphlops angeli, A. angolensis, A. anomalus, A. congestus, A. decorosus, A. elegans, A. liberiensis, A. lineolatus, A. manni, A. punctatus, A. schlegelii, A. schmidti, A. steinhausi.* Blind Snakes of the genus *Afrotyphlops* have a distinct eye (compared to the indistinct eye of *Letheobia*). The dorsum is usually dark in color. Although their size is variable, *Afro-*

FIGURE 5.2. *Afrotyphlops lineolatus* (Typhlopidae), Ethiopia. By S. Spawls

---

Key to the Families Typhlopidae and Leptotyphlopidae in Central and Western Africa

| 1 | More than 16 scale rows midbody ....Typhlopidae |
| 1' | 16 or fewer scale rows midbody .... Leptotyphlopidae |

Key to Central and Western African Genera of the Family Typhlopidae

| 1 | Eye indistinct .... *Letheobia* |
| 1' | Eye clearly visible .... 2 |

| 2(1) | Upper labials not in contact with the preocular or, at most, the second upper labial covering the preocular .... *Afrotyphlops* |
| 2' | Third upper labial overlaps the preocular .... 3 |

| 3(2) | 24 dorsal scale rows; 2 postoculars .... *Xerotyphlops etheridgei* |
| 3' | 20 dorsal scale rows; 1 postocular .... *Indotyphlops braminus* |

typhlops tend to have large, robust bodies. Schlegel's Giant Blind Snake, *A. schlegelii*, can reach almost 1 m in length, making it the largest typhlopid species in the world.

African Gracile Blind Snakes: Genus *Letheobia* Cope (1869)

Letheobia is a genus of 30 species, distributed in a broad band across sub-Saharan Africa, including central Africa, eastern Africa, and western Africa, and extending as far south as the Democratic Republic of the Congo and Tanzania. Eighteen species occur within our zone: *Letheobia acutirostrata, L. caeca, L. coecatus, L. crossi, L. debilis, L. feae, L. gracilis, L. graueri, L. kibarae, L. leucosticta, L. newtoni, L. pauwelsi, L. praeocularis, L. rufescens, L. stejnegeri, L. sudanensis, L. wittei, L. zenkeri*. *Letheobia* have an indistinct eye (compared to the distinct eye seen in *Afrotyphlops*). They are generally slender bodied and lack pigmentation so that they appear pink. Note that Pyron and Wallach (2014) assign the African species *L. coecatus* and *L. zenkeri* to the otherwise entirely West Indian genus *Typhlops*, on the basis of visceral morphology.

Desert Blind Snakes: *Xerotyphlops* Hedges, Marion, Lipp, Marin, and Vidal (2014)

The genus *Xerotyphlops* includes four species found in dry zones in Africa, Europe, and the Middle East. One species, *Xerotyphlops etheridgei* from Mauritania, occurs within our zone.

South Asian Blind Snakes: *Indotyphlops* Hedges, Marion, Lipp, Marin, and Vidal (2014)

Indotyphlops is a genus of 22 species found in south and east Asia. One species, *Indotyphlops braminus*, has been introduced to many parts of the world, including Africa, where it is now widespread in our zone. *I. braminus* is known by the common name "Flower Pot Snake," a reference to one of the ways it has reached so many continents, through the commercial trade in tropical potted plants. This species is particularly effective as a colonist because females can reproduce on their own by parthenogenesis (Wynn et al. 1987).

FIGURE 5.3. *Letheobia sp.* (Typhlopidae), Democratic Republic of Congo. By J. Kielgast

TABLE 5.1. Typhlopid species of central and western Africa

GENUS	SPECIES	ORIGINAL DESCRIPTION	DISTRIBUTION WITHIN OUR ZONE
Afrotyphlops	angeli	Guibé (1952)	Guinea
	angolensis	Bocage (1866a)	not in West Africa, but broad distribution in the rest of our zone
	anomalus	Bocage (1873)	southern Angola
	congestus	Duméril and Bibron (1844)	from Nigeria east to central and eastern Democratic Republic of Congo
	decorosus	Buchholz and Peters (1876)	Cameroon, Central African Republic
	elegans	Peters (1868)	Príncipe
	liberiensis	Hedges et al. (2014)	Guinea, Sierra Leone, Ivory Coast,
	lineolatus	Jan (1864)	West Africa to eastern Democratic Republic of Congo, Rwanda, and Burundi
	manni	Loveridge (1941)	Liberia, Guinea
	punctatus	Leach (1819)	West Africa to northern Democratic Republic of Congo
	schlegelii	Bianconi (1847)	Angola, Democratic Republic of Congo
	schmidti	Laurent (1956)	southeastern Democratic Republic of Congo
	steinhausi	Werner (1909a)	Nigeria to northern Democratic Republic of Congo
Letheobia	acutirostrata	Andersson (1916)	Democratic Republic of Congo
	caeca	Duméril (1856)	West Africa to eastern Democratic Republic of Congo
	coecatus	Jan (1864)	Ghana, Ivory Coast
	crossi	Boulenger (1893)	Nigeria
	debilis	Joger (1990)	Central African Republic
	feae	Boulenger (1906a)	São Tomé, Príncipe
	gracilis	Sternfeld (1910)	southern Democratic Republic of Congo
	graueri	Sternfeld (1912)	eastern Democratic Republic of Congo, Rwanda, Burundi
	kibarae	Witte (1953)	southern Democratic Republic of Congo (Katanga Province)
	leucosticta	Boulenger (1898)	Liberia
	newtoni	Bocage (1890)	São Tomé, Príncipe
	pauwelsi	Wallach (2005)	Gabon
	praeocularis	Stejneger (1893)	Nigeria to western Democratic Republic of Congo to Angola
	rufescens	Chabanaud (1916)	Central African Republic, northern Democratic Republic of Congo
	stejnegeri	Loveridge (1931)	Congo, west and central Democratic Republic of Congo
	sudanensis	Schmidt (1923)	eastern Democratic Republic of Congo
	wittei	Roux-Estève (1974)	central Democratic Republic of Congo
	zenkeri	Sternfeld (1908a)	Cameroon
Xerotyphlops	etheridgei	Wallach (2002)	Mauritania
Indotyphlops	braminus	Daudin (1803a)	introduced to Africa from Asia, now widespread

Key to Central and Western African Genera of the Family Leptotyphlopidae

1   16 midbody scale rows; striped pattern .... *Rhinoleptus*
1'   14 midbody scale rows; no striped pattern 2

2(1) Dark brown above and below .... *Leptotyphlops*
2'   Underside paler than above 3

3(2) Light brown above, underside white .... *Myriopholis*
3'   Brown above, underside light brown 4

4(3) 1 or 2 upper labials; maximum length 322 mm .... *Namibiana*
4'   3 upper labials; maximum length 188 mm .... *Tricheilostoma*

African Thread Snakes: Genus *Leptotyphlops* Fitzinger (1843)

Leptotyphlops is a genus of 22 species found in southeastern Africa. Five species occur within our zone: *L. conjunctus*, *L. emini*, *L. kafubi*, *L. nigricans*, *L. scutifrons*.

Many-scaled Thread Snakes: Genus *Myriopholis* Hedges, Adalsteinsson and Branch in Adalsteinsson, Branch, Trape, Vitt, and Hedges (2009)

This genus includes 24 species distributed throughout Africa and extending through the Arabian Peninsula and into southwestern Asia. Eight species occur within our zone: *Myriopholis adleri, M. albiventer, M. algeriensis, M. boueti, M. macrorhyncha, M. narirostris, M. natatrix, M. perreti.* The genus name *Myriopholis* (meaning "many scales") refers to the numbers of subcaudals and middorsal scale rows, which are typically high in members of this genus compared with other leptotyphlopid genera.

FIGURE 5.4. *Leptotyphlops scutifrons* (Leptotyphlopidae), Botswana. By S. Spawls

FIGURE 5.5. *Myriopholis macrorhyncha* (Leptotyphlopidae), Ethiopia. By S. Spawls

TABLE 5.2. Leptotyphlopid species of central and western Africa

GENUS	SPECIES	ORIGINAL DESCRIPTION	DISTRIBUTION WITHIN OUR ZONE
Leptotyphlops	*conjunctus*	Jan (1861)	southeastern Democratic Republic of Congo
	emini	Boulenger (1890)	eastern Democratic Republic of Congo
	kafubi	Boulenger (1919a)	Angola, southern Democratic Republic of Congo
	nigricans	Schlegel (1839)	Angola, southeastern Democratic Republic of Congo
	scutifrons	Peters (1854)	Angola
Myriopholis	*adleri*	Hahn and Wallach (1998)	West Africa extending as far east as Chad
	albiventer	Hallermann and Rödel (1995)	West Africa
	algeriensis	Jacquet (1896)	northern West Africa
	boueti	Chabanaud (1917)	West Africa
	macrorhyncha	Jan (1860)	extending from West Africa to Egypt to the Middle East
	narirostris	Peters (1867)	West Africa
	natatrix	Andersson (1937)	Gambia
	perreti	Roux-Estève (1979)	Cameroon, Gabon, Congo?
Tricheilostoma	*bicolor*	Jan (1860)	West Africa, possibly reaching as far east as Cameroon
	broadleyi	Wallach and Hahn (1997)	Ivory Coast (known only from type specimen)
	greenwelli	Wallach and Boundy (2005)	southwestern Nigeria
	sundewalli	Jan (1861)	Ghana to Cameroon to Central African Republic
Rhinoleptus	*koniagui*	Villiers (1956)	West Africa
Namibiana	*labialis*	Sternfeld (1908a)	southern Angola

West African Forest Thread Snakes: Genus *Tricheilostoma* Jan (1860)

The genus *Tricheilostoma* includes four species found in rainforest habitat in west Africa. All four species occur within our zone: *Tricheilostoma bicolor, T. broadleyi, T. greenwelli, T. sundewalli.*

Note that the genus name *Tricheilostoma* was formerly applied to leptotyphlopids of the New World genus now known as *Trilepida.* African *Tricheilostoma* were formerly assigned to the genus *Guinea.* Hedges (2011) replaced the genus *Guinea* with *Tricheilosto-*

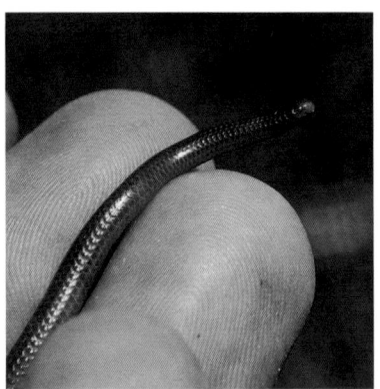

FIGURE 5.6. *Tricheilostoma bicolor* (Leptotyphlopidae), Ghana. By S. Spawls

ma in order to rectify an historical error of nomenclature.

Slender-snouted Thread Snakes: Genus *Rhinoleptus* Orejas-Miranda, Roux-Estève, and Guibé (1970)

The genus *Rhinoleptus* includes two species, one from West Africa and one from east Africa. The West African species, *Rhinoleptus koniagui*, occurs in our zone, with a distribution including Guinea, Mali, and Senegal. Snakes of this genus get their name from the distinctively shaped rostral scale, which is usually narrow and pointed.

Namibian Thread Snakes: Genus *Namibiana* Hedges, Adalsteinsson and Branch in Adalsteinsson, Branch, Trape, Vitt, and Hedges (2009)

The genus *Namibiana* includes five species found in southwest Africa. One of these, *Namibiana labialis*, occurs within our zone, with its range extending into southern Angola.

6

Families Boidae and Pythonidae

The Alethinophidia (true snakes) other than the Caenophidia (recent snakes) are sometimes called the "Henophidia." This term refers to a paraphyletic assemblage of 210 or so species worldwide. Henophidians often have common names "boa" and "python," and the characteristics many of them share include retained primitive features such as spurs (pelvic vestiges), teeth on the premaxilla, and presence of a left lung. Representatives of this assemblage occurring within our zone include members of the families Boidae (*sensu* Zheng and Wiens 2016) and Pythonidae. Recent molecular and paleontological studies indicate that these two families diverged between 109 and 64 mya and are not each other's closest relatives (Noonan and Chippindale 2006; Vidal et al. 2009; Hsiang et al. 2015; Zheng and Wiens 2016).

Pythons (family Pythonidae) have an Old World distribution (sub-Saharan Africa, south and Southeast Asia, Australia) and lay eggs. Within our zone, they are represented by four species in the genus *Python*. The family Boidae has proven more complicated. We use the family Boidae here in the sense used by recent authors (Pyron et al. 2013; Zheng and Wiens 2016) (see Fig. 6.1) to include lineages collectively distributed in the New World (Boinae, Ungaliophiinae): Madagascar (Sanziniinae), southwest Pacific Islands (Candoiinae), central Africa (Cal-abariinae), and northern Africa eastward through central Asia (Erycinae). Within our zone, the Boidae are represented by the Calabar Boa, *Calabaria reinhardtii*, the only member of the monotypic genus *Calabaria* (subfamily Calabariinae), and by two species of Sand Boa (genus *Eryx*, subfamily Erycinae).

Family: Boidae

Genus *Calabaria* Gray 1858

Calabar Boa: *Calabaria reinhardtii* (Schlegel 1851)

Calabaria is a monotypic genus and the sole representative of the subfamily Calabariinae (family: Boidae). *Calabaria reinhardtii* is a burrowing species found in forested areas of central and western Africa. Its range extends from Liberia to Lake Kivu in eastern Democratic Republic of Congo and includes Bioko Island, Equatorial Guinea. The type locality is "Gold Coast" (meaning Ghana).

C. reinhardtii are nocturnal and crepuscular in their habits, found in forest leaf litter and most often encountered above ground after rain. A population inhabiting forest and plantation mosaic habitat in Nigeria were found to feed almost entirely on small mammals, rodents, and shrews, including nestlings. The only exception found in this same study was one example of snake eggs (probably *Psammophis phillipsii*) as stomach contents. Both nestling rodents and snake eggs are dietary items that reflect the fossorial foraging habits of *C. reinhardtii*. Females were found to lay clutches of 2–9 eggs in alternate years, with larger females producing larger clutches (Luiselli et al. 2002).

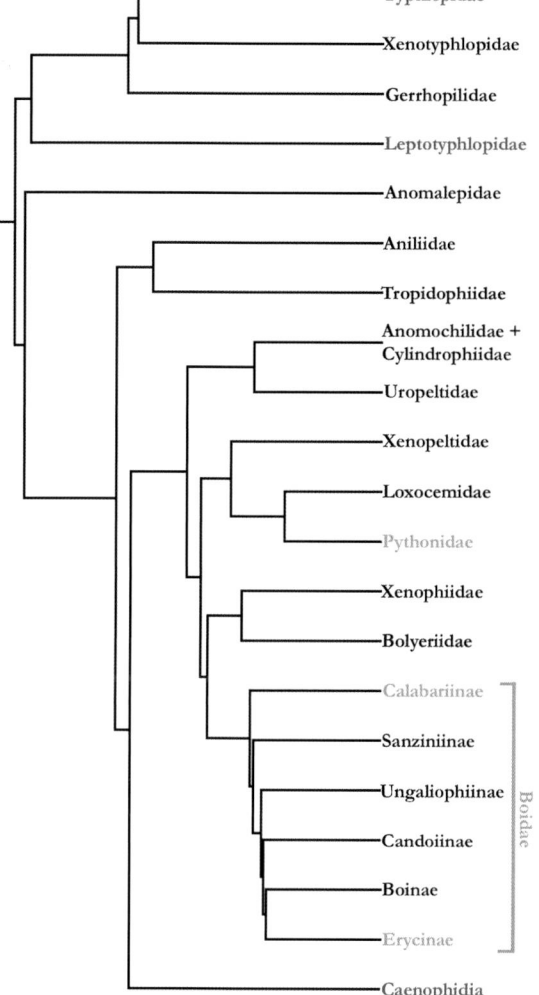

FIGURE 6.1. Higher-level phylogeny of snake families and subfamilies, with families and subfamilies occurring within our zone highlighted (red or orange). The family Pythonidae and the subfamilies of the family Boidae occurring within our zone are highlighted in orange. Modified by K. Jackson from Zheng and Wiens (2016)

When threatened, *C. reinhardtii* will often coil the body with the head protected inside the coils and the blunt, rounded tip of the tail held above the body so that it resembles the head.

The common name "Ground Python" is sometimes applied to *C. reinhardtii*, reflecting historical confusion about its taxonomic affiliations. Whereas all pythons (Pythonidae) lay eggs, boas (Boidae) usually give birth to live young. Thus *C. reinhardtii* was once thought to be a python because it lays eggs. It is now recognized as a boid.

The scales of the head and body are glossy brown or reddish above and below, with a scattering of iridescent white scales. The tail tends to be darker than the rest of the body. The head is short and blunt with no distinct neck separating it from the thick, cylindrical body. The tail is short and thick with a blunt end, resembling the head. The eye is small with a vertically elliptical pupil. The head scales consist of large platelike scales. The nasal scale is single. There is one pair of internasals and two pairs of prefrontals, sometimes fused to one another or further subdivided. There is a loreal scale. There are 2 postoculars, 1 preocular, and 2 supraoculars. The frontal scale is broader than it is long. There are 8(3,4) or sometimes 7(3,4) upper labials and 8–11(0) lower labials. There are no sensory pits on the upper or lower labials scales. There may be one small pair of chin shields or no chin shields. There is no submandibular groove. The dorsal scales are smooth, without apical pits, and arranged in 29–37 straight rows. The ventral scales are narrower than the ventral surface in contact with the ground. There are 218–242 ventrals and 19–28 single subcaudals. The cloacal scale is single. There is a spur on either side of the cloaca.

FIGURE 6.2. *Calabaria reinhardtii* in defensive posture, Republic of Congo. The individual shown here is getting ready to shed, hence the opaque eye and overall faded appearance. By K. Jackson

FIGURE 6.3. *Calabaria reinhardtii*, captive. By J. Murphy

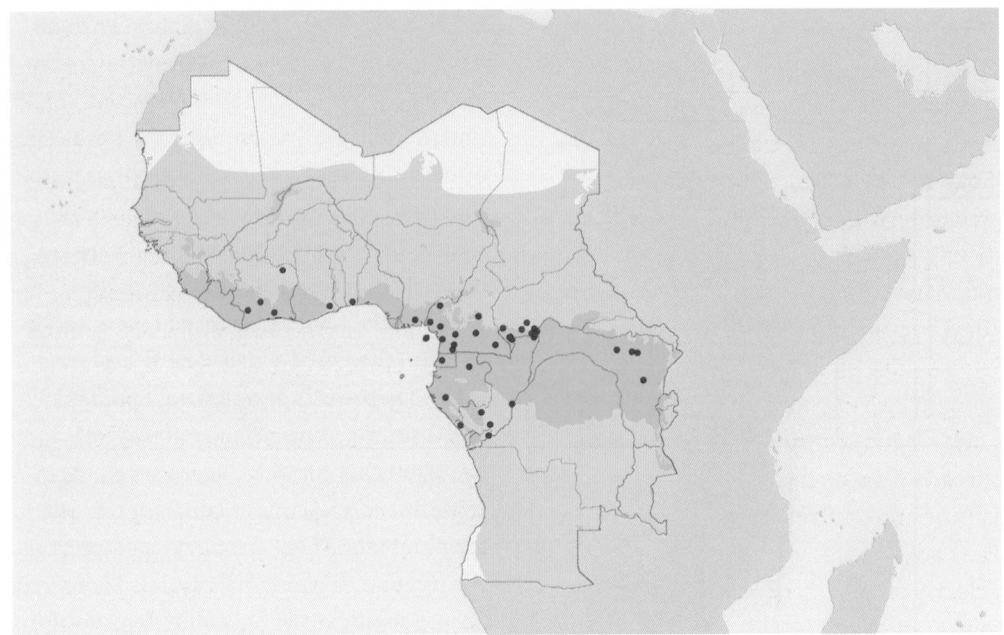

FIGURE 6.4. *Calabaria reinhardtii* (blue). By K. Jackson

FIGURE 6.5. *Calabaria reinhardtii* RMCA 29853. By T. Giri

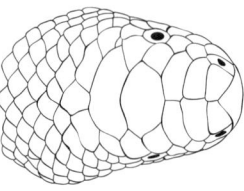
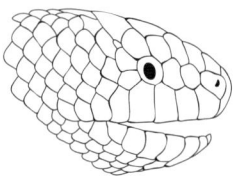
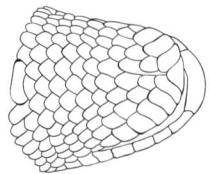

The maximum length recorded for this species is 1,030 mm (Villiers 1966), but the average adult length is closer to 600 mm. The hemipenes are unforked, and the sulcus spermaticus is not forked. The maxillary dentition consists of 12–19 short, thick unspecialized teeth, decreasing in size posteriorly.

Sand Boas: Genus *Eryx* Daudin 1803a

Belonging to the subfamily Erycinae (family: Boidae), *Eryx* is a genus of 12 species of small, primarily nocturnal, desert-adapted boas, distributed through desert areas of north Africa, Asia, and the Middle East. Two species occur within our zone. The maxillary dentition of Sand Boas occurring within our zone consists of 13–18 unspecialized teeth, decreasing in size posteriorly. The diet of adults consists mainly of rodents and birds, while juveniles may feed on lizards (Spawls et al. 2004). Sand Boas are ambush predators, hiding themselves in sand and striking at their prey as it passes by (Trape and Mané 2006a). Once caught, the prey is killed by constriction before swallowing. Females give birth to litters of up to 20 live young. The head is small and short with a blunt, rounded snout, which is used for burrowing into sand. The nostrils are dorsally oriented to avoid breathing sand in. The body is thick and cylindrical without a clearly defined neck. The tail is short and cone shaped with a pointed tip. The eye is small with a vertically elliptical pupil. The head scales are small and relatively undifferentiated. There are no sensory pits on the upper or lower labial scales. The dorsal scales may be smooth or keeled, lack apical pits, and are arranged in 37–59 straight rows. The vertebral row is not enlarged relative to the other dorsal scale rows. The ventral scales are narrower than the ventral surface of the snake. The subcaudals are single. The cloacal scale is single. There is a spur on either side of the cloaca. The hemipenes are unforked. The sulcus spermaticus is forked for the distal two-thirds of the organ.

East African Sand Boa: *Eryx colubrinus* (Linnaeus 1758)

Eryx colubrinus is found in desert, sahel, and savannah habitats from Niger to Egypt. The type locality is Egypt. The overall coloration is pale yellowish or pinkish with darker dorsal blotches that fade into the background coloration at the edges. The underside is pale pinkish or cream. There are 12–15 perioculars and 9–12 interoculars. There are 12–13 upper labials, none of which is in contact with the eye, and 11–14 lower labials. There are no chin shields. There is no submandibular groove. The dorsal scales are weakly keeled and arranged in 39–59 straight rows. There are 162–198 ventrals and 20–28 single subcaudals. The cloacal scale is single. The maximum length recorded for this species is 520 mm (Villiers 1950).

Key to Central and Western African Species of the Genus *Eryx*

1   Fewer than 7 interocular scales .... *E. muelleri*
1' More than 8 interocular scales .... *E. colubrinus*

FIGURE 6.6. *Eryx colubrinus*, Tanzania. By S. Spawls

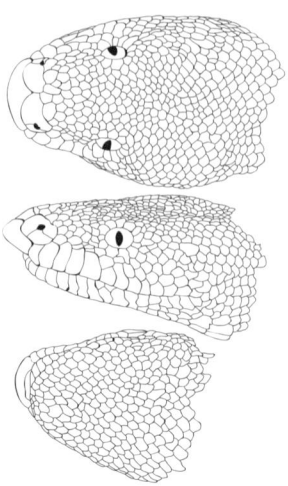

FIGURE 6.7. *Eryx colubrinus* 1914.4.29.2. By T. Giri

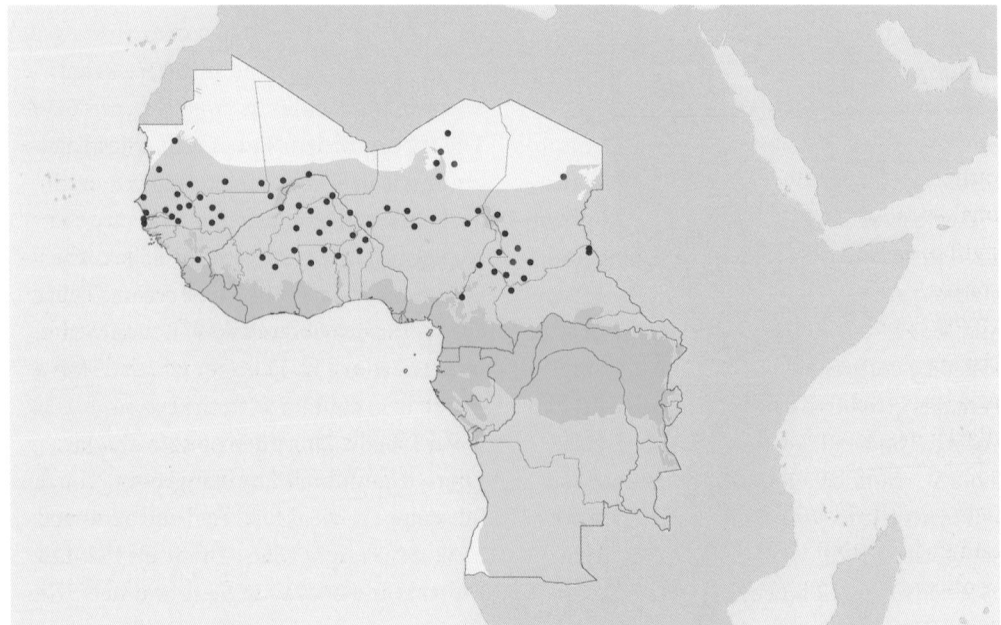

FIGURE 6.8. *Eryx muelleri* (blue), *Eryx colubrinus* (red). By K. Jackson

Sahara Sand Boa: *Eryx muelleri* (Boulenger 1892)

The range of *Eryx muelleri* extends from Mauritania to Sudan and south as far as Ivory Coast. Type locality is Sennar Sudan. This is a savannah species, found in gen-erally less dry habitats than *E. colubrinus*. This species lives in vegetated areas where the young feed primarily on lizards and the adults on small rodents (Vignoli et al. 2016). The overall coloration is pale yellowish or pinkish with darker, clearly delineated, dorsal blotches. The underside is pale pinkish or cream. The markings of adults fade as the animal ages. There are 8–11 perioculars and 5 interoculars. There are 8–10 upper labials, none of which is in contact with the eye, and 10–13 lower labials. There are no chin shields. There is no submandibular groove. The dorsal scales are smooth and arranged in 37–48 straight rows. There are 172–188 ventrals and 14–24 single subcaudals. The cloacal scale is single. The maximum length recorded for this species is 700 mm (Villiers 1950). One subspecies, *E. m. subniger* Angel 1938, was described but is no longer recognized (Tokar 1995).

Family: Pythonidae

Pythons: Genus *Python* Daudin 1803b

The genus *Python* includes ten species in Africa and Asia, four of which occur within our zone. This genus includes the largest snakes in Africa. Because of their large size, pythons are in demand by local populations for their meat and their skin. Akani et al. (1999) considered *Python sebae* to be one of the species most affected by human hunting and loss of habitat, leading to rapid population declines. Internationally, they are sought out, often in large numbers, by the pet trade (Buffrénil 1995; Ineich 2006; Reed and Rodda 2009). According to the Convention on International Trade in Endangered Species (CITES), based on information obtained from official importers and exporters of animals, there were 28,600 specimens of *P. sebae* exported between 1995 and 2000 from just 10 countries in sub-Saharan Africa. Exports are currently decreasing sharply and have not exceeded 2,000 specimens per year since 2002 (Luiselli et al. 2012). This sharp decline was probably due to market conditions, such

as exchange rates, which prompted buyers to fall back on captive-bred animals, rather than political pressure by the European Union and the United States to restrict trade on live or dead animals. The average sizes reported of pythons in the wild are diminishing, when compared to the older literature, presumably reflecting the impact of hunting pressures, or other human environmental impacts, on populations.

The maxillary dentition consists of 15–30 long, recurved teeth decreasing in size posteriorly. Although not venomous, a defensive bite by one of the larger *Python* species (e.g., *P. sebae, P. natalensis*) can cause a serious wound to a human, simply as a result of the mechanical damage caused by the teeth. Large Boidae and Pythonidae do attack primates, including humans, under certain circumstances. This has been studied by Headland and Greene (2011) among Agta Negritos hunters in the Philippines. In Africa, accidents have not been documented systematically, but the literature reports several deaths, especially of young children, attributable to *P. sebae* (Branch and Håcke 1980).

The head is long and triangular, and the neck well defined. The body is stout and the tail is short. The eye is small with a vertically elliptical pupil. The head scales include some large platelike scales and others that are small and undifferentiated. The nasal may be single, divided, or semi-divided. The internasals and prefrontals are paired. The loreal is broken up into several smaller scales. There are 2 or more preoculars, 2 or more postoculars, 1 or more supraoculars, 2 or more frontals, and a variable number of suboculars. The scales of the temporal region are small and undifferentiated. There are 9–17 upper labials and 15–23 lower labials. Some of the labial scales bear sensory pits. The chin shields are absent. The submandibular groove is pronounced. The dorsal scales are smooth, without apical pits, and are arranged in straight rows. The vertebral row is not enlarged relative to the other dorsal scale rows. The ventral scales are narrower than the ventral surface of the snake. The subcaudals are divided. The cloacal scale may be single or divided. A pair of spurs, which represent the vestiges of the hind legs, are present on either side of the cloaca and are larger in males than in females. The hemipenes are unforked with the sulcus spermaticus divided for the distal two-thirds.

Angolan Python / Dwarf Python: *Python anchietae* Bocage 1887a

The range of this small python extends from southern Angola to northern Namibia. The type locality is Catumbela, near Lobito, Angola.

Python anchietae lives in areas of grassy savanna, sometimes more or less rocky. This species is similar to *P. regius* both in size and behavior. Rather shy and fearful, it feeds mainly on small mammals, such as gerbils, birds (Branch 1994), and sometimes lizards or arthropods (Colclough 2016). In the wild, females usually lay clutches of fewer than 6 eggs, but in captivity, clutch size can reach as many as 10 eggs (Colclough 2016). Hatchlings measure about 450 mm in length.

The overall coloration above ranges from light to dark brown with large round or oval yellow blotches outlined in black. The dorsum of the head is the same color as the body. Two stripes, a yellow one with a black one beneath it, run along the sides of the head on each side from the nostril through the eye to the neck. The upper and lower labials are pale. The venter is whitish. The nasal may be single or divided. There are 4–5 preoculars, 2 or more postoculars, 1 or more supraoculars, and 2–3 suboculars. There are 15–18 perioculars and about 10 interoculars. There are 12–15 upper labials, none of which is in contact with the eye. Four or five of the upper labials have sensory pits. There are 15–16 lower labials. The dorsal scales are arranged in 55–61 straight rows. There are 253–267 ventrals and 46–57 divided subcaudals. The cloacal scale may be single or divided. The maximum length recorded for this species is 1,800 mm (Branch 1994).

Key to Central and Western African Species of the Genus *Python*

1	Fewer than 65 dorsal scale rows	2
1'	More than 65 dorsal scale rows	3
2(1)	More than 250 ventrals	*P. anchietae*
2'	Fewer than 210 ventrals	*P. regius*
3(1)	Fewer than 5 large scales between the eyes	*P. sebae*
3'	More than 8 small scales between the eyes	*P. natalensis*

FIGURE 6.9. *Python anchietae*, captive. By J. Murphy

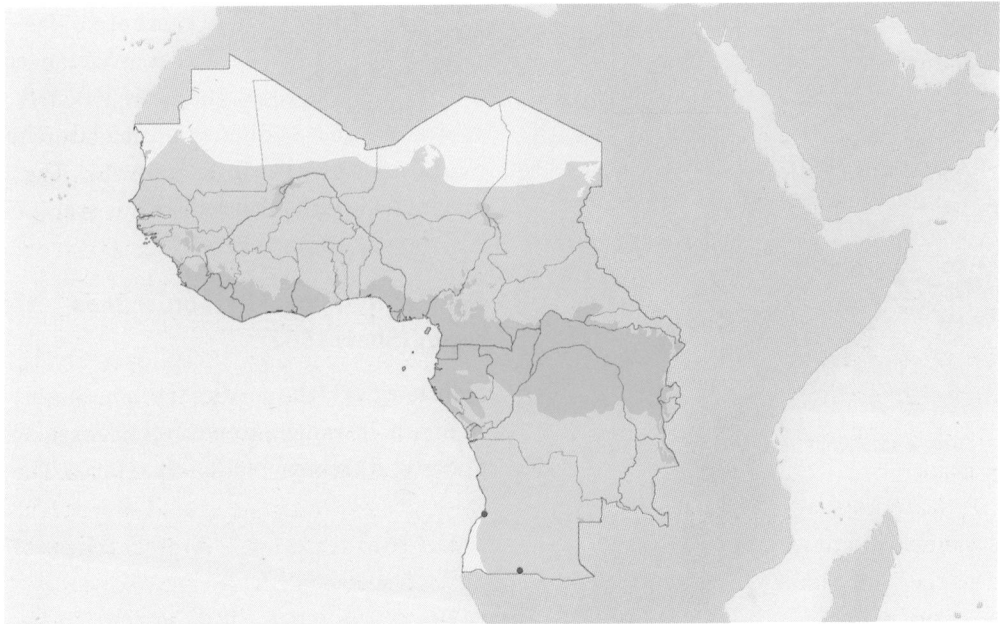

FIGURE 6.10. *Python anchietae* (red). By K. Jackson

Southern African Python: *Python natalensis* Smith 1840

The range of *Python natalensis* extends from southern Angola east into Kenya and south to South Africa. The type locality is Port Natal (=Durban, South Africa). *P. natalensis* was formerly considered a subspecies of *P. sebae*, which it resembles in overall appearance, though *P. natalensis* is probably smaller than *P. sebae* (Reed and Rodda 2009). *P. natalensis* differs from *P. sebae* in its distribution, which extends farther south than that of *P. sebae*, but the ranges of the two species overlap slightly in Kenya, Tanzania, and Zimbabwe. *P. natalensis* lives in drier (10 mm of monthly rainfall in months) and colder (5°C to 7°C lower) habitats than *P. sebae* (Reed and Rodda 2009). The mating and egg-laying seasons of *P. natalensis* are reversed relative to those of *P. sebae* owing to the northern distribution of *P. sebae* and southern one of *P. natalensis*.

P. natalensis can be distinguished from

P. sebae in the pattern on the side of the head: in *P. natalensis*, the dark patch in front of and posterior to the eye is paler and narrower than in *P. sebae*, giving the appearance of a dark stripe as opposed to a yellow stripe at the level of the eye. The overall coloration of *P. natalensis*, though similar to *P. sebae*, is drabber. The scales on the dorsum of the head are small and medium sized in *P. natalensis* as opposed to medium and large sized in *P. sebae*, and there are small differences in some of the other head scale counts. The nasal may be single or divided. There are 2–3 preoculars, 3–4 postoculars, 2–3 supraoculars, and a variable number of suboculars. There are 8–13 perioculars and 6–10 interoculars. There are 10–16 upper labials, none of which is in contact with the eye. There are sensory pits on two of the upper labials, as well as smaller ones on three of the lower labials. There are 17–20 lower labials. The dorsal scales are arranged in 76–99 straight rows. There are 260–291 ventrals and 63–84 divided subcaudals. The cloacal scale may be single or divided. The maximum length recorded for this species is 5,600 mm (Branch 1994).

Ball Python / Royal Python: *Python regius* (Shaw 1802)

Python regius is the smallest python species in Africa. Its range extends from Senegal to Sudan and Uganda, and south as far as Con-

FIGURE 6.11. *Python natalensis*, Botswana. By S. Spawls

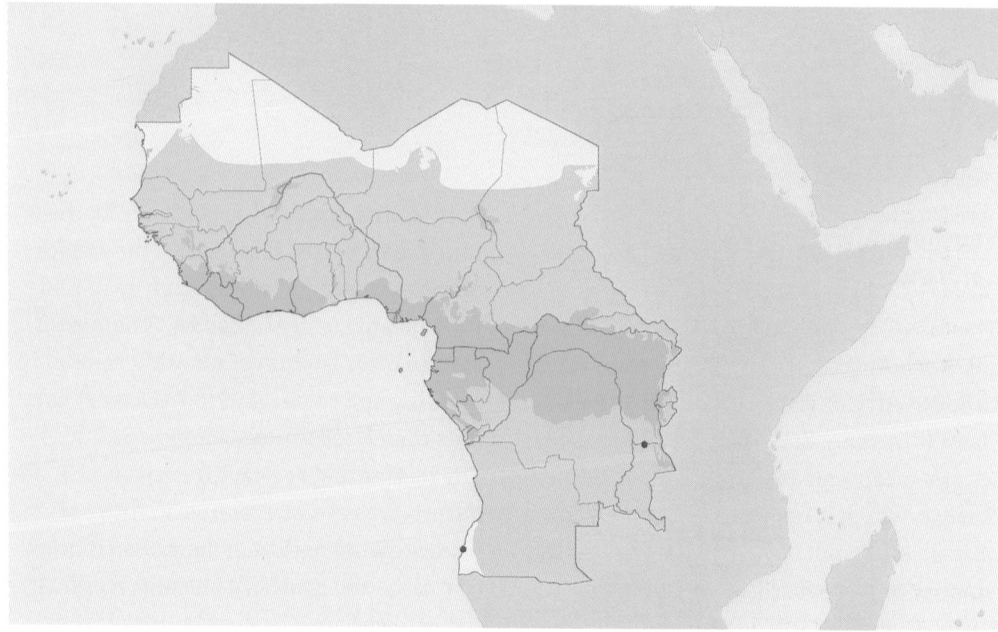

FIGURE 6.12. *Python natalensis* (red). By K. Jackson

go and the Democratic Republic of Congo. The type locality is unknown.

Ball Pythons are typically extremely tame and get their common name from their habit of rolling themselves into a tight ball with the head protected inside the coils. This tameness of disposition has contributed to their enormous popularity as pet snakes, but their timidity can also lead them to be reluctant to eat, so they often don't do well in captivity. Wild-caught Ball Pythons are imported from West Africa in enormous numbers for the international pet trade. Their diet consists of mammals and sometimes birds. Ball pythons lay small clutches (up to 10) of relatively large eggs (60 mm in length by 40 mm in width) (Chippaux pers. obs.).

The dorsal coloration is dark overall with a pattern of large round or oval yellow spots outlined in black. The dorsum of the head is black. There is a yellow stripe from nostril to eye on each side of the head and two yellow stripes posterior to the eye. The upper and lower labials are pale. The venter is whitish. The nasal is single. There are 2–4 preoculars, 3–4 postoculars, 1–2 supraoculars, and no suboculars. There are 9–12(5) or (5,6) or (6) upper labials. Three or four of the upper labials have sensory pits. There are 15 or so lower labials. The dorsal scales are arranged in 51–63 straight rows. There are 191–207 ventrals and 28–37 divided subcaudals. The cloacal scale may be single or divided. The maximum length recorded for this species is 1,440 mm (Trape pers. comm.).

African Rock Python: *Python sebae* (Gmelin 1789)

The African Rock Python is a broadly distributed species found in all sorts of habitats ranging from forests to savannah and sahel. In the rivers and flooded forests of Congo, juveniles are frequently caught in traps and gill nets set to catch fish (Jackson pers. obs.). Reports from the early twentieth century indicate its presence in southern Algeria, in the Sahara Desert (Angel and Lhote 1938). Its range extends from Senegal to Somalia and as far south as Zimbabwe. In the United States, this species has become an introduced pest in Florida as a result of escaped snakes imported for the pet trade. The type locality is Kadugli, Sudan. African Rock Pythons are mostly nocturnal and terrestrial, but they are often aquatic and may sometimes climb trees. Juveniles eat birds, frogs, fishes, and small mammals. The diet of adults consists mainly of mammals, but *Python sebae* is versatile and adaptable in its diet. Luiselli et al. (2001a) found that *P. sebae* in forested areas of Nigeria ate rodents, bats, genets, and monkeys, as well as some reptiles (stomach contents of the largest individuals included a dwarf crocodile, *Osteolaemus tetraspis*, and a monitor lizard, *Varanus niloticus*), while *Python sebae* in semi-urban areas ate dogs, goats, rats, and poultry. Rock Pythons lay clutches of up to 100 eggs.

The overall coloration is dark, made up of a dorsal pattern in black, brown, and yellow of large round, oval, or irregular blotches with dark centers and pale edges. On the sides of the head there is a yellow stripe at the level of the eye, with a broad dark blotch on the side of the head in front of and posterior to the eye. The upper labials are generally dark with pale edges. The lower labials are pale. The venter is whitish with dark spots. The scales on the dorsum of the head are medium and large sized. The nasal is semi-divided. There are 2–4 preoculars, 3–4 postoculars, 1–2 supraoculars, and

FIGURE 6.13. Close-up of the head of a captive specimen of *Python regius*. By M. Rubio

2–3 suboculars. There are 9–12 perioculars and 4 interoculars. There are 11–17 upper labials, none of which is in contact with the eye. There are sensory pits on two of the upper labials, as well as smaller ones on three of the lower labials. There are 15–23 lower labials. The dorsal scales are arranged in 71–99 straight rows. There are 259–294 ventrals and 55–80 divided subcaudals. The cloacal scale is usually single but sometimes divided. The maximum length recorded for this species is 9,800 mm (Bérart 1932 cited by Roux-Estève 1969), but the average length of adults is 3,500 mm.

FIGURE 6.14. *Python regius*, rolled into a ball in defensive posture. Ghana. By M. Fujita

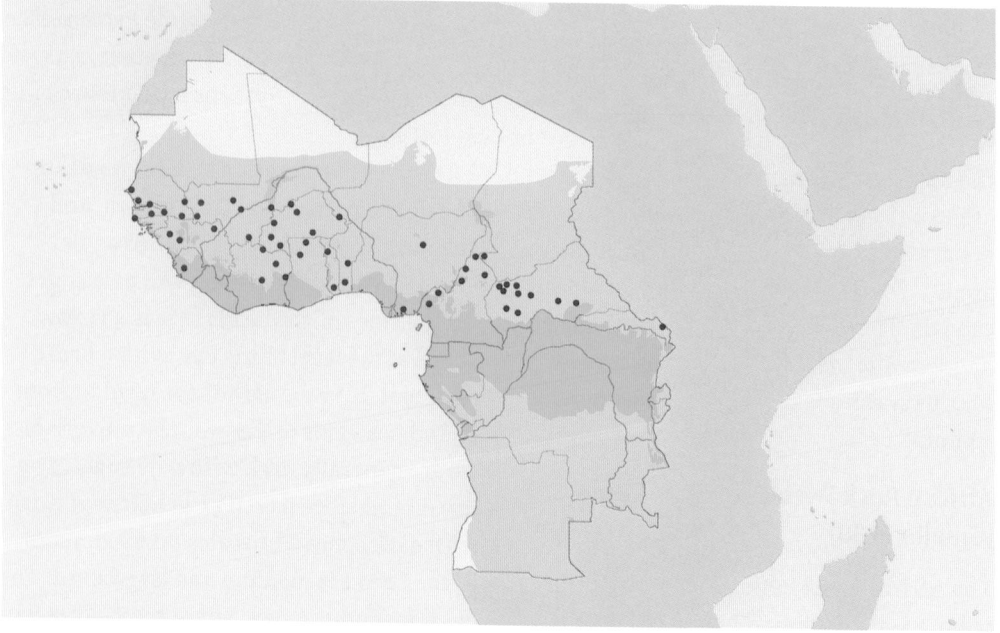

FIGURE 6.15. *Python regius* (blue). By K. Jackson

FIGURE 6.16. *Python sebae*, Kenya. By S. Spawls

FIGURE 6.17. *Python sebae* juvenile, Benin. By M.-O. Rödel

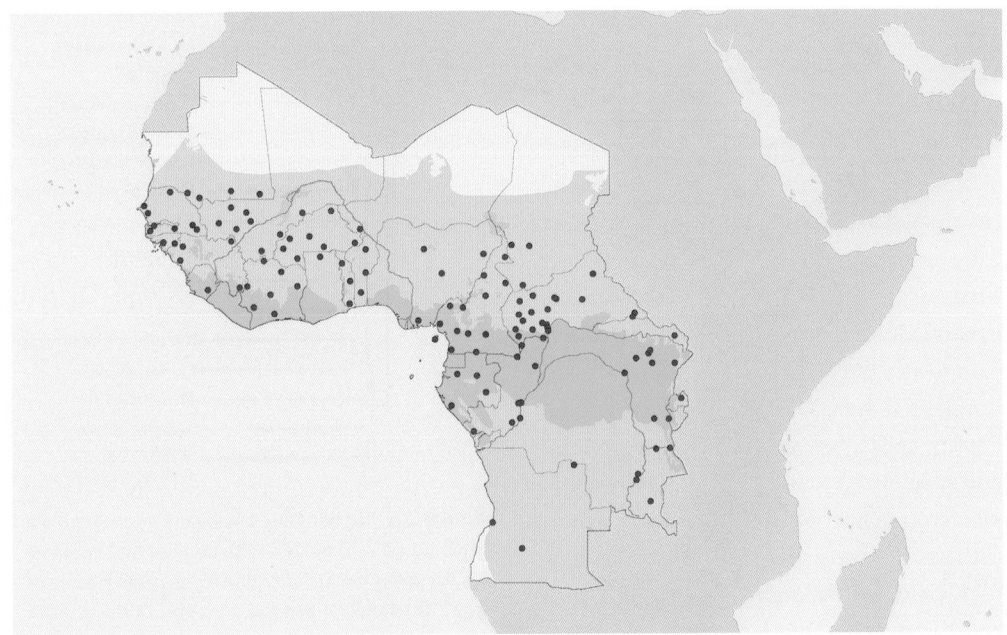

FIGURE 6.18. *Python sebae* (blue). By K. Jackson

7

Family Viperidae

The family Viperidae includes just over 300 species and has an almost worldwide distribution (absent from Madagascar and Australia). The Viperidae divided approximately 47 mya into two main lineages: the subfamily Crotalinae (New World and Asian Pit Vipers) and the subfamily Viperinae (African and Eurasian Vipers). All viperids occurring in our zone originated with the major radiation of the Viperinae in Africa 38 mya. The five viperid genera now present within our zone are thought to have originated with the diversification of this African branch of the Viperinae into two African lineages: one giving rise to the genera *Bitis* and *Atheris*, the other to *Causus*, *Cerastes*, and *Echis* (Wüster et al. 2008).

The viperid genera found within our zone represent a high degree of ecological diversity. The Bush Vipers, genus *Atheris*, found in forested areas of sub-Saharan Africa, are fully arboreal with prehensile tails. Desert Vipers, genus *Cerastes*, are adapted for life in sand and hunt as sit-and-wait predators, completely buried except for their dorsally oriented eyes. The Night Adders, genus *Causus*, are largely nocturnal, terrestrial, and specialized toad-eaters. Like the solenoglyph lamprophiid genus *Atractaspis* (Stiletto Snakes), some *Causus* species have highly elongate venom glands, extending one-third of the length of the body. *Causus* are unique among African vipers in having

large platelike head scales similar to those of colubrids and lamprophiids rather than the small, relatively undifferentiated head scales seen in most viperids. Giant vipers of the genus *Bitis*, such as the broadly distributed Puff Adder, *Bitis arietans*, can produce litters of up to 100 live young. One species of Saw-scaled Viper, genus *Echis* (*Echis ocellatus*), is responsible for more human deaths than any other African snake because of its potent hemorrhagic ven-

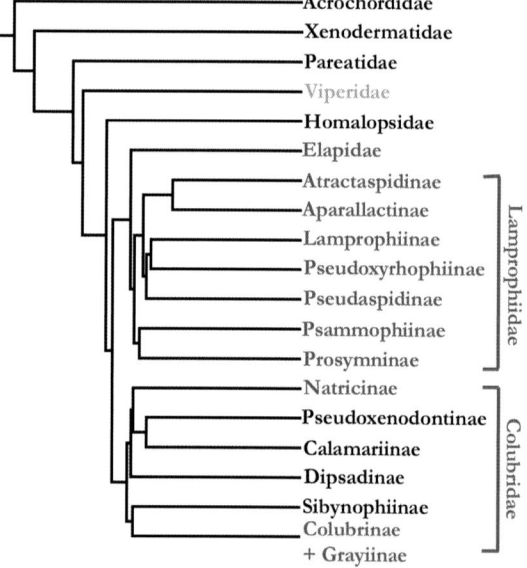

FIGURE 7.1. Higher-level phylogeny of caenophidian families and subfamilies, with families and subfamilies occurring within our zone highlighted (red or orange). The family Viperidae, the topic of this chapter, is highlighted in orange. Modified by K. Jackson from Zheng and Wiens (2016)

om as well as its small size and ability to adapt well to human-altered and densely human-populated habitats.

Bush Vipers: Genus *Atheris* Cope 1862

The Bush Vipers, *Atheris*, are a genus of 16 species of small, arboreal viper found in tropical Africa. Nine species occur within our zone. *Atheris* are nocturnal and primarily arboreal. They give birth to litters of about 7–9 live young. Bush vipers are small and usually brightly colored. Juveniles may be especially bright and quite different in coloration from adults. For example, *Atheris chlorechis* adults are uniformly bright green, while neonates are bright yellow. Sometimes the tip of the tail may be especially brightly colored and used as a lure, the snake wiggling the tail tip so that it resembles a maggot or caterpillar, in such a way as to entice potential prey to approach within striking distance. Luiselli et al. (2000a) studied habitat use and diet in *A. squamigera* in southeastern Nigeria. *A. squamigera* fed primarily on mammals but also on lizards and birds. Juveniles ate mammals (mostly shrews) and skinks (genus *Trachylepis*), while adults ate mammals (mostly rodents) and birds. *A. squamigera* were found in a variety of natural habitat types, including primary and secondary dry forest, swamp forest, and shrubland, but were rarely observed in human-altered habitats, suggesting that this species (and perhaps this genus in general) may be less able to adapt to human-altered habitats than other snake species.

Although bites by Bush Vipers to humans are rare, they often result in serious symptoms. The venom of *Atheris* contains enzymes that cause a local and systemic hemorrhagic syndrome. Some enzymes convert fibrinogen into fibrin, and integrins cause strong aggregation of the blood platelets (Mebs et al. 1998). In addition to this defibrination syndrome, which can result in severe anemia (Hatten et al. 2013), acute renal insufficiency has been reported (Top et al. 2006). There has been no published report of tissue necrosis, but many deaths have resulted from *Atheris* envenomation, particularly from bites by *A. squamigera* (Chippaux unpubl. data).

The head is triangular, with a rounded snout and a well-defined neck. The eye is small to medium sized with a vertically elliptical pupil. The tail is prehensile. The head scales are small and undifferentiated. The rostral is small. The nasal is single or semi-divided. The eye is separated from the nostril by 2–4 scales. There are 9–19 perioculars and 6–12 interoculars. There are 0–3 rows of scales between the eye and the upper labials. There are 8–13 upper labials and 8–15 lower labials. There is 1 pair of chin shields. The sublingual groove is pronounced. The dorsal scales are keeled, without apical pits, and arranged in 14–37 straight rows. The vertebral row is not enlarged relative to the other dorsal scale rows. The cloacal scale is single. The hemipenes are bilobed for the distal half, with a forked sulcus spermaticus.

Mocquard's Bush Viper: *Atheris anisolepis* Mocquard 1887

The range of *Atheris anisolepis* extends from Gabon to Angola. The type locality is Alima-Lékéti, Congo.

The body is usually uniformly leaf green, sometimes with dark yellow spots toward the posterior end. The venter is yellow or pale green. The nasal is undivided. There

Key to Central and Western African Species of the Genus *Atheris*

1	Eye in contact with one or more upper labials	*A. subocularis*
1'	Eye separated from the upper labials by suboculars	2
2(1)	Paraventral row narrower than the other dorsal scale rows	3
2'	Paraventral row as broad or broader than the other dorsal scale rows	35
3(2)	From 6 to 8 interocular scales	*A. anisolepis*
3'	9 or more interocular scales	4
4(3)	Fewer than 145 ventrals	*A. katangensis*
4'	More than 150 ventrals	*A. chlorechis*
5(2)	Gulars smooth	*A. nitschei*
5'	Gulars keeled	6
6(5)	Keels on scales of the head and neck exaggerated, elongate and spine-like, giving a hairy or bristly appearance	7
6'	Scales of the head and neck strongly keeled	8
7(6)	2 rows of (usually keeled) scales between the eye and the nasal	*A. hispida*
7'	1 row of (always smooth) scales between the eye and the nasal	*A. hirsuta*
8(6)	A black stripe along the side of the head extending from the eye and broadening| posteriorly	*A. broadleyi*
8'	No black stripe on the head	*A. squamigera*

are 2–4 scales separating the eye from the nostril. There are 12–17 perioculars and 6–8 interoculars. There are 2 rows of suboculars. There are 9–13 upper labials and 10–14(2) lower labials. The gular scales are keeled. There are 19–25 dorsal scale rows. The paraventrals are smaller than the other dorsal scales. There are 150–162 ventrals and 46–55 subcaudals. The maximum length recorded for this species is 650 mm (Mocquard 1887).

Some authors question the validity of this species. Lawson and Ustach (2000) synonymized *A. anisolepis* with *A. squamigera* on the grounds that the number of rows of subocular scales, an important character used to distinguish these two species from one another (2 rows of suboculars in *A. anisolepis* versus 1 row of suboculars in *A. squamigera*), is in fact a variable sexually dimorphic character rather than a difference between species. But these authors offered no comment on the other scale characteristics used to distinguish *A. anisolepis* from *A. squamigera* (size of paraventral scales relative to other dorsal scale rows, shape of the rostral scale, and number of suprarostrals).

FIGURE 7.2. *Atheris broadleyi*, Cameroon. By V. Gvoždík

Broadley's Bush Viper: *Atheris broadleyi* Lawson 1999

The range of *Atheris broadleyi* extends from Nigeria to Central African Republic and possibly Congo (Brazzaville). The type locality is Lipondji, Cameroon.

The body is yellow anteriorly, becoming gradually greenish posteriorly. The flanks are patterned with a series of 30 or so incomplete bands of sulfur yellow outlined in black. The tail is marked with 9 or so dark bands, and the tip of the tail is black. The venter is light blue with diffuse whitish blotches on the anterior part of the body, becoming darker posteriorly. The head is the same color as the dorsum of the body. One each side of the head, a darker triangle, with its point toward the eye, extends from the eye to the corner of the mouth. The nasal is undivided. There are 2 or occasionally 3 scales separating the eye from the nostril. There are 12–16 (usually 15) perioculars and 3–8 (usually 6) interoculars. There is 1 row of suboculars. There are 9–12 upper labials and from 9(2) to 12(4) lower labials. The gular scales are keeled. There are 17–23 dorsal scale rows. The paraventrals are the same size as the other dorsal scales. There are 157–169 ventrals and 45–61 subcaudals. The maximum length recorded for this species is 768 mm (Lawson 1999).

Green Bush Viper: *Atheris chlorechis* (Pel 1851)

This species is found in forests from Guinea to Togo. The type locality is Gold Coast, restricted to Butre, Ghana (Hughes and Barry 1969).

This snake is usually uniformly leaf green. The venter is the same color as the dorsum but a paler shade. Some specimens have an irregular pattern of black spots. Others have a pair of yellow paravertebral stripes, sometimes with black spots arranged along them. The nasal is undivided. There are 3–4 scales separating the eye from the nostril. There are 14–20 perioculars and 9–13 interoculars. There are 2 rows of suboculars. There are 9–12 upper labials and 9–14(2) lower labials.

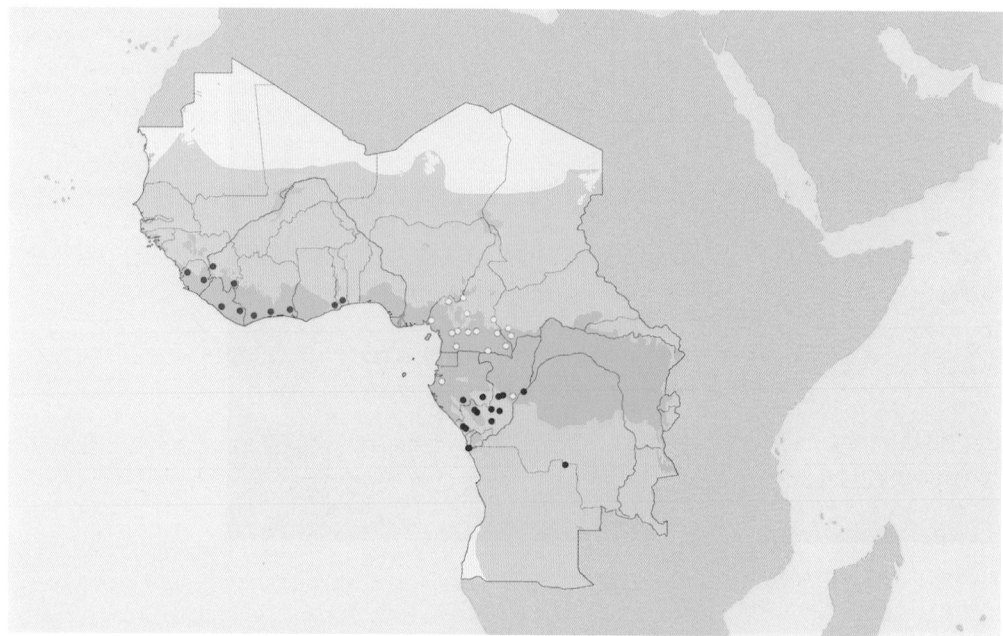

FIGURE 7.3. *Atheris anisolepis* (blue), *Atheris broadleyi* (yellow), *Atheris chlorechis* (red). By K. Jackson

FIGURE 7.4. *Atheris chlorechis*, Ghana. By P. Naskrecki

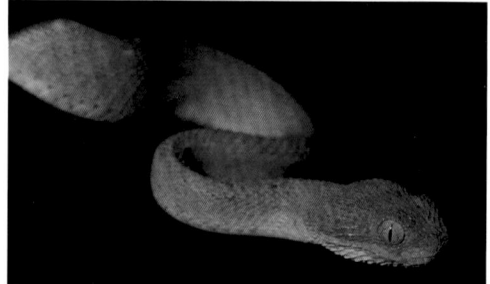

FIGURE 7.5. *Atheris chlorechis*, Ivory Coast. By M.-O. Rödel

FIGURE 7.6. *Atheris chlorechis* juvenile, Ivory Coast. By M.-O. Rödel

The gular scales are keeled. There are 25–37 dorsal scale rows. The paraventrals are smaller than the other dorsal scales. There are 151–165 ventrals and 48–64 subcaudals. The maximum length recorded for this species is 650 mm (Villiers 1975).

West African Hairy Bush Viper: *Atheris hirsuta* Ernst and Rödel 2002

Atheris hirsuta is endemic to the Taï Forest, southwestern Ivory Coast. The type locality is Taï National Park, Ivory Coast.

A. hirsuta is one of two *Atheris* species (the other being *A. hispida* in east Africa) that have a bizarre and highly distinctive "hairy" appearance, which results from the keels on dorsal scales being greatly exaggerated and lengthened. The dorsum is bronze, with the keels and distal edges of some dorsal scales dark brown, forming a pattern of discontinuous bands. The iris of the eye is golden yellow. The venter is cream or pale yellow. The nasal is undivided. There are 2 scales separating the eye from the nostril. There are 14–15 perioculars and 9 interoculars. There is 1 row of suboculars. There are 9–10 upper labials and 8(3) or 9(3) lower labials. The gular scales are keeled. There are 16 dorsal scale rows. The paraventrals are the same size as the other dorsal scales. There are 160 ventrals and 58 subcaudals. The maximum length recorded for this species is 480 mm (Ernst and Rödel 2002).

African Hairy Bush Viper: *Atheris hispida* Laurent 1955

The range of *Atheris hispida* extends from eastern Democratic Republic of Congo to

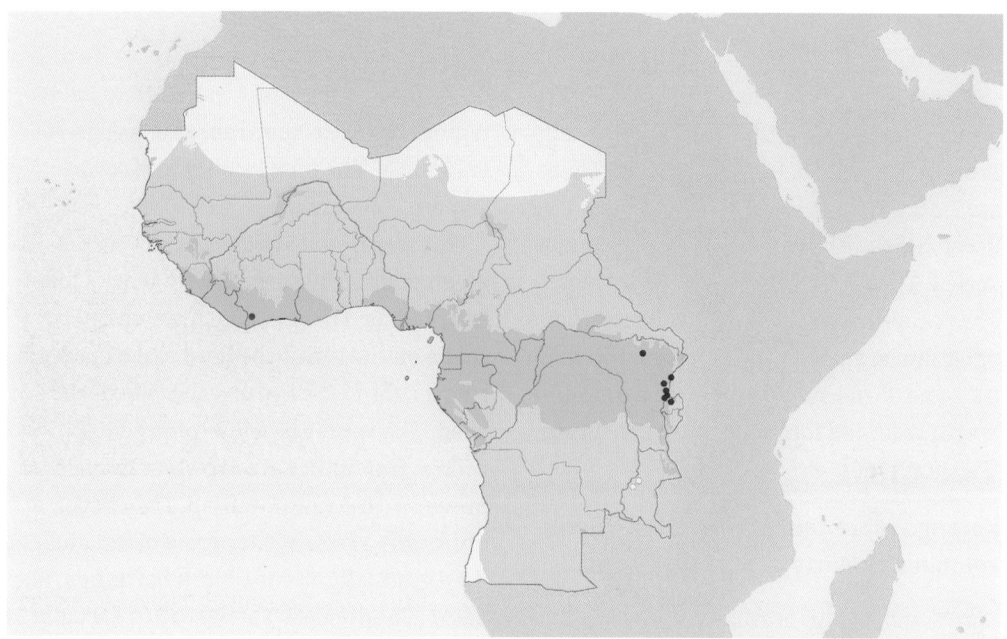

FIGURE 7.7. *Atheris hirsuta* (red), *Atheris hispida* (blue), *Atheris katangensis* (yellow). By K. Jackson

FIGURE 7.8. *Atheris hirsuta*, Ivory Coast. By M.-O. Rödel

Kenya and Tanzania. The type locality is Lutunguru, Kivu Province, Democratic Republic of Congo.

Like *A. hirsuta* in West Africa, *A. hispida* has a distinctive hairy or bristly appearance, resulting from the keels on the dorsal scales being greatly exaggerated and lengthened. In males, the dorsum is usually olive or greenish brown, and the venter greenish, becoming darker posteriorly. Sometimes there is a dark stripe behind the eye. In females, the dorsum is usually yellowish brown or olive, and the venter yellowish brown. Both sexes have a black mark on the nape. The nasal may be divided or undivided. There are 2 scales separating the eye from the nostril. There are 9–16 perioculars and 7–9 interoculars. There is usually 1 row of suboculars (occasionally 2). There are 7–10 upper labials and from 8(2) to 10(3) lower labials. The gular scales are strongly keeled. There are 15–19 dorsal scale rows. The paraventrals are the same size as the other dorsal scales. There are 149–166 ventrals and 35–64 subcaudals. The maximum length recorded for this species is 735 mm (Laurent 1956).

Katanga Mountain Bush Viper: *Atheris katangensis* Witte 1953

Atheris katangensis is known only from the type locality of Mubale-Munte, Upemba National Park, Democratic Republic of Congo.

The dorsum is pale brown to olive or purple brown above, with a vertebral series of dark-bordered and dark-centered yellowish rhombic markings. The tail tip is yellow. The venter is yellow anteriorly, sometimes becoming gray-green posteriorly, about every third ventral with a yellow lateral spot and a few ventrals with short black transverse bars. The nasal is undivided. There are 2–3 scales separating the eye from the nostril. There are 14–17 perioculars and 9–11 interoculars. There is 1 row of suboculars. There are 9–12 upper labials and 11 lower labials. The gular scales are weakly keeled. There are 23–31 dorsal scale rows. The paraventrals are smaller than the other dorsal scales. There are 133–144 ventrals and 38–59 subcaudals (fewer than 43 in females, more than 44 in males). The maximum length recorded for this species is 397 mm (Witte 1953).

Great Lakes Bush Viper: *Atheris nitschei* Tornier 1902

Atheris nitschei is found in the Great Lakes region of Africa from Tanzania to Uganda to Zambia. The type locality is Mpororo Swamp, Tanzania.

This large, stout-bodied Bush Viper is green or olive, patterned with black blotches or bands. There is a dark blotch or V shape on the top of the head and a black stripe behind the eye on each side of the head. The venter is yellow or greenish yellow. Hatchlings are brown or grayish brown with a yellow tail tip. They become uniformly green as they grow older, and then gradually add the black markings. The nasal is undivided. There are 2–4 (usually 3) scales separating the eye from the nostril. There are 12–19 perioculars and 8–12 interoculars. There is usually 1 row of suboculars, sometimes 0 or 2. There are 8–13 upper labials and from 9 to 15 lower labials. The gular scales are smooth. There are 23–34 dorsal scale rows. The paraventrals are the same size as the other dorsal scales. There are 140–164 ventrals and 35–59 subcaudals. The maximum length recorded for this species is 750 mm (Witte 1941).

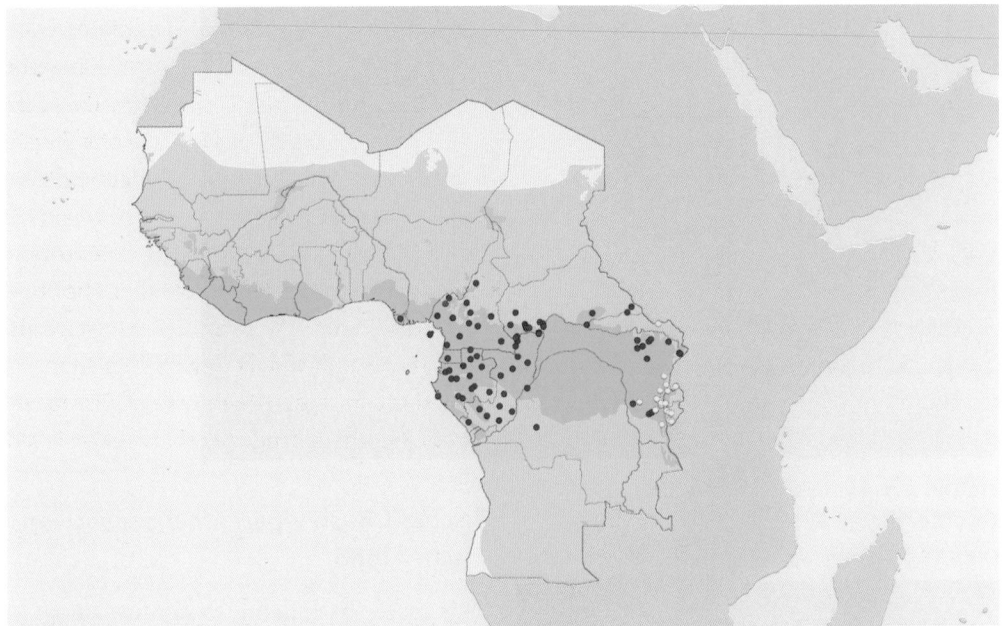

FIGURE 7.9. *Atheris nitschei* (yellow), *Atheris squamigera* (blue), *Atheris subocularis* (red). By K. Jackson

Rough-scaled Bush Viper: *Atheris squamigera* (Hallowell 1854a)

The range of *Atheris squamigera* extends from Nigeria to Kenya to Tanzania to Angola. The type locality is Gabon, restricted to Libreville (Lawson and Ustach 2000).

This species is dark leaf green, sometimes uniformly so, but usually with a pattern of black and yellowish blotches arranged in bands. In some individuals, the yellowish blotches are the most prominent so that the overall coloration tends toward yellow or dark yellow. The venter is the same color as the dorsum but a paler shade. The head is usually the same color as the body, sometimes with dark spots on the sides. The nasal is undivided. There are 2 or occasionally 3 scales separating the eye from the nostril. There are 10–18 (usually 14) perioculars and 6–9 interoculars. There are usually 2 rows of suboculars (occasionally 1 or more than 2). There are 9–12 upper labials

FIGURE 7.10. A captive *Atheris nitschei* feeding. By R. Carmichael

and from 9(2) to 12(4) lower labials (usually 10 or 11(3)). The gular scales are keeled. There are 15–25 (usually 19 or 21) dorsal scale rows. The paraventrals are the same size as the other dorsal scales. There are 152–165 ventrals and 45–67 subcaudals. The maximum length recorded for this species is 799 mm (Roux-Estève 1965).

FIGURE 7.11. *Atheris squamigera*, Democratic Republic of Congo. By J. Kielgast

FIGURE 7.12. *Atheris squamigera* juvenile, Democratic Republic of Congo. By J. Kielgast

Fischer's Bush Viper: *Atheris subocularis* Fischer 1888

This species is endemic to southwestern Cameroon. The type locality is Cameroon.

The dorsum is olive green or yellowish. The dorsum of the body is patterned with 30 or so greenish bands, edged anteriorly with darker green or black. The dorsum of the head bears a dark marking in the shape of a partial chevron and/or other dark blotches. The sides and the front of the head are a paler shade of dark green than the top of the head. The venter is uniformly a dull pale yellow speckled with black. The nasal is semi-divided. There are 2 (occasionally 3) scales separating the eye from the nostril. There are 11–14 perioculars and 6–7 interoculars. There are no suboculars, a feature unique among *Atheris* species within our zone. There are 8(4,5) to 10(4,6) upper labials, or sometimes only 1 upper labial in contact with the eye. There are 8(2) to 9(4) lower labials. The gular scales are keeled. There are 14–16 dorsal scale rows. The paraventrals are larger than the other dorsal scales. There are 152–163 ventrals and 58–65 subcaudals. The maximum length recorded for this species is 491 mm (Lawson et al. 2001).

African Adders: Genus *Bitis* Gray 1842

This genus includes 17 species of stout-bodied, terrestrial vipers distributed throughout sub-Saharan Africa. Included within our zone are the four broadly distributed giant species of *Bitis*, as well as three smaller southern African species. The Giant Vipers (*Bitis arietans, B. gabonica, B. rhinoceros, B. nasicornis*) are heavy bodied, slow moving, and primarily nocturnal. They hunt by ambush, feeding primarily on small mammals (primarily rodents), but also birds, amphibians, and lizards. *Bitis* give birth to live young, with some of the giant members of the genus producing enormous litters. The Puff Adder, *B. arietans*, is known to give birth to litters of as many as 100 young, though 30–50 is more usual (Pitman 1974). In forest habitat in Nigeria, Rhinoceros Vipers (*B. nasicornis*) produced litters of 13–42 (mean = 25), while Gaboon Vipers (*B. gabonica*) produced litters of 5–32 (mean = 18). Although the species with the larger average body size (*B. gabonica*) produced smaller litters on average than *B. nasicornis*, the species with the smaller average body size, within each species, larger females had larger litters on average (Luiselli et al. 1998a). Among the small species of *Bitis*, litter sizes of 4–10 for the Dwarf Puff Adder (*B. peringueyi*) and 14–19 for the Horned Adder (*B. caudalis*) are reported (Spawls and Branch 1995).

The head is broad and triangular with a well-defined neck. The body is broad and flat with a short tail. Some species have scales modified into horns above the eyes or at the tip of the rostrum. The eye is small with a vertical pupil. The head scales are small and undifferentiated. The rostral is small. There are 10–21 perioculars and 6–17 interoculars. There are from 2 to 5 rows of scales between the eye and the upper labials. There are 10–18 upper labials and 10–22 lower labials. There is 1 pair of chin shields. The sublingual groove is pronounced. The dorsal scales are keeled, with apical pits, and arranged in 23–46 slightly oblique rows. The subcaudals are divided. The cloacal scale is single. The hemipenes are bilobed from the base, and the sulcus spermaticus is forked.

The venoms of *Bitis* differ significantly depending on the species (Calvete et al. 2007). In general, all *Bitis* venoms contain enzymes that act on blood coagulation and break down tissues. The venoms of *B. gabonica* and *B. rhinoceros* appear to be the most complex. They have strong hemorrhagic effects owing to the presence of serine proteases (gabonase, rhinocerase), which can break down both the α and β chains of fibrinogen (like natural thrombin) but can also dissolve a clot once it has formed (Pirkle et al. 1986; Vaiyapuri et al. 2010). These enzymes are not inhibited by antithrombin III, heparin, or hirudin (Pirkle et al. 1986; McNally et al. 1993). The inflammatory and hemorrhagic syndromes associated with *Bitis* envenomations (edema, hemorrhage, petechiae, purpura, blisters) can result in dramatic symptoms, but usually blood coagulation is quickly restored with early administration of antivenom (Marsh et al. 1997; Wildi et al. 2001). The venom of *B. arietans* produces few hemorrhagic symptoms—although blood coagulation is disrupted and bleeding resulting primarily from the presence of metalloproteases may be observed—but it can lead to shock and to extensive and severe tissue necrosis (Le Dantec et al. 2004; Currier et al. 2010; Fernandez et al.

Key to Central and Western African Species of the Genus *Bitis*

1　Nasal separated from the first upper labial by at least 4 scales, and from the rostral by at least 3 scales 2

1'　Nasal separated from the first upper labial by not more than 3 scales, and from the rostral by 1 or 2 scales 4

2(1)　A group of several narrow, pointed scales between the two nostrils .... *B. nasicornis*

2'　A single scale between the nostrils 3

3(2)　A single dark triangle on each side of the head with its point extending to the eye ....
.... *B. rhinoceros* (Fig. 7.13)

3'　Two dark triangles on each side of the head, with their points extending toward the eye .... *B. gabonica* (Fig. 7.13)

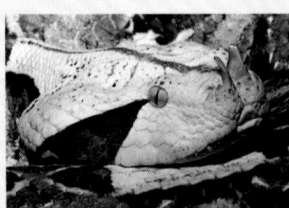 

FIGURE 7.13. *Bitis rhinoceros* (*left*) and *Bitis gabonica* (*right*) showing characteristic markings on the side of the heads: one triangle behind the eye in *B. rhinoceros*, two triangles behind and below the eye in *B. gabonica*. By K. Jackson using photos of captive individuals by D. Williams

4(1)　A scale shaped like a horn above each eye .... *B. caudalis*

4'　No scale shaped like a horn on the head 5

5(4)　23–27 dorsal scale rows and fewer than 10 interoculars .... *B. peringueyi*

5'　27–41 dorsal scale rows and more than 10 interoculars 6

6(5)　12 lower labials or fewer .... *B. heraldica*

6'　13 or more lower labials .... *B. arietans*

2014), which may necessitate amputations. In addition, phospholipase A$_2$ in the venom of *B. caudalis* functions as a presynaptic neurotoxin (neurotoxin-β) capable of causing muscular paralysis similar to that seen with elapid envenomations (Viljoen et al. 1982).

Puff Adder: *Bitis arietans* (Merrem 1820)

The Puff Adder is the most widely distributed viperid in Africa, with a range extending from the Sahara Desert to southern Africa. The type locality is Cape of Good Hope, South Africa. Though primarily a savannah species, *Bitis arietans* is the most adaptable of the giant *Bitis* in terms of habitat, also making its way into forested areas and disturbed habitats. Juveniles especially can be found in unexpected places, including swimming in the water in flooded forest habitat in Congo (Jackson pers. obs.) and basking in a tree at a height of 4 m off the ground in a suburban garden in South Africa (Branch and Branch 2004). The body is dark beige or sandy brown above with a pattern of pale chevrons along the body with the point extending posteriorly. The top of the head is dark with a pale band across the head between the eyes and a pale stripe from the posterior margin of each eye to the corner of the mouth. The venter is pale with darker mottling. There are 10–16 perioculars and 7–11 interoculars. There are 3–4 rows of subocular scales between the eye and the upper labials. There are 12–17 upper labials. There are 13–19(4), or occasionally (3) or (5), lower labials. There are 28–41 dorsal scale rows. There are 123–148 ventrals.

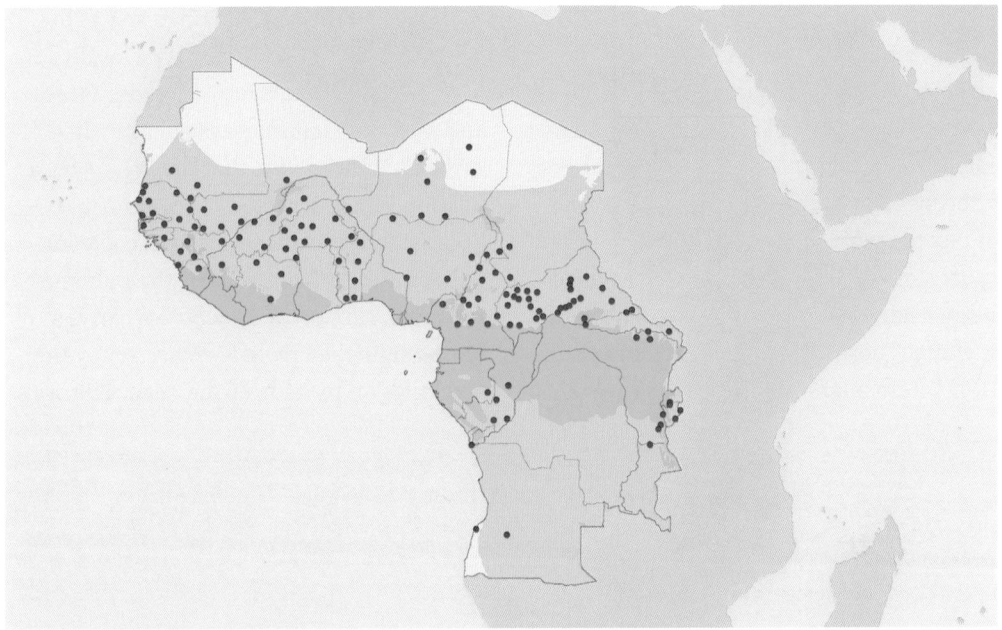

FIGURE 7.14. *Bitis arietans* (blue). By K. Jackson

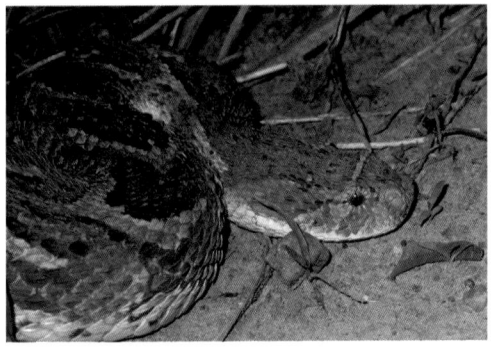

FIGURE 7.15. *Bitis arietans*, Ivory Coast. By M.-O. Rödel

FIGURE 7.16. *Bitis arietans* juvenile, Ghana. B M. Fujita

There are 14–38 subcaudals (fewer than 25 in females). The maximum length recorded for this species is 1,905 mm (Pitman 1974).

Horned Adder: *Bitis caudalis* (Smith 1839)

This species is found in arid regions throughout southwestern Africa, with a distribution extending east as far as Zimbabwe. Within our zone, it occurs in southwestern Angola. The type locality is "sandy districts north of the Cape Colony" (South Africa).

Bitis caudalis is a desert species, though less specialized for sand than *B. peringuey*. It can sidewind on sand and has a horn above each eye. The dorsal coloration is light gray or brown with a variable darker pattern of blotches along the back, and pale-edged spots along the sides. The top of the head bears a dark V-shaped marking. The venter is white. There are 9–20 perioculars and 11–17 interoculars. There are 2–5 rows of subocular scales between the eye and the upper labials. There are 10–14 upper labials. There are 10–15(2) or (3) lower labials. There are 25–31 dorsal scale rows. There are 120–155 ventrals. There are 18–34 subcaudals. The maximum length recorded for this species is 500 mm (FitzSimons 1974).

Gaboon Viper: *Bitis gabonica* (Duméril, Bibron and Duméril 1854a)

The range of the Gaboon Viper, *Bitis gabonica*, extends from Benin to Angola to the Democratic Republic of Congo and as far east as western Kenya. The type locality is Gabon. *B. gabonica* is found in both forest and savannah habitats as well as in disturbed habitats. Note that the West Africa Gaboon Viper, *B. rhinoceros*, was formerly considered a subspecies of *B. gabonica*. Gaboon Vipers from west of the Dahomey gap are *B. rhinoceros*. *B. gabonica* can easily be distinguished from *B. rhinoceros* by the markings on the side of the head. The over-

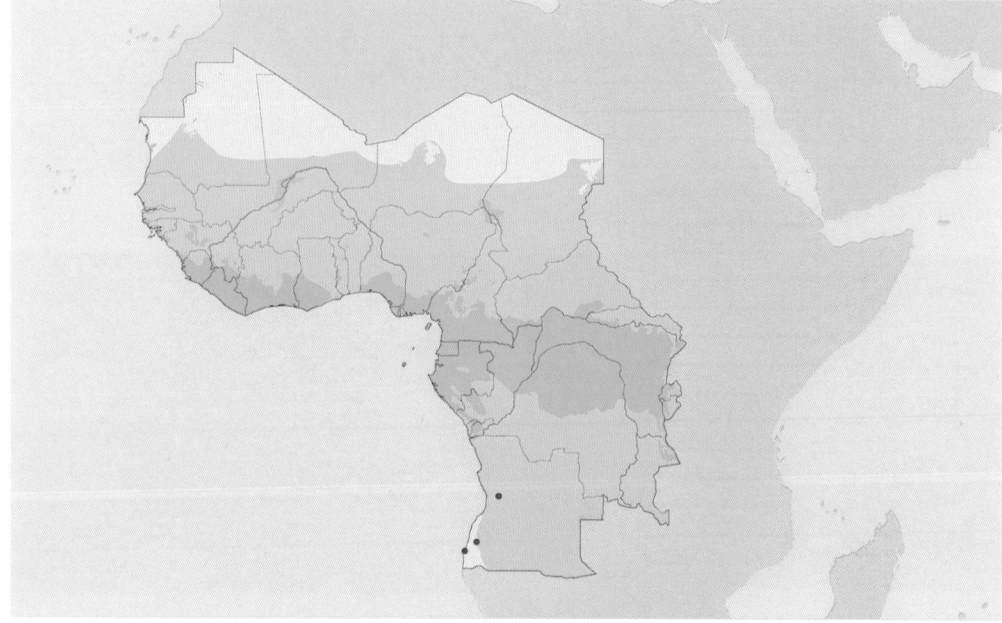

FIGURE 7.17. *Bitis caudalis* (red). By K. Jackson

FIGURE 7.18. *Bitis caudalis*, Namibia. By W. Branch

all dorsal color is brown with markings in a dark geometrical pattern of X's outlined in yellow or beige. The top of the head is pale with a black stripe along the midline, dividing into three branches at the level of the eyes in some specimens. On each side of the head are two black triangles, one extending vertically down from the eye to the upper labials, the other extending diagonally backward from the eye to the corner of the mouth. These markings distinguish *B. gabonica* from *B. rhinoceros* since *B. rhinoceros* lacks the first of these triangles. The venter is yellow mottled with black. There are no horns at the tip of the snout in *B. gabonica*, another feature that distinguishes this species from *B. rhinoceros*. There are 14–21 perioculars and 12–16 interoculars. There

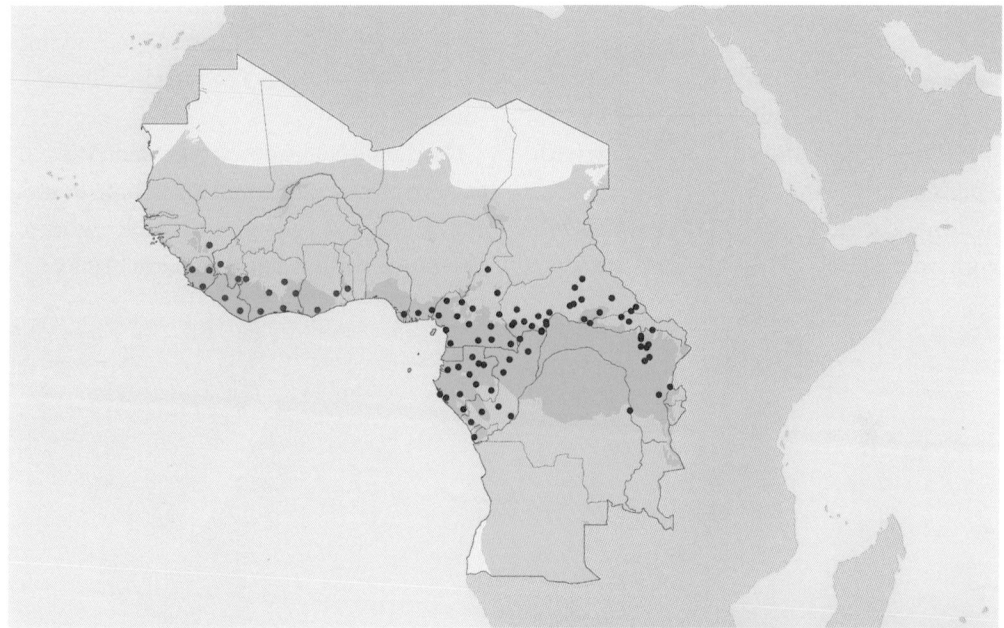

FIGURE 7.19. *Bitis gabonica* (blue), *Bitis rhinoceros* (red). By K. Jackson

FIGURE 7.20. *Bitis gabonica*, Democratic Republic of Congo. By J. Kielgast

are 4–5 rows of subocular scales between the eye and the upper labials. There are 12–18 upper labials. There are 16–22(4) or (5) lower labials. There are 28–46 dorsal scale rows. There are 124–140 ventrals (more than 130 in females, fewer than 133 in males). There are 17–33 subcaudals (more than 24 in males, fewer than 24 in females). The maximum length recorded for this species is 1,740 mm (Grasset 1946).

Angolan Adder: *Bitis heraldica* (Bocage 1889)

This small viper is known only from a few specimens and is probably endemic to the high plateau of central Angola. The type locality is tributary of the Cunene River, between 13° and 14° south, to the east of Caconda, Huila, Angola.

The dorsal coloration is light brown with a dark pattern of oval blotches along the midline. The sides of the head are marked with a dark stripe from the eye to the cor-

ner of the mouth. The venter is white with dark gray spots. There are 12–14 perioculars and 11–13 interoculars. There are 2–3 rows of subocular scales between the eye and the upper labials. There are 13–14 upper labials. There are 11–12(3), or occasionally (4), lower labials. There are 27–31 dorsal scale rows. There are 124–131 ventrals. There are 19–27 subcaudals. The maximum length recorded for this species is 325 mm (Bocage 1895a).

Rhinoceros Viper: *Bitis nasicornis* (Shaw 1802)

The Rhinoceros Viper is a forest species, found in forested habitats in central and western Africa, from Guinea to Gabon to the Democratic Republic of Congo, and into Rwanda and Zambia. The type locality is "inner Africa."

The dorsal coloration is a geometrical pattern of rectangles and X's in blue or lavender, red, and yellow, with black outlines. The top of the head bears a large black

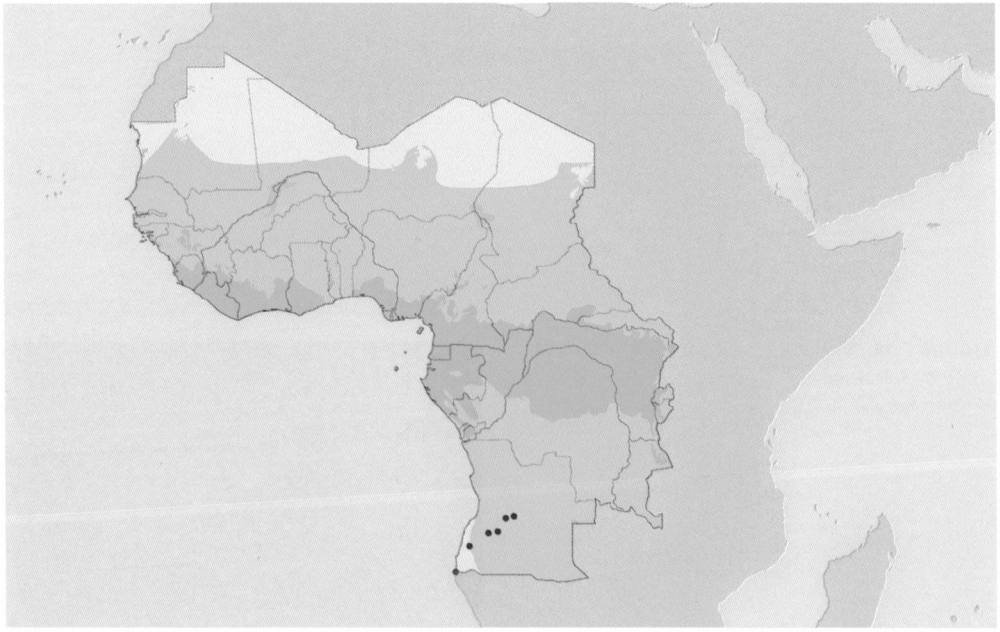

FIGURE 7.21. *Bitis heraldica* (blue), *Bitis peringueyi* (red). By K. Jackson

arrow with the tip pointing anteriorly. The venter is pale with dark mottling. Improbable as this coloration may sound, or even appear when the animal is seen in isolation, it provides remarkably effective camouflage against the leaf litter of the forest floor. At the tip of the snout are 2–3 pairs of horns formed by enlarged scales. There are 19–20 perioculars and 12–16 (usually 14) interoculars. There are 4–5 rows of subocular scales between the eye and the upper labials. There are 15–17 upper labials. There are 15–20(5) lower labials. There are 31–43 dorsal scale rows. There are 117–140 ventrals. There are 16–34 subcaudals (more than 24 in males, fewer than 20 in females). The maximum length recorded for this species is 1,300 mm (Ota et al. 1987).

Dwarf Puff Adder: *Bitis peringueyi* (Boulenger 1888)

This small species is found in the Namib Desert from western Namibia into southwestern Angola. The type locality is 10 miles east of Walvis Bay, Namibia.

Bitis peringueyi is highly adapted for life in

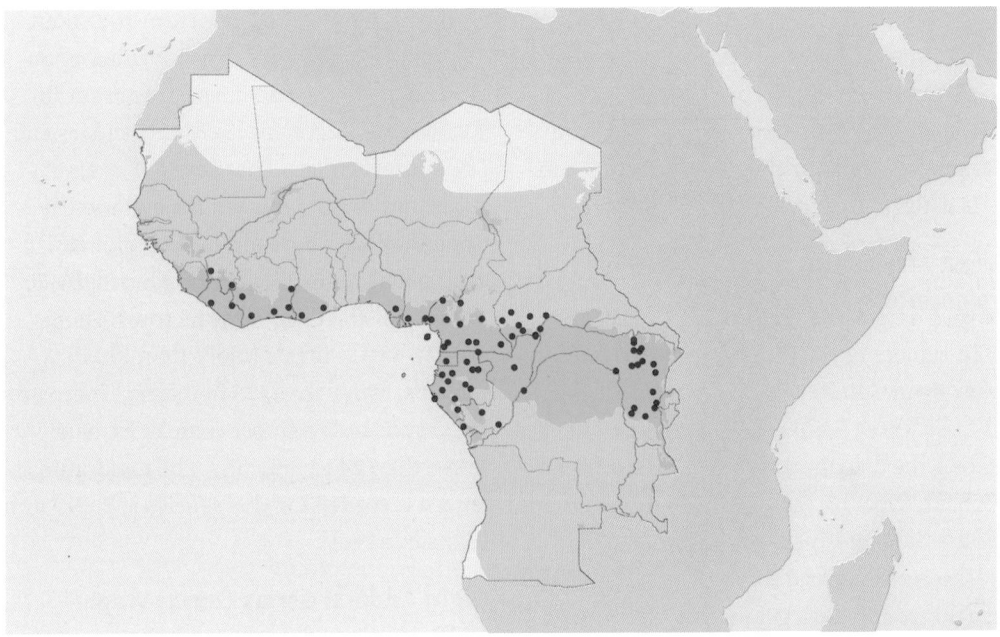

FIGURE 7.22. *Bitis nasicornis* (blue). By K. Jackson

FIGURE 7.23. *Bitis nasicornis*, Ghana. By P. Naskrecki

sand. Like the Saharan Sand Viper, *Cerastes vipera*, it lacks horns above the eyes and has its eyes dorsally oriented. It buries itself in sand and can move rapidly across loose sand by sidewinding. Spawls and Branch (1995) report that its diet consists primarily of sand dune lizards (*Meroles anchietae*). *B. peringueyi* is orange in color, like the sand of the Namib Desert. There are 10–13 perioculars and 6–9 interoculars. There are 2–4 rows of subocular scales between the eye and the upper labials. There are 10–14 upper labials. There are 10–13(2) to (4) lower labials. There are 23–27 dorsal scale rows. There are 117–144 ventrals (more than 124 in females, fewer than 139 in males). There are 15–30 subcaudals (more than 21 in males, fewer than 26 in females). The maximum length recorded for this species is 320 mm (Haacke 1975).

West African Gaboon Viper: *Bitis rhinoceros* (Schlegel 1855)

Bitis rhinoceros is found in West African forests, from Guinea to Togo. The type locality is Cape Three Points, Ghana. Figure 7.19 shows the distribution of the West African Gaboon Viper, *B. rhinoceros*, relative to that of the Gaboon Viper, *B. gabonica*.

The overall dorsal color is brown with markings in a dark geometrical pattern of X's outlined in yellow or beige. The top of the head is pale with a black stripe along the midline, dividing into three branches at the level of the eyes in some specimens. On each side of the head is a black triangle, extending diagonally backward from the eye to the corner of the mouth. These markings distinguish *B. rhinoceros* from *B. gabonica* since *B. gabonica* also has a second black triangle extending vertically downward from the eye to the upper labials. The venter is yellow

FIGURE 7.24. *Bitis rhinoceros*, Ivory Coast. By M.-O. Rödel

mottled with black. At the tip of the snout there is a pair of horns, each formed by an enlarged scale. These horns are absent in *B. gabonica*. There are 15–21 perioculars and 12–16 interoculars. There are 4–5 rows of subocular scales between the eye and the upper labials. There are 13–18 upper labials. There are 16–22(4) or (5) lower labials. There are 28–46 dorsal scale rows. There are 124–140 ventrals (more than 130 in females, fewer than 133 in males). There are 17–33 subcaudals (more than 24 in males, fewer than 24 in females). The maximum length recorded for this species is 2,050 mm (Cansdale 1961).

Night Adders: Genus *Causus* Wagler 1830

Causus is a genus of six species of small, terrestrial vipers found in sub-Saharan Africa. Five species occur within our zone. Night Adders are nocturnal and feed primarily on anurans, especially toads (genus *Sclerophrys* = formerly *Bufo*). But it is not unusual to observe some *Causus*, including *Causus maculatus* in west Africa especially in forest areas, out hunting in the late afternoon well before dark (Chippaux pers. obs.). They lay large clutches (up to 26) of relatively small eggs (Ineich et al. 2006). Although bites to humans are common, Night Adder envenomations generally result in only mild symptoms. Pain and swelling may occur, but there has never been a documented case involving a hemorrhagic syndrome or tissue necrosis resulting from a *Causus* bite.

The genus *Causus* is unique among African viperids in having platelike head scales similar to those of elapids, colubrids, and lamprophiids rather than the generally small, undifferentiated head scales characteristic of viperids. The presence of such characteristic colubroid head scales in a viper has been interpreted as a primitive characteristic and as an indication that *Causus* represent a basal branch among the Old World vipers (Viperinae). But there is increasing agreement from recent molecular phylogenies (e.g., Wüster et al. 2008; Pyron et al. 2011) that *Causus* originated within the Viperinae, a clade that originated in Africa and a branch of which has since diversified into the genera *Atheris*, *Bitis*, *Causus*, and *Cerastes* in Africa, and *Echis* in Africa and Asia. The condition of the head scales must therefore be considered a derived feature. *Causus* are morphologically unusual in other ways as well. For example, *Causus* resemble the Stiletto Snakes (genus *Atractaspis*, Lamprophiidae) in that while some species in the genus possess normal venom glands, similarly proportioned to those of other viperids, other species possess elongate venom glands extending along the sides of the body far beyond the head (Phisalix 1922; Shayer-Wollberg and Kochva 1967).

The head is short and broad. The body is wide and somewhat dorsoventrally flattened. The tail is short. The eye size is medium or large with a round pupil. Because the head scales of *Causus* are platelike, resembling those of elapids, colubrids, and lamprophiids, the same scale counts that are used to describe the arrangement of head scales as in those families can be used. The nasal is divided. The internasals and prefrontals are paired. There are 1 or 2 loreals. There are 5–7 perioculars, including a supraocular; 1–3 preoculars; 1–3 postoculars; and 1–2 suboculars, the latter arranged in a single row between the eye and the upper labials. There are 6 or 7 upper labials, none of which are in contact with the eye (because of the suboculars), and 7–11 lower labials. There are 1 or 2 (occasionally 3) anterior temporals and 2–4 posterior temporals. There is 1 pair of chin shields. The submandibular groove is deep. The dorsal scales bear 1 apical pit, may be smooth or weakly keeled, and are arranged in 15–23 oblique rows. The vertebral row is not enlarged relative to the other dorsal scale rows. The subcaudals may be single or divided. The cloacal scale is single. The number of ventrals and subcaudals is sexually dimorphic in some species, with females having more ventrals

FIGURE 7.25. Head scales of Night Adders (genus *Causus*) are large and platelike, resembling the head scales of elapids and of nonvenomous snakes, rather than small and undifferentiated like the head scales of other African viperids (*Causus rhombeatus*, Kenya). By W. Wüster

and fewer subcaudals than males. The hemipenes are bilobed, and the sulcus spermaticus is forked.

Two-striped Night Adder: *Causus bilineatus* Boulenger 1905

The range of *Causus bilineatus* extends from southern Angola to western Tanzania and north to southern Democratic Republic of the Congo and Rwanda. The type locality is Benguela-Bihe, Angola. This is one of the *Causus* species with elongate venom glands.

The dorsum is brown or gray with dark blotches. The species gets its name from the two pale dorsolateral stripes running the length of the body. There is a dark V shape on the dorsum of the head with the tip pointing toward the snout. The venter is pale gray to black.

The eye is large. The internasals are shorter than or equal to in length and narrower than the prefrontals. There is 1 loreal. There are usually 6 perioculars. These include the supraocular, 2 preoculars of equal size, usually 1 postocular, and 2 suboculars. There are usually 6, sometimes 7, upper labials, and usually 9(4), sometimes 10(4), lower labials. The temporal formula is usually 2+3, sometimes 2+2 or 2+4. The dorsal scales are weakly keeled and usually arranged in 17 (but sometimes 15 or 19) oblique rows. There are 119–144 ventrals (fewer than 142 in males, more than 127 in females), and 18–35 divided subcaudals. The maximum length recorded for this species is 650 mm (Spawls et al. 2004).

Forest Night Adder: *Causus lichtensteini* (Jan 1859)

Causus lichtensteini is a forest species with a range extending from Guinea to Kenya and the Democratic Republic of Congo. The type locality is Ghana. *C. lichtensteini* has short venom glands, as opposed to the elon-

---

Key to Central and Western African Species of the Genus *Causus*

1	Subcaudals single	*C. lichtensteini*
1'	Subcaudals divided	2
2(1)	Coloration green, with no markings on the head	*C. resimus*
2'	Coloration brown or gray, with a V-shaped marking on the back of the head with the point directed anteriorly	3
3(2)	Dorsal pattern consists of a middorsal line of dark spots between a pair of pale dorsolateral stripes that extend the full length of the body	*C. bilineatus*
3'	Dorsal pattern of distinct or indistinct spots without dorsal stripes	4
4(3)	The suture between the internasals is as long or longer than the suture between the prefrontals; no contact, or a point of contact only between the loreal and the internasal	*C. maculatus*
4'	The suture between the internasals is distinctly shorter than the suture between the prefrontals; contact between the loreal and the internasal is as great or greater than the contact between the loreal and the prefrontal	*C. rhombeatus*

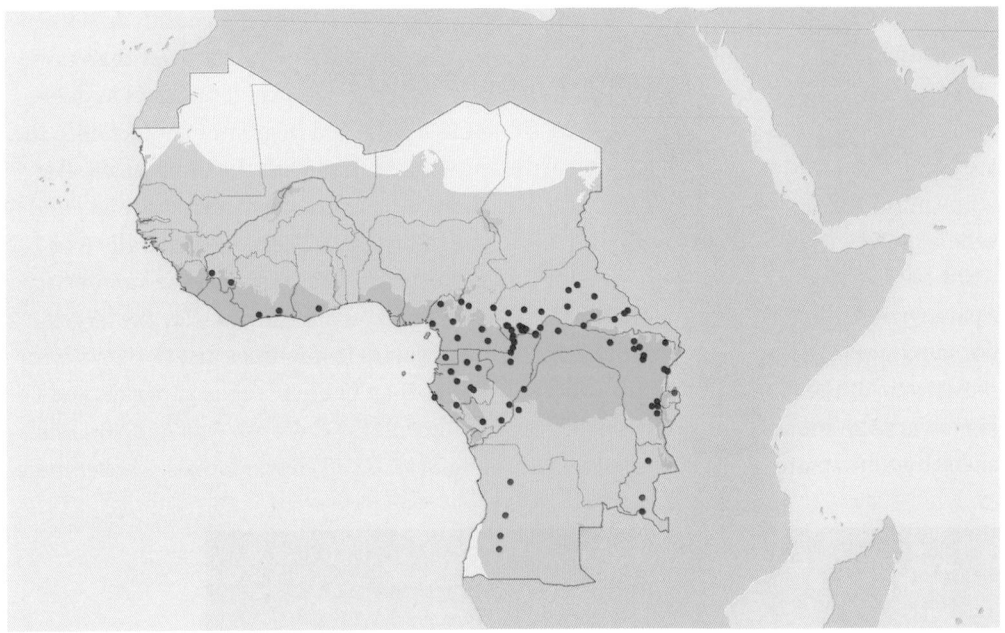

FIGURE 7.26. *Causus bilineatus* (red), *Causus lichtensteini* (blue). By K. Jackson

gate venom glands seen in some species of *Causus.*

The dorsum is brown or greenish, with or without dark spots. There is a black V shape that is outlined in white on the dorsum of the head with the tip pointing toward the snout. The venter is pale with some dark bands across the neck. The eye is large. The internasals are longer and narrower than the prefrontals. There is 1 loreal, sometimes 2, as deep as or a bit deeper than long. There are 5–7 perioculars. These include the supraocular, 1–2 preoculars, the inferior larger than the superior, 2–3 postoculars, the superior usually the smallest, and 2 suboculars. There are 6 or 7 upper labials, and 8–12(4), usually 9(4), lower labials. The temporal formula is usually 2+3, sometimes 2+2 or 2+4. The dorsal scales are weakly keeled and usually arranged in 15 oblique rows. There are 128–156 ventrals and 14–23 (more than 17 in males, fewer than 20 in females)

FIGURE 7.27. *Causus lichtensteini*, Ivory Coast. By M.-O. Rödel

single subcaudals. The maximum length recorded for this species is 700 mm (Spawls et al. 2004).

Spotted Night Adder: *Causus maculatus* (Hallowell 1842)

Causus maculatus is a broadly distributed species, common throughout central and western Africa. Its range extends from Mauritania to Angola to eastern Democratic Republic of Congo. The type locality is

Liberia. *C. maculatus* is often mistaken for *C. rhombeatus*, an eastern and southern African species that it resembles. Reports in the literature of *C. rhombeatus* in West Africa are usually attributable to *C. maculatus*. This is one of the *Causus* species with elongate venom glands.

The dorsum is light or dark brown with a pattern of black oval spots that fade as the snake ages. There is a dark V shape on the dorsum of the head with the tip pointing toward the snout. The venter is pale and sometimes patterned with dark bands.

The eye is large. The internasals are longer than or as long as and narrower than the prefrontals. There is 1 loreal. It is as deep as it is long, and in greater contact with the prefrontal than with the internasal (often there is no contact between the loreal and the internasal). There are generally 6 or 7 perioculars. These include the supraocular, 2 or 3 preoculars (the inferior the largest), 2 postoculars (sometimes just 1), the inferior larger than or equal to the superior, and 1 or 2 suboculars. There are 6 or 7 upper labials and 8(4) to 11(4) lower labials. The temporal

FIGURE 7.28. *Causus maculatus*, Guinea. By P. Naskrecki

FIGURE 7.29. *Causus maculatus*, Republic of Congo. By K. Jackson

FIGURE 7.30. *Causus maculatus*, uniform phase, Ghana. By S. Spawls

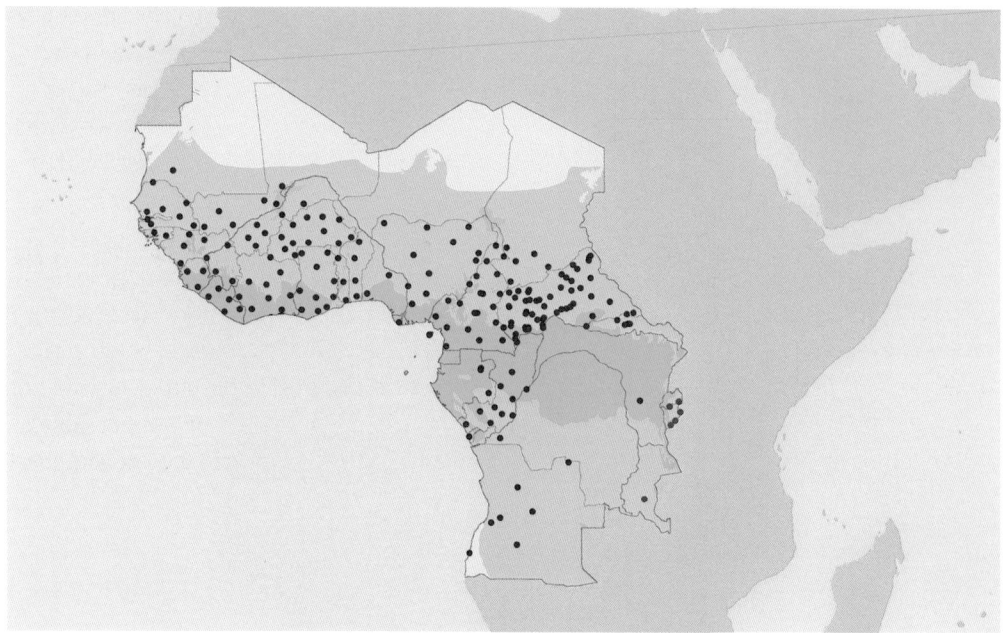

FIGURE 7.31. *Causus maculatus* (blue), *Causus rhombeatus* (red). By K. Jackson

formula is usually 2+3 or 2+4, occasionally 1+3. The dorsal scales are smooth or weakly keeled and usually arranged in 19 (but sometimes 17 or 21) oblique rows. There are 118–159 ventrals (fewer than 138 in males, more than 124 in females), and 14–27 (more than 14 in males, fewer than 24 in females) divided subcaudals. Hughes (1977) noted a higher average ventral count in savannah populations of *C. maculatus* relative to forest populations. The maximum length recorded for this species is 700 mm (Stucki-Stirn 1979).

Green Night Adder: *Causus resimus* (Peters 1862)

The range of *Causus resimus* extends from Cameroon to Somalia in the east and to Angola and Mozambique in the south. The type locality is Jebel Ghule, Sennar, Sudan. *C. resimus* inhabits woodland savanna and open gallery forest habitats. This is one of the *Causus* species with elongate venom glands.

The dorsum is usually uniformly green, though some specimens are light brown. There is sometimes a dark V shape on the dorsum of the head with the tip pointing toward the snout, as in other *Causus* species. The venter is uniformly light green or white. The eye is medium sized. The internasals are longer and narrower than the prefrontals. There is 1 loreal, as deep as or a bit deeper than long. There are usually 6–7 perioculars. These include the supraocular, 2 preoculars of equal size, 2 postoculars, the inferior larger than or equal to the superior, and usually 1 but sometimes 2 suboculars. There are usually 7, but sometimes 6, upper labials, and 8–11(4), usually 10(4), lower labials. The temporal formula is 2+3, occasionally 2+4. The dorsal scales are smooth or weakly keeled and usually arranged in 19 (but sometimes 17 or 21) oblique rows.

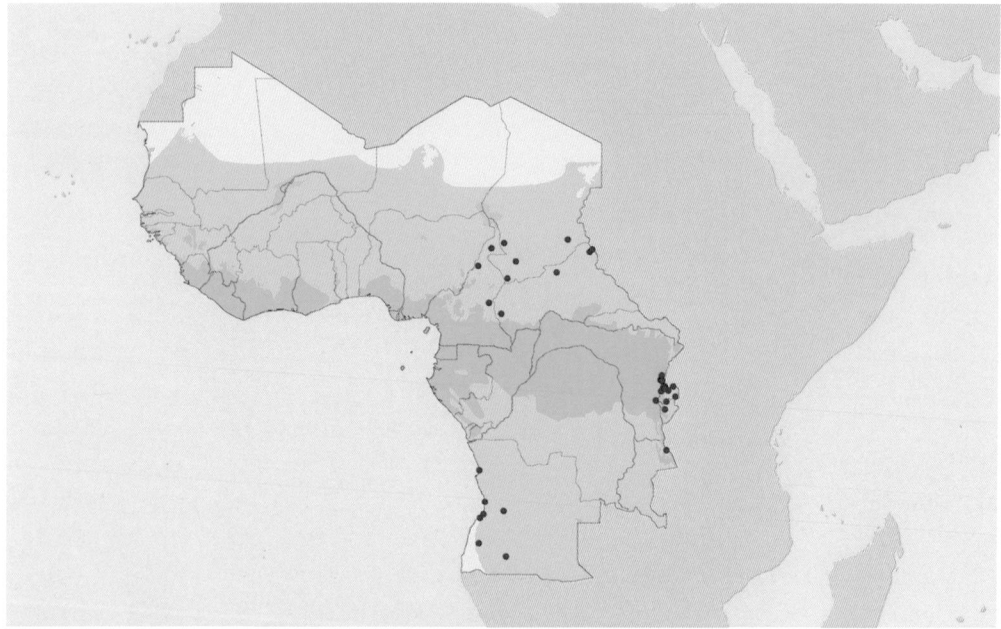

FIGURE 7.32. *Causus resimus* (blue). By K. Jackson

FIGURE 7.33. *Causus resimus*, Kenya. By W. Wüster

There are 131–155 ventrals and 15–27 divided subcaudals. The maximum length recorded for this species is 750 mm (Spawls et al. 2004).

Rhombic Night Adder: *Causus rhombeatus* (Lichtenstein 1823)

Causus rhombeatus is found in eastern and southern Africa. Its range extends from southeastern Sudan and Somalia to South Africa. The type locality is Cape of Good Hope, South Africa. *C. rhombeatus* is often mistaken for the central and western African species *C. maculatus*. This is one of the *Causus* species with elongate venom glands.

The dorsum is light or dark brown with a pattern of black oval spots that fade as the snake ages. There is a dark V shape on the dorsum of the head with the tip pointing toward the snout. The venter is pale and sometimes patterned with dark spots. The eye is large. The internasals are shorter and narrower than the prefrontals. There is 1 loreal. It is deeper than it is long, and in greater contact with the internasal than with the prefrontal. There are generally 6 or 7 perioculars. These include the supraocular; 2 or 3 preoculars, the inferior the largest; 1 or 2 postoculars, the inferior larger than or equal to the superior; and 1 or 2 suboculars. There are 6 or 7 upper labials, and 7(3) to 10(4) lower labials, occasionally 11(5) or 12(5). The temporal formula is usually 2+3, sometimes 2+2 or 2+4, and occasionally 3+3. The dorsal scales are weakly keeled and usually arranged in 19 (but sometimes 17, 21, or 23) oblique rows. There are 135–166 ventrals (with no sexual dimorphism) and 20–36 (more than 20 in males, fewer than 31 in females) divided subcaudals. The maximum length recorded for this species is 950 mm (Spawls et al. 2004).

FIGURE 7.34. *Causus rhombeatus*, Kenya. By S. Spawls

Desert Vipers: Genus *Cerastes* Laurenti 1768

Cerastes is a genus of four species of nocturnal, terrestrial vipers found in the deserts of north Africa and the Middle East. Two species occur within our zone.

Desert vipers of this genus are well adapted for their sandy habitats. They are able to bury themselves in loose sand, often with just their eyes showing above the sand. From this position they hunt by ambush, feeding on rodents, birds, lizards, and invertebrates. The tip of the tail, which is often black, may be used as a lure. *Cerastes* are also able to move rapidly across loose sand by sidewinding. Of the two species found in our zone, *Cerastes vipera* is more specialized than *C. cerastes* for life in sand, with eyes more dorsally oriented on the head, which allows the eyes to remain above the surface when most of the rest of the snake is submerged in sand. *C. cerastes* lays 10–23 eggs under rocks and in holes. The natural history of *C. vipera* is less well known. Like the Saw-scaled Vipers (genus *Echis*), *Cerastes* will rub the keeled scales of the sides of their body against each other, producing a hissing sound, as part of a defensive display.

The head is triangular, broad, and flat. The body is stout with a distinct neck and a short tail. The eye is small or medium sized with a vertical pupil. The eyes are dorsally or dorsolaterally positioned on the head. There may be a horn above each eye. The head scales are small and undifferentiated. There is 1 pair of chin shields. The submandibular groove is pronounced. The dorsal scales are keeled, with 1 apical pit, and are arranged in straight or slightly oblique rows. The subcaudals are divided. The cloacal scale is single. The hemipenes are forked starting from the base, with a forked sulcus spermaticus.

The venom of *Cerastes* contains many enzymes that act at different levels in the blood-clotting process and on the breakdown of tissues. Serine proteases, including the prothrombin activator, Factor X, and thrombin-like enzymes activate the fibrinogen-fibrin transformation, leading to hypofibrinogenemia (Laraba-Djebari et al. 1995; Marrakchi et al. 1997a; Chérifi and Laraba-Djebari 2013). Integrins promote platelet aggregation, leading to thrombocytopenia and the release of endothelial activators of inflammation and coagulation (Chérifi et al. 2014). However, disintegrins inhibit platelet aggregation (Marrakchi et al. 1997b). Along with hemorrhagins, which destroy the vascular endothelia, they are responsible for localized hemorrhaging. Finally, phospholipase A2 molecules are associated with tissue necrosis (Oukkache et al. 2012). The clinical picture is characterized by a hemorrhagic syndrome (bleeding, purpura, blisters, and extensive edema), microangiopathic hemolytic anemia, and renal insufficiency that sometimes results in death (Schneemann et al. 2004; Chani et al. 2008). A few cases of arterial thrombosis leading to infarction have been reported (Mounir et al. 2009; Chani et al. 2012). Tissue necrosis is usually localized and not extensive (Chippaux 1982), although it can manifest itself far from the site of the bite in certain major organs, such as the pancreas (Valenta et al. 2010).

Desert Horned Viper: *Cerastes cerastes* (Linnaeus 1758)

Cerastes cerastes is found in the Sahara Desert from Mauritania to Egypt, in all types of habitats except moving dunes. The type locality is "Oriente," which probably means Egypt.

The dorsum is sand colored, yellowish or light brown, with a pattern of 30 or so dark blotches or crossbands. As the snake ages, these darker markings tend to fade and to merge, darkening the overall color of the individual. The venter is ivory or iridescent white without markings. The tip of the tail is sometimes black. There are 14–18 perioculars and 15–21 interoculars. A prominent feature usually seem in this species is a horn above each eye. Some local populations of *C. cerastes* lack horns above the eye, for example, in the Aïr Mountains of Niger (Trape and Mané 2006a). The horn above each eye is a modified single scale.

There are 4–5 rows of sublabials. There are 12–15 upper labials, none of which is in contact with the eye, and 12(3) to 15(4) lower labials. The dorsal scales are arranged in 27–35 straight or slightly oblique rows. There are 130–165 ventrals and 25–42 divided subcaudals. The cloacal scale is single. The maximum length recorded for this species is 730 mm (Le Berre 1989).

Key to Central and Western African Species of the Genus *Cerastes*

1   Fewer than 129 ventrals; not more than 27 dorsal scale rows .... *C. vipera*
1'   More than 129 ventrals; at least 27 dorsal scale rows .... *C. cerastes*

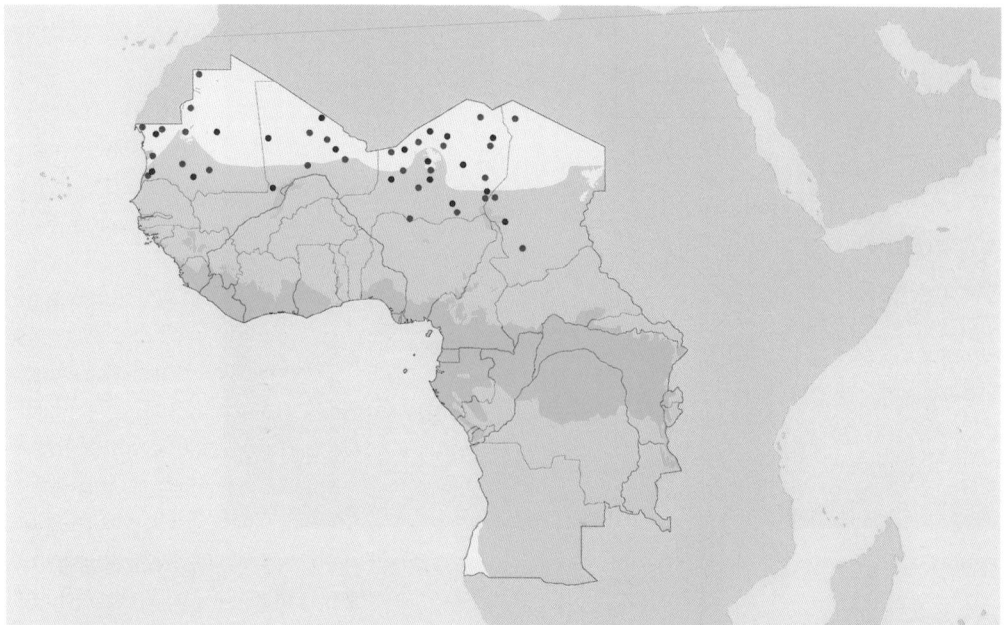

FIGURE 7.35. *Cerastes cerastes* (red), *Cerastes vipera*
(blue). By K. Jackson

Sahara Sand Viper: *Cerastes vipera* (Linnaeus 1758)

Cerastes vipera is found in sandy habitats
and sand dunes of the Sahara Desert from
Mauritania to Egypt. The type locality is
Egypt.

The dorsum is sand colored, yellow,
brown, or sometimes slightly reddish. The
dorsal pattern consists of 30 or so pairs of
brown or gray spots arranged in staggered
rows or merging together. The venter is
uniformly whitish. The tip of the tail is
sometimes black. There are 9–14 perioculars and 7–13 interoculars. This species lacks
horns. The eyes are dorsally oriented. There
are 3–4 rows of sublabials. There are 8–13
upper labials, none of which is in contact
with the eye, and 9(3) to 13(4) lower labials. The dorsal scales are arranged in 23–27
straight or slightly oblique rows. There are
99–128 ventrals and 16–26 divided subcau-

FIGURE 7.36. *Cerastes cerastes*, captive. By D. Williams

dals. The cloacal scale is single. The maximum length recorded for this species is 490
mm (Le Berre 1989).

Saw-scaled Vipers / Carpet Vipers: Genus *Echis* Merrem 1820

Echis is a genus of 10 species of small, nocturnal, terrestrial vipers, found in dry sahel
and savannah habitats of north Africa, central Asia, and the Middle East. Three or four
species occur within our zone. The com-

FIGURE 7.37. *Cerastes vipera*, captive. By W. Wüster

FIGURE 7.38. *Cerastes vipera*, captive, buried except for the eyes. By W. Wüster

mon name, "Saw-scaled Viper," refers to the characteristic threat display of *Echis*, which involves vigorous rubbing of the sides of the body against one another, producing a rasping sound. *Echis* feed on a variety of prey, especially small mammals and invertebrates such as scorpions and centipedes. The species occurring within our zone lay clutches of up to 20 eggs, but some *Echis* from other parts of the range of the genus give birth to live young.

In the savannah belt of west Africa, Saw-scaled Vipers are responsible for approximately 80% of envenomations (Chippaux 2011). This corresponds with a population of 250 million people at risk (rural population in the savannah zone of countries where *Echis* are found) and between 150,000 and 200,000 envenomations, resulting in 10,000–15,000 deaths as well as 5,000–100,000 amputations or other serious sequelae each year. *Echis ocellatus* is responsible for more human deaths per year than any other African snake.

The venom of *Echis* is complex and rich in a large variety of enzymes (Wagstaff et al. 2009). It contains metalloproteases (also called hemorrhagins), which cause destruction of the vascular endothelium (Howes et al. 2003, 2005), and prothrombin-activating enzymes. The prothrombin-activating enzyme of *E. ocellatus* (a west African species) shows about 90% homology with ecarin isolated from the venom of the Asian species, *E. carinatus* (Hasson et al. 2003). Finally, there are several different disintegrins that inhibit platelet aggregation (Smith et al. 2002).

The hemorrhagins are responsible for persistent bleeding at the site of the bite wound as well as from the mucous membranes and sometimes also from other recently healed wounds. The prothrombin-activating enzymes and the disintegrins lead to an afibinogenemia, which can in turn lead to hemorrhagic syndromes (peritoneal, meningeal, or cerebral bleeding) and be complicated by anemia and by hypovolemic shock. Extensive tissue necrosis has been observed in some cases, particularly following bites to the hand in children bitten while trying to catch small animals in their burrows for food, leading to serious symptoms and to amputation of the hand or arm (Gras et al. 2012). There is considerable variation in venom composition between different *Echis* species (Gillissen et al. 1994). Nonetheless, antivenoms appear to be effective against the hemorrhagic symptoms (Meyer et al. 1997; Chippaux et al. 1998, 1999, 2007b, 2015b), provided that they are made using venoms from African *Echis*. Antivenoms manufactured using *E. carinatus* from India have been ineffective for the treatment of bites by African *Echis* species (Visser et al. 2008; Warrell 2008).

The head is oval, with a well-defined neck. The eye is medium to large sized with a vertical pupil. The head scales are small and undifferentiated. The rostral is small. The nasal is divided. The internasals are paired. There are 10–20 perioculars and 7–15 keeled interoculars. There are 1 or 2 rows of scales between the eye and the upper labials. There are 9–13 upper labials, none of which is in contact with the eye, and 8–14 lower labials. There is 1 pair of chin shields with a pronounced submandibular groove. The dorsal scales are strongly keeled, with apical pits, and arranged in 23–33 straight rows. The vertebral row is not enlarged relative to the other dorsal scale rows. The cloacal scale is single. The hemipenes are bilobed from the base, and the sulcus spermaticus is forked.

Joger's Saw-scaled Viper: *Echis jogeri* Cherlin 1990

See *Echis ocellatus* below.

White-bellied Saw-scaled Viper: *Echis leucogaster* Roman 1972

Echis leucogaster is a sahel species. Its range extends from Mauritania east as far as Chad, in sahel habitat. The type locality is Boubon, Niger.

Key to Central and Western African Species of the Genus *Echis*

1	Venter uniformly white	2
1'	Venter speckled with black	3
2(1)	More than 155 ventrals and more than 30 subcaudals	*E. leucogaster*
2'	Fewer than 140 ventrals and fewer than 31 subcaudals	*E. jogeri* (=*E. ocellatus*)
3(1)	Rostral two times as broad as deep; at least 27 subcaudals	*E. pyramidum*
3'	Rostral equal in breadth and depth; not more than 30 subcaudals	*E. ocellatus*

The dorsum is grayish to rust brown with a mid-vertebral line of white blotches, each surrounded by a darker brown area that extends onto the sides of the body. The venter is uniformly white. There are 13–19 smooth perioculars and 7–15 keeled interoculars. There are 2 rows of smooth scales between the eye and the upper labials. There are usually 10 or 11 upper labials. There are 10–14 lower labials, usually 11–13(3 or 4). The dorsal scales are arranged in 25–33 rows. There are 158–189 ventrals (fewer than 180 in males, more than 169 in females) and 25–39 subcaudals (more than 30 in males, fewer than 26 in females). The maximum length recorded for this species is 830 mm (Trape and Mané 2006a).

FIGURE 7.39. *Echis leucogaster*, captive. By S. Spawls

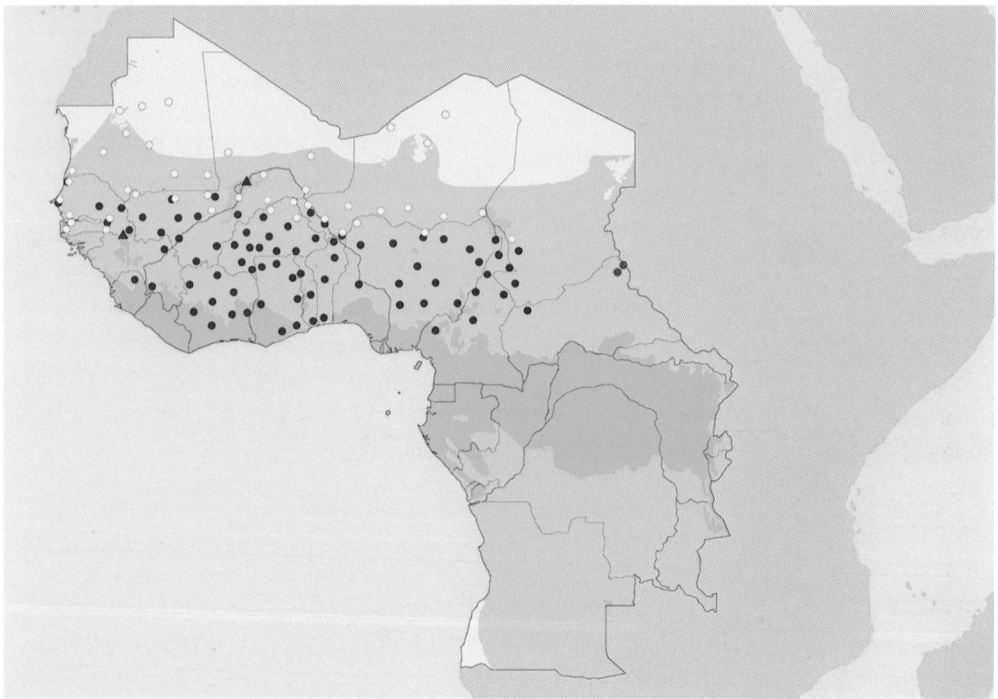

FIGURE 7.40. *Echis leucogaster* (yellow), *Echis ocellatus* (blue), *Echis pyramidum* (red). Blue triangle in Mali indicates the type locality of *E. jogeri*. Blue triangle in Senegal indicates capture locality of other proposed *E. jogeri* (Pook et al. 2009). By K. Jackson

West African Viper: *Echis ocellatus* Stemmler 1970

Echis ocellatus is a savanna species. Its range extends from Senegal east as far as western Central African Republic. The type locality is Garango, Burkina Faso.

The dorsum is grayish to rust brown with a vertebral stripe of alternating black and white blotches, and, along the sides of the body, bright ivory spots outlined in dark brown. The venter is pale beige or cream speckled with dark brown flecks. There are 12–17 smooth perioculars and 8–15 keeled interoculars. There are 1 or 2 rows of smooth scales between the eye and the upper labials. There usually 9 or 10 upper labials. There are 8–13(3 or 4) lower labials. The dorsal scales are arranged in 23–33 rows. The number of dorsal scale rows is geographically variable, increasing from west to east across the range of the species. There are 121–167 ventrals (fewer than 150 in males, more than 127 in females). The number of ventrals increases from west to east across the range of the species like the number of dorsal scale rows (Trape and Mané 2006a). There are 17–30 subcaudals (more than 19 in males, fewer than 26 in females). The maximum length recorded for this species is 550 mm (Roman 1973).

Some authors recognize *E. jogeri* Cherlin 1990 as a separate species closely related to *E. ocellatus*. *E. jogeri* resembles *E. ocellatus*, differing from it in having generally lower ventral counts. Ventral counts in *E. ocellatus* vary across its range, however, increasing from west to east (Trape and Mané 2006a). Molecular studies by Pook et al. (2009) found that *E. ocellatus* collected in eastern Senegal represented a lineage that had diverged long ago from all other *E. ocellatus*

FIGURE 7.41. *Echis ocellatus*, captive. By D. Williams

FIGURE 7.42. *Echis ocellatus*, Ghana. By S. Spawls

they collected (from Cameroon, Niger, Nigeria, and Togo). They interpret this result as evidence in support of the validity of *E. jogeri* as a separate species from *E. ocellatus*.

In this scenario, *E. ocellatus* with low ventral counts, in the western part of its range (e.g., Senegal), represent a separate species (*E. jogeri*), rather than the low end of a morphocline of ventral counts in a single species (*E. ocellatus*) with a large distribution across West Africa. Thus morphological observations of possible *E. jogeri* may correlate with molecular evidence that *E. ocellatus* from Senegal represent a basal branch of *E. ocellatus* that diverged long ago from all other *E. ocellatus* and that this branch represents a separate species, but further work is needed to clarify the status of *E. jog-*

eri. Adding to the confusion surrounding the status of *E. jogeri* is the fact that the type locality, which is Timbuktu, Mali, is farther east than the range of *E. jogeri* is thought to extend and farther north than the normal range of *E. ocellatus*, well into the range of the sahel species, *E. leucogaster*. One possible explanation for this anomaly is that the type specimen may have been an individual washed down the Niger River from a location farther south and west. A further confounding factor is an error in the original description. The type specimen of *E. jogeri* (MNHN 1993 144 A-144), a female, is erro-neously reported in the species description (Cherlin 1990) as having a ventral count of 123, a lower ventral count than the lowest known for female *E. ocellatus* (Trape and Mané 2006a). Examination of the type specimen shows the ventral count in fact to be 132, which is well within the range of ventral counts known from *E. ocellatus* from Senegal and Mali. Further morphological study is needed to make sense of the variation seen in the *E. ocellatus* complex across its distribution.

FIGURE 7.43. *Echis jogeri*, Senegal. By W. Wüster

FIGURE 7.44. *Echis jogeri* juvenile, Senegal. By W. Wüster

FIGURE 7.45. *Echis pyramidum*, captive. By D. Williams

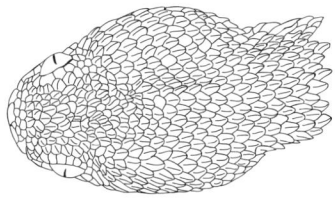

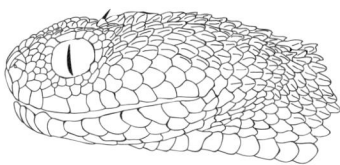

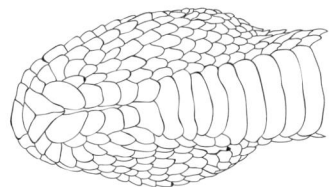

FIGURE 7.46. *Echis pyramidum* USNM 134479. By T. Giri

Egyptian Saw-scaled Viper: *Echis pyramidum* (Geoffroy-Saint-Hilaire 1827)

Echis pyramidum is a savanna species. Its range extends from northeastern Central African Republic east to Kenya and Egypt. The type locality is Egypt.

The dorsum is grayish to rust brown with a mid-vertebral line of white blotches, each surrounded by a darker brown area that extends onto the sides of the body. The venter is white with black flecks. There are 10–20 smooth perioculars and 10–15 keeled interoculars. There are 2 rows of smooth scales between the eye and the upper labials. There are 9–13 upper labials, usually 10 or 11. There are 11–14 lower labials, usually 12 or 13 (3 or 4). The dorsal scales are arranged in 25–31 rows. There are 155–182 ventrals and 27–43 subcaudals. The maximum length recorded for this species is 700 mm (Schleich et al. 1996).

FIGURE 7.47. *Echis pyramidum*, Kenya. By S. Spawls

8

Family Elapidae

The Elapidae is a family with an almost worldwide distribution that includes species with reputations as the "deadliest" snakes in the world. The family is particularly diverse and well represented in Australia, where the largest number of elapid genera is concentrated. Also included among the Elapidae are the True Sea Snakes (Hydrophiinae) and Sea Kraits (*Laticauda*), as well as the New World Coral Snakes (*Micrurus, Micruroides*). Africa boasts a rich elapid fauna, including cobras, mambas, African Garter Snakes, and others described here. Elapids are thought to have originated in Asia in the early Oligocene, with today's African elapids representing independent colonizations of that continent in three lineages of the elapids represented in our zone: (1) cobras, genera *Naja, Pseudohaje*, and *Aspidelaps*; (2) mambas, genus *Dendroaspis*; and (3) African Garter Snakes, genus *Elapsoidea* (Kelly et al. 2009; Pyron et al. 2011).

African elapids can be difficult to distinguish from nonvenomous species (e.g., Colubridae, Lamprophiinae), which they resemble in overall body shape and in their large head scales. The pupil is always round. Elapids lack a loreal scale, which, though not a diagnostic character, helps to distinguish them from most other snake genera within our zone. Though morphologically somewhat conservative, African elapids fill a range of ecological niches, with habitats ranging from arboreal (e.g., *Dendroaspis jamesoni, D. viridis*) to semiaquatic (e.g., *Naja annulata*), and diets equally variable. All are egg-layers.

Elapids are much feared for their formidable venom-delivery system. The elapid dentition is proteroglyph: a tubular fang at the anterior end of a short maxilla, sometimes with a few small solid or grooved teeth posterior to the fang, and a muscularized venom gland located posterior to the eye and connected by a duct to the base of the fang. Bites by elapids are primarily neurotoxic and act by paralyzing muscles, including ultimately, without treatment, the muscles involved in breathing. The victim may die of suffocation due to respiratory paralysis without prompt treatment. The mechanisms of action and effects of elapid venom are described in detail in chapter 4.

Coral and Shield-nose Snakes: Genus *Aspidelaps* Smith 1849

The genus *Aspidelaps* comprises two species from southern Africa, one of which, the Angolan Coral Snake, *Aspidelaps lubricus*, occurs within our zone. *Aspidelaps* are short, relatively stout snakes as elapids go, with a broad head and an enlarged, shield-like rostral scale. The tail is short. The eye is small with a round pupil. The hemipenes are bilobed, and the sulcus spermaticus is divided for the distal half. Found in rocky and dry savanna habitats, they are active at night and are often underground during the day. They are egg-layers and feed on lizards, snakes, and small mammals. When threatened, they will rear up like a cobra. The Coral Snake, *A. lubricus*, has a narrow hood, while the Shield-nosed Snake, *A. scutatus*, rears up but without hooding. Little is known about the venom.

Coral Snake: *Aspidelaps lubricus* (Laurenti 1768)

The range of *Aspidelaps lubricus* extends from Cape Province, South Africa, to Namibia and southern Angola. The type locality is Cape of Good Hope. Three sub-

species are recognized. One of these, *A. l. cowlesi* Bogert 1940, occurs within our zone, its distribution being northern Namibia and southern Angola. The type locality for *A. l. cowlesi* is Munhino, Angola. This northern subspecies differs from the other subspecies in the top of the head being pale and in the higher dorsal scale row count at midbody.

The head and body are light brown above, with the neck a darker brown, followed posteriorly by faint crossbands. The venter is yellowish with 3 or 4 black crossbands on the throat. The nasal is divided. The internasals and prefrontals are paired. The loreal is absent. The rostral is enlarged and shield-like. There is 1 preocular (rarely 2), 2 or 3 postoculars, and 1 subocular. The frontal is longer than broad. There are 7(3,4) upper labials in *A. l. cowlesi*, more often 6(3,4) in the species as a whole. The temporal formula is usually 1+3. There is 1 pair of chin shields. There are 8(3) or 8(4) lower labials. The submandibular groove is shallow. The dorsal scales are smooth, without apical pits, and arranged in oblique rows, most often 21 or sometimes 23 in *A. l. cowlesi*, usually 19 in the species as a whole. The vertebral row is not enlarged relative to the other dorsal scale rows. In *A. l. cowlesi*, there are 144–146 ventrals in males and 158–161 in females, and 31–33 divided subcaudals in males and 30 in females. The cloacal scale is single. The maximum length recorded for this species is 750 mm (Broadley 1983).

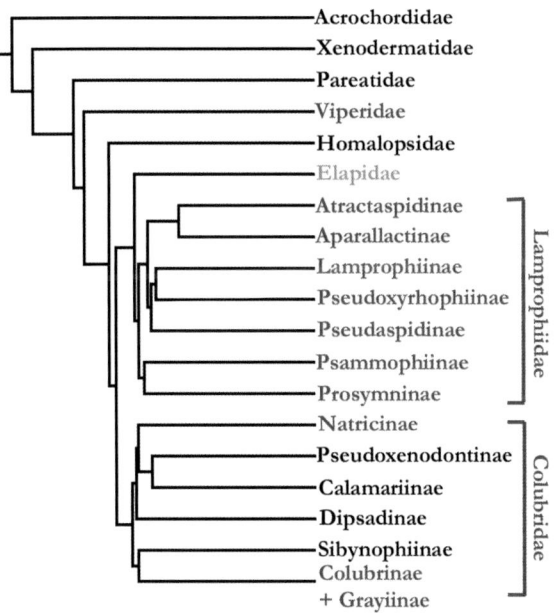

FIGURE 8.1. Higher-level phylogeny of caenophidian families and subfamilies, with families and subfamilies occurring within our zone highlighted in red or orange. Family Elapidae, the subject of this chapter, is highlighted in orange. Modified by K. Jackson from Zheng and Wiens (2016)

Mambas: Genus *Dendroaspis* Schlegel 1848

Mambas are a genus of large elapids, endemic to sub-Saharan Africa. Of four species of *Dendroaspis*, three occur within our zone: the Black Mamba, *Dendroaspis polylepis*, a

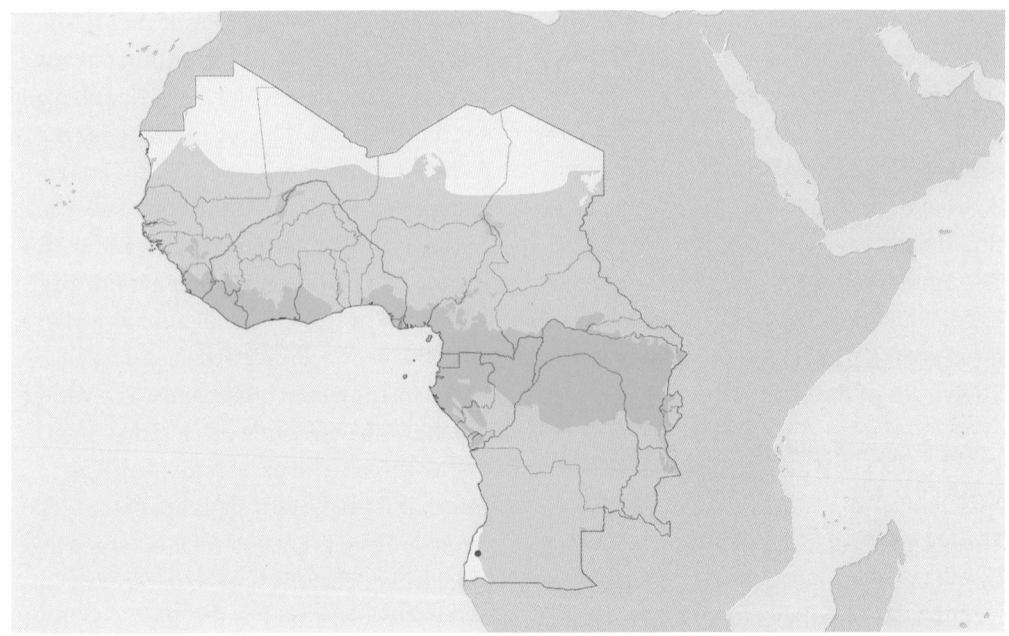

FIGURE 8.2. *Aspidelaps lubricus* (red). By K. Jackson

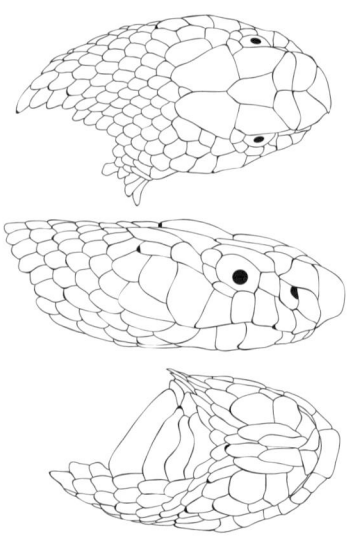

FIGURE 8.3. *Aspidelaps lubricus* RMCA 10668. By T. Giri

FIGURE 8.4. *Aspidelaps lubricus cowlesi*, captive. By W. Wüster

terrestrial species, and two Green Mambas, arboreal species, *D. jamesoni* and *D. viridis*. The species outside our zone is the Eastern Green Mamba, *D. angusticeps*.

Black Mambas feed almost exclusively on terrestrial mammalian prey, the few exceptions including birds that are unable to fly (e.g., chicks). Prey items are small relative to the snakes (1.9% to 7.8% body mass of the snake) (Branch et al. 1995). Surprisingly, for a snake familiar even to the general public, the distribution of the Black Mamba is uncertain. Håkansson and Madsen (1983) reviewed what was known from the literature about the geographical distribution of *D. polylepis*. For a long time (until 1946), the species was confused with the Eastern Green Mamba, *D. angusticeps*, making many locality records from the older literature unreliable. Not until the 1950s was *D. polylepis* documented in West Africa. The Black Mamba is a species associated with open savanna habitats, and it might reasonably be expected to have a distribution similar to that of the Spitting Cobra, *Naja nigricollis*. But *D. polylepis* is not known from the large area encompassed by Chad, Central African Republic, Mali, and Nigeria. It is unclear whether this represents an actual gap in its distribution or is simply an artifact of sampling effort in those countries. These factors explain the sparse dots on the distribution map presented here.

The ecology of Jameson's Mamba, a central African forest species, has been extensively studied in southern Nigeria (Luiselli et al. 2000b). Adults feed almost exclusively on birds, while juveniles also eat lizards and toads. The species was abundant in the region studied, active during both wet and dry seasons, and tolerated a wide range of habitat types, including secondary forest and forest-plantation mosaic. Males are on average larger than females and were observed in combat during the dry season (December–February). Eggs are laid in the wet season (April–June). The clutch size is correlated with the length of the mother,

with the number of eggs laid ranging from 7 to 16.

In addition to the characteristics typical of elapid venoms in general, mambas possess a suite of toxins specific to their genus. These include dendrotoxins and fasciculins, both of which increase the amount of free acetylcholine at neuromuscular junctions, leading to paralysis and tremor, but achieving this by different mechanisms: dendrotoxins act directly to increase the amount of free acetylcholine at neuromuscular junctions, while fasciculins inhibit acetylcholinesterase, which would normally break down acetylcholine, resulting again in an increase in free acetylcholine. Finally, muscarinic toxins block muscarinic receptors, resulting in parasympathetic syndromes (e.g., tears, intolerance of light, muscular cramps, drooling, sweating, vomiting, and diarrhea), some of which may also occur in envenomations by True Cobras, *Naja*, but which are particularly intense and present soon after envenomation in mambas (Chippaux et al. 1977).

Dendroaspis have narrow heads, elongate, somewhat laterally compressed bodies, and long slender tails. The eye is relatively small and the pupil round. The loreal is absent and the internasals and prefrontals paired. The nasal is divided. There are 2 or 3 preoculars, from 2 to 4 postoculars and 1 or 2 suboculars. There are 1 or 2 anterior temporals. There are from 7 to 9 upper labials, 1 or 2 of which are in contact with the eye, and from 8 to 13 lower labials. There are 2 pairs of chin shields. The submandibular groove is pronounced. There are from 13 to 25 oblique rows of smooth dorsal scales without apical pits. There are 210–282 ventrals, a divided cloacal scale, and 94–131 divided subcaudals. The maxilla bears just the fang with no addi-

Key to Central and Western African Species of the Genus *Dendroaspis*

1   13 dorsal scale rows midbody .... *D. viridis*
1'   15 or more dorsal scale rows midbody 2

2(1)   15 or 17 dorsal scale rows midbody .... *D. jamesoni*
2'   21–25 dorsal scale rows midbody .... *D. polylepis*

tional teeth posterior to it. The hemipenes are bilobed, and the sulcus spermaticus is divided at the apex.

Jameson's Mamba: *Dendroaspis jamesoni* (Traill 1843)

Dendroaspis jamesoni is a central African forest species with a distribution extending from Togo to Kenya. Two subspecies are recognized: *D. j. jamesoni* (Traill 1843); *D. j. kaimosae* Loveridge 1936. The range of *D. j. jamesoni* extends from Togo to Central African Republic to south Sudan. The type locality has been corrected to "West Africa" by Mertens (1938). The range of *D. j. kaimosae* extends from eastern Democratic Republic of Congo to Kenya and Uganda. The type locality is Kaimosi Forest, Kakamega, Kenya.

The body, including the head, is bright green, becoming progressively yellow or orange toward the tail. The scales have black edges, making a distinct grid-like pattern, especially noticeable on the tail. There are usually 8(4) upper labials, but this may vary from 7 to 9 and (4) or (4,5). The temporal formula is 1+2 or 2+2. There are usually 9(4) lower labials, but this may sometimes

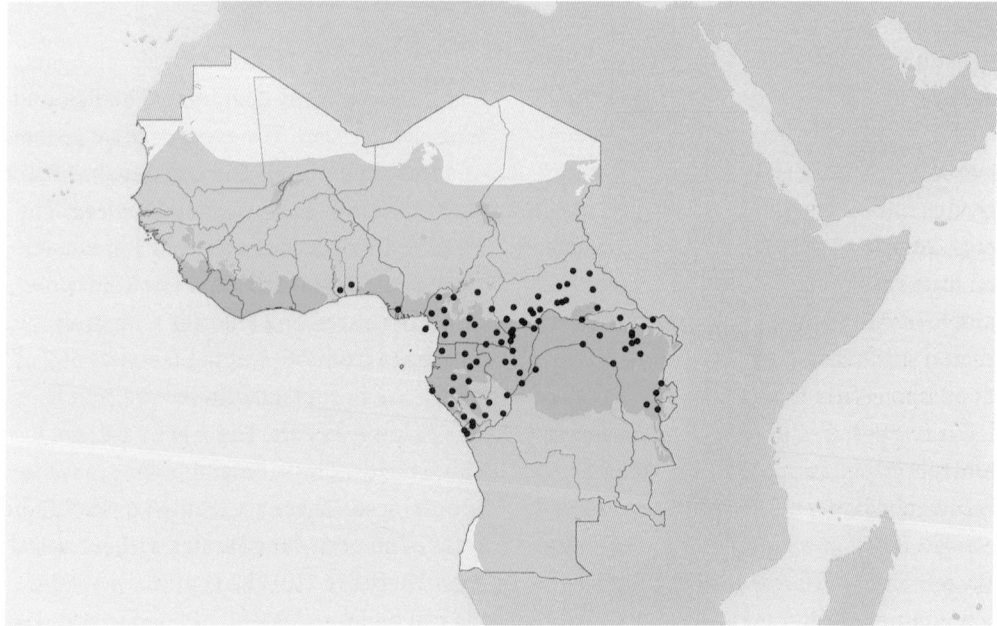

FIGURE 8.5. *Dendroaspis jamesoni* (blue). By K. Jackson

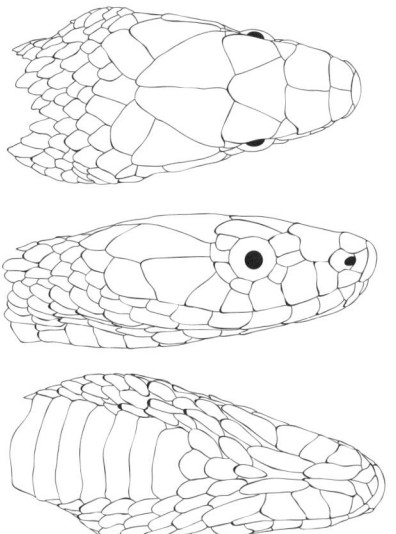

FIGURE 8.6. *Dendroaspis jamesoni* RMCA A7-028-R-0085. By T. Giri

FIGURE 8.7. *Dendroaspis jamesoni*, Kenya. By S. Spawls

vary from 8 to 10(4). There are 15 or 17 dorsal scale rows. There are 202–236 ventrals and 94–122 divided subcaudals. The maximum length recorded for this species is 2,550 mm (Witte 1953).

The two subspecies differ in number of ventrals (210–236 in *D. j. jamesoni*, 202–227 in *D. j. kaimosae*) and number of subcaudals (94–122 in *D. j. jamesoni*, 94–113 in *D. j. kaimosae*). The coloration of the two subspecies is the same except for the tail, which is completely black in *D. j. kaimosae*.

Black Mamba: *Dendroaspis polylepis* Günther 1864

Dendroaspis polylepis is found in savanna habitat throughout most of our zone (but see comments in the genus description about historical confusion between *D. polylepis* and *D. angusticeps*). The type locality is Zambezi, Mozambique.

The body is uniformly brown above, with black edges to the dorsal scales. The underside is pale gray. Young individuals may be olive green but darken with age. The outside of the Black Mamba is not black, but the inside lining of the mouth is black, which may account for the common name. There are usually 8(4) upper labials, but this may vary from 7 to 9 and (4) or (4,5). The temporal formula is 2+3. There are 9(4) to 12(4) lower labials. There are 21–25 dorsal scale rows. There are 242–282 ventrals and 105–131 divided subcaudals. The maximum length recorded for this species is 4,250 mm (Broadley 1983).

Green Mamba: *Dendroaspis viridis* Hallowell 1844

Dendroaspis viridis is a west African forest species with a range extending from Senegal to Togo. The type locality is Liberia.

The body, including the head, is bright green, becoming progressively yellow or orange toward the tail. The scales have black edges, making a distinct grid-like pattern, which is especially noticeable on the tail. There are usually 8(4) upper labials, but this may vary from 7 to 9 and (4) or (4,5). The temporal formula is 2+3. There are 9(4) or 10(4) lower labials. There are 13 dorsal scale rows. There are 211–225 ventrals and 105–128 divided subcaudals. The maximum length recorded for this species is 2,390 mm (Doucet 1963a).

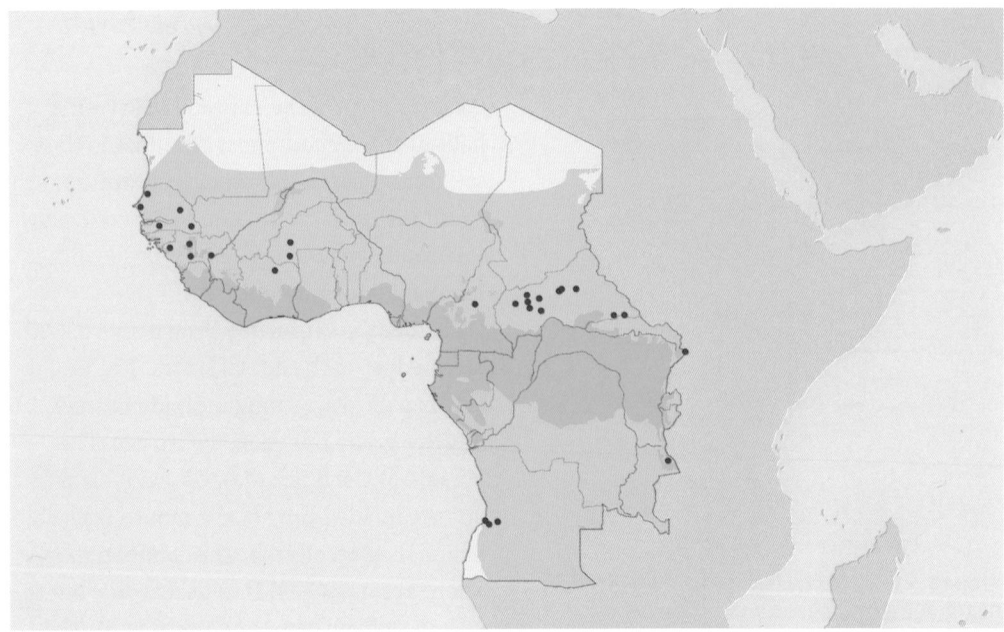

FIGURE 8.8. *Dendroaspis polylepis* (blue). By K. Jackson

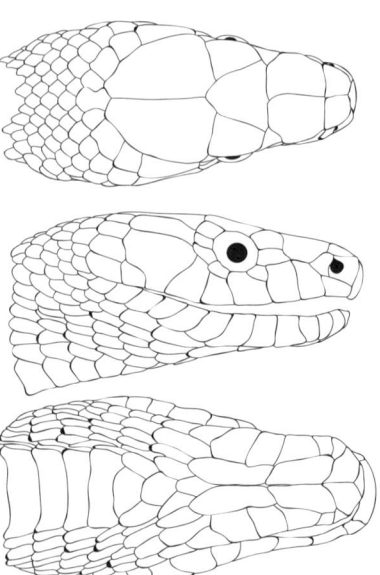

FIGURE 8.9. *Dendroaspis polylepis* RMCA 74-13-R-28. By T. Giri

FIGURE 8.10. *Dendroaspis polylepis*, captive. By S. Spawls

FIGURE 8.11. *Dendroaspis polylepis*, Swaziland. Note the black lining of the mouth, and also the extreme elevation of the snout that is possible in *Dendroaspis*. By A. Barlow

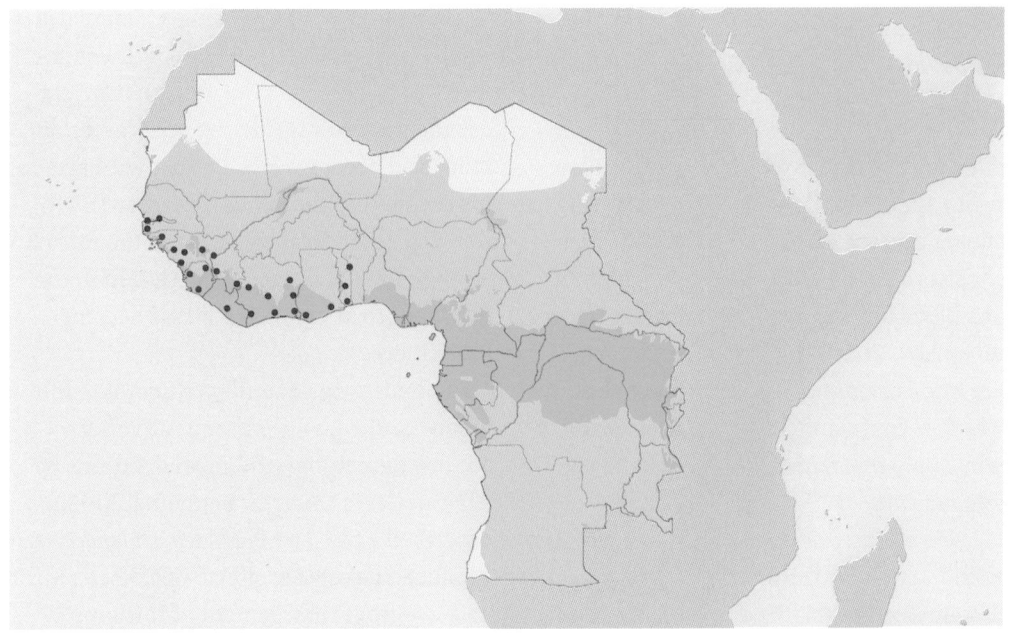

FIGURE 8.12. *Dendroaspis viridis* (blue). By K. Jackson

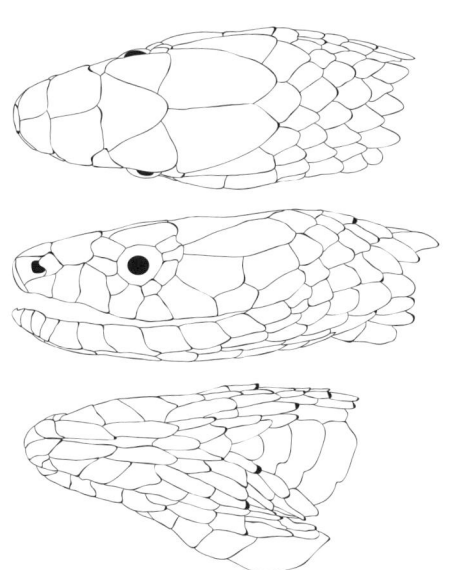

FIGURE 8.13. *Dendroaspis viridis* BMNH 1960.1.5.46. By T. Giri

FIGURE 8.14. *Dendroaspis viridis*, captive. By M. Ponce

FIGURE 8.15. *Dendroaspis viridis*, Ghana. By W. Wüster

African Garter Snakes: Genus *Elapsoidea* Bocage 1866a

Elapsoidea is a genus of seven small elapids, five of which occur within our zone. *Elapsoidea* are small, slow-moving terrestrial snakes and are active at night. Their diet consists of lizards, small snakes, amphibians (Jakobsen 1997), and even insects (Mertens 1937). They are egg-layers. *Elapsoidea* are known commonly as "garter snakes," which is confusing for North Americans, who apply this name to the natricine genus, *Thamnophis*.

The maxillary dentition is proteroglyph, with 1–2 tubular front fangs followed by 3–4 small, unspecialized posterior teeth. The hemipenes are bilobed distally. The sulcus spermaticus is divided for the distal three-fourths.

The head is small, the neck indistinct. The body is cylindrical, with a short tail. The eye is medium sized with a round pupil. The rostral scale is rounded viewed from above. The nasal is divided. The internasals and prefrontals are paired. The loreal is absent. There is 1 preocular, 2 postoculars, and no suboculars. The temporal formula is usually 1+2+3. There are 6–7 upper labials, 2 of which are in contact with the eye. There are 6–7 lower labials. There are 2 pairs of chin shields. The submandibular groove is shallow. The dorsal scales are smooth, without apical pits, and arranged in 13–15 slightly oblique rows. The number of dorsal scale rows tends to decrease posteriorly along the body, so three dorsal scale counts are given: for the neck, for midbody, and for the posterior part of the body (anterior to the cloaca). The vertebral row is not enlarged relative to the other dorsal scale rows. There are 131–171 ventrals and 13–30 divided subcaudals. The cloacal scale is single. *Elapsoidea* have a banded pattern, useful in the identification of species. The pattern is brightest and most distinct in juveniles and then fades and becomes less distinct with age. In his revision of the genus, Broadley (1971b) noted geographical variation in the rate of fading of the pattern (i.e., variation in the body length at which the bands become indistinct).

Elapsoidea are generally reluctant to bite when handled, and consequently there are few descriptions of envenomations by African Garter Snakes. Bennefield (1982) described a bite by *Elapsoidea boulengeri* that produced negligible effects (localized pain and swelling, enlargement of the lymph nodes) but no neurological symptoms. The author also mentioned transitory nasal congestion immediately following the bite. One must not conclude from this single case history that the bite of an African Garter Snake is trivial, however. The venom contains a neurotoxin with the potential to cause symptoms characteristic of elapid envenomations.

Günther's Garter Snake: *Elapsoidea guentheri* Bocage 1866a

The range of *Elapsoidea guentheri* extends from Congo to Zimbabwe. The type locality is Cabinda, Angola.

The body is patterned with 14–24 light brown bands, as broad as or broader than the dark space that separates them, and 2–3 similar bands on the tail. The venter is pale and may range from whitish to gray or light brown. The temporal formula is 1+2+3. There are 6(2,3), sometimes 7(3,4), upper labials. There are usually 7(4) but occasionally 7(3) lower labials. The anterior and posterior pairs of chin shields are equal in

Key to Central and Western African Species of the Genus *Elapsoidea*

1   Prefrontal in contact with an upper lab... *E. laticincta*
1'   Prefrontal not in contact with an upper labial 2

2(1) Underside dark or patterned 3
2'   Underside uniformly pale 4

3(2) Upper labials 6(2,3) .... *E. trapei*
3'   Upper labials 7(3,4) .... *E. loveridgei*

4(3) Length of frontal greater than or equal to 1.5 times its breadth .... *E. guenther*
4'   Length of frontal less than 1.5 times its breadth .... *E. semiannulata*

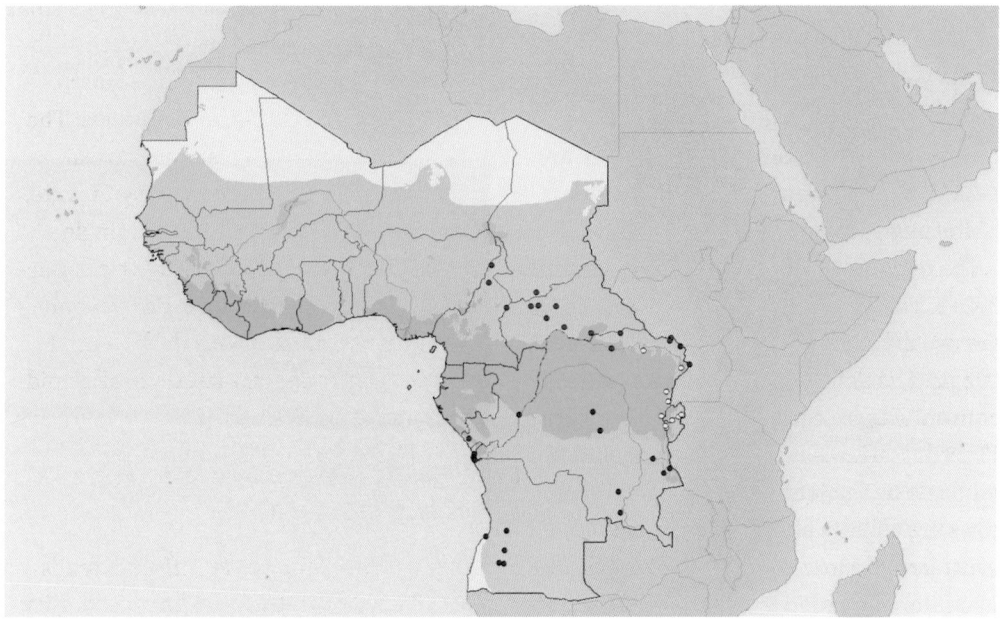

FIGURE 8.16. *Elapsoidea guentheri* (blue), *Elapsoidea laticincta* (red), *Elapsoidea loveridgei* (yellow). By K. Jackson

length. There are 13–15 dorsal scale rows at the neck, and 13 midbody and posteriorly. The dorsal scale rows are slightly oblique and may appear straight. There are 131–160 ventrals and 15–26 divided subcaudals. The maximum recorded length for this species is 630 mm (Loveridge 1944).

Werner's Garter Snake: *Elapsoidea laticincta* (Werner 1919)

The range of *Elapsoidea laticincta* extends from Chad to Sudan to the Democratic Republic of Congo. The type locality is Kadugli, Kordofan, Sudan.

The body is patterned with 8–17 light brown bands, separated by dark bands

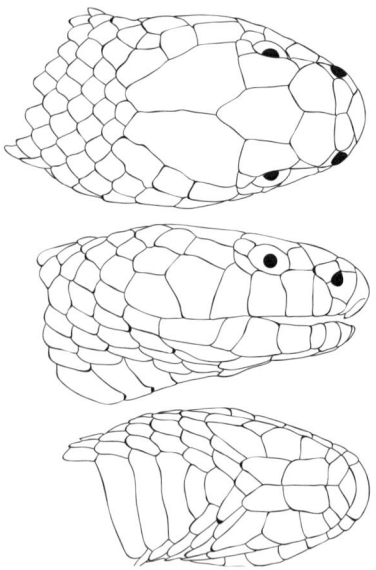

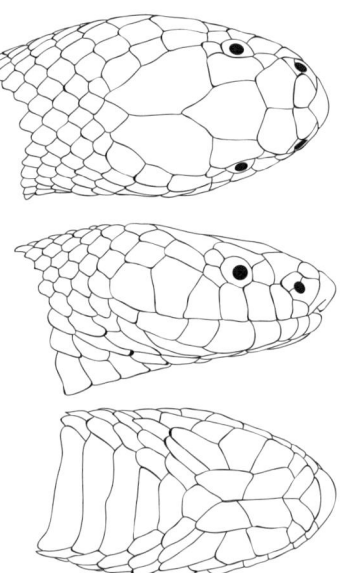

FIGURE 8.17. *Elapsoidea guentheri* RMCA 1482. By T. Giri

FIGURE 8.18. *Elapsoidea laticincta* RMCA 30846. By T. Giri

twice as thick as the light bands. There are 1–2 similar bands on the tail. The venter is uniformly yellow or light brown.

The temporal formula is 1+2+3, sometimes 2+2+3. There are 7(3,4) upper labials. There are usually 7(3), but occasionally 7(4), lower labials. The anterior and posterior pairs of chin shields are equal in length. There are 13–15 dorsal scale rows at the neck, and 13 midbody and posteriorly. The dorsal scale rows are slightly oblique. There are 140–150 ventrals and 13–26 divided subcaudals. The maximum recorded length for this species is 560 mm (Jakobsen 1997).

East African Garter Snake: *Elapsoidea loveridgei* Parker 1949

Three subspecies are recognized, two of which occur within our zone: *Elapsoidea loveridgei colleti* Laurent 1956 and *E. l. multicincta* Laurent 1956. The range of *E. l. colleti* extends from the Democratic Republic of Congo to Burundi, Rwanda, and Uganda. The type locality is Astrida, Rwanda. The

FIGURE 8.19. *Elapsoidea laticincta*, Central African Republic. By S. Spawls

range of *E. l. multicincta* extends from Ethiopia to Tanzania to the Democratic Republic of Congo. The type locality is Uélé, Democratic Republic of Congo.

The dorsum is dark brown or black with a pattern of paler bands, which differs between the two subspecies. The pattern of bands usually persists throughout life but is faded or reduced in older individuals. *E. l. colleti* has 17–24 white-bordered gray to pale brown bands on the body and 2–4 on the tail, most narrower than the brown or black interspaces, while *E. l. multicincta* has 23–35 pale brown bands on the body and 0–6 on the tail, all of which are narrower than the brown or black interspaces. The head is gray to brown in both subspecies, but in *E. l. colleti*, the dark nuchal band extends anteriorly onto the frontal or prefrontals, with bands sometimes replaced by dark spots. In both subspecies, the chin and throat are white to cream. The venter is uniformly gray or brown.

The temporal formula is 1+2+3. There are usually 7(3,4) upper labials. There are usually 7(3), but occasionally 7(4), lower labials. The anterior and posterior pairs of chin shields are equal in length. There are 13–15 dorsal scale rows at the neck, and 13 midbody and posteriorly. The dorsal scale rows are slightly oblique. There are 147–171 ventrals (151–171 for *E. l. multicincta*, 161–170

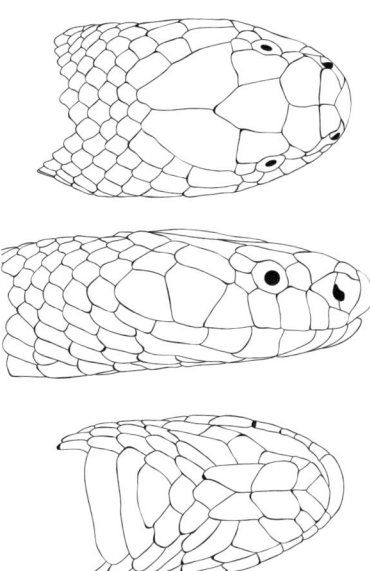

FIGURE 8.20. *Elapsoidea loveridgei* RMCA 78-29-R-44. By T. Giri

for *E. l. colleti*) and 17–30 divided subcaudals (17–26 for *E. l. multicincta*, 18–26 for *E. l. colleti*). The maximum recorded length for this species is 645 mm (Broadley 1971a).

FIGURE 8.21. *Elapsoidea loveridgei*, Tanzania. By S. Spawls

Angolan Garter Snake: *Elapsoidea semiannulata* Bocage 1882

The range of *Elapsoidea semiannulata* extends from Senegal to the Democratic Republic of Congo. Two subspecies are recognized: *E. s. semiannulata* Bocage 1882; *E. s. moebiusi* (Werner 1897). The type localities are Caconga, Angola, for *E. s. semiannulata* and Kete, Ghana, for *E. s. moebiusi*.

The body is patterned with alternating dark large bands and gray bands, the darkest being twice as thick as the gray ones. The bands are separated by a row of pearly white scales. There are 10–21 white bands along the body and 1–3 on the tail. The pattern fades and becomes less distinct as the snake grows. The venter is uniformly whitish. The temporal formula is usually 1+2+3, sometimes 1+1+2 or 2+3+3. There are 7(3,4) upper labials. There are usually 7(3), but occasionally 7(4), lower labials. The anterior chin shields are longer than the posterior pair. There are 13–15 dorsal scale rows at the neck, and 13 midbody and posteriorly. The dorsal scale rows are slightly oblique. There are 136–167 ventrals (136–161 for *E. s. semiannulata* and 145–167 for *E. s. moebiusi*) and 13–28 divided subcaudals. The maximum recorded length for this species is 670 mm (Mané 1999).

Senegal Garter Snake: *Elapsoidea trapei* Mané 1999

Elapsoidea trapei is known only from the type locality of Ndébou, Senegal, and from a few nearby villages in eastern Senegal.

The dorsum is patterned with alternating light and dark bands of roughly equal thickness, separated by thin white crossbands that are one scale thick. There are 20 or so dark bands on the body. The pattern fades

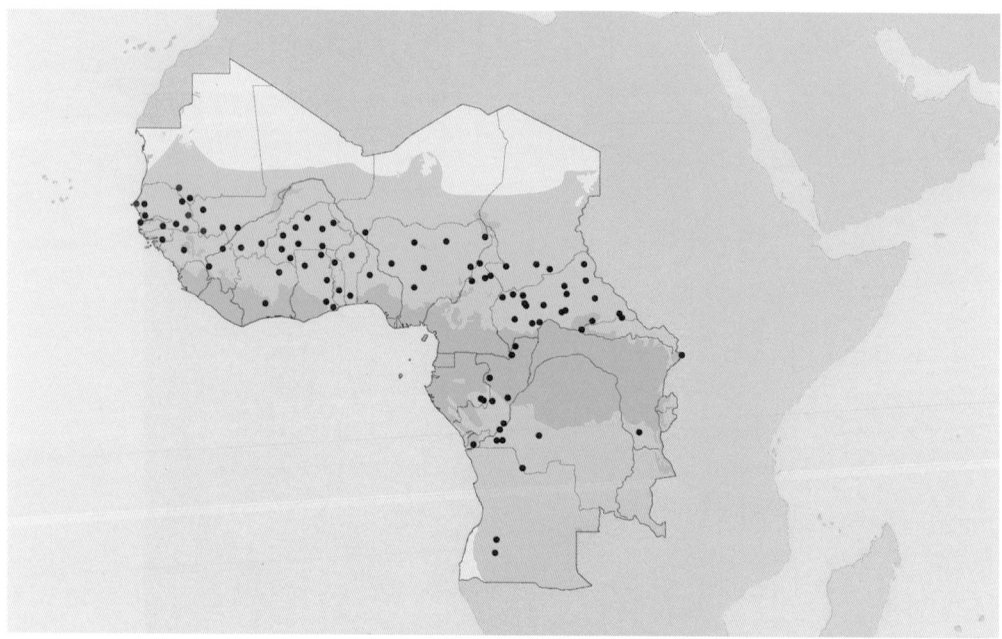

FIGURE 8.22. *Elapsoidea semiannulata* (blue), *Elapsoidea trapei* (red). By K. Jackson

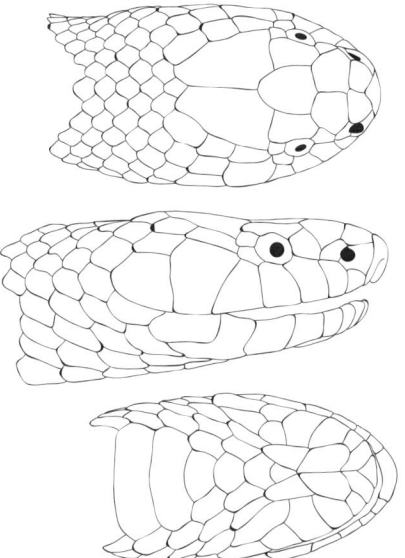

FIGURE 8.23. *Elapsoidea semiannulata* RMCA A7-003-R-0014. By T. Giri

FIGURE 8.24. *Elapsoidea semiannulata moebiusi*, Ivory Coast. By M.-O. Rödel

as the snake grows. The venter is uniformly dark brown, like the dark bands on the dorsum.

The temporal formula is 1+2+3. There are 6(2,3) upper labials. There are 7(4) lower labials. The anterior pair of chin shields is equal to or longer than the posterior pair. There are 13 dorsal scale rows at the neck, at midbody and posteriorly. The dorsal scale

rows are slightly oblique. There are 155–170 ventrals and 18–27 divided subcaudals. The maximum recorded length for this species is 680 mm (Trape and Mané 2006a).

True Cobras: Genus *Naja* Laurenti 1768

Among African snakes, the true cobras (genus *Naja*) are perhaps the best known and most easily recognizable to the general public for their classic defensive posture, with the front third of the body raised ver-

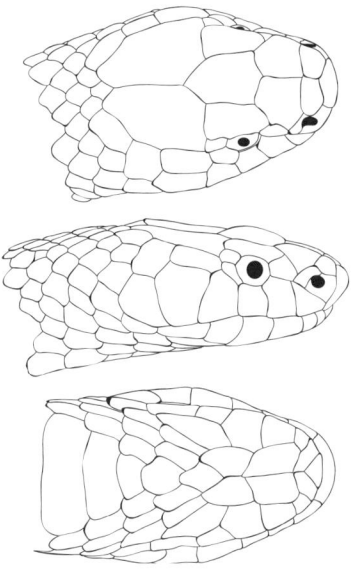

FIGURE 8.25. *Elapsoidea trapei* MNHN 1999.6505. By T. Giri

FIGURE 8.26. *Elapsoidea trapei*, Senegal. By W. Wüster

tically off the ground and the hood spread. *Naja* is an Old World genus, present in both Africa and Asia. The taxonomy of the genus is currently in a state of flux, with numbers of recognized species steadily increasing, but the genus includes a total of approximately 26 species: 11 in Asia and 15 in Africa, with 11 or so occurring in our zone.

Adult *Naja* are often large snakes, with some species attaining lengths of more than 3 m. Species adapted to different habitats occur in almost all areas within our range. These include species restricted to rainforest and others who inhabit open habitats. Among the latter, some species "spit," meaning that, in addition to biting, they are able to project their venom from a distance into the eyes of a potential attacker.

Wallach et al. (2009) proposed subgenus names for the four major clades that make up the genus *Naja*. Asiatic *Naja* are the sister group to the three African clades and retain the subgenus name *Naja*. The subgenus *Afronaja* comprises African Spitting Cobras of the genus *Naja* (*Naja katiensis, N. mossambica, N. nigricollis, N. nigricincta, N. nubi-*

ae). The subgenus *Boulengerina* comprises African (non-spitting) rainforest cobras of the genus *Naja* (*N. annulata, N. christyi, N. melanoleuca, N. multifasciata*). Note that these include Water Cobras formerly assigned to the genus *Boulengerina* (*N. annulata, N. christyi*) and the Burrowing Cobra formerly assigned to the genus *Paranaja* (*N. multifasciata*). The subgenus *Uraeus* comprises African (non-spitting) open formation habitat cobras of the genus *Naja* (*N. anchietae, N. haje, N. senegalensis*).

Reconstructing the history of spitting in elapids (the only group including species that spit), Wüster et al. (2007) proposed that spitting had evolved three times: (1) once in the *Afronaja* clade, (2) once in the Rinkhals (genus *Haemachatus*: African but not within our zone), and (3) once in a clade of Asiatic *Naja* descended from a non-spitting African lineage that had colonized Asia. The common wisdom is that spitting performs a warning function, analogous to rattling by rattlesnakes in the New World. Only if the snake continues to be harassed will it resort to biting. It was

FIGURE 8.27. Mozambique Spitting Cobra, *Naja mossambica*, spitting, Swaziland. By W. Wüster

therefore predicted that spitting in African *Naja* had probably evolved along with the evolution of open grassland habitats and large, potentially threatening mammals. Wüster et al.'s (2007) molecular clock dating put the basal divergence of African spitting *Naja* 16 million years earlier than predicted by the above scenario, however, coinciding with the earliest open grassland formations but predating the large mammals. This leaves the original adaptive function of spitting unclear.

Spitting is accomplished by a specialization of the discharge orifice (the opening at the fang tip of the venom canal that runs through the shaft of the tubular fang). In non-spitting species, the discharge orifice is an elongate opening extending toward the tip of the fang, whereas in spitting species the discharge orifice is roughly tear shaped, opening a bit above the tip of the fang, and with its lower edge representing the abrupt termination of the venom canal inside (Figs. 8–27 and 8–28). Bogert (1943) experimented with fangs, injecting water under pressure into the venom canal of the fangs of spitting and non-spitting cobras. While water injected into the venom canal of non-spitting species exited the fang in the same direction the fang pointed (i.e., down), the jet of water injected into the venom canal of spitting species hit the end of the venom canal and was propelled outward through the discharge orifice at an angle of almost 45°. Spitting cobras are known to aim for the eyes of those who threaten them, and they maximize the chance of some of the venom hitting its target by a rapid rotational movement of the head during spitting, which distributes the venom over a wider area (Young et al. 2008).

When the venom of spitting cobras comes

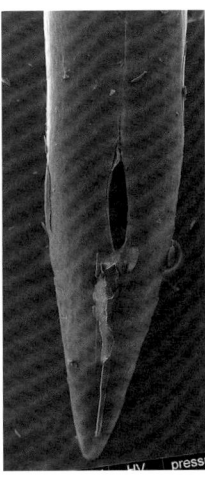

FIGURE 8.28. Tips of the fangs of a (non-spitting) Forest Cobra, *Naja melanoleuca* (*left*), and a Black-necked Spitting Cobra, *Naja nigricollis* (*right*). In non-spitting species, the discharge orifice is an elongate opening extending toward the tip of the fang, whereas in spitting species, the discharge orifice is shorter and rounded, opening slightly basal to the tip of the fang, and with its lower edge representing the abrupt, forward-inflected termination of the venom canal inside. By K. Jackson using scanning electron micrographs by A. Ripley. *Naja melanoleuca* fang, MCZ R-48464, imaged with permission from the Museum of Comparative Zoology, Harvard University

into contact with the eyes of a victim, the result is painful, as cytotoxins in the venom break apart cell membranes on the surface of the eyeball. This type of envenomation is nonetheless not life-threatening or even especially serious if treated correctly. But inadequate or inappropriate treatment can result in the serious complication of lesions on the eyeball, potentially causing permanent blindness (Warrell and Ormerod 1976; Yaya and Danai 2007). The correct treatment is thorough rinsing of the eyes with large quantities of water, followed by the application of anesthetic eye drops. The use of corticosteroids is not recommended because it can increase the risk of infection.

With variations of coloration and patterns

of scutellation between species, *Naja* resemble other elapids in general morphological characteristics, with heads that appear foreshortened, accommodating the proteroglyph dentition, absence of a loreal scale, large platelike scales on the head, round pupils, and smooth scales without apical pits. Scale characters that are particularly noteworthy in cobras are nuchal scales, which are sometimes useful in distinguishing species, and cuneates, which can confuse the counting of lower labial scales (see chapter 1 for details). Within our range, all non-spitting cobras (as well as the spitting species, *Naja nubiae*) have a third upper labial scale that is in contact with both eye

Key to Central and Western African Species of the Genus *Naja*

1	Fewer than 40 subcaudals	*N. multifasciata*
1'	40 or more subcaudals	2
2(1)	At least 2 suboculars separating the eye partially or completely from the upper labials	3
2'	No subocular or a single subocular; at least 1 upper labial in contact with the eye	5
3(2)	17 dorsal scale rows midbody (occasionally 19); fewer than 20 rows at the neck	*N. anchietae*
3'	21 dorsal scale rows midbody (occasionally 19); more than 20 rows at the neck	4
4(3)	21 (occasionally 23) dorsal scale rows at the neck; pattern on head scales	*N. haje*
4'	25 or 27 (occasionally 23) dorsal scale rows at the neck; no pattern on head scales	*N. senegalensis*
5(2)	1 upper labial in contact with the eye	6
5'	2 upper labials in contact with the eye	9
6(5)	Alternating light and dark crossbands along the back and tail	*N. nigrincincta*
6'	Lack of transverse bands along the body and tail	7
7(6)	Frontal approximately equal in length and breadth	*N. nigricollis*
7'	Frontal narrow, clearly longer than broad	8
8(7)	West Africa; 160–186 ventrals and 42–59 subcaudals	*N. katiensis*
8'	Southern Angola; 177–205 ventrals and 52–69 subcaudals	*N. mossambica*
9(5)	Dorsal scale rows oblique	10
9'	Dorsal scale rows straight	11
10(9)	1 anterior temporal	*N. melanoleuca*
10'	2 anterior temporals	*N. nubiae*
11(9)	17 dorsal scale rows midbody	*N. christyi*
11'	21 or more dorsal scale rows midbody	*N. annulata*

and nasal scale, with just 1 preocular above, whereas in all the other spitting species there are 2 preoculars, the lower of which occupies part of the position of the third upper labial in the other group but does not reach the lip (Wüster and Broadley 2003).

Subgenus *Afronaja*: African Spitting Cobras

Katian Spitting Cobra: *Naja katiensis* Angel 1922

The distribution of *Naja katiensis* extends from Senegal to Cameroon. The type locality is Kati, Mali, giving the species its name.

  N. katiensis is uniformly rusty brown above. The head and underside are lighter brown. There are two dark bands across the throat. *N. katiensis* has 2 elongate preoculars, 3 postoculars that are approximately equal in size, and no suboculars. The upper labials are usually 6(3), but sometimes 7(3) or 8(3). The third upper labial is the deep-est. There are 8–13 nuchals, usually more than 10. The temporal formula is usually 2+4, but sometimes 3+4 or 3+5. There is 1 pair of chin shields, and the lower labials range from 7(4) to 10(5). The fifth lower labial is the longest. There are 21–27, but usually 25, oblique dorsal scale rows. There are 160–186 ventrals (fewer than 174 in males, more than 163 in females). The cloacal scale is single, and there are 42–59 (more than 45 in males, fewer than 55 in females) divided subcaudals. The maximum length recorded for this species is 1,070 mm (Trape and Mané 2006a).

Mozambique Spitting Cobra: *Naja mossambica* Peters 1854

The range of *Naja mossambica* extends from southern Angola to South Africa to Tanzania. The type locality is Sena, Tete, Mozambique. Within our zone, this species is known only from a single individual in southern Angola.

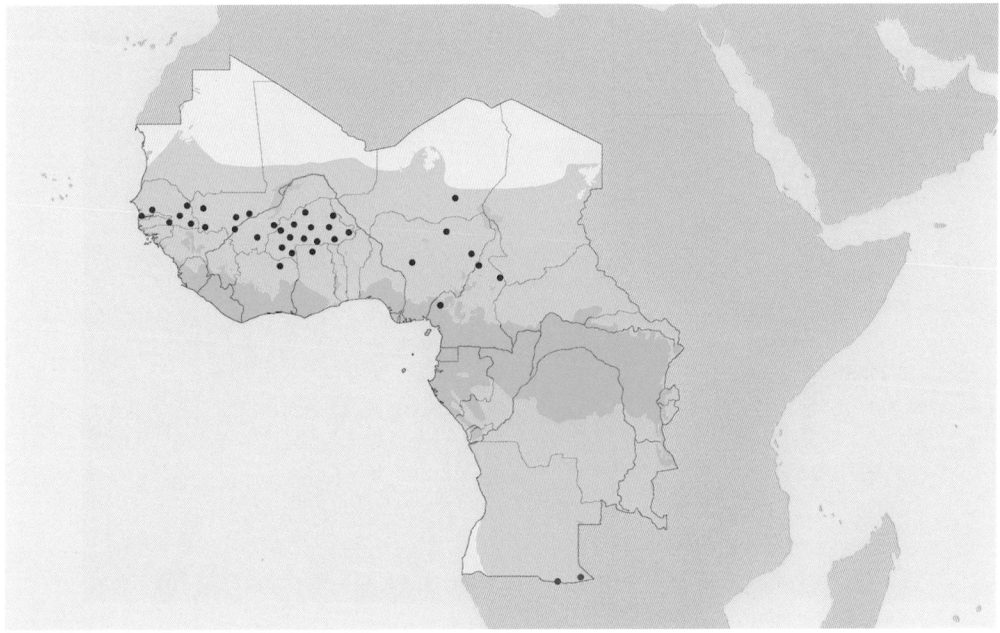

FIGURE 8.29. *Naja katiensis* (blue), *Naja mossambica* (red). By K. Jackson

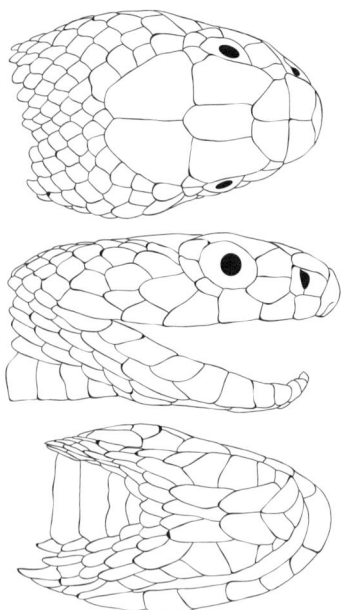

FIGURE 8.30. *Naja katiensis* RMCA 29286. By T. Giri

The body is gray to brown above, with black edges to some or all of the dorsal scales. The venter is pinkish to yellowish with irregular black bars across the throat. *N. mossambica* has 2 elongate preoculars, 3 postoculars of approximately equal size, and no suboculars. The upper labials are usually 6(3), but sometimes 7(3) or 8(3). The third upper labial is the deepest. There are 11–14 nuchals, usually more than 12. The temporal formula is usually 2+4, sometimes 3+4 or 5. There is 1 pair of chin shields, and the lower labials may range from 7(5) to 10(5). The fifth lower labial is the longest. There are 21–27, but usually 23 or 25, oblique dorsal scale rows midbody. There are 177–205 ventrals. The cloacal scale is single, and there are 52–69 divided subcaudals. The maximum length recorded for this species is 1,500 mm (Spawls et al. 2004).

FIGURE 8.31. *Naja katiensis*, Ghana. By S. Spawls

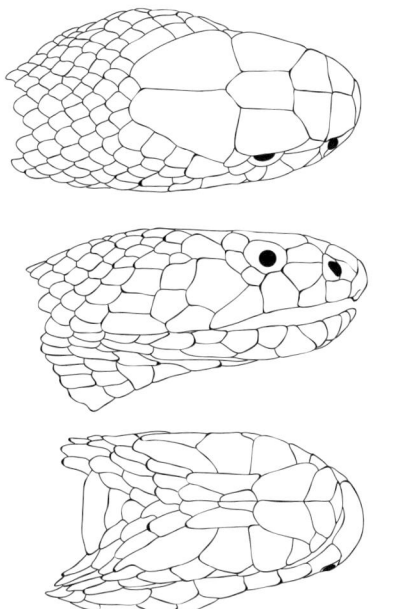

FIGURE 8.32. *Naja mossambica* FMNH 17658. By T. Giri

Western Barred Spitting Cobra / Zebra Cobra: *Naja nigricincta* Bogert 1940

Formerly treated as a subspecies of *Naja nigricollis*, *N. nigricincta* is now recognized as a separate species, and it is actually more closely related to *N. mossambica* than to *N. nigricollis* (Wüster et al. 2007). The range of the Western Barred Spitting Cobra extends from southwestern Angola to southwestern South Africa and western Botswana. The type locality is Munhino, Angola. This species differs from *N. nigricollis* in coloration and in number of ventrals and subcaudals.

   N. nigricincta has 6(3) upper labials. The dorsal scales are smooth, without apical pits, and arranged in 21, or sometimes 23, oblique rows. *N. nigricincta* has a generally higher (though overlapping) ventral and subcaudal count than *N. nigricollis*. There are 192–226 ventrals and 60–73 divided sub-

caudals. The Western Barred Spitting Cobra may be dirty white, olive, or reddish above, with numerous black crossbands on the body and tail, earning it its common name. There is a broad black band at the throat, and the chin may also be black. The maximum length recorded for this species is 1,100 mm (Bogert 1940).

Black-necked Spitting Cobra: *Naja nigricollis* Reinhardt 1843

Naja nigricollis is a widely distributed species with a range extending from Mauritania to Kenya to Angola. The type locality is "Guinea," but presumably meaning "Guinea Coast" rather than the country of Guinea. It is likely that the true type locality is somewhere in Ghana.

   The most thorough ecological studies of *N. nigricollis* were carried out in Nigeria, where the snake occupies a diverse range of habitats. Most active in the wet season (June–July) and least active during the hottest, driest periods (December–February), egg laying nonetheless takes place over a broad time span, and there is a positive correlation between maternal size and clutch size. Adults feed mainly on lizards and to a lesser extent on mammals and frogs. Juveniles feed with equal frequency on lizards, frogs, and fish (Luiselli and Angelici 2000). The past three decades have seen considerable destruction of the pristine rainforest of eastern Nigeria, which is the habitat of *N. melanoleuca*. *N. nigricollis* has colonized this newly opened habitat, where it occurs in sympatry with *N. melanoleuca*. In its original habitat in the arid savannas of central Nigeria, *N. nigricollis* aestivates during the driest months of the year. In the newly colonized former forest habitat, food is available year-round, and there is no need to aestivate. *N. melanoleuca* continues to feed all year, but *N. nigricollis* reduces feeding rates during what would be the dry months in its old habitat, an atavistic behavior that originally served a purpose (Luiselli 2002).

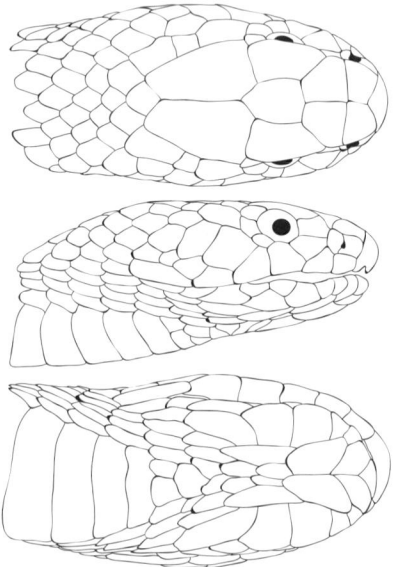

FIGURE 8.33. *Naja nigricollis* RMCA R.G.1254. By T. Giri

The coloration of *N. nigricollis* is variable. Uniformly brown, reddish brown, or black above, it is speckled with black. The underside is black with pink to red bands across the anterior part of the body in specimens from central and western Africa. *N. nigricollis* has 2, or occasionally 1, elongate preoculars, 2 or sometimes 3 postoculars, and no suboculars. The upper labials are usually 6(3), but sometimes 7(3) or 8(3). The third upper labial is the deepest. Sometimes the sixth and seventh upper labials are fused, making the sixth the longest. There are 7–11 nuchals, usually fewer than 10. The temporal formula is usually 2+3 or 2+4. There is 1 pair of chin shields, and the lower labials are usually 9(4) but may range from 8(4) to 11(4). The fourth lower labial is the longest. There are 17–23, but usually 21, oblique dorsal scale rows midbody. There are 176–219 ventrals. The cloacal scale is single, and there are 52–69 divided subcaudals. The maximum length recorded for this species is 2,200 mm (Villiers 1975).

FIGURE 8.34. *Naja nigricollis*, Benin. By L. Naudin

Nubian Spitting Cobra: *Naja nubiae* Wüster and Broadley 2003

The distribution of *Naja nubiae* extends from Egypt to the Sudan and to Saharan oases in Niger and Chad. The type locality is Kom Ombo, Aswan, Egypt.

The overall coloration of *N. nubiae* is similar to that of *N. katiensis*. The body is uniformly rusty brown above. There is a pair of broad, dark rings on the neck. The underside is uniformly paler brown. There is a teardrop-shaped black spot on the upper labials immediately below the eye.

N. nubiae has 1, or sometimes 2, elongate preoculars, 2 postoculars of approximately equal size, and 0 or 1 suboculars. The upper labials are usually 7(3,4) but sometimes 6 or 8(3,4), and occasionally 6 or 7(4). The third or fourth upper labial is the longest. There are 8-12 nuchals, usually more than 12. The temporal formula is usually 2+4 or 2+5, but sometimes 3+4 or 3+5. There is 1 pair of chin shields, and the lower labials may range from 7(4) to 11(4). The fourth lower labial is the longest. There are 17-27 oblique dorsal scale rows midbody. There are 179-200 ventrals. The cloacal scale is single, and there are 56-69 divided subcaudals. The maximum length recorded for this species is 1,489 mm (Wüster and Broadley 2003).

Subgenus *Boulengerina*: African (Non-spitting) Rainforest Cobras

Ringed Water Cobra: *Naja annulata* Buchholz and Peters 1876

The range of *Naja annulata*, the more commonly encountered of the two species of water cobra, extends from Cameroon to the Democratic Republic of Congo to Gabon and Central African Republic. The type locality is Mbusa, Gabon. It is strictly aquatic and feeds mostly on fishes, maybe amphibians (Spawls et al. 2004).

N. annulata is yellowish to light brown above, including the head. There are 21-23 black rings along the length of the body, the first 5 single, and subsequent ones double. The first ring is around the neck, and the last is before the tail. The tail is dark. Usually the rings completely encircle the body, but occasionally the underside is uniformly brown or blackish, darkening posteriorly. There is 1 elongate preocular, 2 postoculars of roughly equal size, and no suboculars. There are 7(3,4) or sometimes 8(3,4) upper labials, of which the sixth is the longest. The temporal formula is 1+2 or 1+3, and there are 7-9 nuchals. There is just 1 pair of chin shields, and 8(4), or occasionally 9 or 10(4) or (5), lower labials, of which the fifth is the longest. The submandibular groove is not pronounced. There are 21-25 straight dorsal scale rows, 199-226 ventrals, and 70-77 divided subcaudals. The maximum length recorded for this species is 1,900 mm (Schmidt 1923).

Two subspecies are recognized: *N. a. annulata* Buchholz and Peters 1876; *N. a. stormsi* (Dollo 1886). These differ from one another in their coloration, *N. a. stormsi* having fewer rings than *N. a. annulata* and having them restricted to the anterior part of the body, as well as in their geographic distribution, with *N. a. annulata* found from Cameroon to Central African Republic and *N. a. stormsi* farther east, from eastern Democratic Republic of Congo to Kenya and Tanzania.

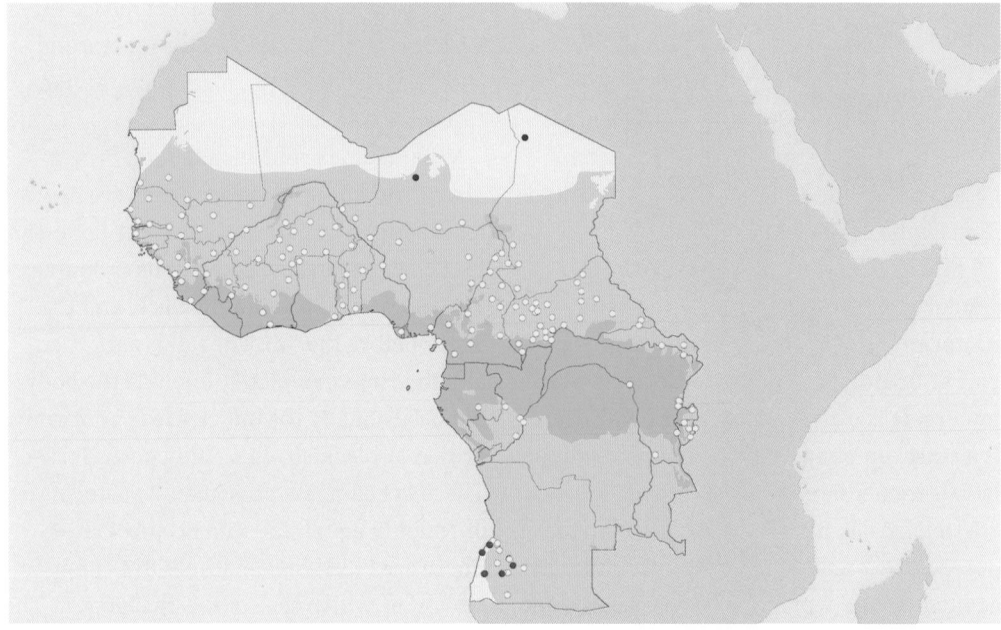

FIGURE 8.35. *Naja nigricollis* (yellow), *Naja nigricincta* (red), *Naja nubiae* (blue). By K. Jackson

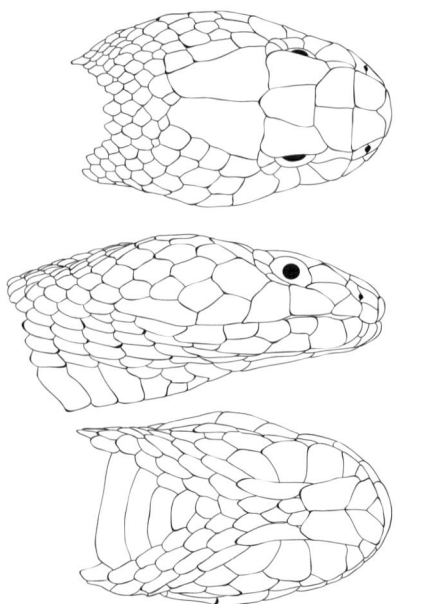

FIGURE 8.36. *Naja nubiae* FMNH 75228. By T. Giri

FIGURE 8.37. *Naja nubiae*, captive. By W. Wüster

Christy's Water Cobra: *Naja christyi* (Boulenger 1904a)

The known range of *Naja christyi* is more restricted than that of *N. annulata*, extending from the Congo to the Democratic Republic of Congo. The type locality is Kinshasa, Democratic Republic of Congo. Like *N. annulata*, it is strictly aquatic and feeds on fishes. *N. christyi* is much less frequently encountered than *N. annulata*, so much so that the species is known only from female

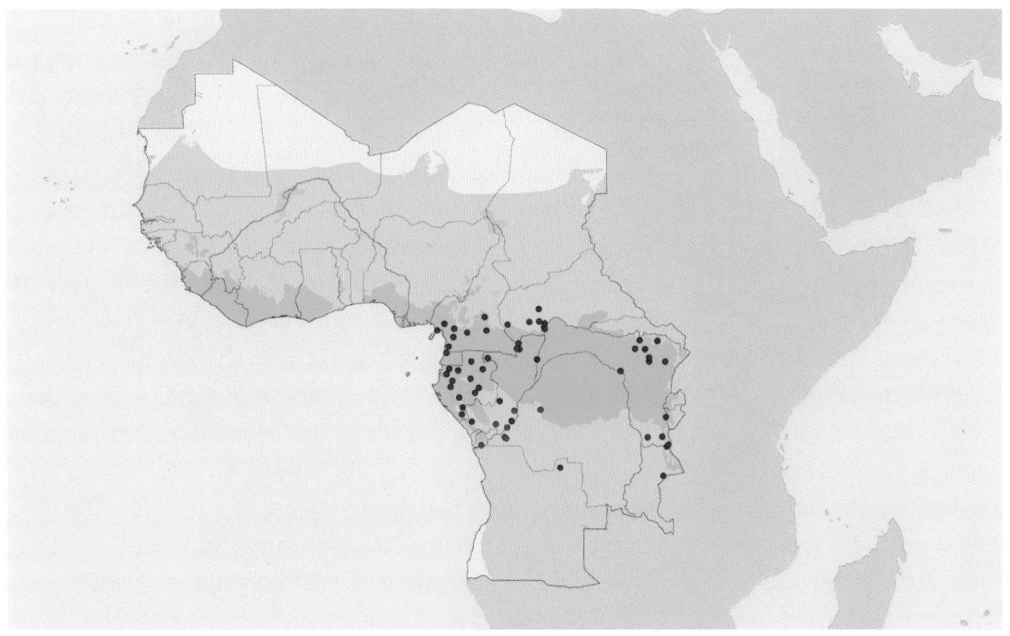

FIGURE 8.38. *Naja annulata* (blue), *Naja christyi* (red). By K. Jackson

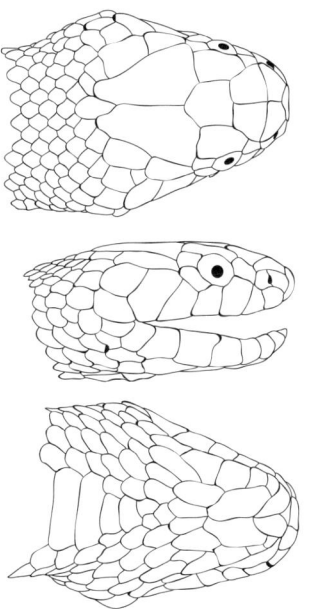

FIGURE 8.39. *Naja annulata* RMCA A7-31. By T. Giri

FIGURE 8.40. *Naja annulata*, juvenile, Republic of Congo. By K. Jackson

FIGURE 8.41. *Naja annulata*, adult, Republic of Congo. By K. Jackson

specimens. Although it resembles *N. annulata* in most scale counts, a notable difference is the number of dorsal scale rows.

N. christyi is dark brown above, including the head. The anterior part of the underside of the body is patterned with narrow yellow bands. There is 1 preocular, 2 postoculars of roughly equal size, and no suboculars. There are 8(3,4) or (4,5) or occasionally 7(3,4) upper labials, of which the sixth is the longest. The temporal formula is 1+2 or 1+3, occasionally 2+3 or 2+4, and there are 7–8 nuchals. There is just 1 pair of chin shields and 9(4), or occasionally 8(4), lower labials, of which the fourth is the longest. The submandibular groove is not pronounced. There are 17 straight dorsal scale rows, 206–209 ventrals, and 69–73 divided subcaudals. The maximum length recorded for this species is 1,340 mm (Trape and Roux-Estève 1990).

Forest Cobra: *Naja melanoleuca* Hallowell 1857

Widely distributed in forest habitat throughout sub-Saharan Africa, *Naja melanoleuca*'s range extends from Senegal to Tanzania to Angola. The type locality is Gabon. In Nigeria, *N. melanoleuca* mainly inhabits primary and secondary forest patches, especially primary swamp forest, though it is also found in plantations and suburbia. Most active in the wet season (June–July) and least active during the hottest, driest periods (December–February), egg laying nonetheless takes place over a broad time span, and there is a positive correlation between maternal size and clutch size. Adults feed equally on mammals, frogs, and fish, while juveniles feed primarily on fish (Luiselli and Angelici 2000).

N. melanoleuca is black above, often (especially in young individuals) with white flecks. The anterior third of the underside is yellow with broad black bands. The rest of the underside is dark gray. There is 1 preocular and usually 3 postoculars (sometimes 2), the lowest of which may be in the position of a subocular. There are 7(3,4) upper labials, of which the sixth is the longest. The temporal formula is 1+2 or 1+3. There are usually 2 pairs of chin shields, but sometimes a pair of false chin shields takes the place of the posterior pair (see definition in chapter 1). There are 8(4), occasionally 7(4) or 9(4), lower labials, of which the fifth is the longest. There are 7–9 nuchals, and 19 (occasionally 17 or 21) oblique dorsal scale rows. There are 198–228 ventrals and 52–76 divided subcaudals. The maximum length recorded for this species is 3,200 mm (Villiers 1975).

Two subspecies are recognized, which differ in certain scale counts. *N. m. melanoleuca* Hallowell 1857 has the temporal formula of 1+2, 211–228 ventrals, and 52–76 subcaudals, whereas *N. m. subfulva* Laurent 1955 has a temporal formula of 1+3, 198–218 ventrals, and 55–67 subcaudals.

Burrowing Cobra: *Naja multifasciata* (Werner 1902)

Secretive and rarely encountered, *Naja multifasciata* burrows under leaf litter in wet areas of primary forest. Its range is restricted to central African forest from Cameroon to western Democratic Republic of Congo. The type locality is Maringa, Democratic Republic of Congo.

The dorsal scales of *N. multifasciata* are pale yellow anteriorly and black posteriorly. The head is black above and yellow on the sides. Posterior to the parietals, a thin pale-yellow band separates the head from

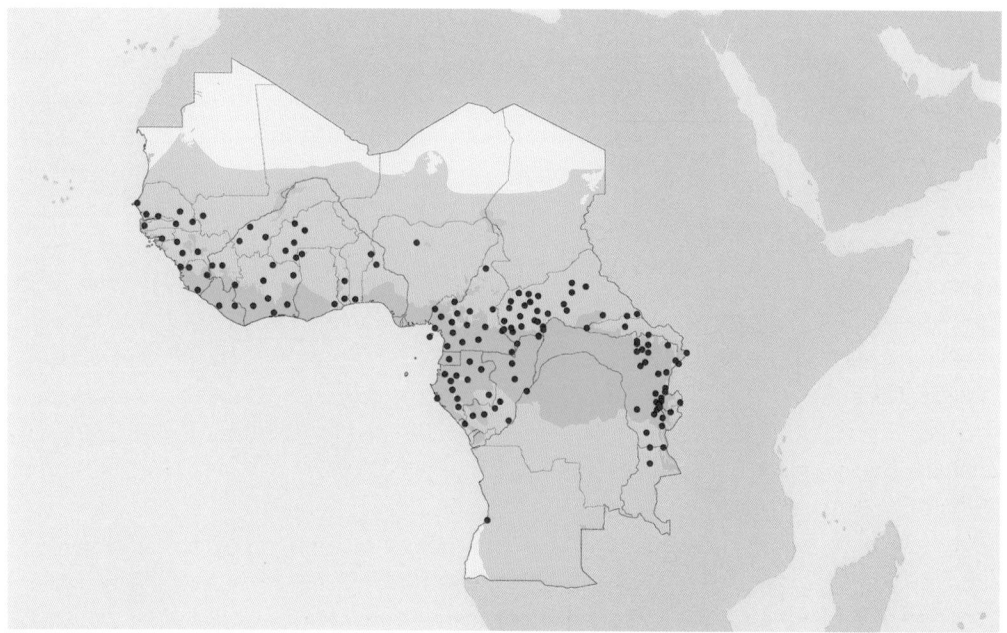

FIGURE 8.42. *Naja melanoleuca* (blue). By K. Jackson

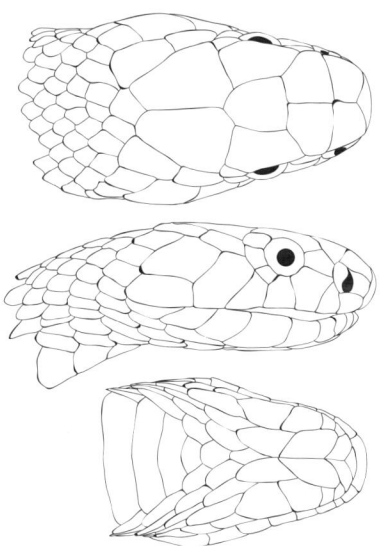

FIGURE 8.43. *Naja melanoleuca* RMCA 21432. By T. Giri

FIGURE 8.44. *Naja melanoleuca*, Benin. By M.-O. Rödel

the body. There is 1 preocular, 2 postoculars, and 1 subocular. The subocular does not completely separate the eye from the upper labials, so the upper labials are usually 6 or 7(3,4). The temporal formula is 1+2 or 1+3.

There are just 15, or less often 17, slightly oblique dorsal scale rows. There are 153–175 ventrals. The cloacal scale is single, and there are just 30–39 divided subcaudals. The maximum length recorded for this species is 760 mm (Witte 1962).

Two subspecies have been proposed, though they are not agreed upon: *N. m. multifasciata* (Werner 1902) and *N. m. anomala* (Sternfeld 1917). These differ in the number

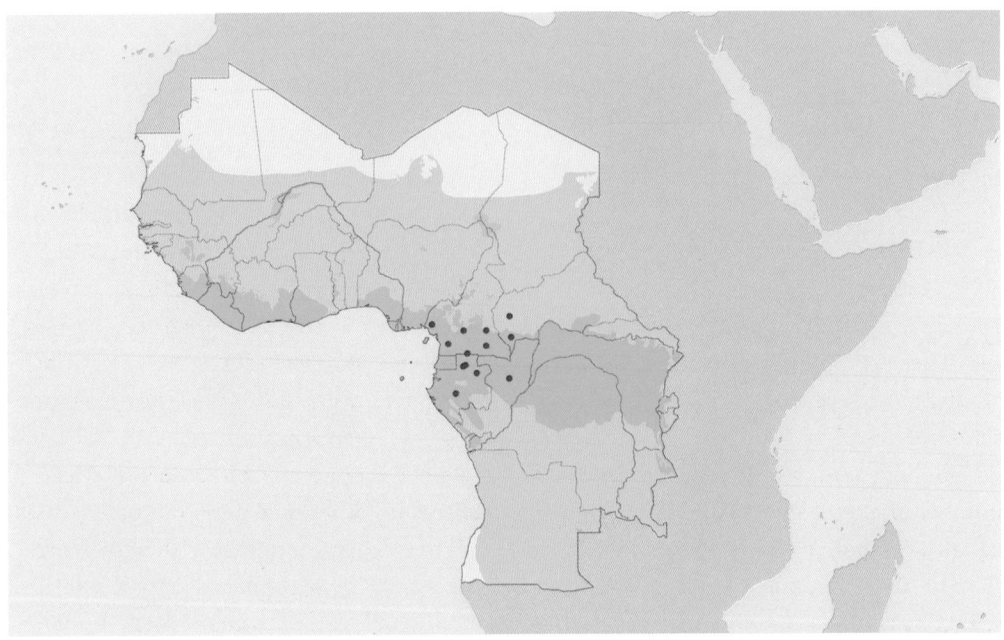

FIGURE 8.45. *Naja multifasciata* (red). By K. Jackson

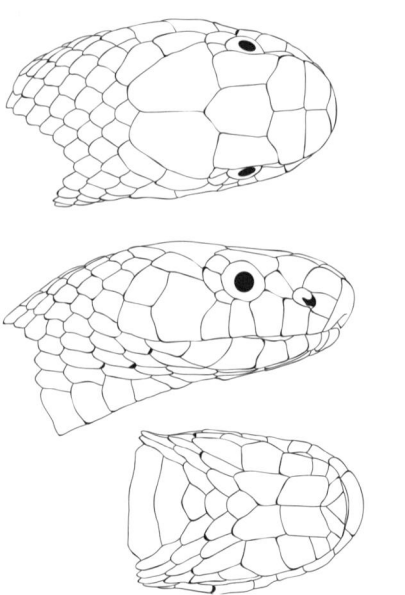

FIGURE 8.46. *Naja multifasciata* RMCA 76-1-R-3. By T. Giri

FIGURE 8.47. *Naja multifasciata*, captive. By W. Wüster

of upper labial scales, which is 7(3,4) in *N. m. multifasciata* and 6(3,4) in *N. m. anomala*.

Subgenus *Uraeus*: African (Non-spitting) Open Formation Cobras

Anchieta's Cobra: *Naja anchietae* Bocage 1879

The range of *Naja anchietae* extends from Angola to Namibia to Zimbabwe. The type locality is Caconda, Angola.

N. anchietae is dark brown above. The underside is yellowish, darkening posteriorly. The head is paler than the body, and there is usually a dark nuchal band across the nape. *N. anchietae* has 6 or 7 perioculars. There are usually 7(0) but occasionally 6 or 8(0) upper labials. The sixth upper labial is the longest. The temporal formula is usually 1+2 or 1+3 but occasionally 2+2 or 2+3. There are 7–9 nuchals. There is 1 pair of chin shields and usually 9 or 10(4), but sometimes 8(4) or 11(4), lower labials, of which the fourth or fifth is the longest. The number of dorsal scale rows at midbody is usually 17 but may occasionally range from 15 to 19. There are fewer than 20 dorsal scale rows at the neck. There are 179–200 ventrals and 51–56 divided subcaudals. The maximum length recorded for this species is 2,180 mm (Broadley 1983).

Egyptian Cobra: *Naja haje* (Linnaeus 1758)

Naja haje was originally a species with an immensely broad distribution in sub-Saharan Africa, which has since been split up into five species (subgenus *Uraeus*), three of which occur within our zone (Trape et al. 2009). The range of *N. haje*, in its current restricted sense, extends from Egypt to Senegal to Zambia. The type locality is northern Egypt.

N. haje is uniformly blackish above scattered with occasional darker or paler scales, and the underside is yellowish with black markings. The scale characters of *N. haje*, *N. senegalensis*, and *N. anchietae* are similar, the clearest difference being in dorsal scale rows, particularly at the neck. Because postoculars are sometimes in the position of suboculars and preoculars extend to the lip and become upper labials, it is useful to refer to all the scales touching the eye

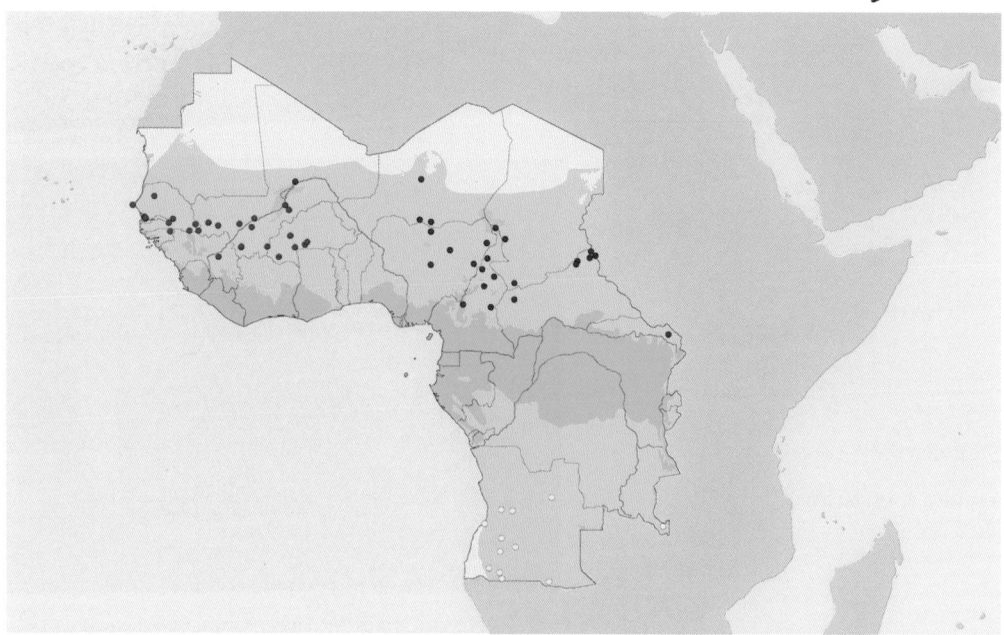

FIGURE 8.48. *Naja anchietae* (yellow), *Naja haje* (blue), *Naja senegalensis* (red). By K. Jackson

(except upper labials) as perioculars in describing snakes of the subgenus *Uraeus*.

N. haje has 6 or 7 perioculars. There are usually 7(0) or sometimes 6 or 8(0) upper labials, but occasionally the third upper labial is in contact with the eye. The sixth upper labial is the longest. The temporal formula is usually 1+2 or 1+3, sometimes 2+3. There are 5–7 nuchals. There is 1 pair of chin shields and usually 9 or 10(4), but sometimes 8(4) or 11(4), lower labials, of which the fourth is the longest. The number of dorsal scale rows at midbody is usually 21 but may range from 19 to 23. At the neck, the number of dorsal scale rows may range from 19 to 23. There are 191–222 ventrals (fewer than 217 in males, more than 201 in females) and 53–68 divided subcaudals. The maximum length recorded for this species is 3,000 mm (Stucki-Stirn 1979).

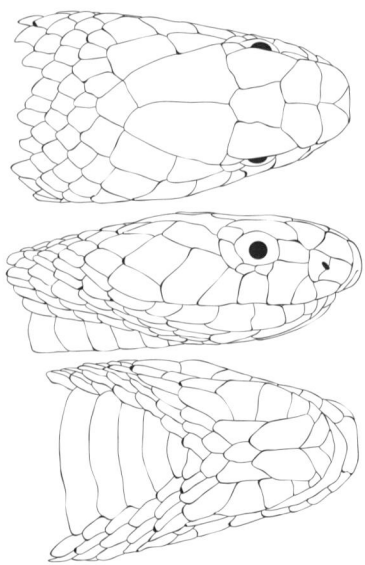

FIGURE 8.49. *Naja haje* RMCA 28509. By T. Giri

Senegalese Cobra: *Naja senegalensis* Trape, Chirio, Broadley and Wüster 2009

The range of *Naja senegalensis* extends from Senegal to Nigeria. The type locality is Diel-mo, Senegal.

N. senegalensis is uniformly grayish brown above, though the paravertebral rows may be paler. The underside of the neck is dark gray with a lighter band across it. The rest of the underside is yellowish, darkening toward the tip of the tail. In juveniles, the overall coloration is more contrasting. There is a white marking on the back of the neck (the back of the hood) similar to the hood markings in Asian *Naja*. This marking is sometimes retained in the adult, although it becomes less clear as the ground color of the back of the neck becomes paler with age (Trape et al. 2009).

N. senegalensis has 5–7 perioculars. There are usually 7(0) but occasionally 8(0) upper labials. The sixth upper labial is the longest. The temporal formula is usually 1+2 or 1+3. There are 6–9 nuchals. There is 1 pair of chin shields and usually 9 or 10(4), but

FIGURE 8.50. *Naja haje*, Kenya. By S. Spawls

sometimes 8(4) or 11(4), lower labials, of which the fourth is the longest. The number of dorsal scale rows at midbody is usually 21 but may occasionally be 23. At the neck, the number of dorsal scale rows is 25 or occasionally 23. There are 205–225 ventrals (fewer than 220 in males, more than 218 in females) and 56–66 divided subcaudals. The maximum length recorded for this species is 2,450 mm (Trape et al. 2009).

Tree Cobras: Genus *Pseudohaje* Günther 1858

The Tree Cobras, *Pseudohaje,* are a rare and elusive genus, of which both species occur within our zone. Their biology is virtually unknown compared to that of other large, arboreal, elapids, such as mambas. Akani et al. (2005), having spent a decade studying thousands of snakes in the field in Nigeria, published their notes on the 62 individuals of Goldie's Tree Cobra captured during that period, recognizing that this was too small a sample size for a serious study but at the same time that these observations would provide the most data available about the natural history of *Pseudohaje*. Their findings are reported here, and, in the absence of comparable data for the Black Tree Cobra, we can only extrapolate from the slightly better-known Goldie's Tree Cobra.

Based on observations of those 62 individuals, *Pseudohaje goldii*'s preferred habitat is primary forest, followed by swamp forest, mangroves, and secondary forest. A few individuals were found in human-altered habitats such as plantations and semi-urban areas. Seventy percent were captured within 25 m of a body of water of some sort. It is therefore not surprising that their diet

consists primarily of amphibians and to a slightly lesser extent fishes. Akani et al. (2005) found a few surprises among *Pseudohaje* stomach contents—two rats and even one juvenile tortoise—but those were aberrations from rare individuals captured near urban areas. Males and females were captured in approximately equal numbers. Males were much larger than females, and male Goldie's Tree Cobras engaged in combat were observed twice. Few small individuals were captured, which the researchers attributed to a fast growth rate through the small body sizes.

The head is short with a well-defined neck. The body and tail are long and slender. The maxilla is short with 2–4 teeth at its posterior end, separated by a diastema from the front fang. The hemipenes are bilobed in their distal half, and the sulcus spermaticus is divided. The nasal is single. The internasals and prefrontals are paired. Although, like other elapids, they lack a loreal, occasionally a very small scale, meeting the definition of a loreal, is present, often on only one side. There are 1 or 2 preoculars, 2 postoculars, and 1 subocular. There are 7(3,4) upper labials. The temporal formula may be 1+1, 1+2, or 1+3. There are 2 pairs of chin shields and 7–10 lower labials, of which the first 4 are in contact with the anterior pair. The submandibular groove is pronounced. The dorsal scales are smooth, lack apical pits, and are arranged in 13 or 15 slightly oblique rows. There are 180–205 ventrals and 76–96 divided subcaudals. The cloacal scale is single.

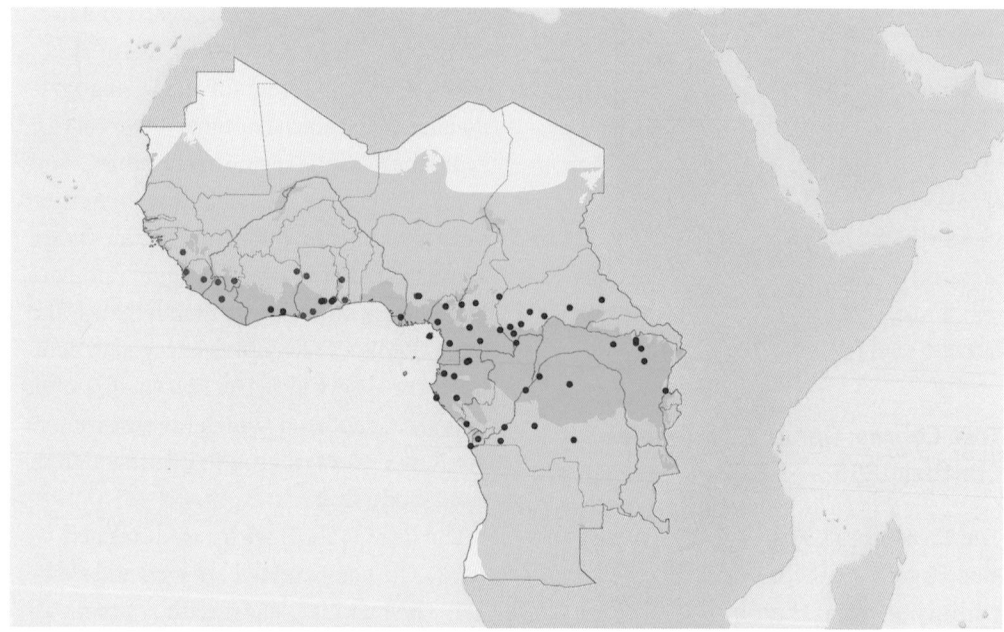

FIGURE 8.51. *Pseudohaje goldii* (blue), *Pseudohaje nigra* (red). By K. Jackson

Key to Central and Western African Species
of the Genus *Pseudohaje*

1   15 dorsal scale rows .... *P. goldii*
1' 13 dorsal scale rows .... *P. nigra*

Goldie's Tree Cobra: *Pseudohaje goldii* Boulenger 1895a

The range of Goldie's Tree Cobra extends from the Ivory Coast to Kenya to Angola. The type locality is Asaba, Nigeria. *Pseudohaje goldii* was named in honor of Sir George Taubman Goldie (1846–1925), who, while governor of the Royal Niger Company, facilitated the collecting expedition that resulted in the description of this new species.

The body is uniformly black above. The underside is white, becoming darker toward the tail. The upper labials are whitish with black edges. There is 1, occa-sionally 2, preocular. There are 15 dorsal scale rows. There are 191–205 ventrals and 81–96 divided subcaudals. The maximum length recorded for this species is 2,570 mm (Stucki-Stirn 1979).

Black Tree Cobra: *Pseudohaje nigra* Günther 1858

The range of the Black Tree Cobra extends from Sierra Leone possibly as far east as Nigeria. The type locality is West Africa.

The body is uniformly black above with a slightly paler (dark brown) underside. There is 1 preocular. There are 13 dorsal scale rows. There are 180–187 ventrals and 76–82 divided subcaudals. The maximum length recorded for this species is 2,130 mm (Boulenger 1896). What little is known of its natural history indicates that *Pseudohaje nigra*, like *P. goldii*, feeds mainly on amphibians (Pauwels and Ohler 1999).

FIGURE 8.52. *Pseudo-haje goldii*, Kenya. By S. Spawls

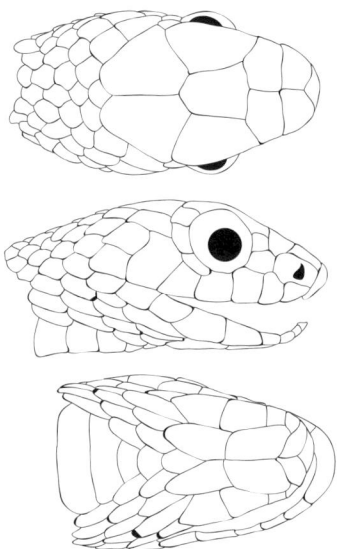

FIGURE 8.53. *Pseudohaje goldii* RMCA 15464. By T. Giri

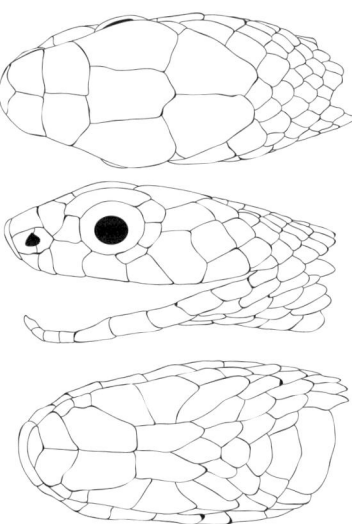

FIGURE 8.54. *Pseudohaje nigra* BMNH 1959.1.2.62. By T. Giri

9

Family Lamprophiidae

Subfamilies Atractaspidinae and Aparallactinae

Snakes of the lamprophiid subfamilies Atractaspidinae and Aparallactinae resemble one another in their ecomorphology—both groups are fossorial or semi-fossorial snakes with glossy scales and usually drab coloration. Their skulls are quite different from one another, however. Atractaspidine snakes found in our zone have a solenoglyph dentition (Stiletto Snakes, genus *Atractaspis*) and were once thought to be viperids, while aparallactines have an opisthoglyph or sometimes aglyph dentition.

The Atractaspidinae include approximately 20 species in 2 genera, occurring primarily in Africa but also extending into the Middle East. Within our zone, the Atractaspidinae are represented by 12 species in the genus *Atractaspis*. The Aparallactinae include 50 species in 10 genera, all found in sub-Saharan Africa. In all, 29 species in 7 genera occur within our zone.

Subfamily Atractaspidinae

The lamprophiid subfamily Atractaspidinae includes two genera, of which one, *Atractaspis*, occurs within our zone. The other genus, *Homoroselaps*, contains two species occurring in southern Africa.

Atractaspis are venomous snakes with a solenoglyph dentition, meaning that the tubular front fang is elongate and the maxilla is reduced to a short stub, allowing the fang and maxilla to rotate so that the fang can lie horizontally in the mouth. Because solenoglyph dentition is otherwise seen only in viperids, *Atractaspis* was for a long time assumed to be a type of viperid. The common names "Mole Viper" and "Burrowing Asp," which are sometimes used for *Atractaspis*, refer to its fossorial habits and also reflect that it was formerly thought to belong to the Viperidae.

The true affiliations of *Atractaspis* were discovered by Bourgeois (1961, 1965) during her studies of the skulls and dentition of African snakes. She noticed that although the solenoglyph skull of *Atractaspis* bore a superficial resemblance to that of vipers, the relationships among the bones making up the palato-maxillary arch (the ectopterygoid, maxilla, and prefrontal) of *Atractaspis* were more reminiscent of those of aparallactine snakes. In aparallactines such as the Centipede-eaters *Aparallactus modestus* and *A. capensis*, she noted, a vertical process of the maxilla extends upward into a notch in the prefrontal, as it does in *Atractaspis*. In viperids, the ectopterygoid extends into a notch on the posterior side of the maxilla, which it does not do in *Atractapis* or in aparallactines (Figs. 9.2 and 9.3). She was the first to propose that *Atractaspis* represented a lineage more closely related to aparallactines than to viperids.

The solenoglyph dentition of viperids allows them to erect the long fangs into a vertical position when striking, and to fold them horizontally against the roof of the mouth when at rest. The solenoglyph dentition of *Atractaspis* allows them to rotate an individual fang sideways out of the almost-closed mouth, stabbing it into a nearby prey item. This is thought to represent an adaptation for envenomating prey in cramped burrows. It also means that *Atractaspis* cannot be safely held behind the head as other snakes can, since *Atractaspis* held in this way can stab one fang sideways out of the mouth and backward into the hand holding its neck. The common names "Side-stabbing Snakes" and "Back-stabbing Snakes" sometimes used for *Atractaspis* refer to this unusual venom injection mechanism, as does "Stiletto Snake," a stiletto being a small, concealed dagger.

If *Atractaspis* was not a viperid, what was it? The similarity of the assemblage of cranial bones of *Atractaspis* to those of aparallactine snakes, in particular the articulation between the maxilla and prefrontal (Fig. 9.3), led Bourgeois (1961, 1963, 1965) to include *Atractaspis* in her subfamily Aparallactinae along with *Aparallactus*, *Xenocalamus*, *Polemon*, *Amblyodipsas*, and *Chilorhinophis*. Subsequent molecular phylogenies have supported the Aparallactinae as the closest relatives of the Atractaspidinae (Nagy et al. 2005; Vidal et al. 2008; Kelly et al. 2009; Pyron et al. 2011, 2013; Zheng and Wiens 2016).

Stiletto Snakes: Genus *Atractaspis* Smith 1849

Atractaspis is a genus containing about 20 species distributed in Africa and the Middle East, 12 of which occur within our zone. Stiletto Snakes are drab-colored, fossorial, nocturnal snakes that are not often encountered. Their most remarkable feature is their venom-delivery system, which is not visible externally. In addition to their pair of long, tubular, independently mobile fangs, used in a side-stabbing motion with the mouth closed, as described above, they possess venom glands that fall into one of two patterns. Underwood and Kochva (1993) recognized two groups within the genus *Atractaspis*: the *bibronii* group with normal-sized venom glands and a sub-Saharan distribution, and the *microlepidota* group characterized by highly elongate venom glands extending backward as much as 20% of the total body length and with a more northerly distribution that extends into the Middle East.

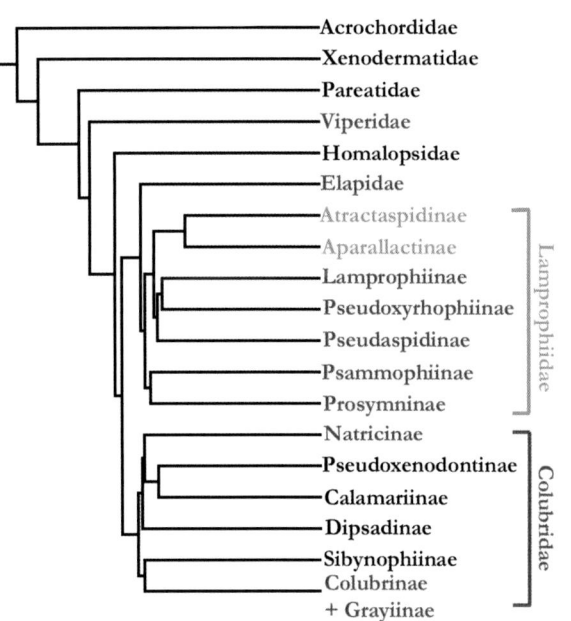

FIGURE 9.1. Higher-level phylogeny of caenophidian families and subfamilies, with families and subfamilies occurring within our zone highlighted (red or orange). The subfamilies Atractaspidinae and Aparallactinae, the topic of this chapter, are highlighted in orange. Modified from Zheng and Wiens (2016)

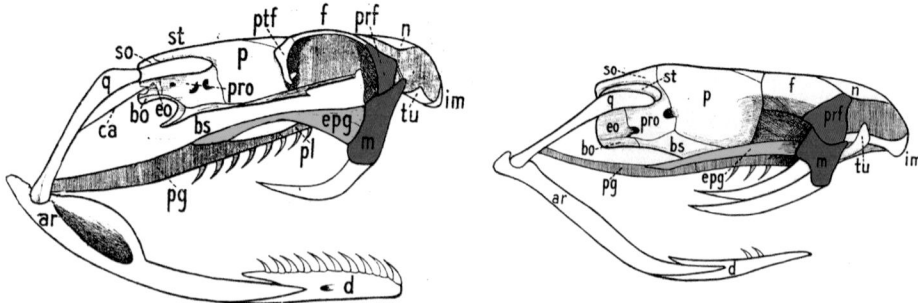

FIGURE 9.2. Skulls of a viperid (*left*) and *Atractaspis* (*right*). M. Phisalix accurately drew the hinge-like articulation between maxilla (red) and ectopterygoid (green) in the viperid, although the ball-and-socket articulation in *Atractaspis* of the maxilla (red) with the prefrontal (blue) has been simplified. This was foresightful at a
time when nobody doubted that *Atractaspis* belonged among the Viperidae (see also Fig. 9.3). Adapted by
K. Jackson from drawings by Phisalix (1922)

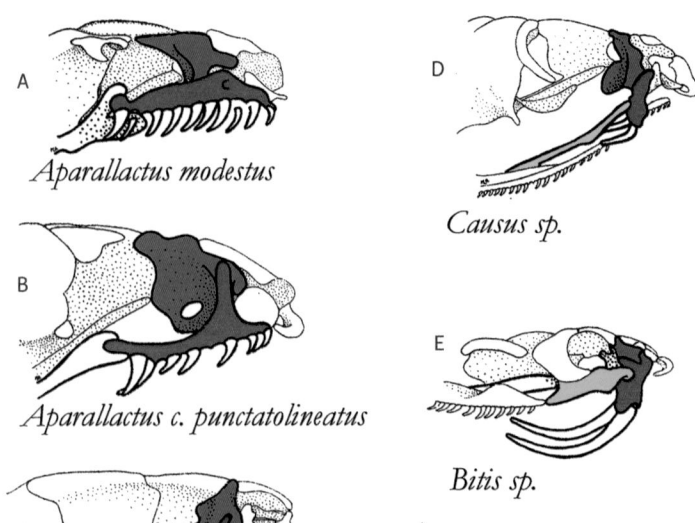

FIGURE 9.3. Illustration explaining the morphocline
leading to the solenoglyph dentition seen in *Atractaspis* (*A*, *B*, and *C*) versus the solenoglyph condition
seen in viperids (*D* and *E*). Maxilla (red), ectopterygoid (green), prefrontal (blue). From Bourgeois (1965);
modified by K. Jackson by the addition of color and
text labeling

FIGURE 9.4. *Atractaspis engaddensis* (a Middle Eastern species) with a single fang protruding through
closed lips. By E. Kochva

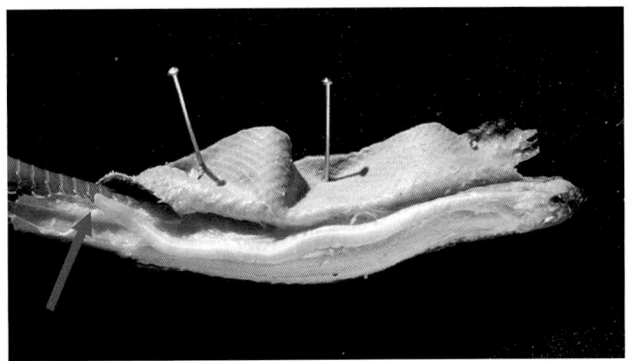

FIGURE 9.5. Dissection of the head of an *Atractaspis* with elongate venom glands. The red arrow points to the posterior end of the venom gland. By E. Kochva

The venom of *Atractaspis* has been less extensively studied than the venoms of elapids and viperids, and most of the studies that do exist are of the Middle Eastern species, *Atractaspis engaddensis*. Although this species does not occur within our zone, in the absence of any data beyond scattered clinical reports for other species in the genus, extrapolating from what is known from *A. engaddensis* may be our best source of information about the venoms of other species of *Atractaspis*. Drop per drop, the venom of *A. engaddensis* is among the most potent of any species of snake (dose killing 50% of mice: = 0.07 mg/kg; Weiser et al. 1984). But the venom yield is small (maximum 0.5 mg; Kochva et al. 1982), even in species with elongate venom glands, which may explain why *Atractaspis* bites are not usually life-threatening, generally resulting only in pain and swelling at the site of the bite wound (Kochva et al. 1982). Corkill et al. (1959) reported 41 cases of bites to humans by a variety of *Atractaspis* species. Four of these 41 envenomations proved fatal. A great variety of symptoms have been reported from *Atractaspis* bites: localized pain, swelling and discoloration, neurological and respiratory symptoms, and others, including cardiovascular effects. Fever, an unusual outcome of snakebite, is a symptom reported from a few victims bitten by *A. microlepidota* (= *micropholis*) and *A. dahomeyensis* (Doucet and Lepesme 1953; Warrell et al. 1976).

Atractaspis venom contains a special toxin not known from the venom of any elapid or viperid. This molecule was isolated by Kochva et al. (1982) from among several other components that made up the venom. Named "sarafotoxin" by Kloog et al. (1988), this toxin powerfully constricts blood vessels. Herpetologist Elazar Kochva survived a bite from *A. engaddensis* sustained while studying it in the lab and wrote an account of the incident in collaboration with clinicians who treated him (Kurnik et al. 1999). The most worrying symptom of Kochva's *A. engaddensis* envenomation was that his blood pressure shot up to 180/110, a symptom he attributes to the effect of sarafotoxin. Sarafotoxin has so far been isolated from the venoms of *A. engaddensis*, *A. bibronii*, and *A. microlepidota*. A new antivenom for *A. engaddensis* has recently been produced (Ismail et al. 2007), and it remains to be seen whether it will be effective against the venoms of other species in the genus.

If threatened, *Atractaspis* will always try to flee, but if forced into a confrontation, the snake assumes a curious posture with the neck arched and the head pointing downward. It may release a foul-smelling substance from the cloaca (perhaps as a deterrent to predators) and may proceed to turn upside down, thrashing from side to side. This display has no known purpose, but is perhaps significant in some way underground (Spawls and Branch 1995).

The feeding mechanism of *Atractaspis* has been described by Deufel and Cundall (2003) as an "evolutionary endpoint." By this they mean that the adaptations of the skull to fossorial life and to feeding underground have come at a cost to other functions, such as prey ingestion. In addition to the fang-erecting mechanism of *Atractaspis* described above, the side-stabbing venom-delivery system of *Atractaspis* may also be advantageous for preying on fossorial lizards, which are able to drop their tails in order to escape a predator. Instead of biting the tail of a fleeing lizard, the side-stabbing venom-delivery system allows *Atractapis* to squeeze forward in the tunnel, past the lizard's tail, in order to inject venom into the body of the prey. Another way in which *Atractaspis* are adapted for life underground is reinforcement of the snout with a network of tight ligaments that immobilize the bones. This rigid snout reduces the *Atractaspis*'s ability to transport prey into the esophagus by alternately ratcheting with the pterygoid-palate teeth on each side as other snakes do.

Atractaspis are drab-colored semi-fossorial snakes with glossy scales. The head is small with small eyes with round pupils, and the neck is not well defined. Most species are fairly thick bodied and slow moving, but there are a few exceptions. The tail is short and comes to a sharp point. The nasal may be single, divided, or semi-divided. Both the internasals and the prefrontals are paired. The loreal is absent. There are 1 or occasionally 2 preoculars and 1 or 2 small postoculars. There is a supraocular but no suboculars. There are 3–7 upper labials, 1 or 2 of which are in contact with the eye. There may be either 1 or 2 anterior temporals. There is a single pair of chin shields with a submandibular groove that is not pronounced and 4–9 lower labials. The dorsal scales are smooth and arranged in 17–37 straight rows, without apical pits. The number of ventrals is highly variable between species, but in some species their number exceeds 300. The cloacal scale may be single or divided. The subcaudals of the short tail may be single, divided, or in some cases a combination of the two, with the subcaudals divided only along part of the length of the tail. The hemipenes are bilobed, and the sulcus spermaticus is forked. *Atractaspis* are egg-layers.

Noteworthy characteristics of the genus *Atractaspis* include the solenoglyph dentition with side-stabbing venom-injection mechanism, which is unique among snakes. *Atractaspis* have dorsal and ventral scale counts that include the highest seen in atractaspidine and aparallactine species. *Atractapis* differ from all aparallactines in having a single pair of chin shields (but see comments on "false chin shields" in the Purple-glossed Snakes, genus *Amblyodipsas*, and in the Quill-snouted Snakes, genus *Xenocalamus*).

Slender Stiletto Snake: *Atractaspis aterrima* Günther 1863a

The range of *Atractaspis aterrima* extends from the Gambia in the west to Uganda and Tanzania in the east. The type locality is West Africa.

Key to Central and Western African Species of the Genus *Atractaspis*

1	Cloacal scale single	2
1'	Cloacal scale divided	9
2(1)	1 anterior temporal	3
2'	2 anterior temporals	7
3(2)	The second pair of lower labials is absent (fused to the chin shields) .... *A. corpulenta*
3'	The second pair of lower labials is present and distinct from the chin shields 4
4(3)	At least 27 dorsal scale rows .... *A. dahomeyensis*
4'	No more than 25 dorsal scale rows 5
5(4)	First upper labial is in contact with the posterior nasal .... *A. aterrima*
5'	First upper labial is not in contact with the posterior nasal 6
6(5)	Mental scale separated from the chin shields by the first pair of lower labials; subcaudals single .... *A. bibronii*
6'	Mental scale in contact with the chin shields; at least some subcaudals divided .... *A. boulengeri*
7(2)	More than 7 lower labials .... *A. watsoni*
7'	Fewer than 7 lower labials 8
8(7)	Fewer than 28 dorsal scale rows; at least 26 subcaudals .... *A. micropholis*
8'	More than 28 dorsal scale rows; no more than 26 subcaudals .... *A. microlepidota*
9(1)	More than 300 ventrals .... *A. reticulata*
9'	Fewer than 260 ventrals 10
10(9)	Fewer than 4 upper labials .... *A. coalescens*
10'	At least 5 upper labials 11
11(10)	23–27 dorsal scale rows (occasionally 21); mental separated from the chin shields by the first pair of lower labials .... *A. irregularis*
11'	19 or 21 dorsal scale rows; mental in contact with the chin shields .... *A. congica*

A. *aterrima* is a relatively slender, fast-moving snake. The body is uniformly black except for the edges of the ventrals, which are paler. The nasal is divided. Both the internasals and the prefrontals are paired.

There is 1 preocular and 1 postocular. There are 5(3,4) upper labials. The temporal formula is 1+2. There are 4–6(3) lower labials (usually 5). There are 3 gulars in contact with the chin shields. The dorsal scales are smooth and arranged in 19–23 straight rows. There are 243–300 ventrals (fewer than 278 in males, more than 262 in females) and 17–26 single subcaudals (more than 20 in males, fewer than 22 in females). The cloacal scale is single. The maximum recorded length is 650 mm (Laurent 1960). The venom gland is short (Underwood and Kochva 1993), and no human fatalities have been recorded for this species (Spawls and Branch 1995).

Characters useful in distinguishing *A. aterrima* from other species within the genus include its relatively slender body form, the presence of 3 gulars in contact with the chin shields, and the single subcaudals and cloacal scale. None of these traits alone is unique among *Atractaspis*, but taken in combination they define the species.

Akani et al. (2001a) carried out a field study of the diets of three species of *Atractaspis* (*A. aterrima, A. corpulenta, A. irregularis*) in Nigeria. Their goal was to determine how these three apparently similar snakes manage to coexist in the same habitat and avoid competing with each other. Their observations suggested that the three species were specializing in different types of prey. The main difference observed was that the three species chose different sizes of prey relative to their own body size. To a lesser extent, they also favored different species of prey. *A. aterrima* specialized in small prey in the form of small snakes and lizards, *A. corpulenta* in medium-sized prey in the form of lizards, and *A. irregularis* took the largest prey items for its body size, feeding on rodents. Gower et al. (2004) came across a specimen of *A. aterrima* in Tanzania whose digestive tract contained a caecilian almost half as large as the predator

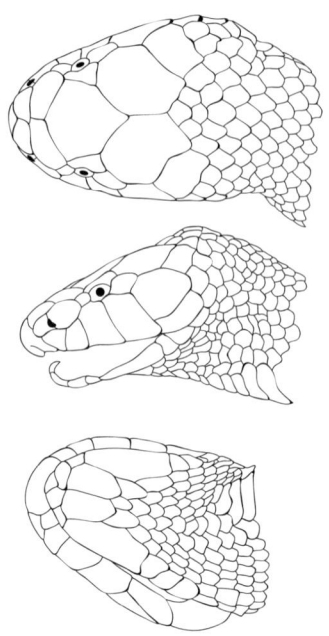

FIGURE 9.6. *Atractaspis aterrima* FMNH 256774.

itself, certainly a deviation from specializing in small-bodied prey, but just a single observation.

Bibron's Stiletto Snake: *Atractaspis bibronii* Smith 1849

Atractaspis bibronii is a southern African species that makes its way into Angola and southern Democratic Republic of Congo. The type locality is Cape Colony, South Africa.

This species is a moderately thick, slow-moving snake. The body is uniformly purple brown to black above. The underside is usually uniformly pale but may be mottled or dark. The nasal is divided. Both the internasals and the prefrontals are paired. There is 1 preocular and 1 postocular. There are 5(3,4) upper labials. The temporal formula is 1+2. There are 5 or 6(3) lower labials. There are 3 gulars in contact with the chin shields. The dorsal scales are smooth

FIGURE 9.7. *Atractaspis aterrima*, Ghana, in its characteristic defensive pose with the head pointed down into the ground. By P. Naskrecki

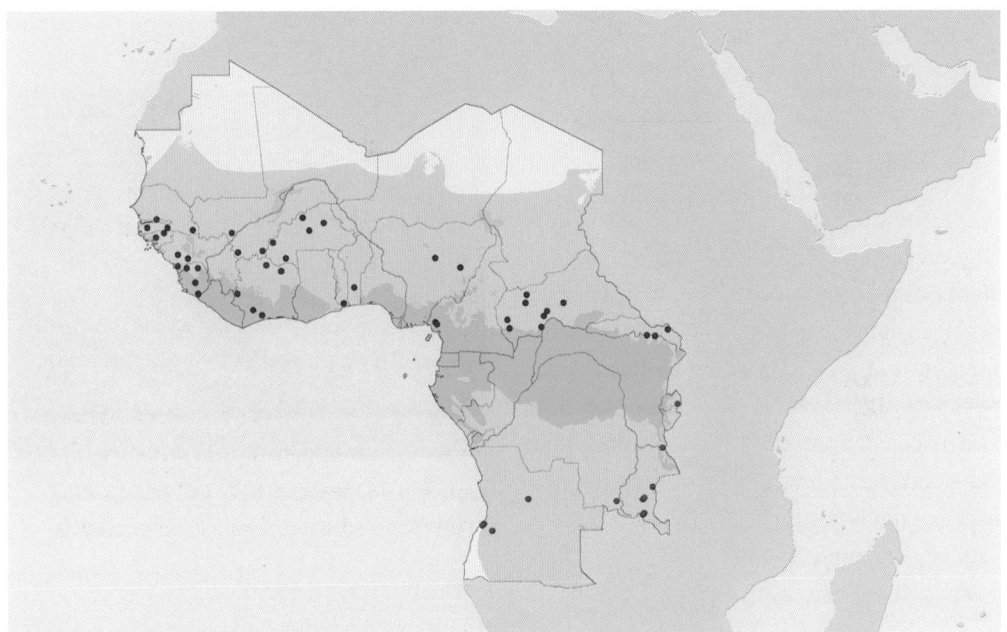

FIGURE 9.8. *Atractaspis aterrima* (blue), *Atractaspis bibronii* (red). By K. Jackson

and arranged in 21–25 straight rows, with no apical pits. There are 212–260 ventrals (fewer than 247 in males, more than 237 in females) and 19–28 single subcaudals. The cloacal scale is single. The maximum recorded length is 646 mm (Laurent 1956). The venom gland is short (Underwood and Kochva 1993), and no human fatalities have been attributed to this species.

Characters useful in distinguishing *A. bibronii* from other species within the genus include its moderately thick body form (though this trait is common in *Atractaspis*), the presence of 3 gulars in contact with the chin shields, and the single subcaudals. None of these traits alone is unique among *Atractaspis*, but taken in combination they define the species.

Shine et al. (2006a) reported that clutch size for this species ranges from 3 to 6 eggs and is not correlated with maternal size. Ionides (1954) listed stomach contents of specimens of *A. bibronii* including four species of snake (the Herald Snake, *Crotaphopeltis hotamboeia*; the Dwarf Sand Snake, *Psammophis angolensis*; the Semiornate Snake, *Meizodon semiornatus*; and the Snouted Night Adder, *Causus defillipi*), lizards, worm lizards (amphisbaenians), frogs, and newborn rats. He did not specify how many of each, but his observations nevertheless provide a general picture of a snake that feeds mostly, though not exclusively, on snakes and lizards. Ionides recommends that enthusiasts search for *Atractaspis* during or after heavy rain, "in heaps of rotting rubbish." He reports a personal experience of being bitten by *A. bibronii* after grabbing hold of a juvenile that he mistook for a harmless Cape Wolf Snake (*Lycophidion capensis*). The result was painful swelling and discoloration at the site of the bite that subsided after a few days. Broadley (1972) also made observations of the natural history of *A. bibronii*. This species, he found, burrowed under stones, often to a depth of a foot or more. The stomach contents of one specimen included a lizard (*Nucra sp.*) and a snake (*Aparallactus capense*).

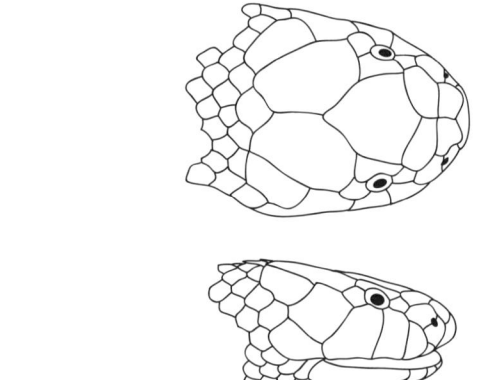
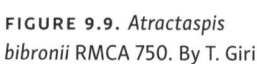

FIGURE 9.9. *Atractaspis bibronii* RMCA 750. By T. Giri

FIGURE 9.10. *Atractaspis bibronii*, Botswana. By S. Spawls

Central African Stiletto Snake: *Atractaspis boulengeri* Mocquard 1897

The distribution of *Atractaspis boulengeri* extends from Cameroon to the Democratic Republic of Congo. The type locality is Lambaréné, Gabon.

The body is uniformly black or slate gray above, with a paler underside. The nasal is divided. Both the internasals and the prefrontals are paired. There is 1 preocular and 1 postocular. There are 5(3,4) upper labials. The temporal formula is 1+2, or occasionally 1+3 or 1+4. There are 4–6(3) lower labials (usually 5). There are 3 gulars in contact with the chin shields. The dorsal scales are smooth and arranged in 21–23 straight rows, without apical pits. There are 192–218 ventrals and 22–27 subcaudals, which are usually divided but occasionally single. The cloacal scale is single. The maximum recorded length is 650 mm (Villiers 1975). The venom gland is short (Underwood and Kochva 1993), but the effect of the bite is not known.

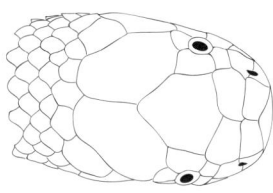

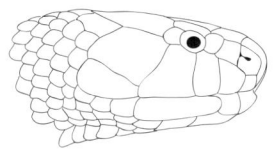

FIGURE 9.11.
Atractaspis boulengeri RMCA 5455. By T. Giri

The single most useful character in identifying *A. boulengeri* within the genus is the presence of 3 gulars in contact with the chin shields. This is a trait shared with *A. aterrima*, *A. congica*, *A. reticulata*, and *A. bibronii*, but each of these four can be distinguished from *A. boulengeri* by other traits.

As many as six subspecies are recognized by some authors: *A. b. boulengeri* Mocquard 1897; *A. b. matschiensis* Werner 1897; *A. b. mixta* Laurent 1945; *A. b. schmidti* Laurent 1945; *A. b. schultzei* Sternfeld 1917; *A. b. vanderborghti* Laurent 1956.

Black Stiletto Snake: *Atractaspis coalescens* Perret 1960

This species is known only from a single specimen from the type locality of Bangwa, Cameroon. Nothing is therefore known about *Atractaspis coalescens* except the external morphology of that one specimen. The body is uniformly shiny black with a bluish sheen to it. Though both paired, the

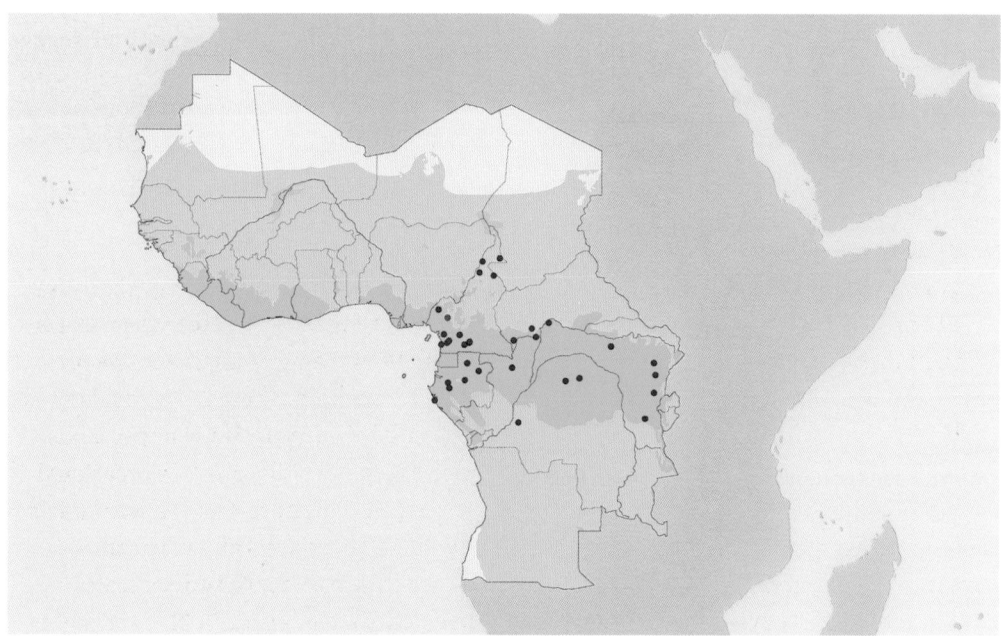

FIGURE 9.12. *Atractaspis boulengeri* (blue), *Atractaspis coalescens* (red). By K. Jackson

internasals are fused to the prefrontals on either side. The nasal is divided. There is 1 preocular and 1 postocular. There are 3(1,2) upper labials. The temporal formula is 1+3. There are 6(3) lower labials. Four gulars are in contact with the chin shields. The dorsal scales are smooth and arranged in 25 straight rows. There are 234 ventrals and 23 divided subcaudals. The cloacal scale is divided. The specimen is 375 mm long (Perret 1960). It is possible that this specimen is not in fact a separate species but rather simply an aberrant specimen of *A. irregularis*. This species is easily distinguished from others within the genus by the fusion of the prefrontal to the internasal on either side.

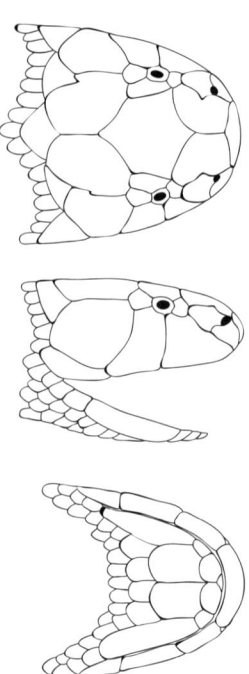

FIGURE 9.13. *Atractaspis coalescens* redrawn from the original description (Perret 1960). By T. Giri

FIGURE 9.14. *Atractaspis congica* RMCA 18701. By T. Giri

Congo Stiletto Snake: *Atractaspis congica* Peters 1877

Atractaspis congica has a broad central African distribution, extending from Cameroon to Zambia. The type locality is Chinchoxo, Cabinda, Angola.

The body is uniformly dark brown. The nasal is divided. Both internasals and prefrontals are paired. There is 1 preocular and 1 postocular. There are 5(3,4) upper labials. The temporal formula is 1+2. There are 4–6(3) lower labials (usually 5), and 3 gulars in contact with the chin shields. The anterior tip of the chin shields shares a point of contact with the mental. The dorsal scales are smooth and arranged in 19–21 straight rows, without apical pits. There are 200–237 ventrals (fewer than 226 in males, more than 219 in females) and 18–23 divided subcaudals (more than 20 in males, fewer than 22 in females). The cloacal scale is divided. The maximum recorded length is 481 mm (Laurent 1950a). This species has a short venom gland (Underwood and Kochva

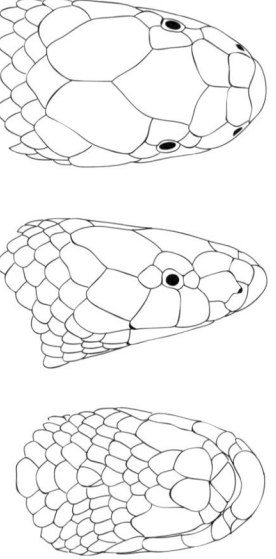

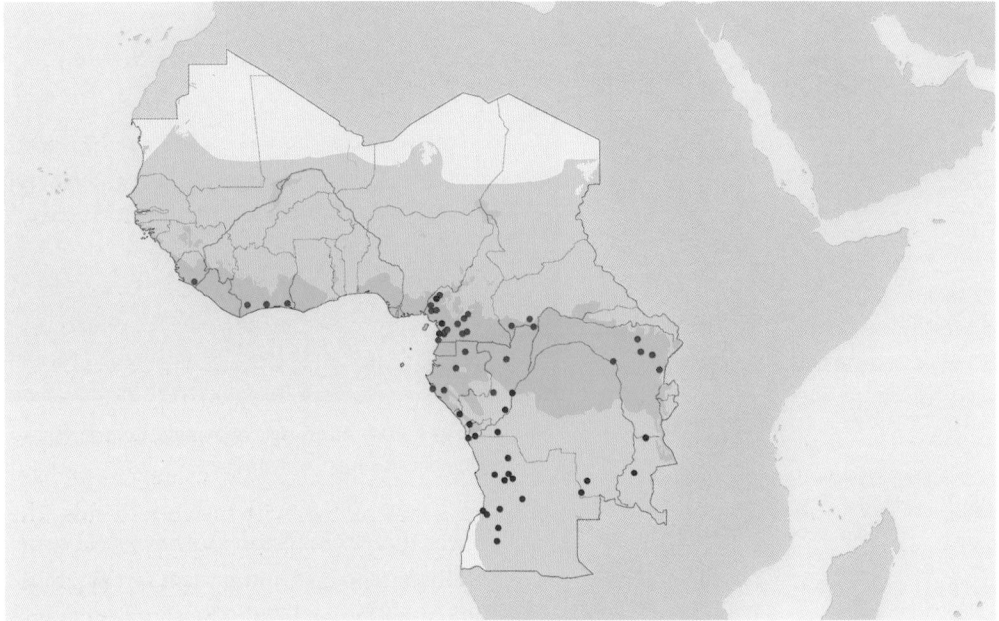

FIGURE 9.15. *Atractaspis congica* (blue), *Atractaspis corpentula* (red). By K. Jackson

1993). There are no reports of serious bites to humans.

Notable in *A. congica* is the presence of a point of contact between the anterior tip of the chin shields and the mental, an arrangement unique to this species.

As many as three subspecies are recognized by some authors: *A. c. congica* Peters 1877; *A. c. orientalis* Laurent 1945; *A. c. leleupi* Laurent 1950a.

Fat Stiletto Snake: *Atractaspis corpulenta* (Hallowell 1854b)

Atractaspis corpulenta is a central and western African forest species with a range extending from Guinea to the Democratic Republic of Congo. The type locality is Gabon.

The body is uniformly black with a dark blue sheen. The edges of the subcaudals are white, and the underside of the tip of the tail is white. The nasal is divided. Both the internasals and the prefrontals are paired.

There is 1 preocular and 1 postocular. There are 5(3,4) upper labials. The temporal formula is 1+2 or 1+3. There are 4–7(3) lower labials. The second lower labials are fused to the chin shields. There are 5 gulars in contact with the chin shields. The dorsal scales are smooth and arranged in 23–29 straight rows, with no apical pits. There are 178–208 ventrals and 23–27 divided subcaudals. The cloacal scale is single. The maximum recorded length is 680 mm (Courtois 1979). This species has a short venom gland (Underwood and Kochva 1993).

This species is unusual among *Atractaspis* in that the second lower labials are fused to the chin shields, a trait shared only with *A. reticulata*. *A. corpulenta* is easily distinguished from *A. reticulata* in having 5 gulars in contact with the chin shields.

As many as three subspecies are recognized by some authors: *A. c. corpulenta* Hallowell 1854b; *A. c. leucura* Mocquard 1885; *A. c. kivuensis* Laurent 1958.

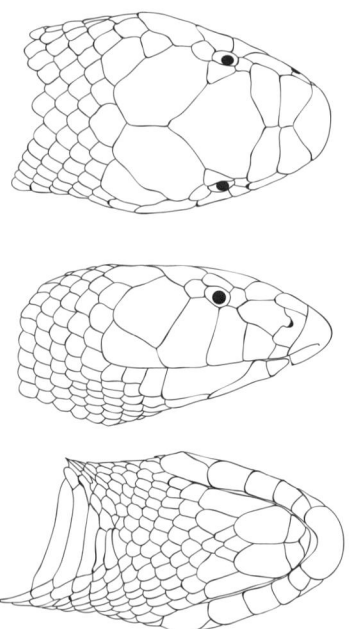

FIGURE 9.16. *Atractaspis corpulenta* RMCA 1685. By T. Giri

FIGURE 9.17. *Atractaspis corpentula*, Democratic Republic of Congo. By J. Kielgast

Dahomey Stiletto Snake: *Atractaspis dahomeyensis* Bocage 1887b

This West African species has a range extending from Senegal to Central African Republic. The type locality is Zomai, Benin.

The body is uniformly black above, with a dark gray underside. The nasal is divided. The internasals and prefrontals are both paired. There is 1 preocular and 1 postocular. There are 5(3,4) upper labials. The temporal formula is 1+2 to 1+4. There are 4–6(3) lower labials (usually 5). There are 5 gulars in contact with the chin shields. The dorsal scales are smooth and arranged in 29–35 straight rows. The number of dorsal scale rows is sexually dimorphic, with fewer than 32 in males, more than 32 in females. There are 210–250 ventrals (fewer than 236 in males, more than 234 in females) and 22–30 subcaudals, some of which may be divided and others not (more than 23 in males,

fewer than 27 in females). The cloacal scale is single. The maximum recorded length is 555 mm (Roman 1973). The venom glands are short (Underwood and Kochva 1993).

Characters useful in distinguishing this species from others within the genus are the presence of a single anterior temporal and the presence of 5 gulars in contact with the chin shields. Taken in combination, these traits are unique to the species.

Variable Stiletto Snake: *Atractaspis irregularis* (Reinhardt 1843)

Atractaspis irregularis is a widely distributed species with a range extending from Sierra Leone to Angola to Tanzania. The type locality is probably Ghana, though it was originally recorded as "Guinea." (In the 1800s, the whole West African coast was referred to as the Guinea Coast.)

The body is uniformly black. The nasal is divided. Both internasals and prefrontals are paired. There is 1 preocular and 1 postocular. There are 5(3,4) upper labials. The temporal formula is 1+2 to 1+4. There are 4–6(3) lower labials. There are 4 gulars in contact with the chin shields. In rare cases, *A. irregularis* may have 3 or 5 gulars in con-

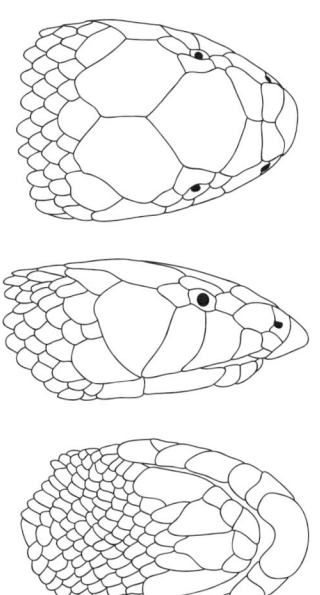

tact with the chin shields; should such a situation occur, the specimen can be identified as an aberrant *A. irregularis* by reference to the rest of the species description. The dorsal scales are smooth and arranged in 23–27 (usually 27; occasionally 21) straight rows, without apical pits. There are 213–251 ventrals (fewer than 236 in males, more than 229 in females) and 20–32 divided subcaudals (more than 20 in males, fewer than 27

FIGURE 9.18. *Atractaspis dahomeyensis* FMNH 170712. By T. Giri

FIGURE 9.19. *Atractaspis dahomeyenis*, Ivory Coast. By M. O. Rödel

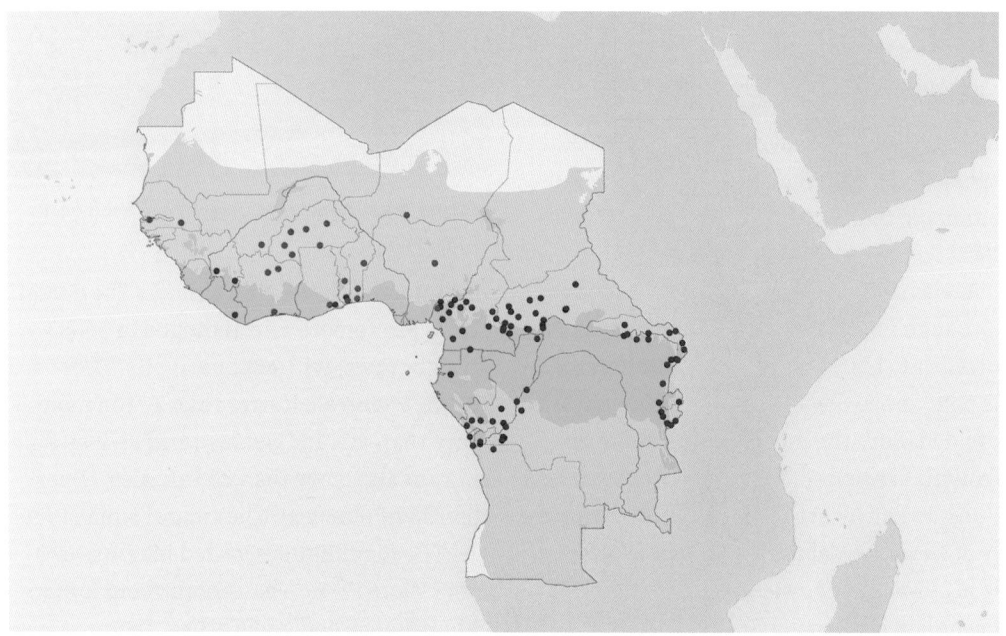

FIGURE 9.20. *Atractaspis dahomeyensis* (red), *Atractaspis irregularis* (blue). By K. Jackson

in females). The cloacal scale is divided. The maximum recorded length is 570 mm (Villiers 1975).

This species is easily distinguished from other *Atractaspis* by the presence of 4 gulars in contact with the chin shields, a trait shared only with *A. coalescens*, from which *A. irregularis* can easily be distinguished by the arrangement of the internasals and prefrontals.

A. irregularis has a short venom gland (Underwood and Kochva 1993). There are anecdotal reports of two fatalities, both associated with unusual circumstances. In one case, a baby was bitten while sleeping and died within a few minutes. The other fatality was a man who while sleeping rolled on a large individual and was bitten nine times (Spawls and Branch 1995).

As many as five subspecies are recognized by some authors: *A. i. irregularis* (Reinhardt 1843); *A. i. conradsi* Sternfeld 1908a; *A. i. parkeri* Laurent 1945; *A. i. uelensis* Laurent 1945; *A. i. loveridgei* Laurent 1945.

Small-scaled Stiletto Snake: *Atractaspis microlepidota* Günther 1866

Atractaspis microlepidota is a West African species with a range extending from Mauritania to the Gambia. The type locality is West Africa.

A. microlepidota is a fast-moving species, although not a slender one. The body is uniformly blackish above. The underside is pale, as are the lower labials. The nasal is divided. Both the internasals and the prefrontals are paired. There is 1 preocular and 1 postocular. There are 6(3,4) or occasionally 5(3,4) upper labials. The temporal formula is 2+3 or 2+4. *A. microlepidota* has 4–6(3) lower labials (usually 5). There are 5 gulars

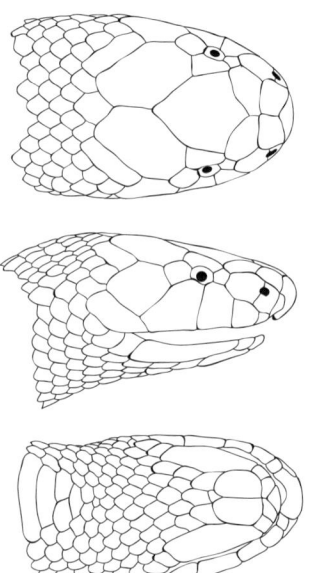

FIGURE 9.21. *Atractaspis irregularis* RMCA 76-3-R-678. By T. Giri

FIGURE 9.22. *Atractaspis irregularis*, Cameroon. By M. Lebreton

in contact with the chin shields. The dorsal scales are smooth and arranged in 29–33 straight rows, without apical pits. There are 198–222 ventrals (fewer than 204 in males, more than 204 in females), and 21–26 single subcaudals (more than 23 in males, fewer than 26 in females). The cloacal scale is single. The maximum recorded length is 670 mm (Mané 1992). The venom gland is elongate (Underwood and Kochva 1993).

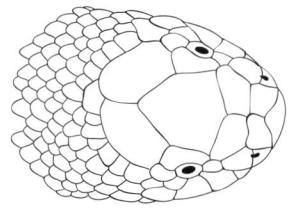

A. *microlepidota* can readily be distinguished from other *Atractaspis* by the presence of 2 anterior temporals and of 5 gulars in contact with the chin shields. Taken in combination, these traits are unique to this species.

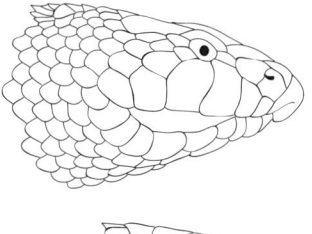

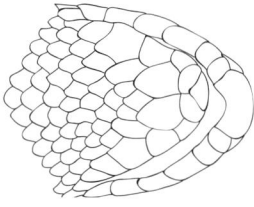

FIGURE 9.23. *Atractaspis microlepidota* RMCA 73-17-R-147. By T. Giri

FIGURE 9.24. *Atractaspis microlepidota*, captive specimen. By J. Benjamin

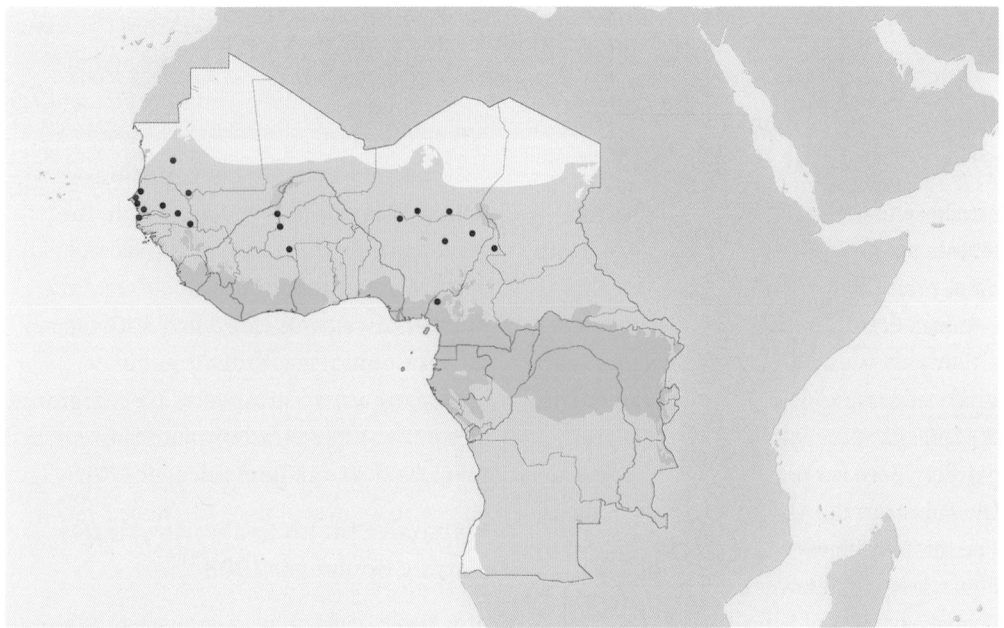

FIGURE 9.25. *Atractaspis microlepidota* (red), *Atractaspis micropholis* (blue). By K. Jackson

Common Stiletto Snake: *Atractaspis micropholis* Günther 1872

The range of *Atractaspis micropholis* extends from Senegal to the Sudan. The type locality is Africa.

The body is brown above with a pale underside. The nasal is divided. Both the internasals and the prefrontals are paired. There is 1 preocular and 1 postocular. There are 6(3,4) upper labials. The temporal formula is 2+3 or 2+4. There are 4–6(3) lower labials (usually 5). There are 7 gulars in contact with the chin shields. The dorsal scales are smooth and arranged in 25–27 straight rows without apical pits. There are 211–237 ventrals and 26–32 single subcaudals (more than 26 in males, fewer than 29 in females). The cloacal scale is single. The maximum recorded length is 913 mm (Trape et al. 2006). The venom gland is elongate (Underwood and Kochva 1993).

The presence of 7 gulars in contact with the chin shields is a trait unique to *A. micropholis*.

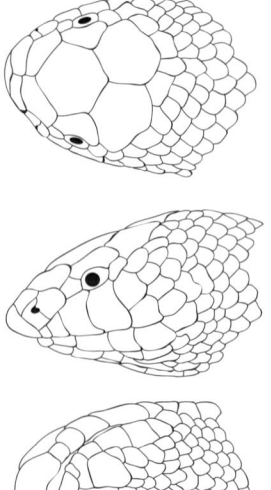

FIGURE 9.26. *Atractaspis micropholis* BMNH 1946.1.18.7 (TYPE). By T. Giri

Reticulate Stiletto Snake: *Atractaspis reticulata* Sjöstedt 1896

Atractaspis reticulata is a central African forest species with a range extending from Cameroon to Tanzania. The type locality is Ekundu, Cameroon.

The body is uniformly brown above, with a pale underside. The nasal is divided. Both the internasals and the prefrontals are paired. There is 1 preocular and 1 postocular. There are 5(3,4) or 6(3,4) upper labials. The temporal formula is 1+(2 or 3)+(4 or 5). There are 4–6(3) lower labials (usually 5). The second pair of infralabials is fused to the chin shields. There are 3 gulars in contact with the chin shields. The dorsal scales are smooth and arranged in 19–23 straight rows, with no apical pits. There are 308–344 ventrals, and 21–26 divided subcaudals. The cloacal scale is divided. The maximum recorded length is 1,135 mm (Laurent 1950b). The venom gland is short (Underwood and Kochva 1993), and the effect of the bite not known.

Characters useful in distinguishing *A. reticulata* from other species within the genus include the fusion of the second pair of infralabials to the chin shields and the number of ventrals exceeding 300, the latter a trait unique within the genus.

As many as two subspecies are recognized by some authors: *A. r. reticulata* Sjöstedt 1896; *A. r. heterochilus* Boulenger 1901.

Watson's Stiletto Snake: *Atractaspis watsoni* Boulenger 1908

Atractaspis watsoni has a range extending from Mauritania to the Sudan. The type locality is Sokoto, Nigeria.

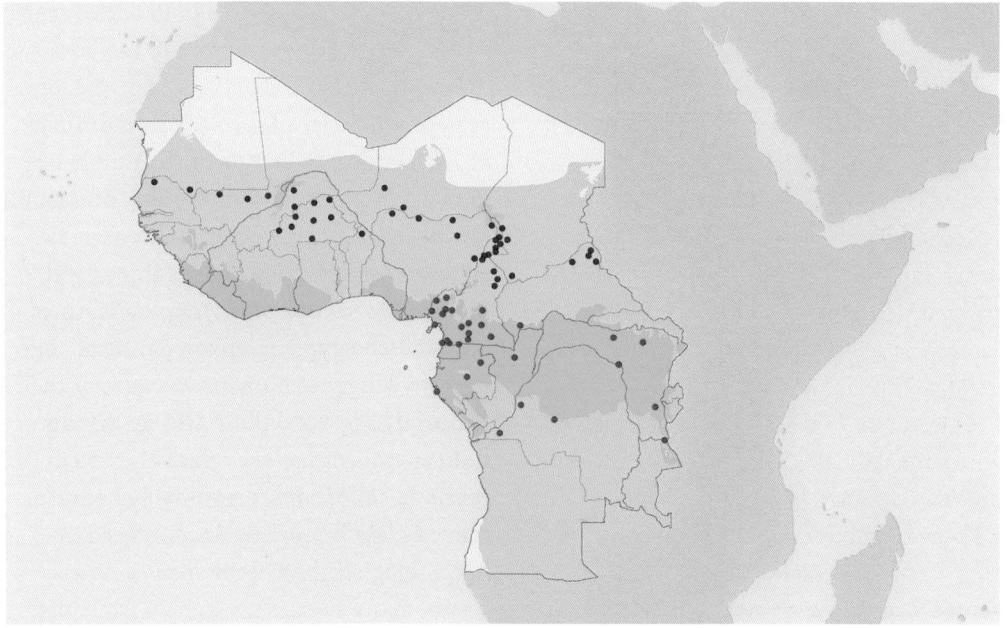

FIGURE 9.27. *Atractaspis reticulata* (red), *Atractaspis watsoni* (blue). By K. Jackson

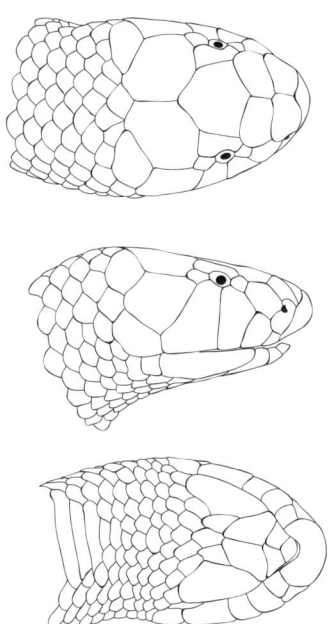

FIGURE 9.28 *Atractaspis reticulata* RMCA 212577. By T. Giri

The body is uniformly black above. The underside is paler. The nasal is divided. Both the internasals and prefrontals are paired. There is 1 preocular and 1 postocular. There are 6(4) or occasionally 6(3,4) upper labials. The temporal formula is 2+3 or 2+4. There are 8(3) or 9(3) lower labials, and 5 gulars in contact with the chin shields. The dorsal scales are smooth and arranged in 29–31 (occasionally 27) straight rows, without apical pits. There are 213–242 ventrals (fewer than 232 in males, more than 219 in females) and 21–30 single subcaudals (more than 23 in males, fewer than 26 in females). The cloacal scale is single. The maximum recorded length is 716 mm (Trape et al. 2006).

This species can be distinguished from others in the genus by the combination of the presence of 2 anterior temporals, with a higher number of lower labials than seen in other *Atractaspis*.

Underwood and Kochva (1993) do not list

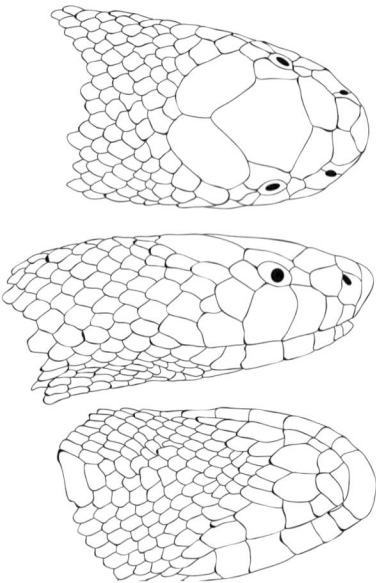

FIGURE 9.29. *Atractaspis watsoni* BMNH 1938.3.1.169. By T. Giri

A. watsoni among either of their groups based on venom gland size, but given its relatively northern distribution and close phylogenetic affiliation with two species with long venom glands, *A. micropholis* and *A. microlepidota* (Trape et al. 2006; Moyer and Jackson 2011), a long venom gland seems possible. The *A. microlepidota* bites reported by Corkill et al. (1959) are presumably attributable to *A. watsoni*, based on the distribution of the species.

Subfamily Aparallactinae

The subfamily Aparallactinae includes 50 species in 10 genera, all found in sub-Saharan Africa. Within our zone, 29 species in 7 genera occur. Aparallactines are secretive, fossorial snakes, with glossy scales, small heads with ill-defined necks, and small eyes with round pupils (with the exception of *Hypoptophis wilsoni*, in which the pupil is slightly vertically elliptical). Their smooth scales lack apical pits and are

most commonly arranged in 15 dorsal scale rows. There is extensive fusion (or "loss") of head scales in aparallactines (e.g., the loreal scale is absent in all species). Loss or fusion of head scales, reduction of the number of dorsal scale rows, and shortening of the tail are all characteristics commonly associated with fossoriality in snakes (Inger and Marx 1965; Savitzky 1983). Aparallactines have opisthoglyph or aglyph dentition and pose no danger to humans. Because of their fossorial habits and their African distribution, aparallactines are relatively rare in museum collections, presumably because they are less frequently encountered by herpetologists than other snakes. As a result, many species are known from just a few specimens, and little is known of their natural history. Witte and Laurent (1947) reviewed the aparallactines and commented that although the systematics of this group is not intrinsically difficult, the rarity of aparallactines in museum collections means that there are simply not enough specimens available to sort out genera, species, and subspecies, and to assess the extent of individual variation within species.

Purple-glossed Snakes: Genus *Amblyodipsas* Peters 1857

The Purple-glossed Snakes are a genus with an African distribution, comprising nine species, four of which occur within our zone. Fossorial, emerging on the surface primarily at night and after heavy rain, *Amblyodipsas* resemble other aparallactines in many ways yet are distinct from them in others. Coloration is generally dark and drab. The head is small, like that of other aparallactines, and the neck is not well defined, but within the genus there is considerable variation in the size and shape of the snout. The eyes are small, with round pupils. The scales are smooth and glossy, the dorsal scales arranged in 15–23 straight rows, without apical pits. The tail is short with a blunt tip. The cloacal scale and subcaudals are divided. The maxilla is relatively short, bearing 3–5 unspecialized teeth in addition to an enlarged and grooved posterior fang (Bogert 1940). The nasal may be single, divided, or semi-divided. The internasals and prefrontals are always paired but may be fused to one another on either side of the head. The loreal is absent, as is the preocular. The supraocular may be present or absent, with "absent" arguably equivalent to "fused to the parietal." There is 1 postocular, which may be fused to the supraocular. The anterior temporal is absent. There are 4–7 upper labials, no suboculars, and 5–7 lower labials. The condition of the chin shields resembles that of *Xenocalamus* in that there is just 1 pair of chin shields, followed posteriorly by gulars that often take the form of "false chin shields." The chin shields of Purple-glossed Snakes have a distinctive shape, however, being broad and rounded anteriorly and narrowing to a point posteriorly. The hemipenes

are unforked with a forked sulcus spermaticus. *Amblyodipsas* are egg-layers. What little natural history information exists indicates that Purple-glossed Snakes subsist on a varied diet of lizards (including amphisbaenians) and snakes (Shine et al. 2006a). Broadley et al. (2003) report that *A. polylepis* kills Blind Snakes (*Typhlops sp.*) by constriction, and this feeding strategy may well apply to other prey types and to other species in the genus.

Of all the aparallactines, Purple-glossed Snakes are the most likely to be mistaken for Stiletto Snakes (*Atractaspis*), though admittedly that is a less dangerous error for the snake collector to make than mistaking a Stiletto Snake for a harmless Purple-glossed Snake has the potential to be.

Katanga Purple-glossed Snake: *Amblyodipsas katangensis* Witte and Laurent 1942

Amblyodipsas katangensis is known in our zone only from savannah habitats in the Katanga Province of the Democratic Republic of Congo, although its range extends south into neighboring Zambia (Broadley et al. 2003). The type locality is N'Gayu, Katanga Province, Democratic Republic of Congo.

Key to Central and Western African Species of the Genus *Amblyodipsas*

1	Prefrontals fused to internasals	*A. katangensis*
1'	Prefrontals distinct from internasals	2
2(1)	19–23 dorsal scale rows	*A. polylepis*
2'	15–17 dorsal scale rows	3
3(2)	6 supralabials, third and fourth in contact with the eye; postocular present (except in rare cases)	*A. unicolor*
3'	5 supralabials, second and third in contact with the eye; no postocular	*A. rodhaini*

This species is uniformly black above and below. The rostral is enlarged but not sharply pointed as in *A. rodhaini*. The shape of the snout is somewhat reminiscent of that of

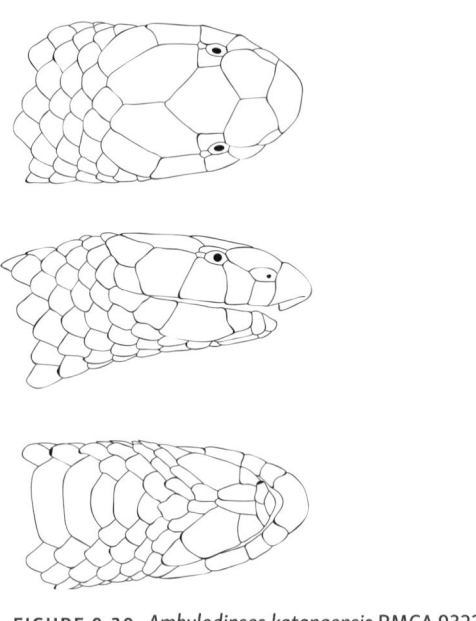

FIGURE 9.30. *Amblyodipsas katangensis* RMCA 9322. By T. Giri

Hypoptophis wilsoni. The nasal is single. The internasals and prefrontals are paired but fused to one another on each side. There is 1 supraocular and 1 postocular. There are usually 5(2,3) upper labials and 5(3) lower labials. The temporal formula is 0+1. The dorsal scales are arranged in 15 straight rows. There are 178–201 ventrals (fewer than 182 in males, more than 195 in females) and 18–27 divided subcaudals (more than 24 in males, fewer than 22 in females). The maximum length recorded for *A. katangensis* is 405 mm (Loveridge 1959).

Common Purple-glossed Snake: *Amblyodipsas polylepis* Bocage 1873

Amblyodipsas polylepis is a southern African species whose range extends from South Africa and Mozambique to Angola. The type locality is Dondo, Angola.

This species is uniformly dark brown above and below. The rostral is rounded and

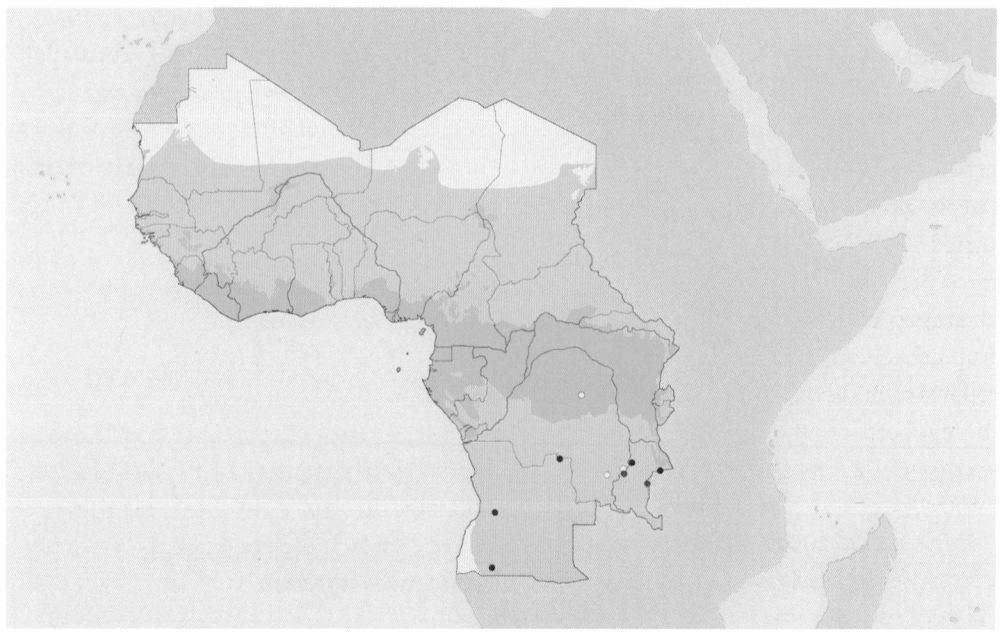

FIGURE 9.31. *Amblyodipsas katangensis* (red), *Amblyodipsas polylepis* (blue), *Amblyodipsas rodhaini* (yellow). By K. Jackson

not enlarged as in some species in the genus. The nasal is single or semi-divided. The internasals and prefrontals are paired and not fused to one another. The supraocular may be either absent or fused to the parietal. There is 1 small postocular, sometimes fused to the supraocular. There are 6(3,4) upper labials. The temporal formula is 0+1. There are 7(4), but sometimes 6(4) or 8(5), lower labials. The dorsal scales are usually arranged in 19 straight rows, although 21 or in rare instances 23 are also possible. There are 154–215 ventrals (fewer than 181 in males, more than185 in females) and 15–31 divided subcaudals (more than 24 in males, fewer than 25 in females. The maximum length recorded for *A. polylepis* is 1,111 mm (Broadley 1971b), making it the largest species of Purple-glossed Snake within our zone.

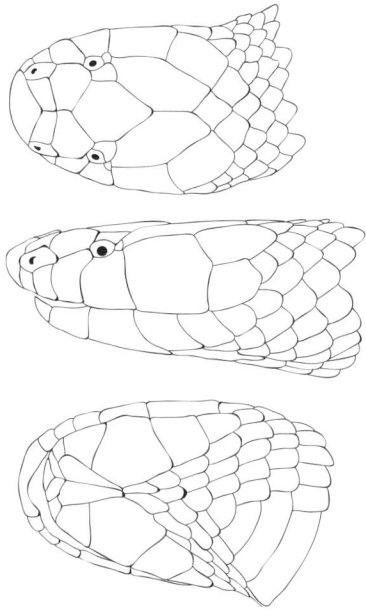

FIGURE 9.32. *Ambylodipsas polylepis* USNM 205568. By T. Giri

Through measurement and dissection of museum specimens and a review of the sparse literature on diet, Shine et al. (2006a) provide natural history information for a sampling of aparallactine species, including *A. polylepis*. A search of the literature for reports of prey items in *A. polylepis* produced the following list, which supports Purple-glossed Snakes' reputation for a varied diet relative to other aparallactines for which such information is available: one report of a caecilian, one report of an amphisbaenian, four reports of lizards, two reports of Blind Snakes (*Typhlops sp.*), and three reports of other snakes (*Lycophidion capense, L. depressirostre, Aparallactus werneri*).

Shine et al.'s (2006a) most peculiar observation of *A. polylepis* is the remarkable sexual dimorphism in size. Females were found on average to be 76% longer than male conspecifics. Although among snakes in general

FIGURE 9.33. *Amblyodipsas polylepis*, Kenya. By S. Spawls

females are usually larger than males, this finding makes *A. polylepis* the most extreme example of sexual size dimorphism known among snakes. The reason for this remarkable sexual dimorphism is not known, and it does not appear to be characteristic of Purple-glossed Snakes in general because in another species measured, *A. ventrimacula-*

ta (a species of *Amblyodipsas* from southern Africa), females were found to be only 20% longer than males on average (Shine et al. (2006a).

Rodhain's Purple-glossed Snake: *Amblyodipsas rodhaini* Witte 1930

Rodhain's Purple-glossed Snake is known only from Katanga Province, Democratic Republic of Congo. The type locality is Lomami, Democratic Republic of Congo.

The ground coloration is uniformly dark above and below. The edges of the dorsal scales are paler, and a pale stripe extends from the side of the head to the tail on both sides of the body. The rostral is enlarged and extended, forming a long, pointed snout reminiscent of *Xenocalamus*. The nasal may be single or divided. The internasals and the prefrontals are both paired and not fused to one another. A single scale above and pos-terior to the eye represents the supraocular and postocular, which are often fused but may be separate. There are 5(2,3) upper labials and 6(3) lower labials. The temporal formula is 0+1. The dorsal scales are arranged in 15 straight rows. There are 196–217 ventrals (fewer than 210 in males and 217 in the only female specimen for which we have data) and 21–29 divided subcaudals (more than 27 in males, 22 in the female specimen previously mentioned). The maximum length recorded for *A. rodhaini* is 438 mm (Broadley 1971b).

Dull Purple-glossed Snake: *Amblyodipsas unicolor* Reinhardt 1843

Amblyodipsas unicolor is a central and western African species, with a distribution extending from Senegal to the Democratic Republic of Congo. The type locality is Ghana.

This species is uniformly dark or medium brown above and below, sometimes with darker markings on the parietals. The snout is rounded with no distinctive specialization of the rostral. The nasal may be single or semi-divided. The internasals and the prefrontals are both paired and not fused to one another. A supraocular is present. A postocular is also present, although it is sometimes fused to the supraocular. There are 6(3,4) or sometimes 5(3,4) upper labials, and 7(4) or occasionally 6(4) lower labials. The temporal formula is 0+1. There are usually 17, but sometimes 15, dorsal scale rows. There are 165–214 ventrals (fewer than 181 in males, more than 189 in females) and 19–41 divided subcaudals (more than 27 in males, fewer than 31 in females). The maximum length recorded for *A. unicolor* is 980 mm (Broadley 1971b).

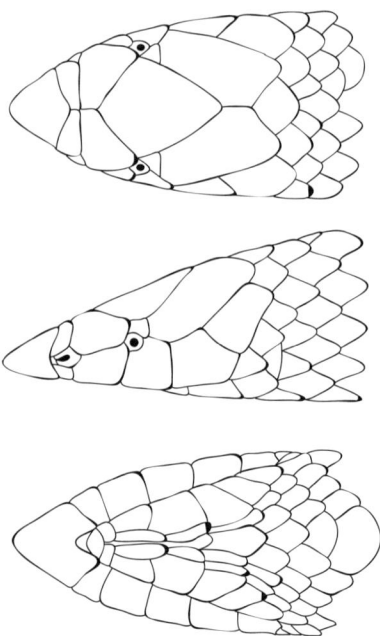

FIGURE 9.34. *Ambylodipsas rodhaini* redrawn from the original description (Witte 1930). By T. Giri

Broadley (1972) reports the stomach contents of this species as including amphisbaenians, Blind Snakes (*Typhlops sp.*), and lizards. He adds that in Zimbabwe, *A. unicolor* are sometimes killed by enterprising cats.

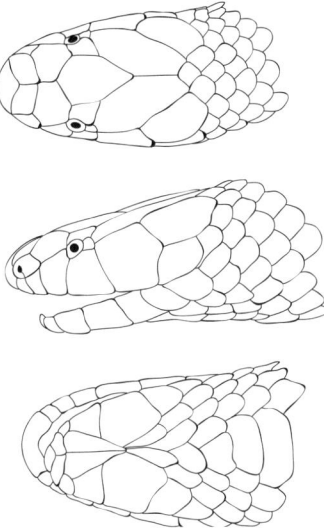

FIGURE 9.35. *Ambylodipsas unicolor* FMNH 3199. By T. Giri

Centipede-eaters: Genus *Aparallactus* Smith 1849

Aparallactus is a genus of 11 species of small, fossorial snakes, restricted to sub-Saharan Africa. Six species occur within our zone. Centipede-eaters are rarely seen above ground (and then most often after rain), and little is known about their natural history.

FIGURE 9.36. *Amblyodipsas unicolor*, Ghana. By S. Spawls

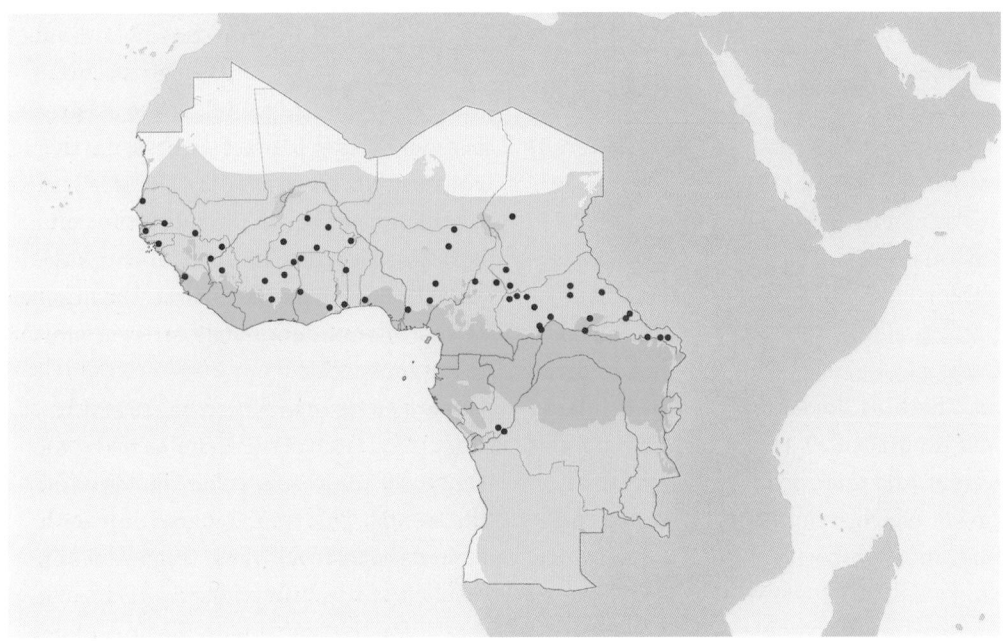

FIGURE 9.37. *Amblyodipsas unicolor* (blue). By K. Jackson

As their name suggests, they are all thought to feed primarily on centipedes, which they subdue with venom. The maxilla is long as aparallactines go, bearing 5–12 unspecialized teeth, and all species but one (*Aparallactus modestus*) have enlarged, grooved rear fangs. The only experimental evidence of the effect of *Aparallactus* venom appears in studies by Andrews (1913) and Christensen (1955), who observed that guinea pigs and mice injected with the venom of *A. capensis* suffered no ill effects. Most *Aparallactus* species, including all within our zone, are egg-layers (Spawls et al. 2004). Most studies of the morphology and taxonomy of *Aparallactus* date back half a century or more (e.g., Bogert 1940; Loveridge 1944; Witte and Laurent 1947).

Centipede-eaters rarely exceed 600 mm in total length, and a more common adult length is half that. Snakes of this genus resemble other aparallactines in many ways associated with adaptation to a fossorial habitat. The head is small and rounded, with small eyes and round pupils, and lacking a well-defined neck. The tail is short and slender. The dorsal scales are generally smooth (but see *A. niger*) and arranged in 15 straight rows, without apical pits. The rostral is small and rounded, the loreal is absent, and the internasals paired. The prefrontals may be single or paired. The nasal may be single, semi-divided, or divided. There is 1 preocular, 1 or 2 postoculars, and 1 supraocular. The anterior temporal is absent, and there are 6 or 7 upper labials, 1 or 2 of which are in contact with the parietal, and no suboculars. There are 2 pairs of chin shields, 6 to 8 lower labials, and a pronounced submandibular groove. The subcaudals and the cloacal scale are single. The number of ventrals and subcaudals is sex-ually dimorphic, with females having more of the former and males more of the latter. The hemipenes could almost be considered unforked, since both the hemipenis itself and the sulcus spermaticus are forked only for the distal 10% of the organ.

The most noteworthy characteristic of the genus is the presence of single subcaudals, a trait shared only (within the Aparallactinae and Atractaspidinae) with *Hypoptophis wilsoni* and a few species of *Atractaspis* and of *Polemon*.

Cape Centipede-eater: *Aparallactus capensis* (Smith 1849)

The range of the Cape Centipede-eater extends from Tanzania to South Africa and into southeastern Democratic Republic of Congo. The type locality is Natal, South Africa.

A small species, not known to exceed 400 mm (Broadley 1983), its body is yellowish or grayish above with a dark head and a pale underside. The nasal may be single, divided, or semi-divided. Both the internasals and the prefrontals are paired. There is 1 preocular, which is in contact with the nasal, and 1 postocular. There are 6(3,4) or 5(3,4) upper labials, depending on the subspecies, and 6(4) lower labials. The anterior chin shields are in contact with the mental. The number of ventrals and subcaudals varies according to the subspecies.

Aparallactus capensis can be readily distinguished from other *Aparallactus* by the contact of the anterior chin shields with the mental. This trait is shared only with *A. moeruensis*, from which it can be distinguished by the differences in size and contact of the parietal of with the upper labials between the two species.

Two subspecies are recognized: *A. c. boca-*

Key to Central and Western African Species of the Genus *Aparallactus*

1	Prefrontal single	2
1'	Prefrontal paired	3
2(1)	Uniformly dark above	*A. niger*
2'	Striped or spotted above	*A. lineatus*
3(1)	Preocular separated from the nasal by a supralabial	*A. lunulatus*
3'	Preocular in contact with the nasal	4
4(3)	Fifth and sixth supralabials both in contact with parietal	*A. moeruensis*
4'	Fifth but not sixth supralabial in contact with parietal	5
5(4)	Chin shields separated from the mental by the first pair of infralabials	*A. modestus*
5'	Anterior chin shields in contact with mental	*A capensis*

gei Boulenger 1895b; *A. c. punctatolineatus* Boulenger 1895b. *A. c. bocagei* has more ventrals (174–191) and more subcaudals (44–63) than *A. c. punctatolineatus* (126–162 ventrals, 36–52 subcaudals). The upper labials also differ between the two subspecies: 6(3, 4) in *A. c. bocagei*, the fifth the largest and in contact with the parietal, whereas *A. c. punctatolineatus* has 5(2, 3) upper labials, and the fourth is the largest and in contact with the parietal, or occasionally both fourth and fifth may be in contact with the parietal.

Observations by Loveridge (1953) and Broadley (1972) demonstrate that Centipede-eaters are prey as well as predators, *A. capensis* having been found in the stomach contents of both a Stiletto Snake (*Atractaspis bibronii*) and an Eagle Owl (*Bubo sp.*). *A. capensis* is found under rocks and logs in savannah as well as other open habitats. *A. capensis* is one of the most studied Centipede-eaters within our zone, but even so, knowledge of its natural history is sparse.

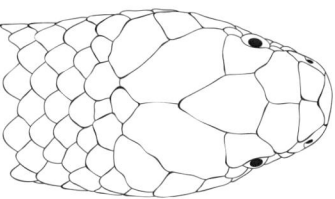
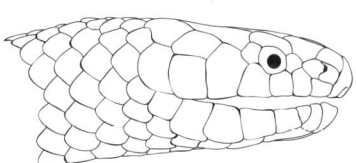
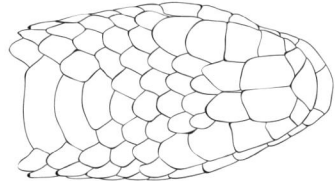

FIGURE 9.38. *Aparallactus capensis* RMCA 15144. By T. Giri

FIGURE 9.39. *Aparallactus capensis*, Tanzania. By S. Spawls

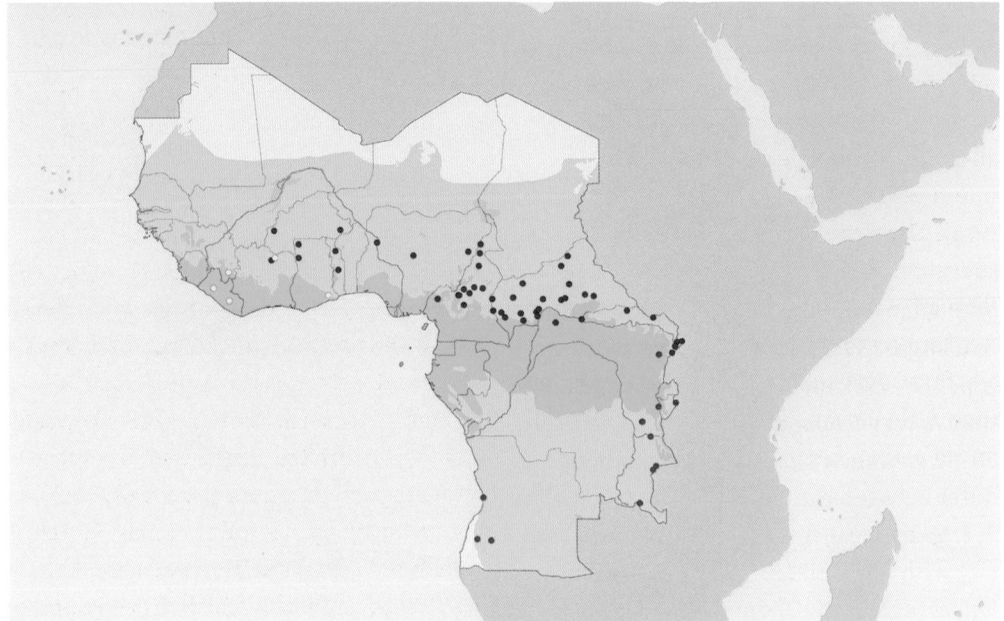

FIGURE 9.40. *Aparallactus capensis* (red), *Aparallactus lineatus* (yellow), *Aparallactus lunulatus* (blue). By K. Jackson

Lined Centipede-eater: *Aparallactus lineatus* Peters 1870

Aparallactus lineatus has a similar distribution to *A. niger*, extending from Guinea to Ghana. The type locality is Kéta, Ghana. These two species differ in coloration. *A. lineatus* is pale brown or gray and gets its name from two or three lines of dark spots that extend the length of the body. The paraventral dorsal scales have pale flecks, and the underside is pale gray mottled with darker gray or black, particularly toward the tail. In terms of head scale morphology, however, *A. lineatus* and *A. niger* are quite similar, most notably in being the only Centipede-eaters within our range to possess a single (rather than paired) prefrontal. The nasal is single. The preocular is in contact with the nasal. The internasals are paired. There is 1 postocular. There are 7(3,4) upper labials, of which the sixth is the largest and both the fifth and sixth are in contact with the parietal. There are

7(3,4) lower labials. Loveridge (1938), knowing that it is common for snakes to darken and their patterns to become indistinct as they grow, synonymized the two species, proposing that small, striped individuals became uniformly dark with age. But later authors (e.g., Witte and Laurent 1947) with access to more specimens of assorted sizes recognized the two species as distinct. Not only are even small *A. niger* uniformly dark whereas large *A. lineatus* retain their stripes, but also there are consistent differences between the species in the shapes of some of the head scales, including that all-important single prefrontal. The maximum recorded length is 600 mm (Villiers 1975). There are 151–170 ventrals and 35–58 subcaudals (more than 51 in males, fewer than 42 in females).

A. lineatus is readily identifiable by its single, fused preocular. Although it shares this trait with *A. niger*, the two species have different color patterns.

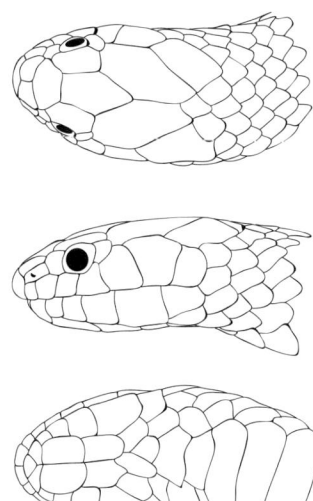

FIGURE 9.41. *Aparallactus lineatus* FMNH 58094. By T. Giri

Reticulate Centipede-eater: *Aparallactus lunulatus* Peters 1854

Aparallactus lunulatus has a broad distribution in sub-Saharan Africa, its range extending from east Africa all the way to Burkina Faso and Ivory Coast in West Africa. The type locality is Tete, Mozambique.

The body is uniformly grayish green, paler on the underside than above, and may have a speckled (or "reticulated") appearance anteriorly resulting from each dorsal scale being darker at its base than at its edges. The head is typically light brown with a dark band across the nape. The number of ventrals is 140–177 (fewer than 163 in males, more than 152 in females) and subcaudals 41–65 (fewer than 52 in males, more than 62 in females). The nasal is single. Both internasals and prefrontals are paired. There is 1 postocular. The preocular is not in contact with the nasal. There are 6(3, 4) upper labials, the fifth the largest and in contact with the parietal. There are usually 6(4) but occasionally 7(4) lower labials. The chin shields are separated from the mental by the first pair of lower labials. The maximum length recorded for this species is 520 mm (Witte and Laurent 1947).

A. lunulatus has a preocular that is not in contact with the nasal, an arrangement unique among *Aparallactus* species within our zone.

Broadley (1972) kept an individual in captivity and observed it feeding on the centipedes he offered it. The snake struck the centipede midbody and manipulated it with its jaws in order to swallow it headfirst. The centipede died quickly from the venom, and its sharp mandibles were not able to puncture the snake's smooth scales. The diet of *A. lunulatus* is not restricted to centipedes, however, and has been known to include scorpions and sometimes smaller snakes (e.g., *Typhlops*, *Leptotyphlops*) (Villiers 1975; Sweeney 1971; Witte and Laurent 1947; Stucki-Stirn 1979). Broadley and Cock (1975) reported that eggs are laid in clutches of just 3 or 4, with each egg 30 mm x 7 mm in size.

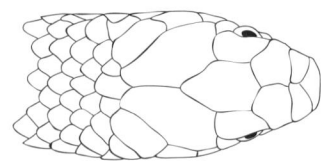

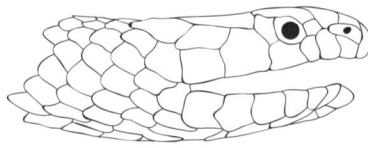

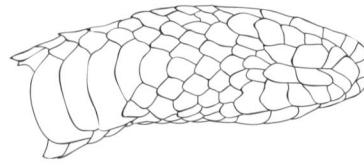

FIGURE 9.42. *Aparallactus lunulatus* RMCA 3862. By T. Giri

FIGURE 9.43. *Aparallactus lunulatus*, Ethiopia. By S. Spawls

Western Forest Centipede-eater: *Aparallactus modestus* Günther 1859

Aparallactus modestus is a forest species found throughout the equatorial forests of central and western Africa. The type locality is West Africa.

The body is uniformly brown or gray, often with a darker band across the nape. Juveniles have a distinctive white or cream band across the nape. The underside is paler, sometimes with dark flecks, and the subcaudals are darker than the ventrals. The ventrals range from 126 to 172 (fewer than 146 in males, more than 147 in females), and subcaudals from 32 to 53 (more than 41 in males, fewer than 49 in females). The nasal may be single, divided, or semi-divided. The preocular is in contact with the nasal. Both the internasals and the prefrontals are paired. There are 2 postoculars, the superior larger than the inferior. There are usually 7(3,4) or occasionally 6(3,4) upper labials, whose size and arrangement with respect to the parietal vary between subspecies. The maximum length reported is 645 mm (Courtois unpubl.), making it the largest species of Centipede-eater within our zone, although 350 mm is a more typical adult length.

A. modestus can be distinguished from other species in the genus by the presence of 2 postoculars. This trait is shared with some individuals of *A. niger*, but the two species are readily distinguishable by the different arrangements of the internasals and prefrontals.

Two subspecies are recognized, *A. m. modestus* Günther 1859 and *A. m. ubangensis* Boulenger 1897a, and can be distinguished by their labial scales. In *A. m. modestus*, only the sixth upper labial is in contact with the

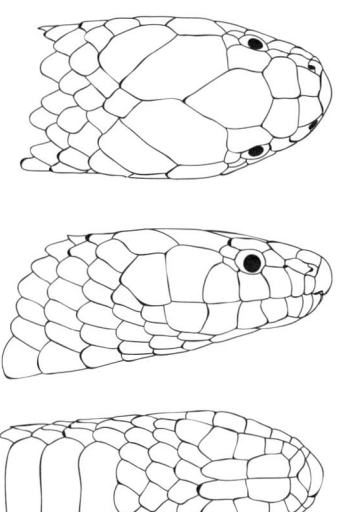

FIGURE 9.44. *Aparallactus modestus* RMCA A7-003-R-0032. By T. Giri

FIGURE 9.45. *Aparallactus modestus*, Guinea. By P. Naskrecki

FIGURE 9.46. *Aparallactus modestus*, Republic of Congo. By K. Jackson

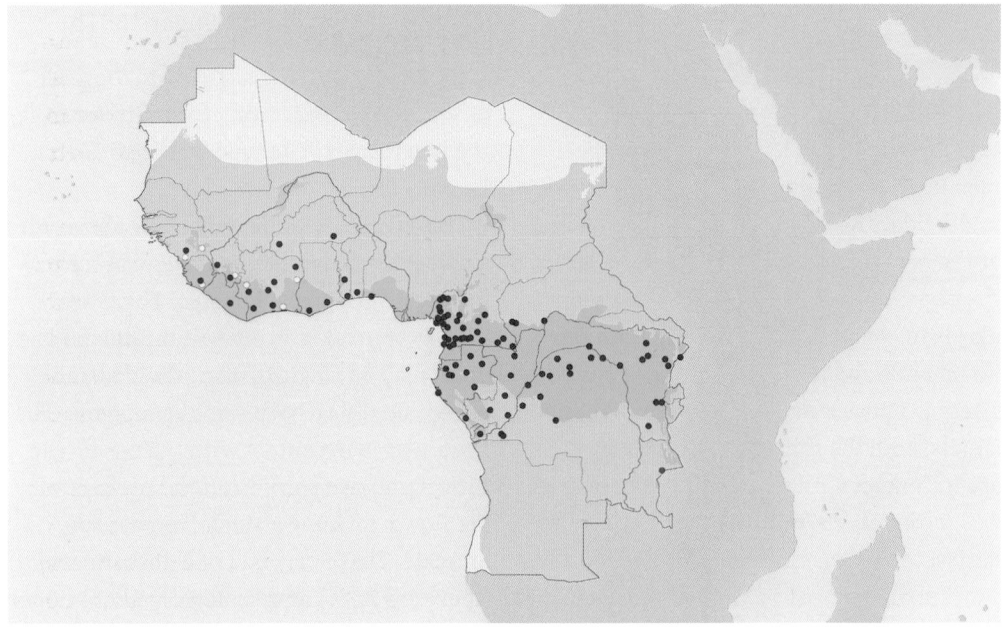

FIGURE 9.47. *Aparallactus modestus* (blue), *Aparallactus moeruensis* (red), *Aparallactus niger* (yellow). By K. Jackson

parietal, and of the lower labials the fifth is the largest, while in *A. m. ubangensis* both the fifth and sixth upper labials are in contact with the parietal, and the fourth lower labial is the largest. *A. m. modestus* has 7(4) or occasionally 8(4) lower labials, while *A. m. ubangensis* has 6(4) lower labials.

A. modestus differs from other Centipede-eaters in lacking grooved posterior fangs (Bogert 1940), though it does possess a Duvernoy's gland, suggesting a possible secondary loss of a grooved fang. Little is known of any influence that an apparently less developed venom-delivery apparatus may have on their diet relative to other species of Centipede-eater. Knoepffler (1966) observed a juvenile *A. modestus* regurgitate an estimated 100 termites.

Zaire Centipede-eater: *Aparallactus moeruensis* Witte and Laurent 1943

This species is known only from the type specimen, collected in the province of Katanga, Democratic Republic of Congo. There is some disagreement as to whether *Aparallactus moeruensis* is a valid species (Witte and Laurent 1947), as in many ways it resembles both *A. capensis* and *A. lunulatus*. Just 389 mm in length, the type specimen is olive gray dorsally with a yellowish underside. The head is darker than the rest of the body, and there is a dark band across the nape. The nasal is divided. The preocular is in contact with the nasal. Both the internasals and the prefrontals are paired. There is 1 postocular. There are 6(3,4) upper labials, both the fifth and sixth of which are in contact with the parietal. There are 7(3) lower labials, of which the sixth is the largest. *A. moeruensis* has 174 ventrals and 57 subcaudals.

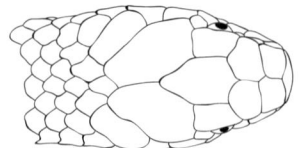
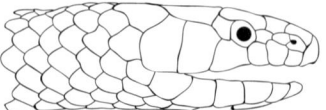
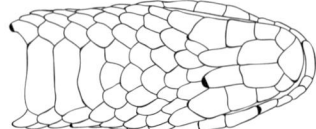

FIGURE 9.48. *Aparallactus moeruensis* RMCA 240 (TYPE). By T. Giri

A distinctive character of *A. moeruensis* is the contact of the chin shields with the mental. It shares this trait with *A. lunulatus*, but the two species can be distinguished by the size and arrangement of the parietals with respect to the upper labials.

Western Black Centipede-eater: *Aparallactus niger* Boulenger 1897b

Aparallactus niger is restricted to the area of West Africa extending from Guinea to the Ivory Coast. The type locality is Sierra Leone.

This species is uniformly black above with a pale gray underside, each ventral scale having a dark posterior edge. There are 151–175 ventrals and 33–64 subcaudals. The nasal may be single or semi-divided and is exceptionally large in some specimens. The preocular is in contact with the nasal. The internasals are paired, but the prefrontals are fused, forming a single, large median scale. There may be 1 or 2 postoculars. There are 7(3,4) or occasionally 6(3,4) upper labials. Of these, the fifth and sixth are in contact with the parietal. There are 6(4) lower labials. *A. niger* differs from all other centipede-eaters by the presence of weakly keeled dorsals on the posterior part of the body and tail in some individuals. The latter may be a sexually dimorphic trait, as keeled dorsals in the anal region are seen in males but not females in some snake species (Loveridge 1938). The maximum recorded length for this species is 605 mm (Witte and Laurent 1947).

The fusion of the prefrontals into a single scale is a distinctive characteristic of *A. niger*. Although it shares this trait with *A. lineatus*, the two species can be distinguished by their color patterns.

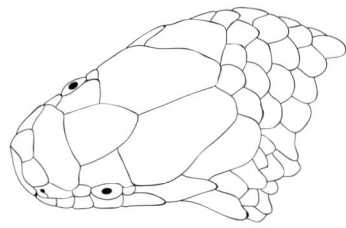

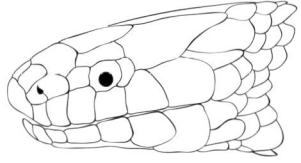

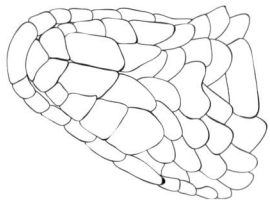

FIGURE 9.49. *Aparallactus niger* BMNH 1946.1.6.89 (TYPE). By T. Giri

Two-headed Snakes: Genus *Chilorhinophis* Werner 1907

Chilorhinophis is a genus comprising three species with an eastern and southeastern African distribution. The range of one species extends into our zone. *Chilorhinophis* are nocturnal burrowing snakes and rarely encountered. When they are, however, these small snakes are so remarkable in appearance that they are easily recognized. The body is elongate and thin, cream or yellow in color, and with three black stripes extending from head to tail. The Two-headed Snakes get their common name from the resemblance between their head and tail. The head is small, black, sometimes with pale spots, with no distinct neck separating it from the cylindrical body. The tail does not taper and its tip is rounded, the same shape and color as the head. When threatened, *Chilorhinophis* will conceal its head in the coils of its body and wave its tail, giving the impression that the tail is in fact the head, presumably to trick a predator into attacking the tail instead of the more vulnerable head.

The small eye on the small head has a round, or in some species slightly vertically elliptical, pupil. The nasal is single and sometimes fused to the first upper labial. Perhaps the most unusual aspect of the head scales of *Chilorhinophis* is the condition of the internasals and prefrontals. There is just 1 pair of scales between the nasals, resulting from a fusion of the internasals with the prefrontals. The loreal is absent. There is usually 1 preocular and 1 postocular. The temporal formula is usually 0+1. There are 4–6 upper labials, one of which is in contact with the eye. There are 2 pairs of chin shields, and of the 5–7 lower labials, 3 are in contact with the anterior pair. The submandibular groove is pronounced. The dorsal scales are smooth, lack apical pits, and are arranged in 15 straight rows. The cloacal scale and subcaudals are divided. The number of ventrals and of subcaudals is sexually dimorphic, females having more ventrals and males more subcaudals.

Gerard's Black and Yellow Burrowing Snake: *Chilorhinophis gerardi* Boulenger 1913

The range of *Chilorhinophis gerardi* extends from northern Zimbabwe to southern Democratic Republic of Congo. The type locality is Kikondja, Katanga Province, Democratic Republic of Congo.

Gerard's Black and Yellow Burrowing Snake burrows in moist savanna habitat and is rarely encountered. Its description is consistent with the description for the genus, with the addition of a few specific details. There are 4(3) or 5(3) upper labials and 5(3) or 6(3) lower labials. The temporal formula is 0+1. There are 263–348 ventrals (fewer than 295 in males, more than 273 in females) and 19–31 divided subcaudals (more than 25 in males, fewer than 27 in females). The maximum length recorded for this species is 569 mm (Witte 1962).

Two subspecies are recognized on the basis of numbers of ventrals and subcaudals: *C. g. gerardi* Boulenger 1913; *C. g. tanganyikae* Loveridge 1951a.

Wedge-snouted Burrowing Snake: *Hypoptophis wilsoni* Boulenger 1908

Hypoptophis is a monotypic genus, the type specimen collected at Inkongo, Democratic Republic of Congo. *Hypoptophis wilsoni* is seldom encountered in the field and is rare in museum collections. It is known from localities in southeastern Democratic Republic of Congo (see Fig. 9.54) and Zambia (Broadley et al. 2003). The species gets its common name from the distinctive shape of its rostral, which is enlarged and pointed, though shorter and sturdier in appearance than that of the Quill-snouted Snakes (*Xenocalamus*). The snout of *Hypoptophis* invites comparison with a spade as much as a wedge, and it is presumably used for digging. In coloration, *H. wilsoni* is uniformly dark brown above, with each dorsal scale outlined in black, and with a lighter brown underside and a paler area in the cloacal region. The head is small relative to the short body, and the neck is not well defined. The eyes are small, and the pupils are vertically elliptical. The maxilla is short, with 2–4 unspecialized teeth and a grooved posterior fang. The nasal is single. Both internasals and prefrontals are paired. The loreal is absent. There is 1 preocular and 2 postoculars, of which the superior is the larger of

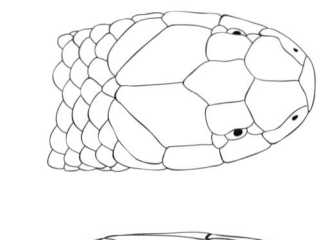

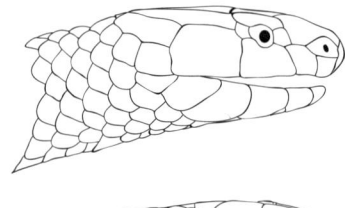

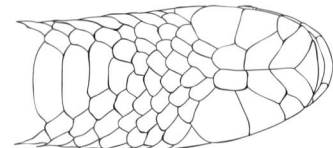

FIGURE 9.50. *Chilorhinophis gerardi* RMCA 748. By T. Giri

FIGURE 9.51. *Chilorhinophis gerardi*, Democratic Republic of Congo. By C. Tilbury

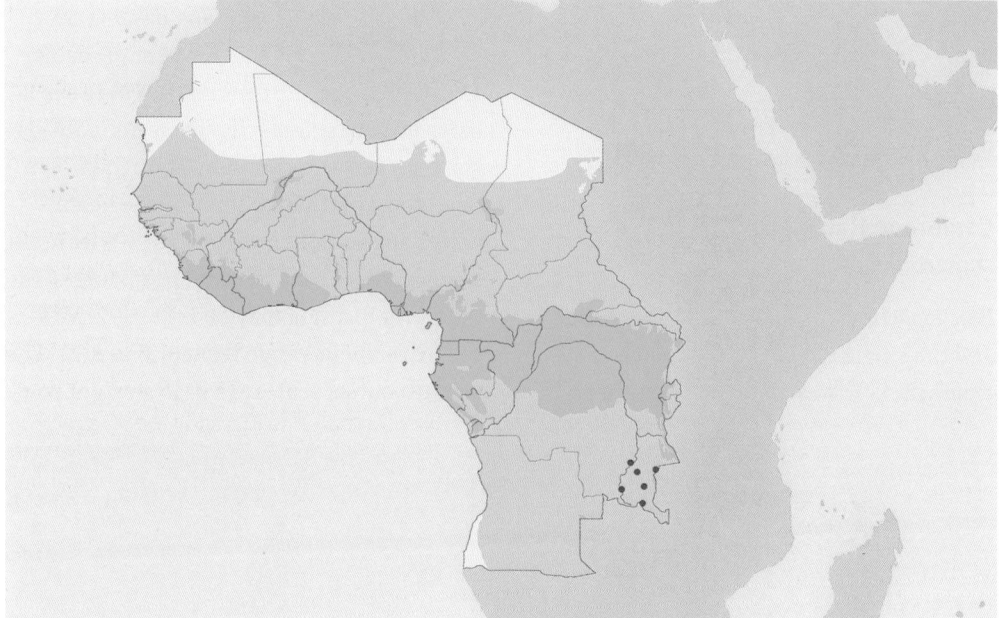

FIGURE 9.52. *Chilorhinophis gerardi* (red). By K. Jackson

the two. There are 7(3,4) upper labials and 6(4) lower labials. The temporal formula is 1+1. There are 2 pairs of chin shields, the anterior pair larger than the posterior pair. The submandibular groove is pronounced. The dorsal scales are arranged in 15 straight rows and lack apical pits. They are smooth anteriorly but keeled posteriorly. There are 102–118 ventrals and 32–45 single subcaudals. The cloacal scale is single. The maximum recorded length is 620 mm (Witte 1962). Information on the structure of the hemipenes is lacking, presumably owing to the small number of specimens in collections. *H. wilsoni* is unique among aparallactines in having a vertical pupil. Another character worthy of note is the condition of the ventral scales, which are keeled posteriorly. The latter is a trait shared only with some individuals of *Aparallactus niger*. Two subspecies are recognized: *H. w. wilsoni* Boulenger 1908; *H. w. katangae* Müller 1911.

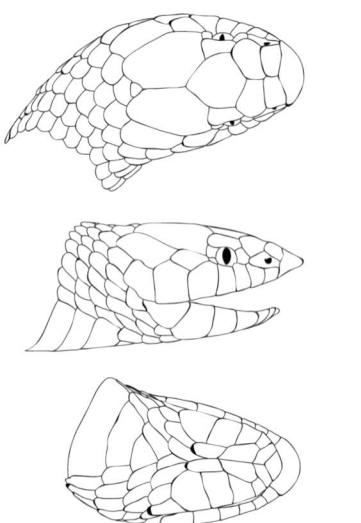

FIGURE 9.53. *Hypoptophis wilsoni* RMCA 74-13-R-128. By T. Giri

Cameroon Racer: *Poecilopholis cameronensis* Boulenger 1903

Poecilopholis is another monotypic genus, and the single species, *Poecilopholis cameronensis*, is known only from southern

Cameroon. The type locality is Efulen, Cameroon.

P. cameronensis resembles other aparallactines in many ways, and in the absence of any natural history information, we can only speculate that like them it is secretive and fossorial in its habits. The body is dark olive green above with a white dot at the center of each dorsal scale, and a white triangle on each side of the head. The underside is uniformly pale. It has a small head without a well-defined neck and small eyes with round pupils. The nasal is single. There is 1 preocular and 1 supraocular, There are 5(3,4) upper labials. There are 2 pairs of chin shields, the posterior pair longer than the anterior pair. Though known from few specimens, some of the head scales are known to vary slightly within the species: there may be 1 or 2 postoculars, the temporal formula may be 1+1 or 1+2, and the infralabials range from 6(3) to 8(3). The smooth dorsal scales are in 15 straight rows

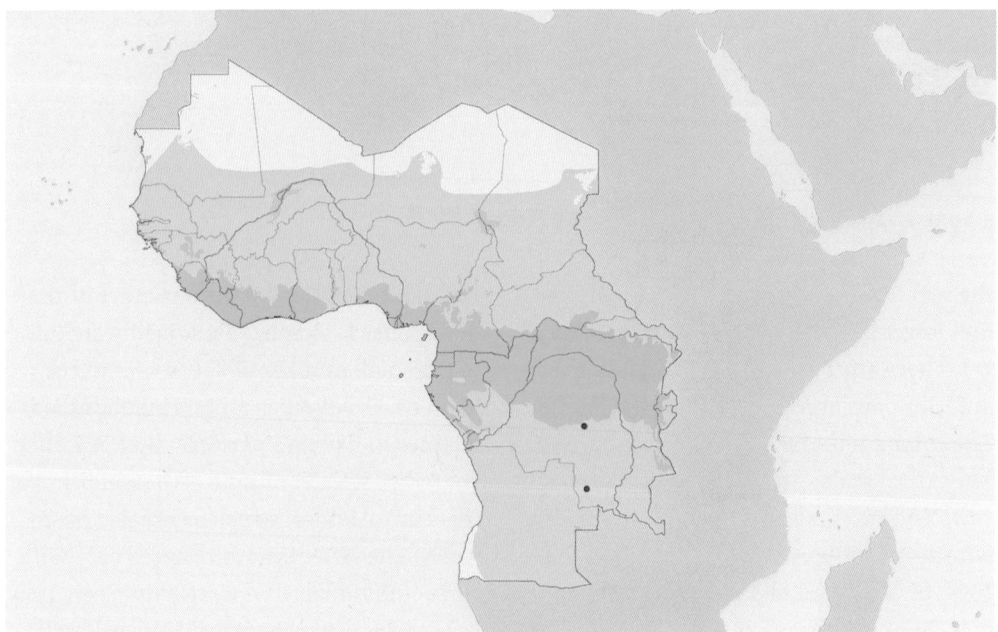

FIGURE 9.54. *Hypoptophis wilsoni* (red). By K. Jackson

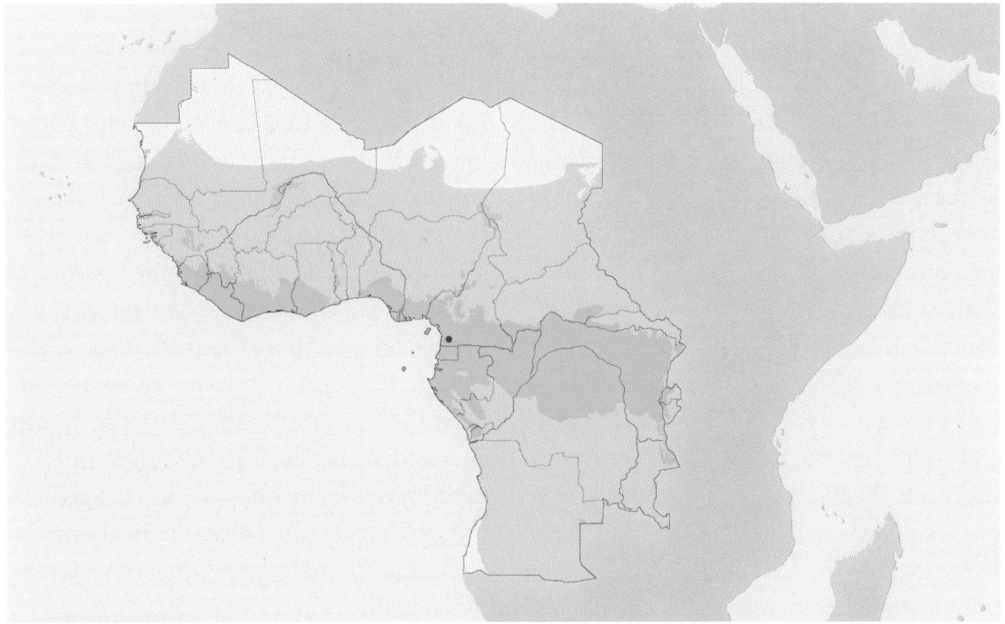

FIGURE 9.55. *Poecilopholis cameronensis* (red). By K. Jackson

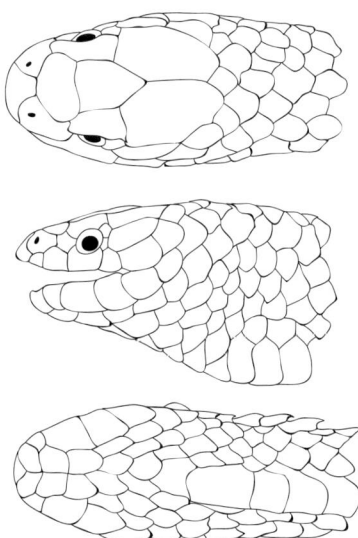

FIGURE 9.56. *Poecilopholis cameronensis* BMNH 1946.1.1.69. By T. Giri

and lack apical pits. There are 175–180 ventrals and a divided cloacal scale. The tail is short with just 20–25 divided subcaudals. The maximum recorded length is 520 mm (Boulenger 1903).

P. cameronensis is unique and easily identifiable by its prefrontals and internasals, which are all fused into a single median scale. But it earns its own genus on the basis of its hemipenes, which are bilobed, a character unique among aparallactines, though shared with the Stiletto Snakes (*Atractaspis*), and by its dentition (having as many as 10 maxillary teeth and lacking any grooved or even enlarged posterior fang).

Snake-eaters: Genus *Polemon* Jan 1858

The Snake-eaters, *Polemon*, are a genus of approximately 14 species, 13 of which can be found within our zone. They are nocturnal, live mostly below ground or in leaf litter, and emerge more frequently after rain. *Polemon* are usually fairly small. Although the maximum length recorded for a Snake-eater is 880 mm, the maximum length for the majority of species is approximately 600 mm, and the average adult length is approximately half that. *Polemon* are therefore are among the few snakes that can be caught in pitfall traps made of standard-sized buckets. The rostral is rounded, with none of the enlargement or specialization of the shape of the snout as seen in some other aparallactines. The eyes on the small rounded head are small, with round pupils. The neck is not strongly defined, nor is it entirely absent, as it is in most aparallactines, and there is often a pale band across the nape. The tail is short and stubby, ending in a sharp point. The maxilla is short, usually bearing just 2 or 3 unspecialized teeth as well as a grooved posterior fang for introducing venom into their prey. *Polemon* are secretive, unaggressive snakes and pose no danger to humans.

The nasal may be single or divided. The loreal is absent. Both the internasals and the prefrontals are always paired. There is never more than 1 preocular, but there may be either 1 or 2 small postoculars. There is a supraocular but no suboculars. There are 6 or 7 upper labials. The temporal formula may be 0+1 or 1+1 and often varies among individuals of the same species. There are 2 pairs of chin shields separated by a pronounced submandibular groove. There are 6–8 lower labials. The dorsal scales are smooth and always arranged in 15 straight rows, without apical pits. The cloacal scale and subcaudals may be single or divided. Both the ventrals and subcaudals are sexually dimorphic, females having more of the former and males more of the latter. In short, scale characters vary comparatively little between species. The hemipenes are almost unforked, divided only at the final 10% of their length, and the sulcus spermaticus is forked. Because the scale characters of the different species in the genus are so

similar, some useful characters to watch for because they are less common within the genus include: (1) single subcaudals, (2) undivided cloacal scale, (3) fewer than 4 lower labials in contact with anterior chin shields, (4) longest upper labial is other than the sixth, and (5) distinctive coloration/patterns. In addition, much of the taxonomic literature on Snake-eaters relies heavily on numbers of ventrals. Where possible, in the species descriptions below, we draw the reader's attention to the particular characters most useful in distinguishing each species from others within the genus, though in most cases these characters are not unique to the species when considered individually.

Although little is known about the natural history of Snake-eaters, some evidence lends support to the accuracy of their common name. Shine et al. (2006a) summarized the known reports of stomach contents of Snake-eaters. Although the sample size was small, it represents the only information available. In all species combined, there were three reports of Blind Snakes (Genus *Typhlops*), one report of a Herald Snake (*Crotaphopeltis hotamboeia*), one caecilian, and one report of lizard eggs. This might seem like a varied diet, but one thing all these fossorial prey items have in common is that they cannot drop their tails as an escape mechanism, as lizards can. Perhaps in the absence of specializations such as the side-stabbing fangs of the Stiletto Snakes (*Atractaspis*) or the elongate and pointed rostral of the Quill-snouted Snakes (*Xenocalamus*), which allow the predator to attack the prey item farther up the body than the tail, *Polemon* are limited to prey that do not drop their tails.

Key to Central and Western African Species of the Genus *Polemon*

1	Cloacal scale single	2
1'	Cloacal scale divided	4

2(1)	Subcaudals single; body not striped	3
2'	Subcaudals divided; body striped	*P. acanthias*

3(2)	Fewer than 215 ventrals	*P. bocourti*
3'	More than 215 ventrals	*P. barthii*

4(1)	3 lower labials contact anterior chin shield	5
4'	4 lower labials contact anterior chin shield	7

5(4)	Dark band across neck; body not striped	*P. notatus*
5'	Pale band or no band across neck; body may be striped	6

6(5)	Pale band across neck	*P. gracilis*
6'	No band across neck	*P. neuwiedi*

7(4)	Ventrals pale	8
7'	Body uniformly dark (ventrals may have pale edges)	*P. christyi*

8(7)	Fewer than 205 ventrals	9
8'	More than 205 ventrals	11

9(8)	Distance from edge of eye to edge of lip less than 2 times the diameter of eye	*P. griseiceps*
9'	Distance from edge of eye to edge of lip more than or equal to 2 times the diameter of eye	10

10(9)	Upper postocular larger than lower postocular	*P. robustus*
10'	Upper postocular equal to or smaller than lower postocular	*P. collaris*

11(8)	Upper postocular equal to or larger than lower postocular	12
11'	Upper postocular smaller than lower postocular	*P. collaris*

12(11)	White or ivory nuchal band	13
12'	No nuchal band	*P. gabonensis*

13(12)	Sixth upper labial the longest; fifth upper labial may or may not contact parietal	*P. fulvicollis*
13'	Fifth upper labial the longest; no upper labial in contact with parietal	*P. graueri*

Reinhardt's Snake-eater: *Polemon acanthias* (Reinhardt 1860)

Polemon acanthias is a West African forest species, with a range extending from Guinea to Nigeria. The type locality is recorded as "Guinea," but presumably meaning "Guinea Coast" rather than the country of Guinea. It is likely that the true type locality is somewhere in Ghana.

The coloration of this species is not typical of the genus, consisting of a reddish-orange ground color above, with five black stripes along the body. There is a white band across the nape, and the underside of body and tail is whitish. The nasal is usually single. There is 1, or sometimes 2, postocular. There are 7(3,4) upper labials, the sixth usually the longest. The temporal formula is 1+1. There may be 6(4) or 7(4) lower labials. There are 182–216 ventrals and 16–24 divided subcaudals. The cloacal scale is single. The maximum length recorded for this species is 585 mm (Loveridge 1944).

The most distinctive characters of this species are the striped color pattern and the single cloacal scale.

Cole (1967) captured a *P. acanthias* in the muddy bank of a stream in Ghana, while it was attempting to eat a caecilian.

Guinea Snake-eater: *Polemon barthii* Jan 1858

The distribution of *Polemon barthii* extends from the Ivory Coast to Gabon and Cameroon. The type locality is recorded as "Guinea," but presumably meaning "Guinea Coast" rather than the country of Guinea. It is likely that the true type locality is somewhere in Ghana.

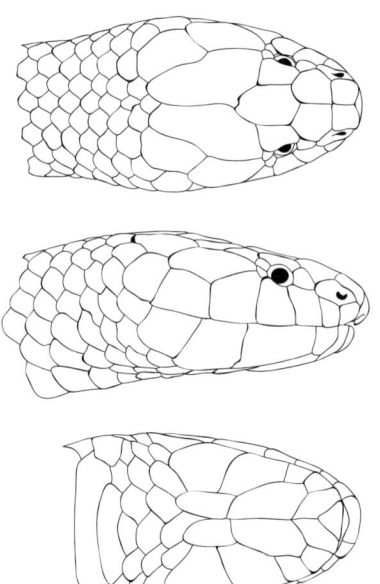

FIGURE 9.57. *Polemon acanthias* RMCA 29550. By T. Giri

It is uniformly black above except for a pale band across the nape and often pale markings along the sides. The underside is pale gray. The nasal is divided. There is 1 postocular. There are 7(3,4) upper labials, of which the fifth is usually the longest. The temporal formula is 0+1. There are 7(4) lower labials. There are 221–229 ventrals and 16–20 single subcaudals. The cloacal scale is single. The maximum length recorded for *P. barthii* is 810 mm (Villiers 1975).

Though not unique to this species, the fact that the fifth rather than the sixth upper labial is the longest places it in a minority within the genus.

Bocourt's Snake-eater: *Polemon bocourti* Mocquard 1897

Polemon bocourti is a central and western African forest species. The type locality is Lambaréné, Gabon.

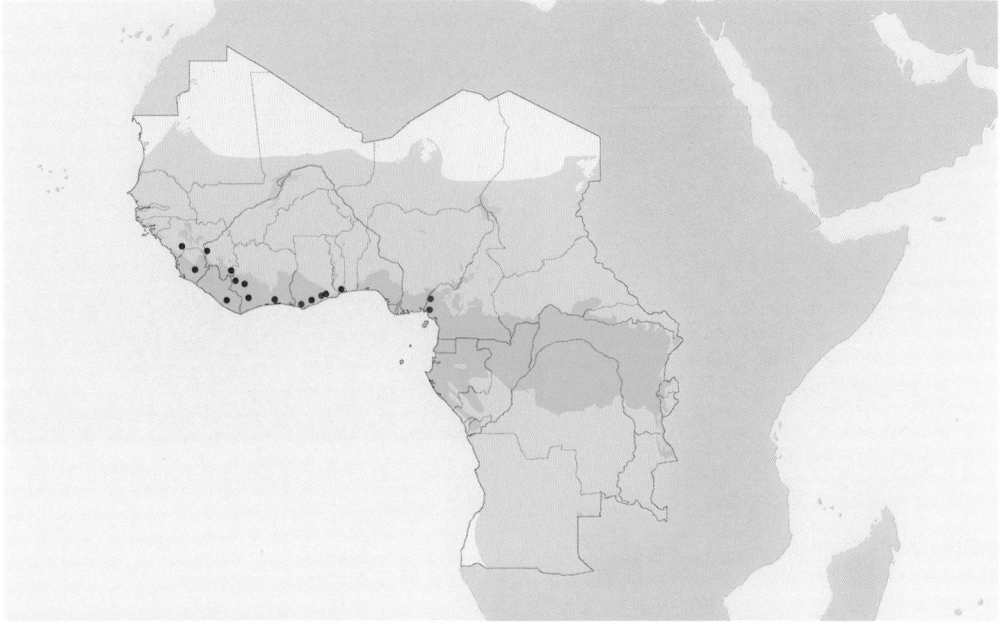

FIGURE 9.58. *Polemon acanthias* (blue), *Polemon barthii* (red). By K. Jackson

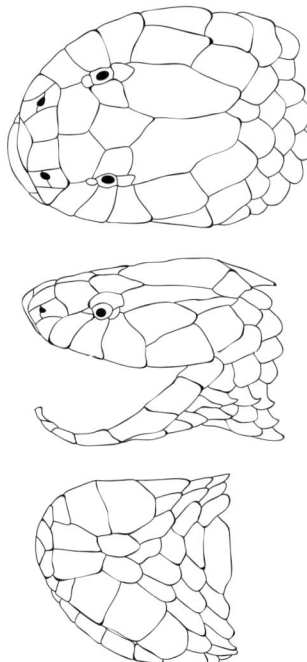

FIGURE 9.59. *Polemon barthii* BMNH 1881.4.9.7 (drawn from a specimen from which the skull had been removed). By T. Giri

The head and body are dark brown or black above with a broad pale band across the nape. The underside is mainly whitish, but with darker edges to the ventrals. The nasal is single. There are usually 2 postoculars. There are 7(3,4) upper labials, of which the sixth is usually the longest. There are 6(4) or 7(4) lower labials. There are 171–210 ventrals and 15–26 single subcaudals. The cloacal scale is single. The maximum recorded length is 980 mm (Witte and Laurent 1947), making it the longest species of *Polemon* ever documented.

The most distinctive features of this species are the single subcaudals and undivided cloacal scale.

Eastern Snake-eater: *Polemon christyi* (Boulenger 1903)

Polemon christyi is an east African species whose distribution extends into Rwanda,

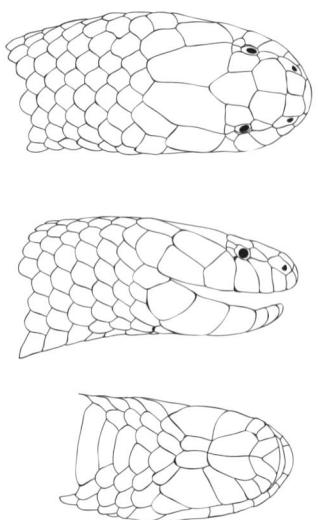

FIGURE 9.61. *Polemon bocourti*, Republic of Congo. By M. Burger

FIGURE 9.60. *Polemon bocourti* RMCA 9588. By T. Giri

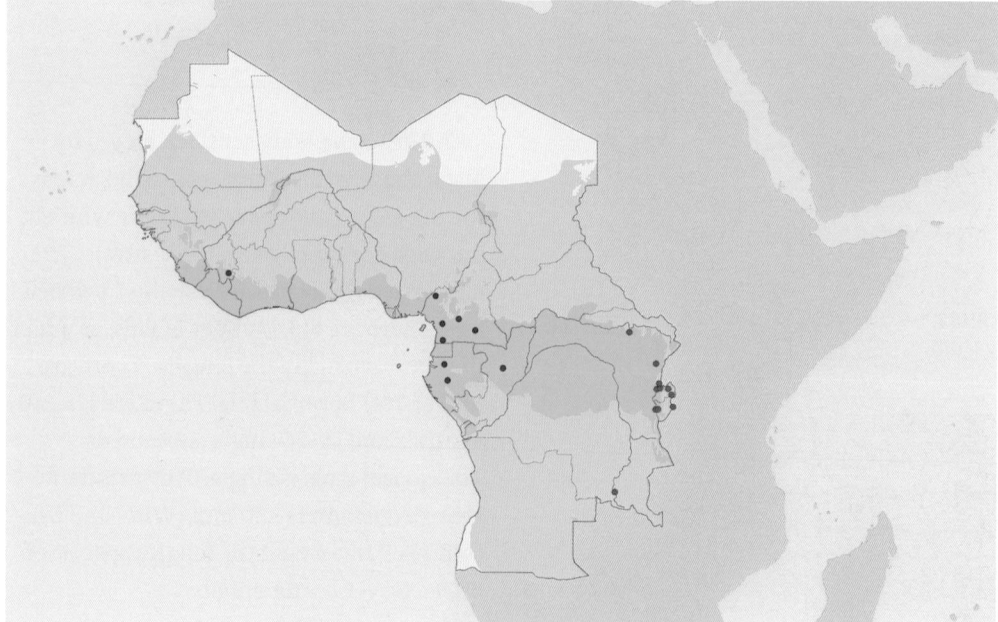

FIGURE 9.62. *Polemon bocourti* (blue), *Polemon christyi* (red). By K. Jackson

Burundi, and eastern Democratic Republic of Congo within our range. The type locality is Uganda.

P. christyi is almost uniformly black above and underneath, though the ventrals may have pale edges. The nasal may be single or divided. There are usually 2 postoculars. There are 7(3,4) upper labials, of which the sixth is usually the longest. The temporal formula is 0+1 or 1+1. There are 7(4) lower labials. There are 199–250 ventrals (fewer than 225 in males, more than 200 in females) and 15–24 (more than 18 in males, fewer than 19 in females) divided subcaudals. The cloacal scale is divided. The maximum recorded length for this species is 841 mm (Laurent 1956).

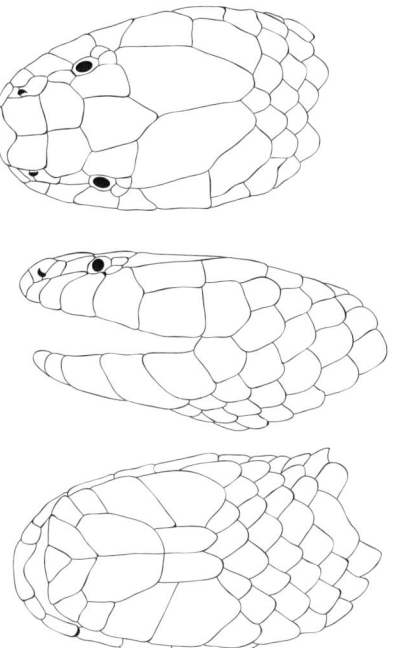

FIGURE 9.63. *Polemon christyi* FMNH 12825. By T. Giri

FIGURE 9.64. *Polemon christyi*, Democratic Republic of Congo. By S. Spawls

This species is readily distinguishable from all other *Polemon* by its uniformly black coloration.

Collared Snake-eater: *Polemon collaris* (Peters 1881)

Polemon collaris has a broad distribution in central Africa, extending from Nigeria to Uganda to Angola. The type locality is Malanje, Cuango, Angola.

This species is brown above with a broad pale band across the nape and a pale underside. The nasal is single. There are usually 2 postoculars. There are 7(3,4) or sometimes 8(3,4) upper labials, of which the sixth is usually the longest. The temporal formula is generally 0+1 but may technically be 1+1 in some cases in which the anterior tip of the first temporal has a point of contact with a postocular. There are 6(3,4) or 7(3,4) lower labials. There are 181–236 ventrals (fewer than 222 in males, more than 219

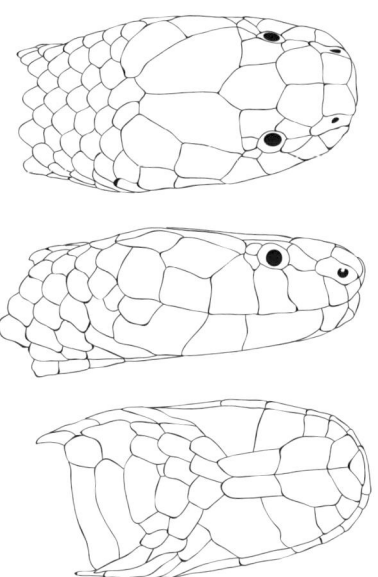

FIGURE 9.65. *Polemon collaris* USNM 570930. By T. Giri

FIGURE 9.66. *Polemon collaris*, Republic of Congo. By M. Burger

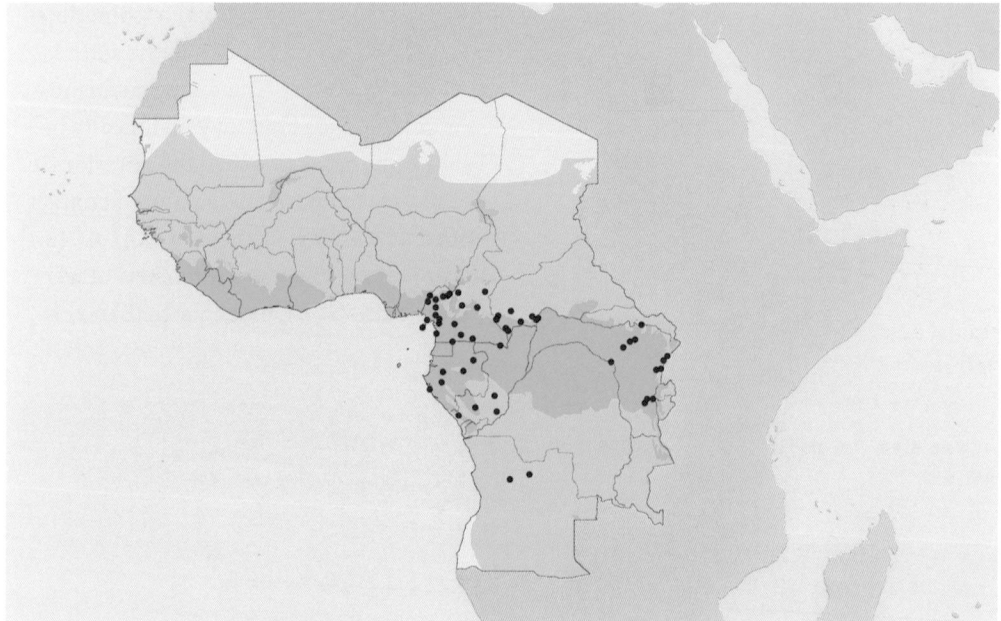

FIGURE 9.67. *Polemon collaris* (blue). By K. Jackson

in females) and 15–24 divided subcaudals (more than 17 in males, fewer than 21 in females). The cloacal scale is divided. The maximum recorded length for the species is 650 mm (Witte and Laurent 1947).

The number of lower labials in contact with the anterior chin shields may prove useful in identifying this species in the case of individuals for which the number is 3.

As many as three subspecies are rec-

ognized by some authors on the basis of the number of ventrals: *Polemon c. collaris* (Peters 1881); *P. c. brevior* (Witte and Laurent 1947); *P. c. longicor* (Witte and Laurent 1947).

African Snake-eater: *Polemon fulvicollis* (Mocquard 1887)

Polemon fulvicollis is a central African forest species with a distribution extending from Gabon to Uganda to the Democratic Republic of Congo. The type locality is Franceville, Gabon.

The body is slate gray above, with pale markings on the tip of the snout and the upper labials. There is a pale band across the nape and the underside is pale. The nasal is single. There are 2 postoculars. There are 7(4) or 7(3,4) upper labials, of which the sixth is usually the longest. The temporal formula is 1+1. There are 7(4) lower labials. There are 242–297 ventrals and 20–29 divided subcaudals. The cloacal scale is divided. The maximum length recorded for *P. fulvicollis* is 506 mm (Resetar and Marx 1981).

One subspecies, *P. f. fulvicollis* Mocquard 1887, is recognized by some authors on the basis of the numbers of ventrals and subcaudals and the ratio of tail length to total length.

Gabon Snake-eater: *Polemon gabonensis* (Duméril 1856)

Polemon gabonensis is a forest species with a distribution extending from Nigeria to Central African Republic to Gabon. The type locality is Gabon.

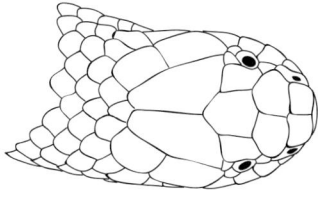
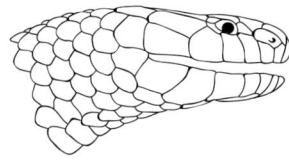
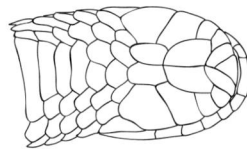

FIGURE 9.68. *Polemon fulvicollis* RMCA 1558. By T. Giri

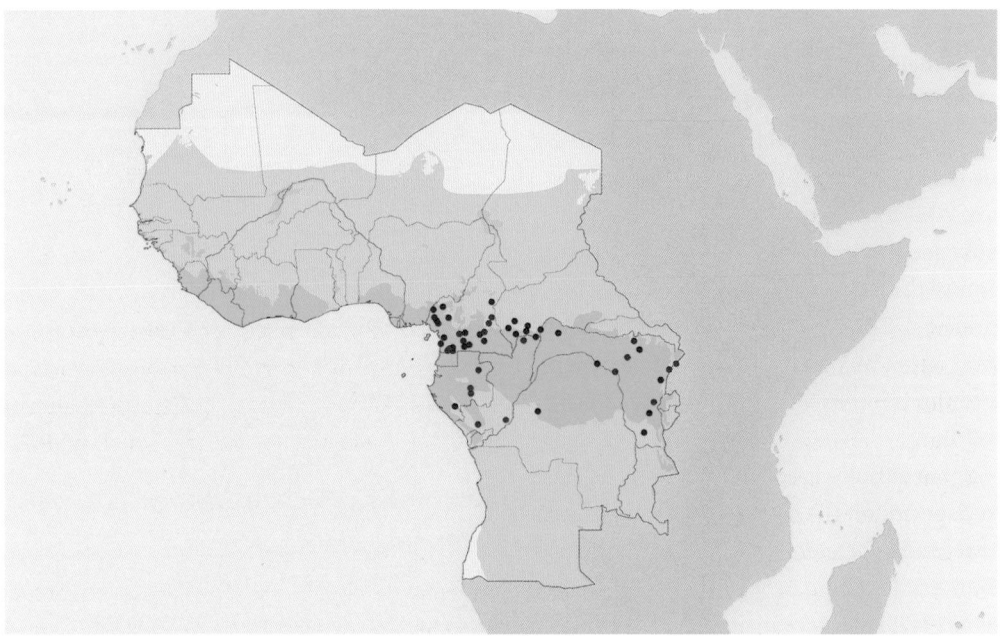

FIGURE 9.69. *Polemon fulvicollis* (red), *Polemon gabonensis* (blue). By K. Jackson

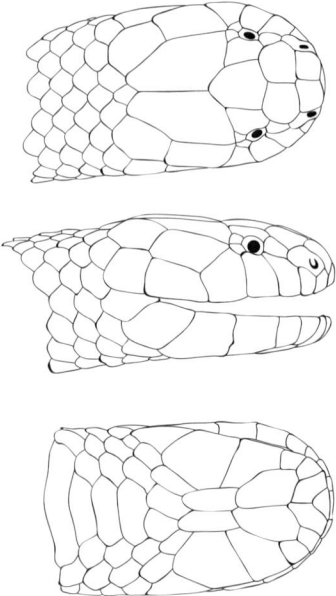

FIGURE 9.70. *Polemon gabonensis* RMCA 10545. By T. Giri

The coloration is dark above except for a pale band across the nape and often pale markings along the sides. The underside is pale gray. The nasal is usually single but may sometimes be semi-divided. There are 2 postoculars. There are 7(3,4) upper labials, of which the sixth is usually the longest. The temporal formula may be either 0+1 or 1+1. There are 7(4) lower labials. There are 208–252 ventrals (depending on the subspecies) and 16–26 divided subcaudals (more than 19 in males, fewer than 21 in females). The cloacal scale is divided. The maximum recorded length for *P. gabonensis* is an impressive 880 mm (Stucki-Stirn 1979).

Some authors recognize as many as three subspecies on the basis of numbers of ventrals and subcaudals: *Polemon g. gabonensis* Duméril 1856; *P. g. schmidtii* Witte and Laurent 1947; *P. g. brachyurus* Laurent 1960.

Slender Snake-eater: *Polemon gracilis* (Boulenger 1911)

Polemon gracilis is a forest species with a distribution extending from Nigeria to Gabon. The type locality is Bitye, Cameroon.

The body is brown above, some individuals having 3 pairs of white stripes the length of the body. There is a pale band across the nape, and the underside is pale. The nasal is single. There is 1 postocular. There are 6(3) upper labials, of which the fifth is usually the longest. The temporal formula is 1+1. There are 6(3) lower labials. There are 247–296 ventrals and 19–29 divided subcaudals. The cloacal scale is divided. The maximum length recorded for *P. gracilis* is 507 mm (Resetar and Marx 1981).

Two potentially useful characters for distinguishing *P. gracilis* from other species in the genus are the fact that the fifth upper labial is the longest, combined with the low number of lower labials in contact with the anterior chin shields.

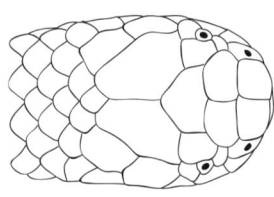

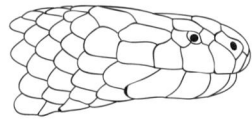

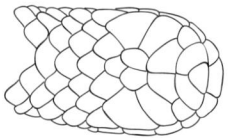

FIGURE 9.71. *Polemon gracilis* RMCA 28318. By T. Giri

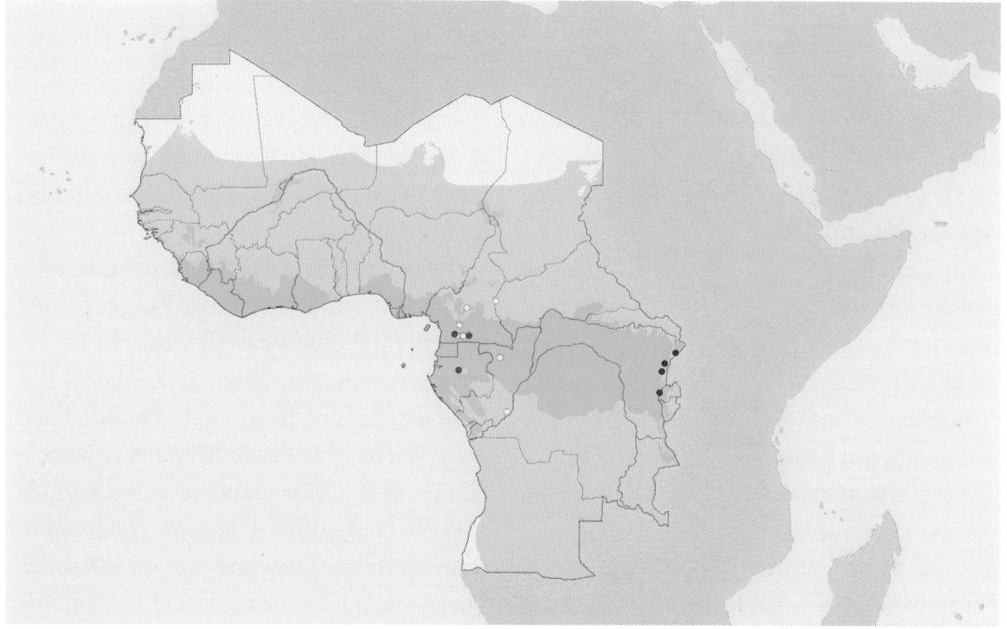

FIGURE 9.72. *Polemon gracilis* (red), *Polemon graueri* (blue), *Polemon griseiceps* (yellow). By K. Jackson

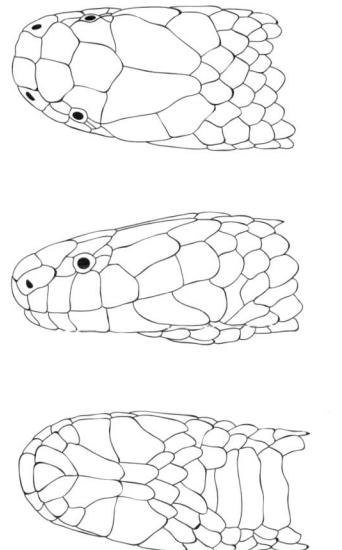

FIGURE 9.73. *Polemon graueri* FMNH 213699. By T. Giri

Central Lake Region Snake-eater: *Polemon graueri* Sternfeld 1908a

Polemon graueri is an east African species whose distribution extends from Uganda and Rwanda in the east to eastern Democratic Republic of Congo. The type locality is Entebbe, Uganda.

The body is uniformly blackish above except for a pale band across the neck. The underside is pale. The nasal may be single or divided. There are 2 postoculars. There are 7(3,4) upper labials, of which the fifth— rather than the sixth—is usually the longest. The temporal formula is 1+1. There are 7(4) lower labials. There are 222–262 ventrals (fewer than 240 in males, more than 228 in females) and 13–21 divided subcaudals (more than 15 in males, fewer than 18 in females. The cloacal scale is divided. The maximum length recorded for *P. graueri* is 560 mm (Spawls et al. 2004).

Cameroon Snake-eater: *Polemon griseiceps* (Laurent 1947)

Polemon griseiceps is a forest species with a distribution extending from Cameroon to the Congo to Central African Republic. The type locality is Bitye, Cameroon

The head and neck are black, while the rest of the body is gray above. The underside and the tip of the tail are whitish. The nasal is divided. There may be either 1 or 2 postoculars. There are 6(3,4) or 7(3,4) upper labials, of which the fourth is usually the longest. The temporal formula is 0+1. There are 7(4) lower labials. There are 178–200 ventrals and 20–25 divided subcaudals. The cloacal scale is divided. The maximum length recorded for *P. griseiceps* is 550 mm (Villiers 1966).

A useful character in distinguishing this species from others in the genus is the fact that the fourth upper labial is the longest.

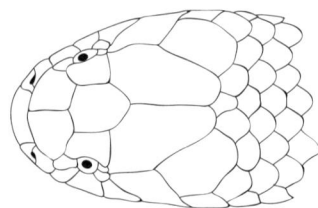

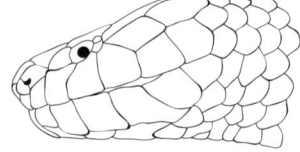

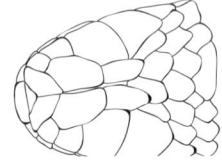

FIGURE 9.74. *Polemon griseiceps* BMNH 1946.1.21.90 (TYPE). By T. Giri

Ivory Coast Snake-eater: *Polemon neuwiedi* (Jan 1858)

Polemon neuwiedi is a West African species with a distribution extending from the Ivory Coast to Nigeria. The type locality is Accra, Ghana.

The coloration of this species is distinctive. The ground color above ranges from cream to light brown with three black stripes running the length of the body. The head and tail are dark, and the underside is pale. The nasal is single. There is 1 postocular. There are 7(3,4) upper labials, of which the fifth is usually the longest. The temporal formula is 0+1. There are 6(3) lower labials. There are 219–261 ventrals and 11–21 divided subcaudals. The cloacal scale is divided. The maximum recorded length for *P. neuwiedi* is 345 mm (Resetar and Marx 1981).

P. neuweidi is distinctive in several ways. Characters useful in identification include the striped color pattern, the fifth rather than the sixth upper labial being the longest, and the low number of lower labials in contact with the anterior chin shields

Peters's Snake-eater: *Polemon notatus* (Peters 1882)

Polemon notatus is a central African forest species with a distribution extending from Cameroon to the Democratic Republic of Congo. The type locality is Cameroon.

The body and head are brown above, with a broad pale band across the nape. The underside is pale. The nasal may be either single or divided. There is 1 postocular. There are 6(3,4) or 7(3,4) upper labials, of which the sixth is usually the longest. The usual temporal formula is 0+1, but it may technically be 1+1 in some cases in which there is a point of contact between the ante-

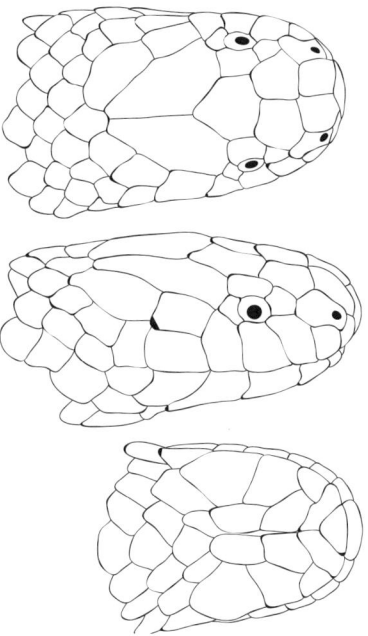

FIGURE 9.75. *Polemon neuweidi* FMNH 19267. By T. Giri

rior tip of the first temporal and the preocular. There are 6(3) or 7(3) lower labials. There are 178–228 ventrals and 17–27 divided subcaudals. The cloacal scale is divided. The maximum length recorded for *P. notatus* is 317 mm (Witte and Laurent 1947). The most useful character for distinguishing this species from most other in the genus is the low number of lower labials in contact with the anterior chin shields.

FIGURE 9.76. *Polemon neuwiedi*, Ghana. By S. Spawls

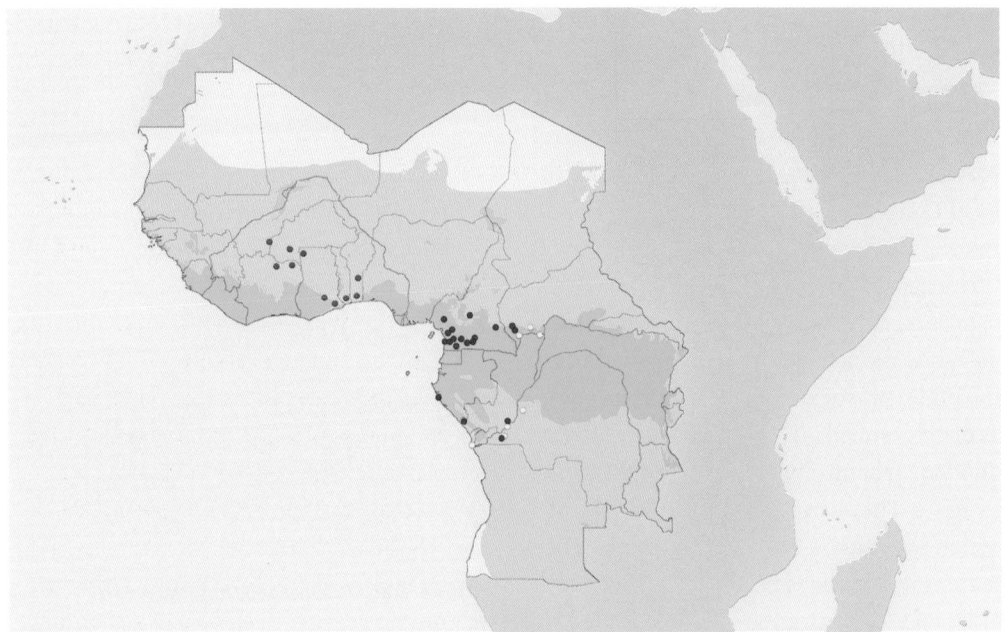

FIGURE 9.77. *Polemon neuwiedi* (red), *Polemon notatus* (blue), *Polemon robustus* (yellow). By K. Jackson

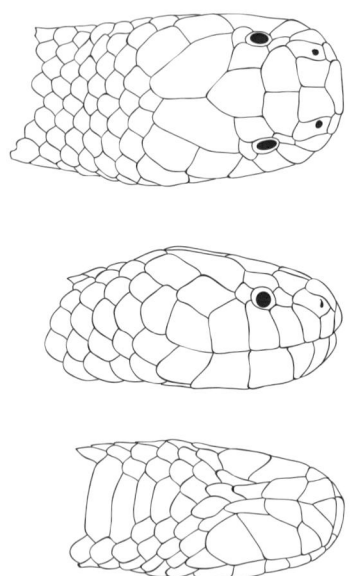

FIGURE 9.79. *Polemon notatus* RMCA A7-003-R-0035. By T. Giri

FIGURE 9.78. *Polemon notatus*, Congo. By M. Burger

Some authors recognize two subspecies on the basis of the number of ventrals: *P. n. notatus* (Peters 1882); *P. n. aemulans* (Werner 1902).

Zaire Snake-eater: *Polemon robustus* (Witte and Laurent 1947)

Polemon robustus is a central African species with a distribution extending from Central African Republic to the Democratic Republic of Congo. The type locality is Bolobo, Democratic Republic of Congo.

The head and body are dark above with a pale band across the nape. The underside is pale, but speckled with a darker shade on the underside of the head and on the edges of the ventrals. The nasal may be single, divided, or semi-divided. There are 2 postoculars. There are 7(3,4) upper labials, of which the sixth is usually the longest. The temporal formula is 1+1. There are 7(4) lower labials. There are 163–189 ventrals (fewer than 171 in males, more than 174 in females) and 17–27 divided subcaudals (more than 21 in males, fewer than 22 in females. The cloacal scale is divided. The maximum length recorded for *P. robustus* is 670 mm (Witte and Laurent 1947).

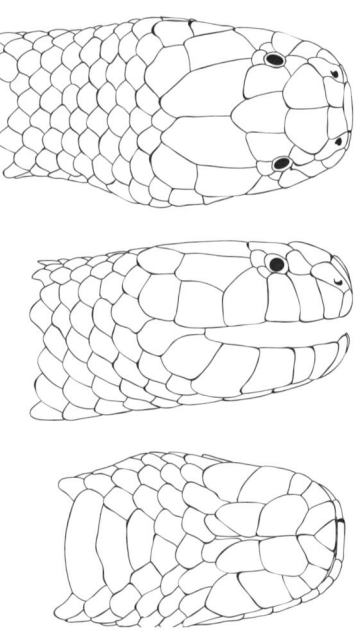

FIGURE 9.80. *Polemon robustus* RMCA 8761. By T. Giri

Quill-snouted Snakes: Genus *Xenocalamus* Günther 1868

Quill-snouted Snakes are a genus comprising five species. Although their distribution is primarily southern African, three species have ranges that extend into our zone. Their appearance is distinctive, with an enormously enlarged rostral scale extended to form a long, pointed snout, with a small, underslung lower jaw. Inhabiting sandy areas (Branch 1994), *Xenocalamus* differ from most other aparallactines in that they truly dig, as opposed to pushing their way through existing tunnels, and digging through sand is probably the primary function of the highly specialized snout. Fossorial and seldom encountered, Quill-snouted Snakes resemble other aparallactines in many ways. The head is small and the neck poorly defined. The eyes are small with round pupils. *Xenocalamus* have 4–6 unspecialized teeth on their relatively short maxilla in addition to an enlarged, grooved rear fang used for injecting venom into the worm lizards (Amphisbaenidae) they feed on. Broadley et al. (2003) report that when captured, *Xenocalamus mechowii* never attempts to bite but may try to stab with its sharp snout.

The head of Quill-snouted Snakes shows a great deal of fusion (or loss) of scales. The frontal is elongate and pushes the paired prefrontals apart, into the position of large preoculars. The internasals are paired. The nasal may be single or divided. The supraocular may be present or absent, and may be fused to one of the 1 or 2 small postoculars. There are no suboculars, and, as with all aparallactines, no loreal. There is no anterior temporal. There are 5 or 6 upper labials and 5–7 lower labials. The submandibular

groove is not pronounced. The chin shields present a puzzle because there is clearly an anterior pair of chin shields on all individuals. But the position of the gulars that follow them varies from individual to individual, sometimes presenting as a pair of scales that meet the definition for posterior chin shields (they are in contact with each other at the midline and with the lower labials on the sides), but sometimes presenting as what are clearly gulars. Quill-snouted Snakes are therefore considered to have just 1 pair of chin shields, followed by a pair of "false chin shields" in some individuals. The subcaudals and cloacal scale are always divided. The hemipenes are unforked with the sulcus spermaticus either unforked (Witte and Laurent 1947) or forked just near the tip. *X. bicolor* lays 2–4 fairly large eggs, and there is no correlation between the size of the mother and the size of the clutch (Shine et al. 2006a).

Most of what little is known of the natural history of Quill-snouted Snakes comes from studies of *X. bicolor*. *X. bicolor* (and probably other species in the genus as well) feeds almost exclusively on worm lizards (amphisbaenians). Like most lizards, worm lizards can drop their tails as a means of escape. Thus a predator that grabs a lizard from behind, by the tail, may end up with nothing but a tail, and the rest of the lizard will escape. Dropping of tails by prey is a problem for aparallactines in general, but worm lizards present a special challenge because they are almost cylindrical in shape rather than tapered, as most lizards are, making it difficult to squeeze past the potential prey in order to grasp it midbody rather than by the tail. It is here that the quill-snout of Quill-snouted Snakes comes into play because its pointed, elon-

Key to Central and Western African Species of the Genus *Xenocalamus*

1	21 dorsal scale rows	*X. michelli*
1'	17 dorsal scale rows	2
2(1)	Supraocular present (may be fused to postocular)	*X. bicolor*
2'	Supraocular absent	*X. mechowii*

gate shape allows *Xenocalamus* to squeeze by a worm lizard even in a narrow passage (Shine et al. 2006a). The quill-snout has other uses as well. Spawls (pers. comm.) reports that *Xenocalamus* can tear through the seams of snake bags with their pointed rostral.

Slender Quill-snouted Snake:
Xenocalamus bicolor Günther 1868

The range of the Slender Quill-snouted Snake extends from Namibia to northern Zimbabwe to Mozambique and northern South Africa. The type locality originally recorded as Zambezi (in error) has subsequently been corrected to Damaraland, Namibia (Broadley 1971b).

The coloration of *Xenocalamus bicolor* is highly variable. Above, the body may be white spotted with black, with or without pale stripes or uniformly black. The underside is white. The nasal is divided. A supraocular, longer than broad, is present and is sometimes fused to the postocular. There is only 1 small postocular, and it may appear absent if fused to the supraocular. There are 6(3,4) upper labials. The temporal formula is 0+1. There are 5(3) or occasionally 4(3) lower labials. The dorsal scales are arranged in 17 straight rows. There are 198–256 ventrals (fewer than 235 in males, more than 215 in females) and 20–36 subcaudals (more than 27 in males, fewer than 30 in females).

The maximum recorded length for *X. bicolor* is 719 mm (FitzSimons 1962).

Elongate Quill-snouted Snake:
Xenocalamus mechowii Peters 1881

The range of the Elongate Quill-snouted Snake extends from the Congo to Namibia and Zimbabwe. The type locality is Malanje, Angola.

The body is white above, with dark brown or black blotches that may be arranged in two distinct lines along the paraventral rows, or that may merge into a less organized pattern. The nasal is divided. The supraocular is absent, a trait unique among the *Xenocalamus* species within our zone, but there are 1 or 2 small postoculars. There are 6(3,4) upper labials. The temporal formula is 0+1. There are 5(3) lower labials. The dorsal scales are arranged in 17 rows. There are 201–248 ventrals (fewer than 231 in males, more than 220 in females) and 25–37 subcaudals (more than 29 in males, fewer than 31 in females. The maximum length recorded for *Xenocalamus mechowii* is 725 mm (Broadley 1971b).

Michell's Quill-snouted Snake:
Xenocalamus michelli Müller 1911

Michell's Quill-snouted Snake is known only from three female specimens, and it is apparently restricted to southern Democratic Republic of Congo. The type locality

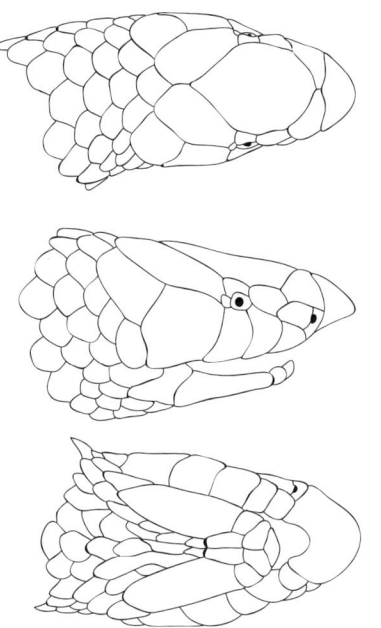

FIGURE 9.81. *Xenocalamus bicolor* FMNH 73477. By T. Giri

FIGURE 9.82. *Xenocalamus bicolor* (spotted phase), Botswana. By S. Spawls

FIGURE 9.83. *Xenocalamus bicolor* (striped phase), Botswana. By S. Spawls

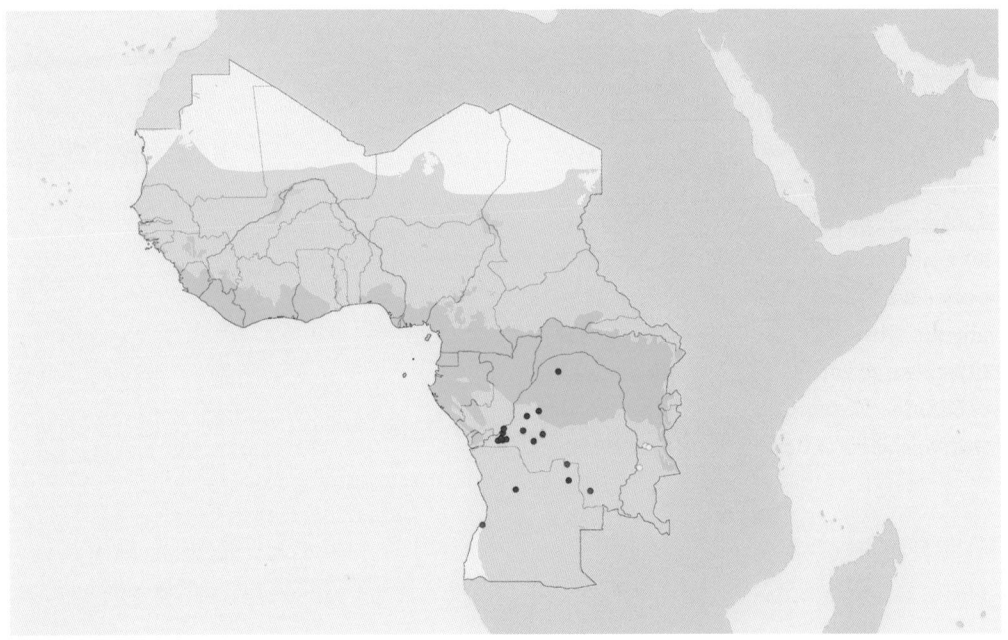

FIGURE 9.84. *Xenocalamus bicolor* (red), *Xenocalamus mechowii* (blue), *Xenocalamus michelli* (yellow). By K. Jackson

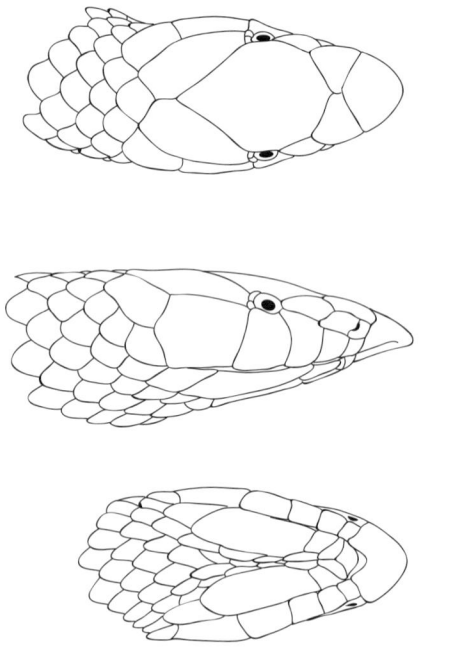

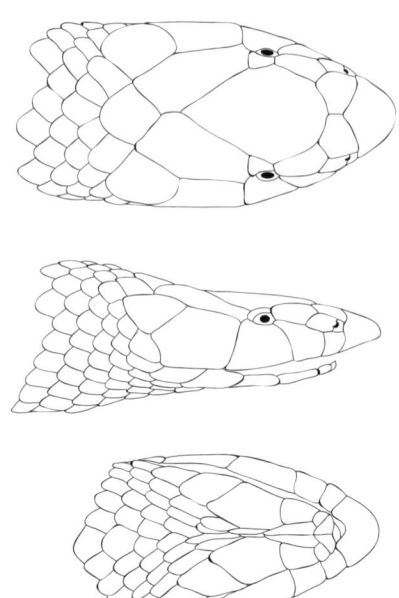

FIGURE 9.85. *Xenocalamus mechowii* RMCA 11763. By T. Giri

FIGURE 9.86. *Xenocalamus michelli* RMCA 18247.

is Kituri, Katanga Province, Democratic Republic of Congo.

Xenocalamus michelli is usually uniformly black, though the underside is sometimes pale. The nasal may be either single or divided. A supraocular is present, and since it is always fused to the single postocular, the postocular is effectively absent. There are 5(2,3) upper labials. The temporal formula is 0+1. There are 7(4) lower labials. The dorsal scales are arranged in 21 rows, this trait setting it apart from the other two species within our range. There are 248–261 ventrals and 27–29 subcaudals. Note that both these counts are based only on female specimens. The maximum length recorded for *X. michelli* is 1,050 mm (Müller 1911).

FIGURE 9.87. *Xenocalamus michelli*, Democratic Republic of Congo. By E. Greenbaum

10

Family Lamprophiidae

Subfamilies Lamprophiinae, Pseudoxyrhophiinae, and Pseudaspidinae

Subfamily Lamprophiinae

The subfamily Lamprophiinae represents a major radiation of nonvenomous snakes in sub-Saharan Africa. Made up of 68 species in 11 genera in total, the Lamprophiinae is represented within our zone by 40 species in 9 genera inhabiting terrestrial, arboreal, and semiaquatic habitats. The Lamprophiinae corresponds to Bogert's (1940) Groups 1 and 2, assigned to those groups on the basis of the presence of vertebral hypapophyses in the posterior part of the body, a forked sulcus spermaticus, and aglyph dentition. Recent molecular phylogenies find the lamprophiids of the subfamily Lamprophiinae to be most closely related to the lamprophiid subfamilies Pseudaspidinae, a lineage of only a few species (Pyron et al. 2013), and Pseudoxyrhophiinae, a large radiation of snakes in Madagascar but with a few representatives occurring within our zone (Vidal et al. 2008; Pyron et al. 2013). We therefore include the subfamilies Pseudoxyrhophiinae and Pseudaspidinae in this chapter with the Lamprophiinae.

African House Snakes: Genus *Boaedon* Duméril, Bibron, and Duméril 1854b

The genus *Boaedon* includes six species, four of which occur within our zone. *Boaedon* have in the past been included within the genus *Lamprophis* (seven species not including *Boaedon*). House Snakes (*Boaedon* + *Lamprophis*) are terrestrial snakes found throughout sub-Saharan Africa in a variety of habitats, including semi-desert, savannah, and forest. They are often common, and many species seem to adapt well to human-altered habitats. House Snakes are nocturnal or crepuscular and often found by day hiding under rocks, logs, debris,

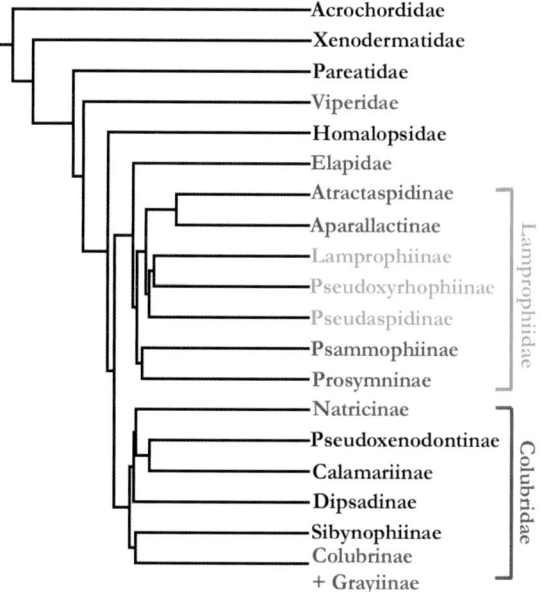

FIGURE 10.1. Higher-level phylogeny of caenophidian families and subfamilies, with families and subfamilies occurring within our zone highlighted (red or orange). The subfamilies Lamprophiinae, Pseudaspidinae, and Pseudoxyrhophiinae, the topic of this chapter, are highlighted in orange. Modified by K. Jackson from Zheng and Wiens (2016)

piles of building materials, and the like. *Boaedon* are dietary generalists, preying on small mammals, lizards, frogs, and nestling birds. Akani et al. (2008) compared diets of the four *Boaedon* species occurring in our zone in the outskirts of towns in Nigeria. Of these four species, they found *B. fuliginosus* to be the most truly generalist in its diet, with *Boaedon lineatus* showing a preference for lizards and frogs, *B. virgatus* a preference for rodents and shrews, and *B. olivaceus* for nestling birds and small mammals. The maxillary dentition is aglyph, with a series of 6 anterior teeth, increasing in size posteriorly, followed after a diastema by a series of 11–16 equal-sized posterior teeth. House Snakes lay eggs, with clutches of 6–20 in *B. fuliginosus* (Nägele 1985).

The head is medium sized with a rounded snout. The neck is not well defined. The tail is short. The eye is small with a vertically elliptical pupil. The nasal is divided or semi-divided. The internasals and prefrontals are paired. The loreal is present. There are 1–2 preoculars and 1–2 postoculars. There is a supraocular but no suboculars. There are 8–11 upper labials and 8–10 lower labials. There are 1–2 anterior temporals. There are 2 pairs of chin shields, the anterior pair longer than the posterior pair. The submandibular groove is pronounced. The dorsal scales are smooth, with 1–2 apical pits (though these may in some cases be difficult to discern), and arranged in 23–35 straight rows. The vertebral row is not enlarged relative to the other dorsal scale rows. There are 168–246 ventrals. There are 38–85 subcaudals, which may be single or divided. The hemipenes are bilobed with a forked sulcus spermaticus.

Brown House Snake: *Boaedon fuliginosus* (Boie 1827)

The Brown House Snake is found throughout Africa, with a distribution extending from Morocco to Somalia to South Africa. The type locality is "Java" (in error).

The dorsum is dark brown, generally without any pattern or stripes. Note, however, that individuals of *Boaedon fuliginosus* from southern Africa have stripes on the head similar to *B. lineatus* Thorpe and McCarthy (1978). The venter is pale gray. The upper labials are paler than the rest

Key to Central and Western African Species of the Genus *Boaedon*

1	Subcaudals single	2
1'	Subcaudals divided	3
2(1)	3 upper labials in contact with the eye	*B. olivaceus*
2'	2 upper labials in contact with the eye	*B. radfordi*
3(1)	23 or 25 dorsal scale rows; preocular not in contact with the frontal	*B. virgatus*
3'	25 to 35 dorsal scale rows; preocular in contact with the frontal	4
4(3)	Frontal distinctly shorter than the parietals	*B. fuliginosus*
4'	Frontal slightly shorter than the parietals	*B. lineatus*

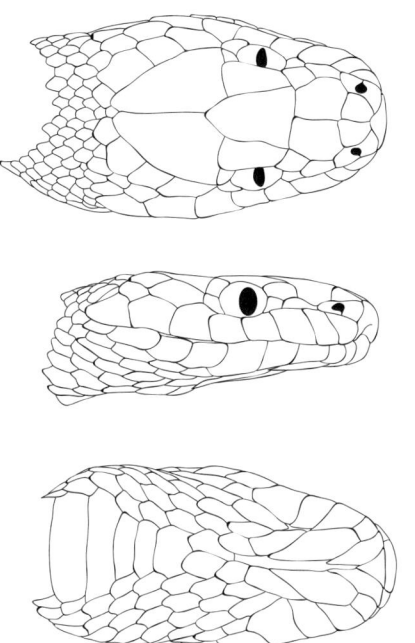

of the head. The nasal is semi-divided. The length of the loreal is 2.5 to 3 times its depth. There are usually 2 preoculars, the superior larger than the inferior, but these are sometimes fused, forming a single preocular. There are 2 postoculars of equal size. There are 8(4,5) or 9(4,5) upper labials, sometimes with the third upper labial also in contact with the eye. The temporal formula is usually 1+2, sometimes 2+2. There are 8(3) to 11(4) lower labials. The dorsal scales are

FIGURE 10.2. *Boaedon fuliginosus* RMCA 78-29-R-1. By T. Giri

FIGURE 10.3. *Boaedon fuliginosus*, Republic of Congo. By K. Jackson

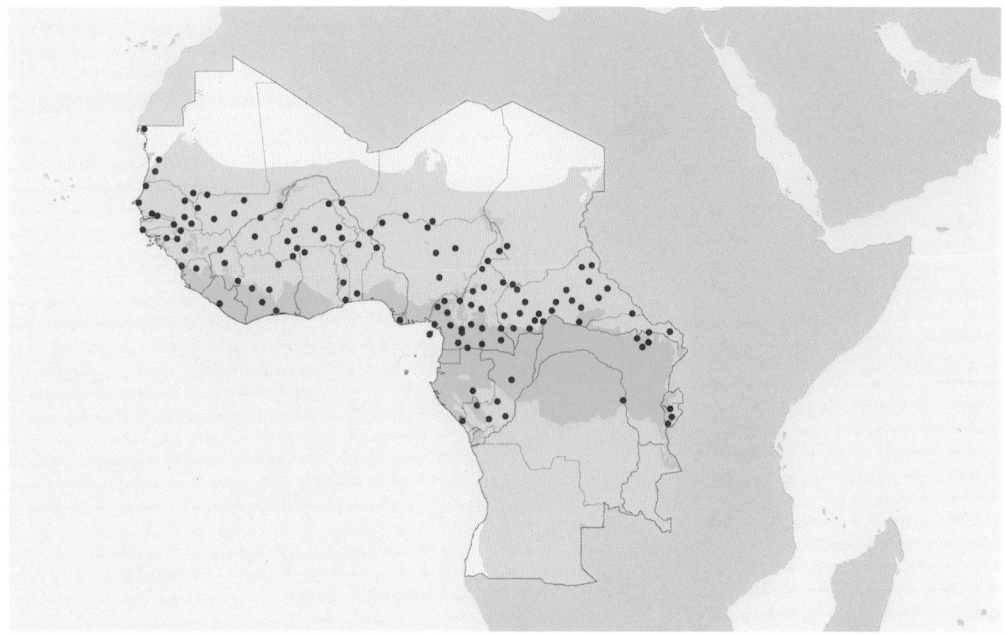

FIGURE 10.4. *Boaedon fuliginosus* (blue). By K. Jackson

smooth, with 2 apical pits, and arranged in 27–33 (occasionally 25) straight rows. There are 185–246 ventrals (fewer than 231 in males, more than 207 in females) and 41–85 divided subcaudals (more than 47 in males, fewer than 60 in females). The cloacal scale is single. The maximum recorded length for this species is 1,110 mm (Laurent 1956).

Three subspecies are recognized by some authors: *B. f. fuliginosus* (Boie 1827); *B. f. mentalis* (Günther 1888); *B. f. bedriagae* (Boulenger 1906). *B. f. mentalis* was described as a distinct subspecies with a range extending from southern Angola to South Africa, distinguished by an unusually pronounced submandibular groove. But this character has been found to be variable, and the subspecies is disputed by Bogert (1940) and Roux-Estève and Guibé (1965a, 1965b). *B. f. bedriagae* is endemic to the island of São Tomé and is distinguished by a higher subcaudal count than *B. f. fuliginosus* (more than 72 in males, more than 66 in females).

Striped House Snake: *Boaedon lineatus* (Duméril, Bibron and Duméril, 1854b)

The Striped House Snake has a large distribution in sub-Saharan Africa, from Senegal in the west to Kenya and Tanzania in the east, and south to South Africa. The type locality is "Gold Coast," meaning Ghana.

The dorsum may range from dark gray

FIGURE 10.5. *Boaedon lineatus*, Senegal. By W. Wüster

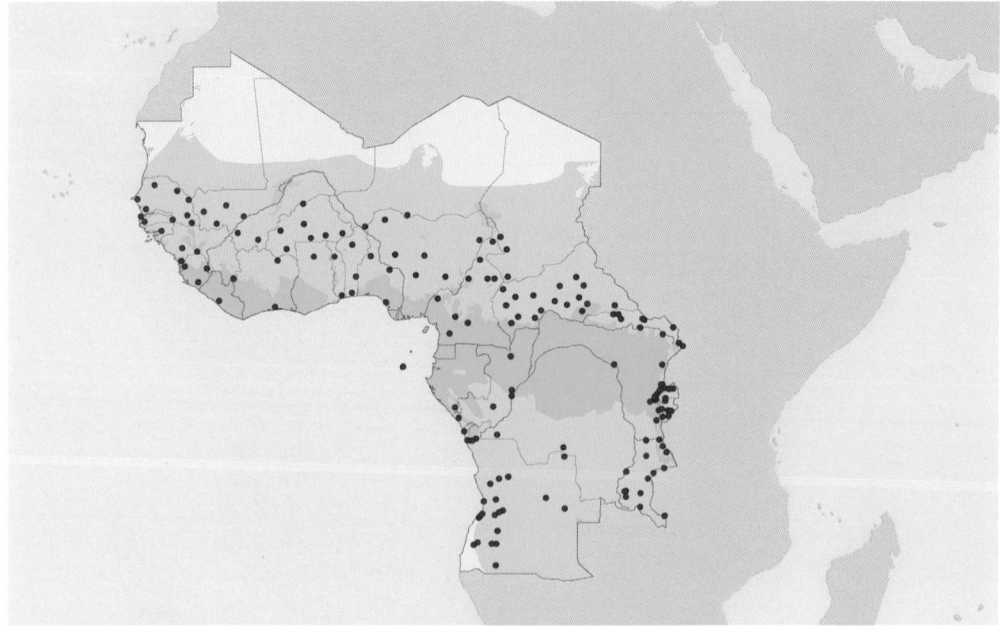

FIGURE 10.6. *Boaedon lineatus* (blue). By K. Jackson

to reddish brown. The venter is pale gray. There are 2 yellow stripes along each side of the head. Thorpe and McCarthy (1978) showed that *Boaedon lineatus* is a distinct species in west Africa rather than a color morph of *B. fuliginosus*, but in southern Africa, *B. fuliginosus* takes on the color pattern of *B. lineatus*. The result of this is that the two species have frequently been confused in the literature, and references to either species may therefore be erroneous.

The nasal is divided. The length of the loreal is roughly twice its depth. There are usually 2 preoculars, the superior larger than the inferior, but these are sometimes fused, forming a single preocular. There are 2 postoculars of equal size. There are 8(4,5) or 9(4,5) upper labials. The temporal formula is 1+2. There are 9(4) to 11(4) lower labials. The dorsal scales are smooth, without apical pits, and arranged in 25–35 straight rows. There are 181–250 ventrals and 40–71 divided subcaudals. The cloacal scale is sin-gle. The maximum recorded length for this species is 1,200 mm (Villiers 1975).

Olive House Snake: *Boaedon olivaceus* (Duméril 1856)

The range of *Boaedon olivaceus* extends from Guinea to Uganda to Bioko Island (Equatorial Guinea). The type locality is Gabon.

The dorsum is uniformly gray or black. The venter is pale, sometimes with black

FIGURE 10.7. *Boaedon olivaceus*, Republic of Congo. By K. Jackson

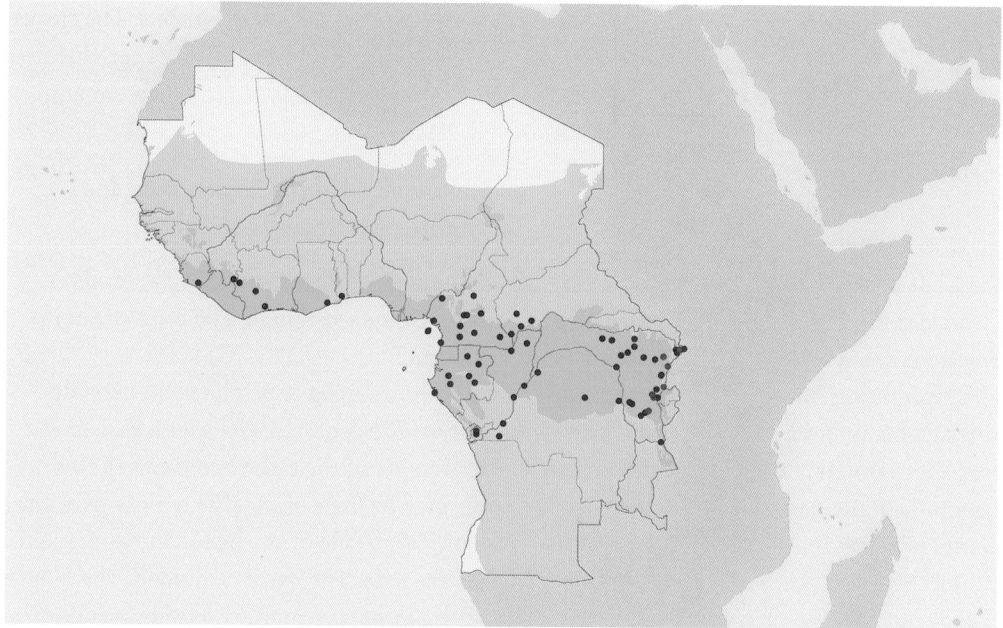

FIGURE 10.8. *Boaedon olivaceus* (blue), *Boaedon radfordi* (red). By K. Jackson

spots. In life, the Olive House Snake is easily recognized by its orange eyes, a feature shared only by *B. radfordi* (see below). The nasal is divided. The length of the loreal is roughly twice its depth. There is usually 1 preocular, sometimes 2, the superior larger than the inferior. The inferior preocular may occasionally be fused to the loreal. There are 2 postoculars, the inferior larger than the superior. There are 8(3,4,5) or 9(3,4,5) upper labials. The temporal formula is usually 1+2, sometimes 2+2. There are 8(4) to 10(4) lower labials. The dorsal scales are smooth, with 0 or 1 apical pits, and arranged in 25–29 (occasionally as many as 31) straight rows. There are 185–222 ventrals (fewer than 211 in males, more than 199 in females). There are 38–63 single subcaudals (more than 49 in males, fewer than 51 in females). The cloacal scale is single. The maximum length recorded for this species is 900 mm (Stucki-Stirn 1979).

FIGURE 10.9. *Boaedon radfordi*, Democratic Republic of Congo. By E. Greenbaum

Radford's House Snake: *Boaedon radfordi* Greenbaum, Portillo, Jackson, and Kusamba 2015

Radford's House Snake is a cryptic species recently described from the Albertine Rift in eastern Democratic Republic of Congo. The type locality is Shatuma-Abis village, Lendu Plateau, Orientale Province, Democratic Republic of Congo.

    Boaedon radfordi most closely resembles *B. olivaceus*, having undivided subcaudals and orange eyes. Molecular evidence indicates that *B. radfordi* diverged from *B. olivaceus* approximately 12 mya. *B. radfordi* has generally higher ventral counts than *B. olivaceus* and 2 (rather than 3) upper labials in contact with the eye. The dorsum is uniformly glossy olive gray, gray, or brown. The grayish dorsal coloration extends ventrally

to the lateral edges of the ventral scales. The ventrals are otherwise yellowish. The underside of the tail is dark gray. There are 1–2 preoculars and 2 postoculars. There are 8(4,5) or 9(4,5) upper labials. There are 9–10 lower labials. The dorsal scales are smooth and arranged in 27–31 straight rows. There are 200–226 ventrals (fewer than 220 in males, more than 211 in females). There are 37–56 single subcaudals (more than 41 in males, fewer than 50 in females). The cloacal scale is single. The maximum length recorded for this species is 801 mm (Greenbaum et al. 2015).

Hallowell's House Snake: *Boaedon virgatus* (Hallowell 1854b)

The range of *Boaedon virgatus* extends from Guinea to Congo. The type locality is Liberia.

    The dorsum is gray. The venter is pale gray at the center, at the edges, and on the underside of the tail. Sometimes the venter may be dark orange. There is a pale line along each side of the head. The upper labials are pale. The nasal is divided. The length of the loreal is roughly 1.5 times its depth. There are usually 2 preoculars, the superior larger than the inferior, but these are sometimes fused, forming a single preocular. There are 2 postoculars, the inferior equal to or larger than the superior. There are 8(4,5) or occasionally 8(3,4,5) upper labials. The temporal formula is 1+2. There are 7(4) to 8(5) lower labials. The dorsal scales are smooth, without apical pits, and arranged in 23 or occasionally 25 straight rows. There are 186–223 ventrals (fewer than 203 in males, more than 203 in females) and 42–64 divided subcaudals (more than 50 in males, fewer than 51 in females). The cloacal scale is single. The maximum length recorded for this species is 915 mm (Villiers 1966).

Genus *Bothrolycus* Günther 1874

Loreal-pitted Snake: *Bothrolycus ater* Günther 1874

Bothrolycus is a monotypic genus. The Loreal-pitted Snake, *Bothrolycus ater*, has a distribution that extends from Cameroon to the Democratic Republic of the Congo and includes Bioko Island (Equatorial Guinea). The type locality is Cameroon.

B. ater is a rarely encountered forest leaf-litter species, and little is known about its natural history. An unusual feature of this species is a deep pit present on each side of the head, between eye and nostril, superficially reminiscent of the thermosensory facial pits of pit vipers and pythons. Al Savitzky (pers. comm.) dissected the loreal region of a specimen to determine whether it resembled that of a rattlesnake, with inconclusive results. He found that in *B. ater* there was no prominent nerve supply to the pit and no rear chamber behind the pit as there is in a rattlesnake. The lining of the pit (the "membrane" in the pit organ of a pit viper) was just the thin preocular scale. Dissection of the gut revealed a large frog in one specimen, which is inconsistent with the thermoreceptive pit hypothesis (one would expect it to be associated with locat-

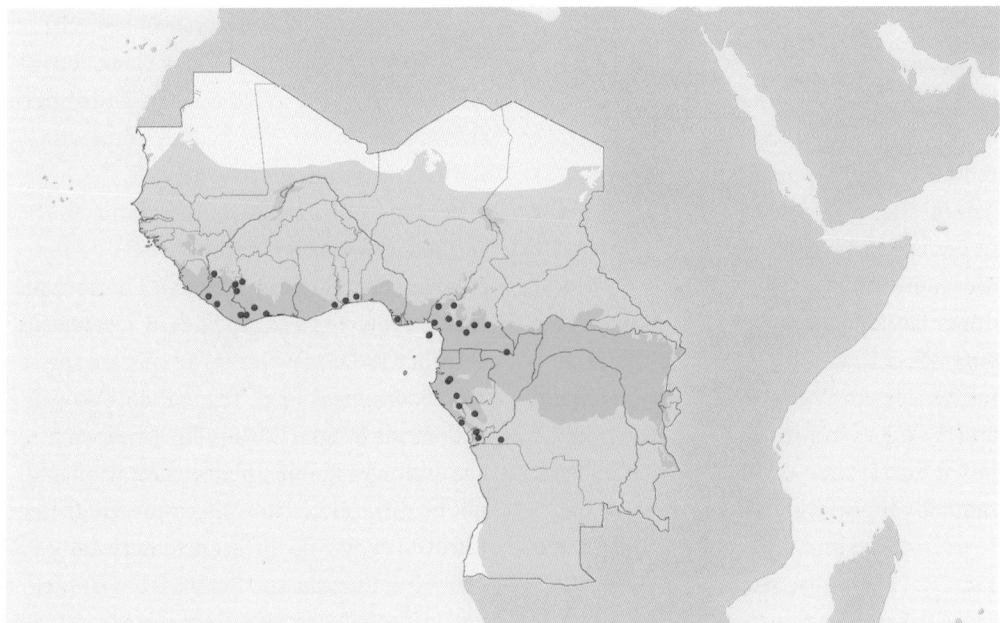

FIGURE 10.10. *Boaedon virgatus* (red). By K. Jackson

ing endothermic prey), though it does not rule out the possibility that the loreal pit of *Bothrolycus* serves a thermosensory function. Within the Lamprophiinae, the genus *Bothrolycus* is thought to be most closely related to the genus *Bothrophthalmus* (Kelly et al. 2011). *Bothrophthalmus* also has a concave loreal region, though to a lesser extent than *Bothrolycus*.

In adults, the body is blackish above, and the underside light brown with pale spots. The head is a lighter color than the body, and the lips and throat are marked with bright white spots outlined in black. In juveniles, the head and nape are cream colored. The body is dark above with pale bands. The head is small and somewhat flattened, with a well-defined neck. The eye is small with a round pupil. The nasal is divided. The internasals and prefrontals are paired. There are 2 preoculars and 2 postoculars. Some confusion surrounds the condition of the loreal because the loreal region is disrupted by the presence of a pit of unknown function between eye and nostril. The loreal scale is usually considered to be absent, though the lower preocular may resemble an elongate loreal in cases in which the upper preocular extends behind the lower preocular and contacts an upper labial. The temporal formula is 1+2 or occasionally 2+2. There are 7(4,5) or 8(4,5) upper labials and 7–9(4) lower labials. The anterior chin shields are longer than the posterior pair. The dorsal scales are smooth and arranged in straight rows. The number of dorsal scale rows is unusual in being sexually dimorphic, with 17 dorsal scale rows in males and 19 in females. There are 132–152 ventrals (fewer than 148 in males, more than 142 in females) and 17–34 divided subcaudals (more than 26 in males, few-

er than 23 in females). The cloacal scale is single. The maximum recorded length for *B. ater* is 702 mm (Witte 1962). The maxillary dentition consists of 20 ungrooved teeth that decrease in size posteriorly. The hemipenes are bilobed, with a forked sulcus spermaticus.

Genus *Bothrophthalmus* Peters 1863

Red-black Striped Snake: *Bothrophthalmus lineatus* (Peters 1863)

Bothrophthalmus is a monotypic genus. The Red-black striped Snake, *Bothrophthalmus lineatus*, has a distribution extending from Guinea to Uganda. The type locality is "Guinea" but presumably meaning "Guinea Coast" rather than the country, Guinea. It is likely that the true type locality is somewhere in Ghana. Among the Lamprophiinae, the genus *Bothrophthalmus* is thought to be most closely related to the genus *Bothrolycus* (Kelly et al. 2011).

B. lineatus is easily recognizable throughout most of its range by its dramatic coloration. The body is brown or black above with 1 to 5 yellow to red stripes. The underside is pink to red. The head is often white with dark stripes on the sides.

In Cameroon, Gabon, and Equatorial Guinea, *B. lineatus* is represented by a unicolored form (lacking stripes). This is sometimes treated as a subspecies, *B. l. brunneus* Günther 1863a, or even as a separate species, *B. brunneus* (e.g., Trape 1985; Pauwels and Vande Weghe 2008). The two forms are distinguishable only by coloration and not by differences in scale count. Hughes (2000) reports no differences between striped *B. lineatus* specimens from Guinea and striped specimens from the eastern part of its distribution, and points out that if

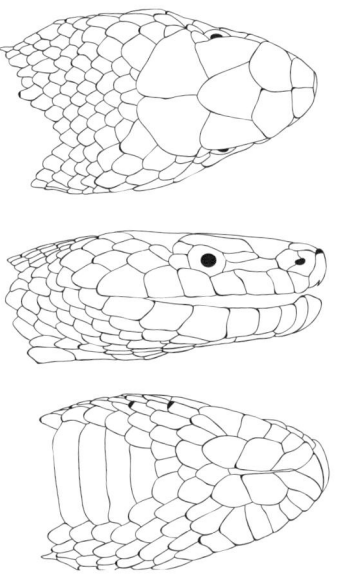

FIGURE 10.11. *Bothrolycus ater* RMCA 9276. By T. Giri

FIGURE 10.12. *Bothrolycus ater*, Republic of Congo. By K. Jackson

FIGURE 10.13. *Bothrolycus ater*, Republic of Congo. By K. Jackson

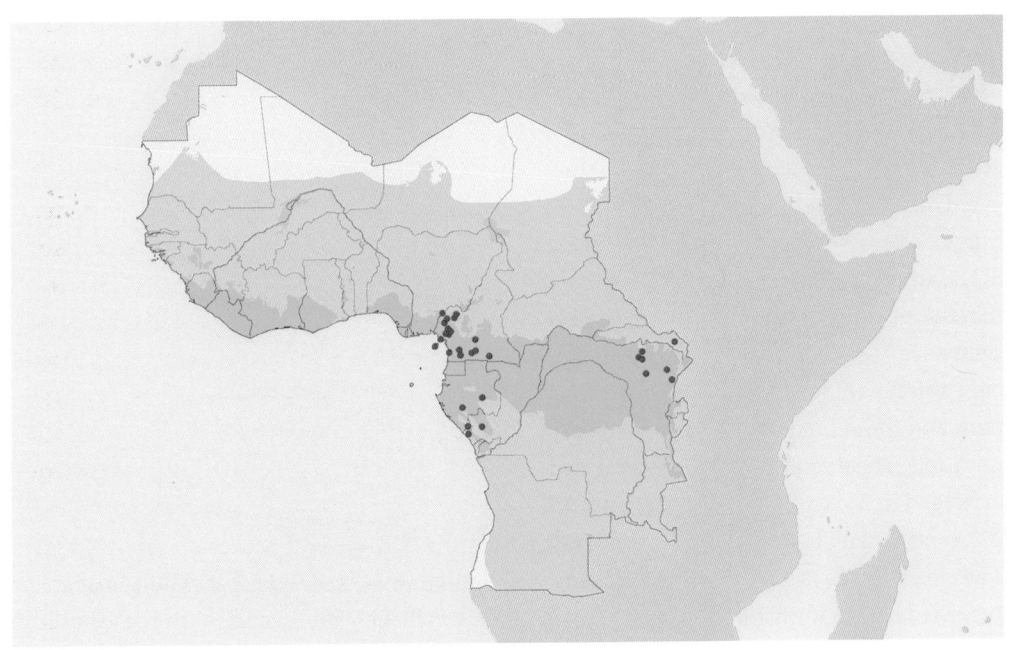

FIGURE 10.14. *Bothrolycus ater* (red). By K. Jackson

the *brunneus* form is considered a separate species, or subspecies, it disrupts the distribution of the striped form, meaning that the nominate subspecies would have to be occurring disjunctly.

Luiselli et al. (1999) extensively surveyed *B. lineatus* in Nigeria, where they found it to be a rare species relative to other snakes, occurring most often in leaf litter on the forest floor in dense primary and secondary forest. *B. lineatus* were found to be more active during the wet season than the dry season. Males were collected more frequently than females. Females lay eggs during the wet season (June in Nigeria). Clutch size ranged from 3 to 5 eggs, with larger females producing larger clutches. Subadult Red-black striped Snakes were found to feed on both scincid lizards and on small mammals, while adults fed exclusively on small mammals, primarily rodents but also insectivores.

The eye is small with a round or slightly vertically elliptical pupil. The nasal is single or semi-divided. The internasals and prefrontals are paired. The loreal is elongate and concave. There are usually 2, but occasionally 3, preoculars and 2 postoculars. Sometimes a tiny subocular is present near the lower anterior corner of the eye. The temporal formula is usually 2+3 but may be 1+3 or 3+3. There are 7(4,5) or 8(4,5) upper labials, with only a point of the fourth upper labial in contact with the eye. There are 7–9(4) lower labials. The anterior chin shields are longer than the posterior pair, and the submandibular groove is pronounced. The dorsal scales are keeled and arranged in 23 straight rows. There are 181–212 ventrals and 62–85 divided subcaudals. The cloacal scale is single. The maximum recorded length for *B. lineatus* is 1,280 mm

(Pitman 1974). Luiselli et al. (1999) found the average length of specimens collected in Nigeria to be 620 mm for males and 700 mm for females, noting that males had shorter bodies but longer tails than females. The maxillary dentition consists of 21 ungrooved teeth, the 6 anteriormost enlarged relative to the posterior maxillary teeth. The hemipenes are bilobed, with a forked sulcus spermaticus.

Banded Snakes: Genus *Chamaelycus* Boulenger 1919b

Chamaelycus is a genus of three species of small snakes that live in forest leaf litter, all of which occur within our zone. Little is known of their natural history, but the diet of *Chamaelycus fasciatus* is reported to include earthworms (Knoepffler 1966), insects (Villiers 1975), reptile eggs (Villiers 1951), and lizards (Ineich 1998).

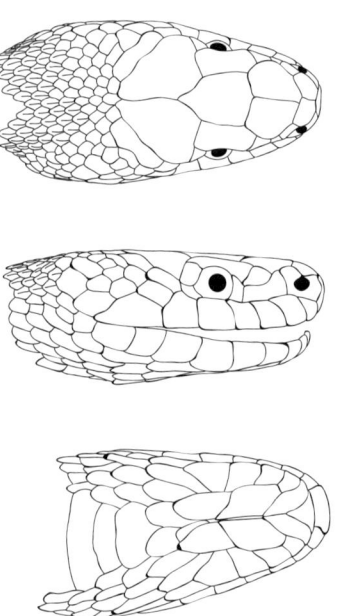

FIGURE 10.15. *Bothrophthalmus lineatus* RMCA 75-8-R-40. By T. Giri

FIGURE 10.16. *Bothrophthalmus lineatus*, Democratic Republic of Congo. By J. Kielgast

FIGURE 10.17. *Bothrophthalmus lineatus*, unicolored form, Republic of Congo. By M. Burger

FIGURE 10.18. *Bothrophthalmus lineatus*, Ivory Coast. By M.-O. Rödel

FIGURE 10.19. *Bothrophthalmus lineatus*, unicolored form, Cameroon. By M.-O. Rödel

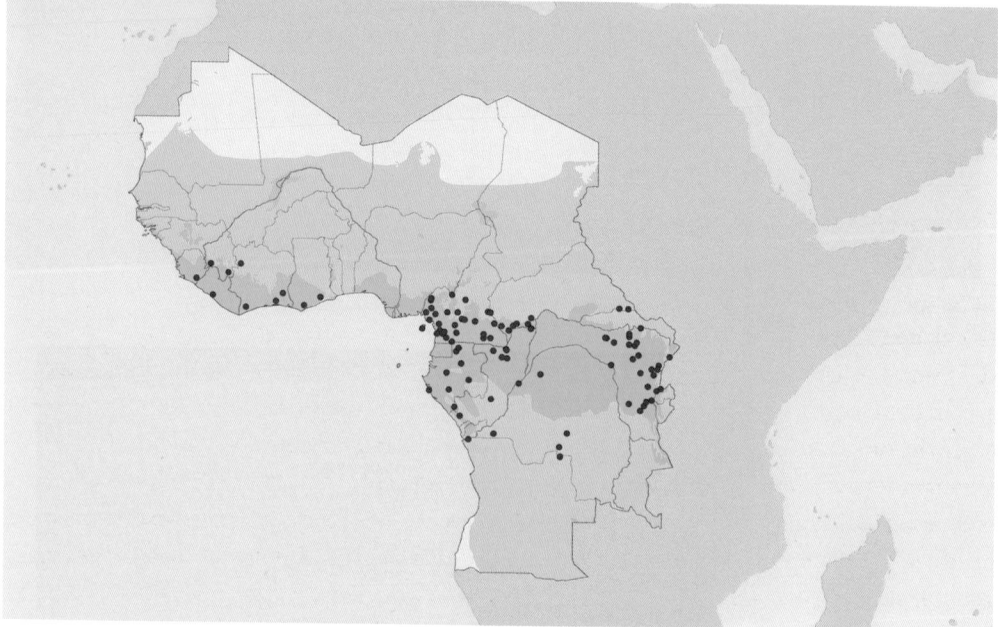

FIGURE 10.20. *Bothrophthalmus lineatus* (blue). By K. Jackson

The head is small and flat, the neck poorly defined. The snout is rounded. The eye is small with a vertically elliptical pupil. The tail is short. The rostral is rounded. The nasal is single. The loreal is present, in a sense, but see below. There is 1 preocular but no suboculars. The internasals and prefrontals are paired. The frontal is broader than it is long, and it is separated from the eye by a supraocular. There are 2 postoculars. The temporal formula is 1+2+ (2 or 3). There are 6–9, usually 7, upper labials, 3 of which are in contact with the eye. There may be 1 or 2 pairs of chin shields. The anterior chin shields are separated from the mental scale by the first pair of lower labials. There are 6–9 lower labials. The dorsal scales are smooth, with 2 apical pits, and arranged in 17 straight rows. The vertebral row is not distinct from the other dorsal rows. The cloacal scale is single. The subcaudals are divided. The numbers of ventrals and subcaudals are thought to be sexually dimorphic in this genus, with males having more subcaudals and fewer ventrals than females. The maxillary dentition is aglyph, consisting of 4–5 anterior maxillary teeth that increase in size posteriorly, followed by 5–10 smaller posterior teeth, all of similar size. The hemipenes are bilobed, with a forked sulcus spermaticus.

Chamaelycus is a genus in which the condition of the loreal and surrounding scales (i.e., the nasal and preocular) is particularly variable. For example, although the nasal scale is single—meaning that it is not divided into anterior and posterior nasal scales, separated from one another by the nostril—it may be fragmented, meaning that the nasal scale is broken up but the line of division does not pass through the nostril (e.g., an additional small scale broken off the posterior edge of the nasal scale). The preocular may sometimes be fused with the loreal, and if one follows the definition of the preocular as the scale anterior to the eye and in contact with it, then one must consider the loreal as absent in such cases. The arrangement of scales in the loreal region is thus variable in Chamaelycus in ways that head scales do not often vary in other snake genera, and different authors have interpreted these variations differently.

Chamaelycus is similar in many ways to Lycophidion and Hormonotus but can be distinguished from them as follows: Chamaelycus can be distinguished from Lycophidion by the condition of the nasal scale, which is single in Chamaelycus and divided in all Lycophidion within our zone, and by the maxillary dentition, the larger anterior teeth being separated from the smaller posterior teeth by a large diastema in Lycophidion and not in Chamaelycus. Chamaelycus can be distinguished from Hormonotus by the number of dorsal scale rows at midbody, which is 17 for Chamaelycus and 15 for Hormonotus; by the vertebral row, which is enlarged relative to the other dorsal scales for Hormonotus and not enlarged in Chamaelycus; and by the shape of the frontal scale, which is broader than long in Chamaelycus and longer than broad in Hormonotus.

Christy's Banded Snake: *Chamaelycus christyi* Boulenger 1919b

The range of Chamaelycus christyi extends from Congo to the Democratic Republic of Congo. The type locality is Madié, Ituri, Democratic Republic of Congo.

The dorsum is uniformly dark olive brown. The ventrals and subcaudals are the same color as the dorsal scales, but with yellow edges. The loreal is small. The preocular is long. There are 2 postoculars, the superior larger than the inferior. The temporal formula is 1+2. The breadth of the frontal is greater than or equal to its length. There are 6(3,4) or 7(3,4) upper labials. The fifth upper labial is the longest. There are 7(4) to 8(5) lower labials. There are usually 2 pairs

---

Key to Central and Western African Species of the Genus *Chamaelycus*

1	3 upper labials in contact with the eye	*C. fasciatus*
1'	2 upper labials in contact with the eye	2
2(1)	At least 40 subcaudals	*C. parkeri*
2'	Not more than 41 subcaudals	*C. christyi*

of chin shields, but sometimes just 1. The submandibular groove is pronounced. The dorsal scales are smooth, with 2 apical pits, and arranged in 17 straight rows. There are 174–195 ventrals and 39–41 divided subcaudals. The maximum length recorded for this species is 370 mm (Boulenger 1919b).

African Banded Snake: *Chamaelycus fasciatus* (Günther 1858)

The range of *Chamaelycus fasciatus* extends from Senegal to the Democratic Republic of Congo. The type locality is West Africa.

The dorsum is brown with or without black bands. The ventrals are brown like the dorsal scales, but they have paler edges. The condition of the loreal and preocular is variable. The loreal is usually but not always present, and there is usually a long preocular. There are 2 postoculars of approximately equal size. The temporal formula is 1+2. The breadth of the frontal is greater than its length. There are usually 7(3,4,5) upper labials, occasionally 8(3,4,5). Occasionally there may be 2 upper labials (4,5) in contact with the eye rather than 3. The sixth upper labial is the longest. There are 8(5) lower labials. There is just 1 pair of chin shields. The submandibular groove is pronounced. The dorsal scales are smooth, with 2 apical pits, and arranged in 17 straight rows. There are 164–198 ventrals and 30–56 divided subcaudals. The maximum length recorded for this species is 380 mm (Bogert 1940).

A fourth species of *Chamaelycus*, *C. werneri* (Mocquard 1902), was described as a different species from *C. fasciatus* on the basis of lacking a loreal scale. We concur with later authors (Angel 1934; Dowling 1969) who synonymize *C. werneri* with *C. fasciatus*. The condition of the loreal and preoculars is unusually variable in *C. fasciatus*, as noted by Witte (1963). Some later authors (e.g., Trape and Roux-Estève 1995) continue to recognize *C. werneri*, however.

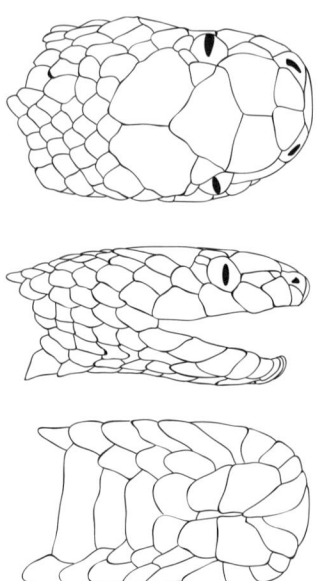

FIGURE 10.21. *Chamaelycus christyi* RMCA 1802. By T. Giri

FIGURE 10.22. *Chamaelycus christyi*, Democratic Republic of Congo. By J. Harvey

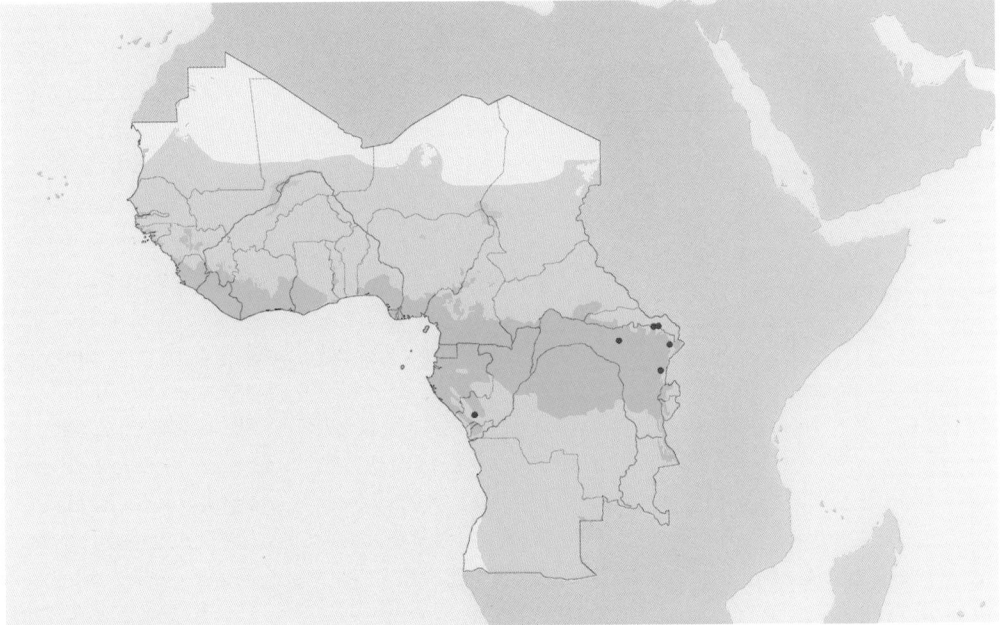

FIGURE 10.23. *Chamaelycus christyi* (red). By K. Jackson

Parker's Banded Snake: *Chamaelycus parkeri* (Angel 1934)

The range of *Chamaelycus parkeri* extends from the Congo to the Democratic Republic of Congo. The type locality is Kabuliré, Democratic Republic of Congo.

The dorsum is uniformly brown with the dorsum of the head slightly darker than the dorsum of the body. The venter is uniformly brown with a pale distal edge to each ventral scale. The loreal is elongate. There is usually 1 preocular, deeper than long, sometimes 2 smaller preoculars. There are 2 postoculars, the inferior larger than the superior. The temporal formula is 1+2. The breadth of the frontal is greater than or equal to its length. There are usually 6(3,4) upper labials. The fifth upper labial is the longest. There are 6–9(3–4) lower labials. There are 2 pairs of chin shields. The submandibular groove is pronounced. The dorsal scales are smooth, with 2 apical pits, and arranged in 17 straight rows. There are 176–178 ventrals and 40–52 divided subcaudals. The maximum length recorded for this species is 312 mm (Trape and Roux-Estève 1990).

Genus *Dendrolycus* Laurent 1956

Cameroon Rainforest Snake: *Dendrolycus elapoides* (Günther 1874)

Dendrolycus is a monotypic genus. The range of *Dendrolycus elapoides* extends from Cameroon to central Democratic Republic of Congo. The type locality is "Cameroon Mountains." Two subspecies are recognized, *D. e. elapoides* (Günther 1874) and *D. e. angusticinctus* (Laurent 1952). These two subspecies differ from one another in color pattern.

The Cameroon Rainforest Snake was first described as a new species of Wolf Snake, *Lycophidion*, since it resembles that genus in most scale characters. But Laurent (1956)

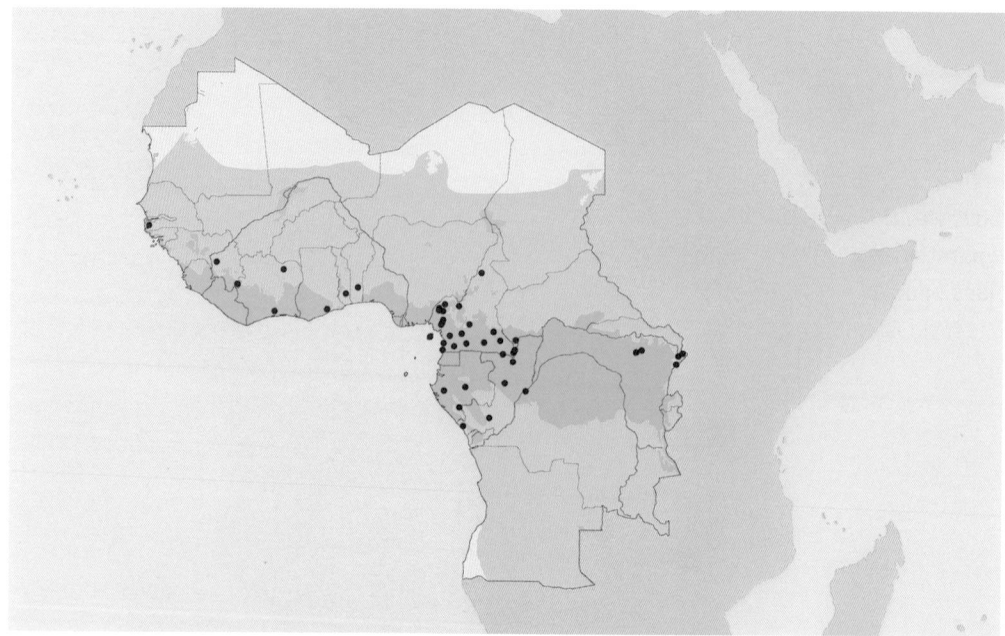

FIGURE 10.24. *Chamaelycus fasciatus* (blue). By K. Jackson

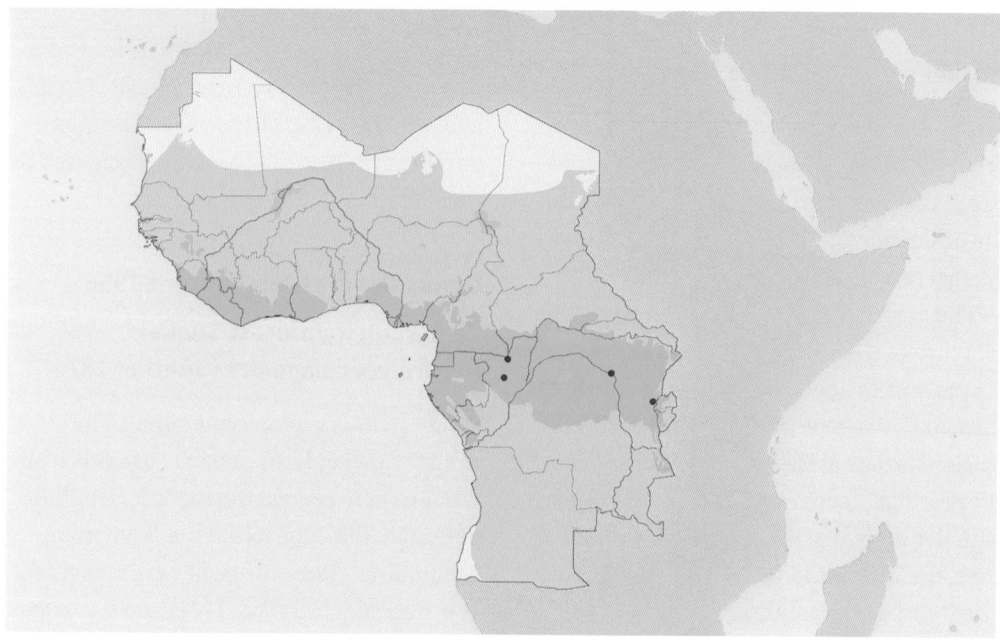

FIGURE 10.25. *Chamaelycus parkeri* (red). By K. Jackson

assigned this species to its own genus because it differs from the terrestrial Wolf Snakes in being arboreal in its habits and in a suite of morphological differences associated with arboreality. The body shape of *Dendrolycus* is laterally compressed, more slender and gracile than *Lycophidion*, with a longer, more slender tail. This difference in shape is reflected in differences in scale counts: 225–253 ventrals and 66–83 subcaudals in *Dendrolycus* versus 144–214 ventrals and 22–58 subcaudals in *Lycophidion*.

The body and tail are patterned with alternating black-and-white or yellowish bands. In *D. e. angusticinctus*, the black bands are narrower than the pale interspaces that separate them, while in *D. e. elapoides*, the black bands are wider than the interspaces. The head is black and the ventrals are white, irregularly spotted with black. The head is small and flat with a well-defined neck. The snout is rounded. The eye is medium sized with a vertically elliptical pupil. The tail is medium length and the body somewhat laterally compressed in shape. The nasal is divided. The internasals and prefrontals are paired. The loreal is present, though in some individuals the loreal is fused to a preocular

FIGURE 10.26. *Dendrolycus elapoides*, Democratic Republic of Congo. By K. Mebert

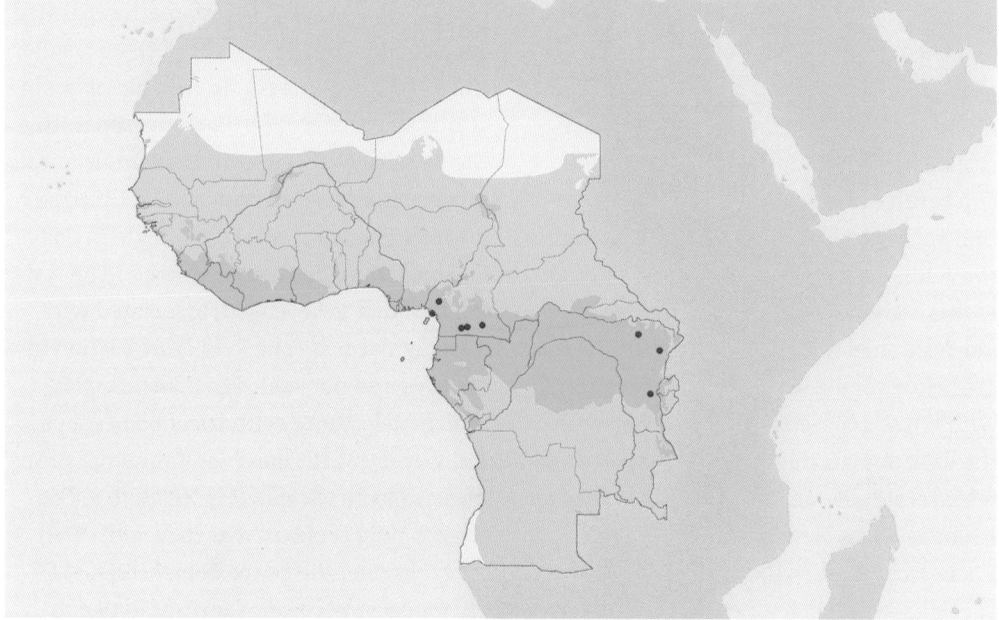

FIGURE 10.27. *Dendrolycus elapoides* (red). By K. Jackson

(so, in a sense, absent). In one of the type specimens of *D. e. angusticinctus*, the loreal is fused to a preocular on one side of the head but is normal on the other. There are 2 preoculars and 2 postoculars, the inferior of which is larger than the superior. The frontal is slightly longer than broad. There are 7(3,4,5) upper labials, of which the seventh upper labial is the largest. There are 8(4) lower labials. The temporal formula is 1+2. There are 2 pairs of chin shields, the anterior pair longer than the posterior pair. The submandibular groove is not pronounced. The dorsal scales are smooth, without apical pits, and arranged in 17 straight rows. The vertebral row is not enlarged relative to the other dorsal scales. There are 225–253 ventrals and 66–83 divided subcaudals. The cloacal scale is single. The maximum recorded length for this species is 410 mm (Witte 1962).

African File Snakes: Genus *Gonionotophis* Boulenger 1893

Gonionotophis is a genus of 15 species from sub-Saharan Africa, 13 of which occur in our zone. Snakes in this genus are distinctive in being triangular in cross section, as a result of hypertrophied dorsal vertebral spines and keeled ventral scales, and the common name, "File Snakes," refers to this body shape, which resembles an old-fashioned three-sided file. File Snakes are also recognizable by their coloration, with white exposed skin visible between the dark dorsal scales, a double-keeled vertebral scale row, and a snout that is square in shape when viewed from above. File Snakes are nocturnal, and many possess a specialized maxillary dentition, similar to that of Wolf Snakes (*Lycophidion*), involving

an arched maxilla and hinged teeth. (See genus description for *Lycophidion* for further detail on this type of specialized maxillary dentition). In File Snakes, this dentition is thought to be associated with a diet of elongate, cylindrical, hard-bodied prey, especially other snakes. Shine et al. (1996) found the diets of *Gonionotophis nyassae* and *G. capensis* to consist primarily of lizards, although snakes were also taken. A literature review included in the same study indicated that *Gonionotophis* in general feed on a wide variety of ectothermic prey, including lizards, snakes (both venomous and nonvenomous), and amphibians. Akani et al. (2001b) found the diet of *G. crossi* in Nigeria to include small mammals as well as lizards.

In the past, two genera of African File Snakes have been recognized, *Mehelya* and *Gonionotophis*, distinguished from one another on the basis of several morphological differences. The most significant of these differences was maxillary dentition, which in *Mehelya* resembled that of *Lycophidion*, the 6–10 anterior maxillary teeth separated by a diastema from a series of 11–26 smaller, hinged posterior teeth. In *Gonionotophis*, by contrast, there was no diastema separating the 24–38 maxillary teeth into anterior and posterior series (Loveridge 1939; Chippaux 2006a). The nasal scale was single in *Gonionotophis* and divided in *Mehelya*. The ventrals tended to be keeled, associated with a triangular body shape in cross section in *Mehelya* and not keeled in *Gonionotophis*, which had a more cylindrical body shape. Kelly et al. (2011), however, found the genus *Mehelya* to be paraphyletic relative to *Gonionotophis*, a problem that they addressed by expanding the genus *Gonionotophis* to encompass species formerly assigned to *Mehelya*. We follow this nomenclature here, using the genus *Gonionotophis* in its new expanded sense. For the File Snake species occurring within our zone, the following ten species of *Gonionotophis* were formerly assigned to the genus *Mehelya*: *G. poensis*, *G. capensis*, *G. nyassae*, *G. crossi*, *G. stenophthalmus*, *G. gabouensis*, *G. guirali*, *G. laurenti*, *G. egbensis*, *G. vernayi*. The remaining three were formerly *Gonionotophis*: *G. brussauxi*, *G. grantii*, *G. klingi*.

The head is broad and flat, distinct from the well-defined neck. The snout is rounded. The eye is small with a pupil that is usually vertically elliptical but may be almost round or round, depending on the species. The tail is medium length or short. The rostral is rounded, visible from above. The nasal may be single or divided. The internasals and prefrontals are paired. The loreal may be present or absent. There are 1 or 2 preoculars but no suboculars. The frontal is broader than or as broad as it is long and is separated from the eye by a supraocular. There are 1 or 2 postoculars. The temporal formula is (0 or 1 or 2)+2+(2 or 3). There are 6–8 upper labials, 2 or 3 of which are in contact with the eye. There are 2 pairs of chin shields. The submandibular groove is pronounced. The anterior chin shields are separated from the mental by the first pair of lower labials. There are 6–10 lower labials. The dorsal scales are keeled, though sometimes so weakly that they appear almost smooth. They lack apical pits and are arranged in 15–21 straight rows at midbody. The vertebral row is enlarged, though sometimes only slightly, and bears a double keel. The ventrals may be rounded or keeled. The cloacal scale is single. The subcaudals are divided. The number of ventrals is not thought to be sexually dimorphic in this genus. The number of subcaudals is, however, with females possessing a lower average number of subcaudals than males. The hemipenes are bilobed, and the sulcus spermaticus is forked.

In a study of *G. capensis* and *G. nyassae*, Shine et al. (1996) found that males and females reached sexual maturity at the same sizes, but that females ultimately grow to much larger sizes than males and differ from males in body proportions. File Snakes are egg-layers, and Shine et al. (1996) found the average clutch size in *G. capensis* (n = 5–11) to be larger than the average clutch size in the smaller species, *G. nyassae* (n = 2–6), though clutch size was not correlated with maternal body size within species.

The genus *Gonionotophis* resembles the genera *Chamaelycus*, *Hormonotus*, and *Lycophidion* but can be distinguished from all these by the presence in *Gonionotophis* of the double-keeled vertebral scale row.

Mocquard's African Ground Snake: *Gonionotophis brussauxi* (Mocquard 1889)

The range of *Gonionotophis brussauxi* extends from Cameroon to the Democratic Republic of the Congo. The type locality is Loudinia-Niari, on the Niari River, Republic of Congo.

The dorsum is uniformly dark brown, sometimes with pale edges to the dorsal scales. The venter is yellowish. The body is cylindrical or slightly laterally compressed. The tail is short. The snout is rounded viewed from above. The pupil is round or slightly vertically elliptical. The nasal is single. The loreal is absent. There is 1 preocular and 2 postoculars, the inferior larger than the superior. The temporal formula is 1+2, or sometimes 2+2. There are usually 8(4,5) but occasionally 7(4,5) upper labials and

8(4) or 9(5) lower labials. The dorsal scales are keeled and arranged in 21, occasionally 23, straight rows. The vertebral row is slightly enlarged and double keeled. There are 167–185 ventrals, not keeled (fewer than 181 in *G. b. brussauxi* and more than 181 in *G. b. prigoginei*). There are 73–95 divided subcaudals (fewer than 93 in *G. b. brussauxi* and more than 93 in *G. b. prigoginei*). The maximum recorded length is 500 mm (Vil-

Key to Central and Western African Species of the Genus *Gonionotophis*

1	15 dorsal scale rows	2
1'	More than 15 dorsal scale rows	10
2(1)	More than 84 subcaudals	*G. poensis*
2'	Fewer than 83 subcaudals	3
3(2)	Fewer than 185 ventrals	
4		
3'	More than 185 ventrals	7
4(3)	Fewer than 155 ventrals	*G. egbensis*
4'	More than 160 ventrals	5
5(4)	Fewer than 50 subcaudals	*G. gabouensis*
5'	More than 50 subcaudals	6
6(5)	Nasal entire; More than 25 maxillary teeth without diastema	*G. grantii*
6'	Nasal divided; Fewer than 26 maxillary teeth with a diastema	*G. nyassae*
7(3)	Frontal almost as long as the parietals	*G. guirali*
7'	Frontal distinctly shorter than the parietals	8
8(7)	3 upper labials in contact with the eye	*G. laurenti*
8'	2 upper labials in contact with the eye	9
9(8)	Frontal as long as broad; dorsal scales strongly keeled	*G. capensis*
9'	Frontal broader than long; dorsal scales weakly keeled	*G. stenophthalmus*
10(1)	21 or 23 dorsal scale rows	*G. brussauxi*
10'	17 or 19 dorsal scale rows	11
11(10)	17 dorsal scale rows	*G. crossi*
11'	19 dorsal scale rows	12
12(11)	2 upper labials in contact with the eye	*G. klingi*
12'	3 upper labials in contact with the eye	*G. vernayi*

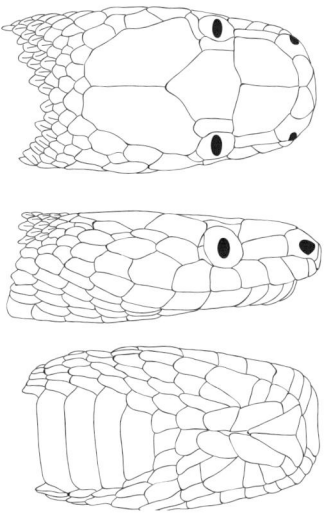

FIGURE 10.29. *Gonionotophis brussauxi*, Republic of Congo. By K. Jackson

FIGURE 10.28. *Gonionotophis brussauxi* RMCA 89-20-R-79. By T. Giri

FIGURE 10.30. *Gonionotophis brussauxi*, Republic of Congo. By M. Burger

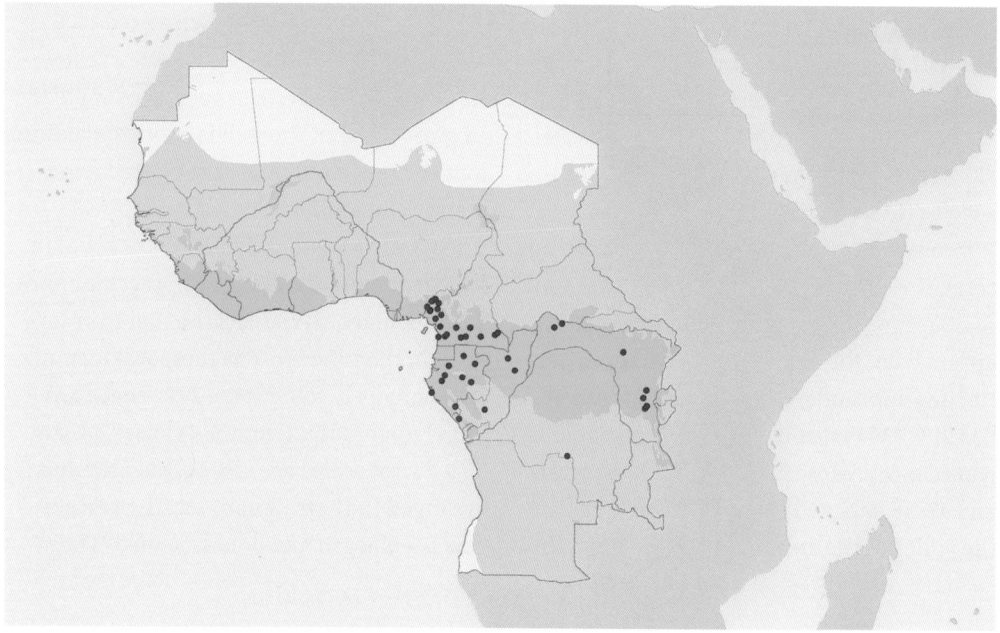

FIGURE 10.31. *Gonionotophis brussauxi* (blue). By K. Jackson

liers 1975). Two subspecies, *G. b. brussauxi* (Mocquard 1889) and *G. b. prigoginei* (Laurent 1956), differing only in numbers of ventrals and subcaudals, are recognized by some, though not most, authors.

Cape File Snake: *Gonionotophis capensis* (Smith 1847)

The range of *Gonionotophis capensis* extends from Cameroon to Somalia to South Africa. The type locality is Cape Province, South Africa.

The dorsum is dark brown, sometimes with a pale spot at the base of each dorsal scale. The venter is paler, with a yellow mid-ventral stripe. The body is triangular in cross section. The tail is short to medium length. The snout is square viewed from above. The pupil is vertically elliptical. The nasal is divided. The loreal is equal in length and depth. There is usually 1 preocular but sometimes 2, the superior larger than the inferior. There are 2 postoculars of roughly equal size. The temporal formula is 1+2. There are 7(3,4) or 8(3,4) upper labials and 7(4) to 9(5) lower labials. The dorsal scales are strongly keeled and arranged in 15 straight rows. The vertebral row is enlarged and double keeled. There are 218–239 keeled ventrals and 53–65 divided subcaudals. The maximum recorded length for this species is 1,403 mm (Loveridge 1939).

Two subspecies are recognized within our zone, *G. c. savorgnani* (Mocquard 1887) and *G. c. capensis* (Smith 1847). *G. c. savorgnani* differs from the nominal subspecies in ventral and subcaudal counts (*G. c. savorgnani* has 218–239 ventrals and 53–65 subcaudals, whereas *G. c. capensis* has 203–224 ventrals and 47–56 subcaudals) and in markings on the scales in the enlarged vertebral row. In *G. c. savorgnani* there is a basal white spot on each scale in the vertebral row, whereas in *G. c. capensis* there is a central white spot on each scale in the vertebral row. The distribution of *G. c. capensis* is generally southern and eastern to that of *G. c. savorgnani*, which extends as far east as the Great Lakes in central Africa. Some authors (e.g., Kelly et al. 2011) elevate these two subspecies to the level of species, and it is possible that further study will provide evidence to support this. For example, the distribution of *G. c. capensis* in central Africa is not known, so it is not known whether the two supposed subspecies occur sympatrically. If they do, it would lend support to the idea that they represent separate species. The subspecies *G. c. unicolor* (Boulenger 1910) was synonymized with *G. c. savorgnani* by Loveridge (1939).

African File Snake: *Gonionotophis crossi* (Boulenger 1895a)

The range of *Gonionotophis crossi* extends from Senegal to Chad. The type locality is Asaba, Nigeria.

The dorsum is dark brown or reddish brown. The venter is paler, with a yellow mid-ventral stripe. The body is triangular in cross section. The tail is short to medium length. The snout is square viewed from above. The pupil is vertically elliptical. The nasal is divided. The loreal is equal in length and depth. There is 1 preocular and 2 postoculars, the superior equal to or larger than the inferior. The temporal formula is usually 1+2, sometimes 2+2. There are 7(3,4) upper labials and 8(4) lower labials. The dorsal scales are strongly keeled and arranged in 17 straight rows. The vertebral row is enlarged and double keeled. There

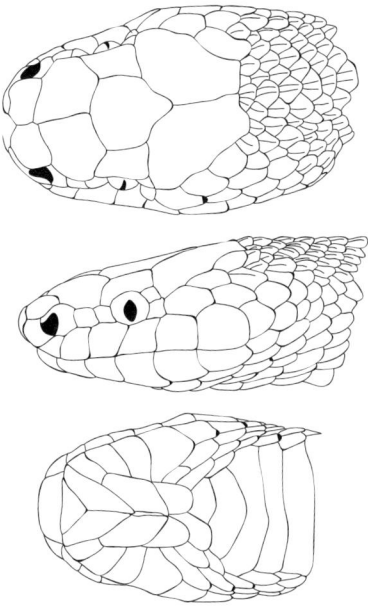

FIGURE 10.32. *Gonionotophis capensis*. Smithsonian Institution catalog number 51628. By T. Giri

FIGURE 10.33. *Gonionotophis capensis*, Cameroon. By M. LeBreton

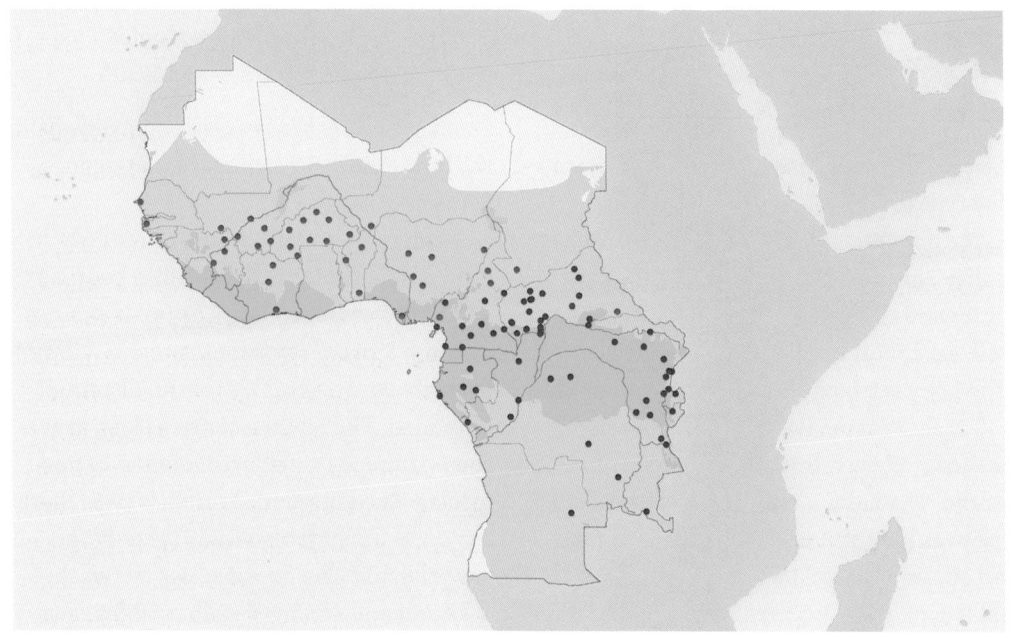

FIGURE 10.34. *Gonionotophis capensis* (blue), *Gonionotopis crossi* (red). By K. Jackson

FIGURE 10.35. *Gonionotophis crossi*, Ghana. By S. Spawls

are 220–240 keeled ventrals and 51–68 divided subcaudals. The maximum length recorded for this species is 1,250 mm (Boulenger 1895a).

Dunger's File Snake: *Gonionotophis egbensis* Dunger 1966

Gonionotophis egbensis is known only from the type locality of Egbe, Nigeria.

The dorsum is dark brown. The venter is paler. The body coloration overall is darker toward the head and lighter toward the tail, except for the snout, which is pale. The body is triangular in cross section. The tail is short to medium length. The snout is square viewed from above. The pupil is round. The nasal is divided. The loreal is 1.5 times as long as it is deep. There is 1 small preocular and 1 postocular. The temporal formula is 1+2. There are 7(3,4) upper labials and 8(5) lower labials. The dorsal scales are weakly keeled and arranged in 15 straight rows. The vertebral row is enlarged and double keeled. There are 147 weakly keeled ventrals and 36 divided subcaudals.

This species is known only from the type specimen, with a total length of 235 mm (Dunger 1966).

Gabou File Snake: *Gonionotophis gabouensis* Trape and Mané 2005

Gonionotophis gabouensis is known only from the type locality of Velingara, Casamance, Senegal.

The dorsum is dark brown or reddish brown. The venter is lighter. The body is triangular in cross section. The tail is short to medium length. The snout is square viewed from above. The pupil is round. The nasal is divided. The loreal is twice as long as it is deep. There is 1 small preocular and 1 postocular. The temporal formula is 0+1. There are 7(3,4) upper labials and 7(4) to 8(5) lower labials. The dorsal scales are weakly keeled and arranged in 15 straight rows. The vertebral row is enlarged and double keeled. There are 172–176 keeled ventrals and 42–43 divided subcaudals. The maximum recorded length is 370 mm (Trape and Mané 2005).

Savanna Lesser File Snake: *Gonionotophis grantii* (Günther 1863a)

The range of *Gonionotophis grantii* extends from Senegal to Central African Republic. The type locality is West Africa.

The dorsum is uniformly dark brown, sometimes with pale edges to the dorsal scales. The venter is yellowish. The body is cylindrical or slightly laterally compressed. The tail is short. The snout is rounded viewed from above. The pupil is round or slightly vertically elliptical. The nasal is sin-gle. The loreal is long and narrow. There is 1 preocular and 1 postocular. The temporal formula is usually 1+2 or 3, but sometimes 0+1, where the fifth upper labial is in contact with the parietal. There are 7(3,4) upper labials and 8(5) or 9(5) lower labials. The dorsal scales are keeled and arranged in 15 straight rows. The vertebral row is slightly enlarged and double keeled. There are 162–178 ventrals, not keeled, and 62–82 divided subcaudals. The maximum recorded length for this species is 500 mm (Villiers 1975).

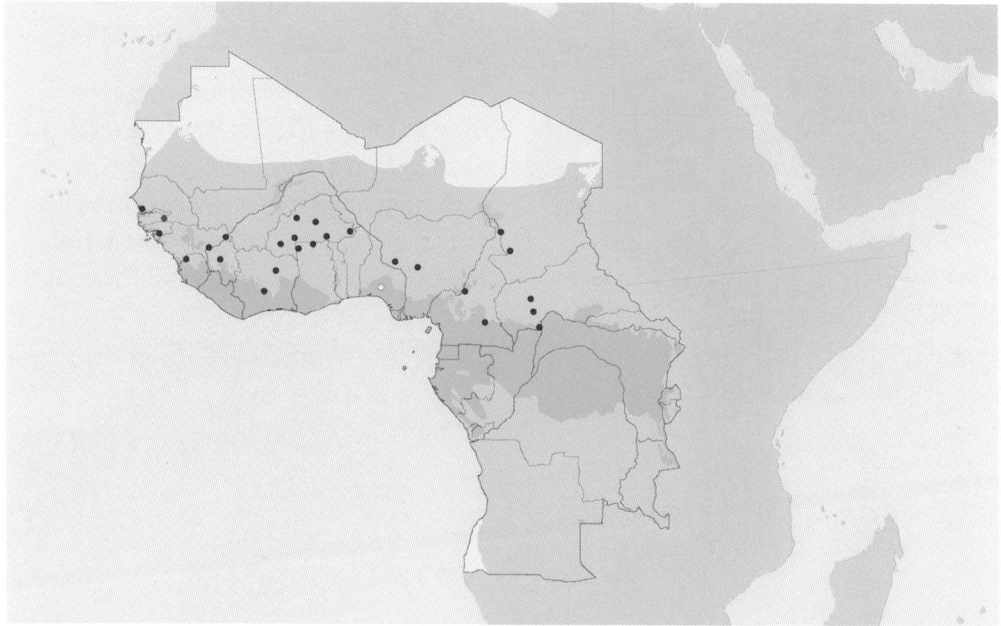

FIGURE 10.36. *Gonionotophis egbensis* (yellow), *Gonionotophis gabouensis* (red), *Gonionotophis grantii* (blue). By K. Jackson

FIGURE 10.37. *Gonionotophis grantii*, Ghana. By S. Spawls

Mocquard's File Snake: *Gonionotophis guirali* (Mocquard 1887)

The range of *Gonionotophis guirali* extends from Sierra Leone to the Democratic Republic of the Congo. The type locality is the Niger River, West Africa.

The dorsum is dark brown, with yellowish tips to the dorsal scales. The venter is pale, with dark edges to the ventral scales. The body is triangular in cross section. The tail is short to medium length. The snout is square viewed from above. The pupil is vertically elliptical. The nasal is divided. The loreal is slightly longer than deep. There is 1 preocular and usually 1 postocular. The temporal formula is 1+2 or occasionally 2+2. There are 7(3,4,5) or sometimes 7(3,4) upper labials and 8(5) lower labials. The dorsal scales are strongly keeled and arranged in 15 straight rows. The vertebral row is enlarged and double keeled. There are 236–262 keeled ventrals and 51–70 divided subcaudals. The maximum length recorded for this species is 1,300 mm (Mocquard 1887).

FIGURE 10.38. *Gonionotophis guirali*, captive specimen. By Y. Tahara

Matschie's African Ground Snake: *Gonionotophis klingi* Matschie 1893a

The range of *Gonionotophis klingi* extends from Guinea to Nigeria. The type locality is Bismarckburg (now Konkoa), Togo.

The dorsum is uniformly dark brown, sometimes with pale edges to the dorsal scales. The venter is yellowish. The body is

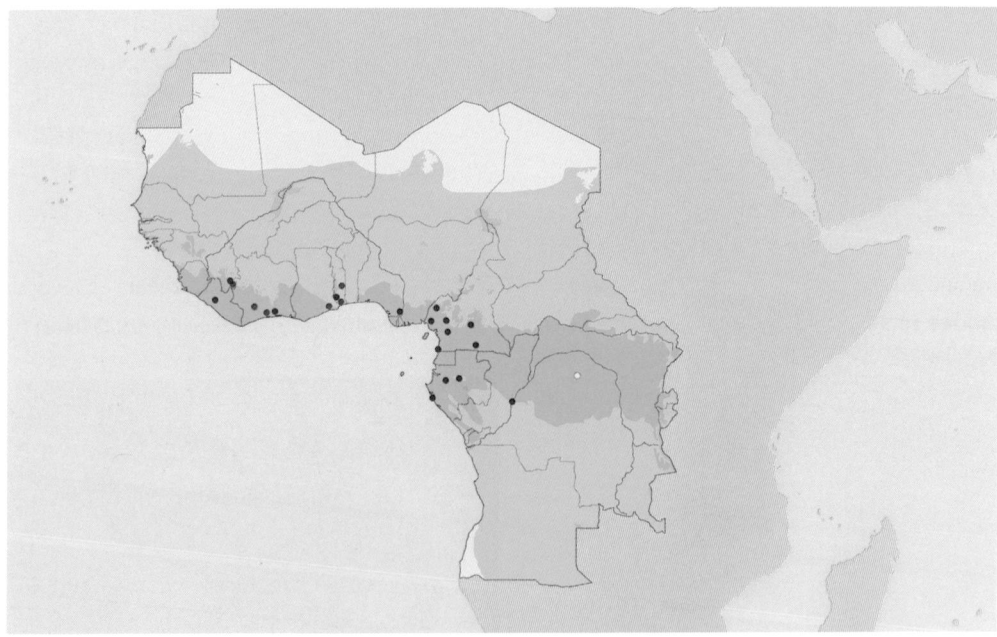

FIGURE 10.39. *Gonionotophis guirali* (blue), *Gonionotophis klingi* (red), *Gonionotophis laurenti* (yellow). By K. Jackson

FIGURE 10.40.
Gonionotophis klingi,
Ghana. By A. Leaché

cylindrical or slightly laterally compressed. The tail is short. The snout is rounded viewed from above. The pupil is round or slightly vertically elliptical. The nasal is single. The loreal is longer than deep and may occasionally be fused to the preocular (in other words, absent). There is 1 preocular and 2 postoculars, the inferior larger than the superior. The temporal formula is 1+2 or sometimes 2+2. There are 7 or 8(4,5), occasionally (4,5,6), upper labials and 7(4) to 9(5) lower labials. The dorsal scales are keeled and arranged in 19 straight rows. The vertebral row is slightly enlarged and double keeled. There are 165–179 ventrals, not keeled, and 79–94 divided subcaudals. The maximum recorded length is 450 mm (Villiers 1975).

Laurent's File Snake: *Gonionotophis laurenti* Witte 1959

Gonionotophis laurenti is known only from the type locality of Ikela, Democratic Republic of Congo.

The dorsum is brown with yellowish tips to the dorsal scales. The venter is yellowish. The body is triangular in cross section. The tail is short to medium length. The snout is square viewed from above. The pupil is vertically elliptical. The nasal is divided. The length of the loreal is less than 1.5 times its depth. There is a small preocular. There is 1 postocular. The temporal formula is 1+2. There are 7(3,4,5) upper labials and 8(5) lower labials. The dorsal scales are weakly keeled and arranged in 15 straight rows. The vertebral row is enlarged and double keeled. There are 202 weakly keeled ventrals and 51 divided subcaudals. This species is known only from the type specimen, which has a total length of 465 mm (Witte 1959).

Dwarf File Snake: *Gonionotophis nyassae* (Günther 1888)

The range of *Gonionotophis nyassae* extends from Kenya and Burundi in the north to northeastern South Africa in the south. The type locality is Lake Nyassa, Malawi.

The dorsum is black or gray with pale skin visible between the scales. The color of the venter may range from white to gray or black with white edges to the ventral scales. The body is triangular in cross section. The tail is short to medium length. The snout is square when viewed from above. The pupil is vertically elliptical. The nasal is divided. The loreal is 1.5 times longer than deep. There is 1 small preocular. There are usually 2 postoculars of equal size. The temporal formula is 1+2. There are 7(3,4) upper labials and 7(4) to 8(5) lower labials. The dorsal scales are keeled and arranged in 15 straight rows. The vertebral row is enlarged and double keeled. There are 165–184 keeled ventrals and 51–73 divided subcaudals. The maximum recorded length for this species is 650 mm (Spawls et al. 2004).

FIGURE 10.41. *Gonionotophis nyassae*, Kenya. By S. Spawls

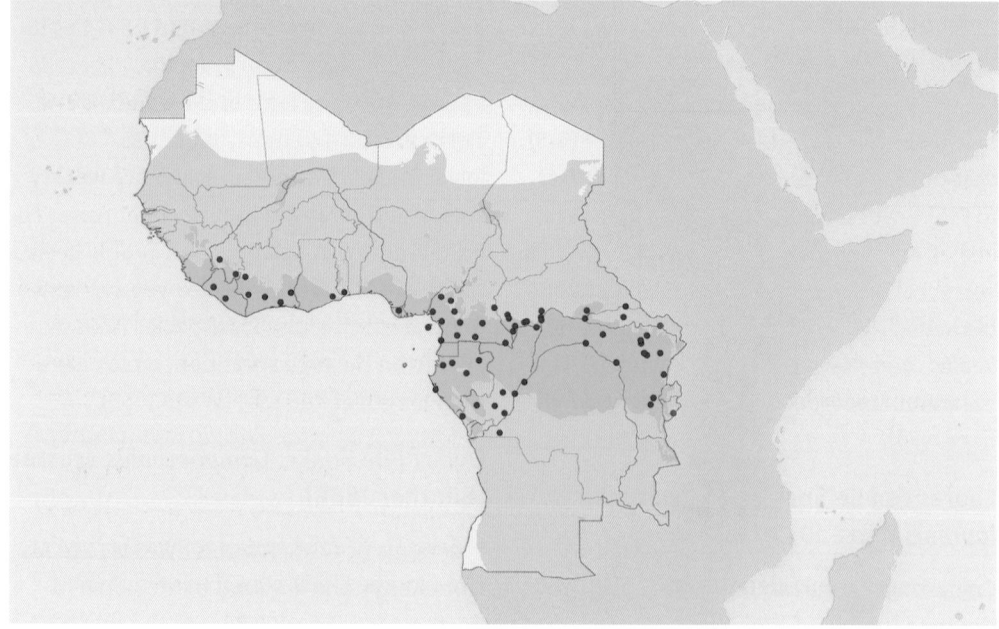

FIGURE 10.42. *Gonionotophis nyassae* (red), *Gonionotophis poensis* (blue). By K. Jackson

FIGURE 10.43. *Gonionotophis poensis*, Democratic Republic of Congo. By E. Greenbaum

Western Forest File Snake: *Gonionotophis poensis* (Smith 1847)

The range of *Gonionotophis poensis* extends from Sierra Leone to Uganda to Angola. The type locality is Bioko, Equatorial Guinea.

The dorsum is dark brown to reddish brown. The venter is paler, with a yellow mid-ventral stripe. The body is triangular in cross section. The tail is short to medium length. The snout is square viewed from above. The pupil may be round or vertically elliptical. The nasal is divided. The loreal is slightly longer than deep. There is 1 preocular and 2 postoculars, the superior equal to or larger than the inferior. The temporal formula is usually 1+2, sometimes 2+2. There are 6–8(3,4) upper labials and 8(4) or (5) lower labials. The dorsal scales are keeled, though sometimes only weakly, and arranged in 15 straight rows. The vertebral row is enlarged and double keeled. There are 234–262 keeled ventrals and 85–124 divided subcaudals. The maximum record-ed length for this species is 1,200 mm (Love-ridge 1939).

Small-eyed File Snake: *Gonionotophis stenophthalmus* (Mocquard 1887)

The range of *Gonionotophis stenophthalmus* extends from Guinea-Bissau to Uganda to the Democratic Republic of the Congo. The type localities are Assini, Ivory Coast, and Cape Lopez, Gabon.

The dorsum is dark brown to reddish brown. The venter is paler. The body is triangular in cross section. The tail is short to medium length. The snout is square viewed from above. The pupil is vertically elliptical. The nasal is divided. The loreal is equal in length and depth. There is 1 preocular and usually 1 postocular. The temporal formula is 1+2. There are 7(3,4), sometimes 8(3,4), upper labials and 7(4) to 8(5) lower labials. The dorsal scales are weakly keeled and arranged in 15 straight rows. The vertebral row is enlarged and double keeled. There are 189–228 keeled ventrals and

47–63 divided subcaudals. The maximum recorded length for this species is 802 mm (Roux-Estève 1965).

Angola File Snake: *Gonionotophis vernayi* (Bogert 1940)

Gonionotophis vernayi is known only from the type locality of Hanha, Angola.

The dorsum is dark olive brown, with pale edges to the dorsal scales. The venter is yellow. The body is triangular in cross section. The tail is short to medium length. The snout is square viewed from above. The pupil is round. The nasal is divided. The length of the loreal is less than 1.5 times its depth. There is 1 preocular and 2 postoculars, the superior equal to or larger than the inferior. The temporal formula is 1+2. There are 7(3,4,5) upper labials and 8(5) lower labials. The dorsal scales are strongly keeled and arranged in 19 straight rows. The vertebral row is enlarged and double keeled. There are 256 weakly keeled ventrals and 65 divided subcaudals. This species is known only from the type specimen, which has a total length of 1,035 mm (Bogert 1940).

FIGURE 10.44. *Gonionotophis stenophthalmus*, Republic of Congo. By M. Burger

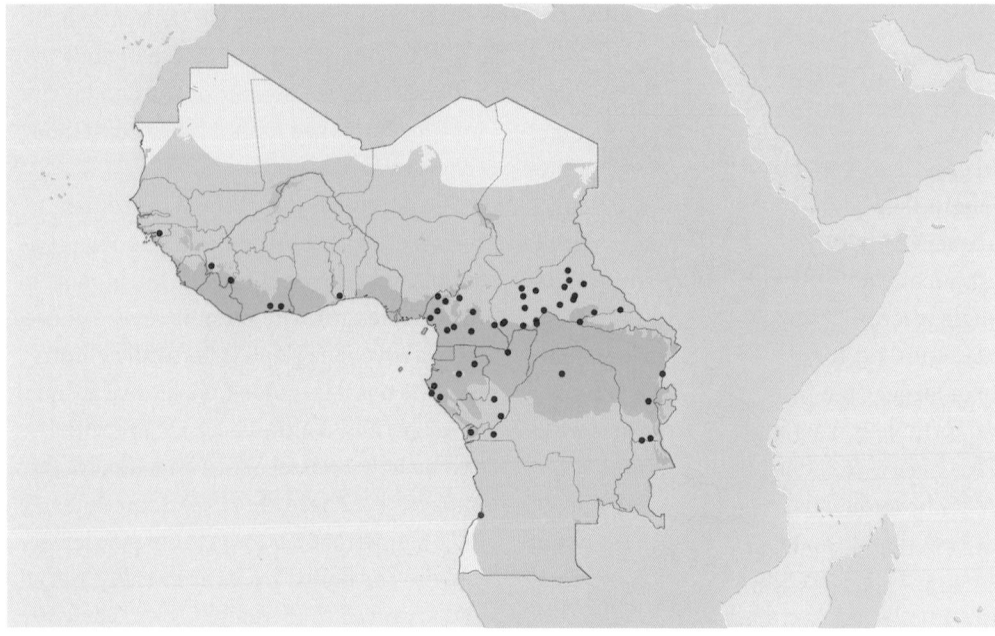

FIGURE 10.45. *Gonionotophis stenophthalmus* (blue), *Gonionotophis vernayi* (red). By K. Jackson

Genus *Hormonotus* Hallowell 1857

Yellow Forest Snake: *Hormonotus modestus* (Duméril, Bibron, and Duméril 1854b)

Hormonotus is a monotypic genus. The range of *Hormonotus modestus* extends from Guinea in the west to Uganda in the east. The type locality is the Guinea coast.

H. *modestus* is nocturnal, terrestrial, and generally found only in forested habitat. Beyond this, little is known of its natural history, as it is rarely encountered by humans. The diet consists of small mammals and lizards. It presumably lays eggs like other lamprophiines. The head is small and flat with a well-defined neck. The snout is square shaped when viewed from above. The body is laterally compressed and the tail medium length to long. The eye is small with a vertically elliptical pupil. The maxillary dentition is aglyph, with a series of 5–6 anterior teeth, increasing in size posteriorly, separated by a diastema from a series of 11–16 equal-sized posterior teeth. The body is brown above, the edges of the dorsal scales darker than the rest. The head is the same color, but with the edges of the scales paler than the rest. The venter is pale. The nasal is usually single but occasionally divided. The internasals and prefrontals are paired. The loreal is twice as long as it is deep. There is 1 preocular and 2 postoculars, the superior larger than or equal to the inferior. The frontal is longer than it is broad. The temporal formula is 1+2, or sometimes 2+2. A temporal formula of 1+2 is more common in specimens from central Africa, while a temporal formula of 2+2 is more common in specimens from West Africa. There are 8(4,5) upper labials, and 9–11, usually 9(4) or 9(5), lower labials. There are 2 pairs of chin shields, the anterior pair longer than the posterior pair. The submandibular groove is pronounced. The dorsal scales are usually smooth but may be weakly keeled. They are arranged in 15 slightly oblique rows. The vertebral row is enlarged and may be smooth or weakly keeled. There are 219–244 ventrals and 81–90 divided subcaudals. The cloacal scale is single. The hemipenes are bilobed at the distal one-third, and the sulcus spermaticus is forked starting at the basal one-third. The maximum length recorded for this species is 875 mm (Laurent 1960).

Water Snakes: Genus *Lycodonomorphus* Fitzinger 1843

Lycodonomorphus is a genus of seven species of nocturnal aquatic snakes distributed throughout central, eastern, and southern Africa. Three species occur within our zone.

FIGURE 10.46. *Hormonotus modestus* RMCA 21564. By T. Giri

Lycodonomorphus are medium-sized aquatic snakes with a cylindrical body, indistinct neck, and short- to medium-length tail. The eye is small. The pupil may be round or vertically elliptical, depending on the species. The nasal may be single, divided, or semi-divided. The internasals and prefrontals are paired. The loreal is present. There are 1–2 preoculars, 1–2 postoculars, and 0–2 suboculars. There is 1 anterior temporal. There are 7–9 upper labials and 7–10 lower labials. There are 2 pairs of chin shields. The submandibular groove is pronounced. The dorsal scales are smooth, with 0 or 2 apical pits, and arranged in 19–25 straight rows. The verte-

FIGURE 10.47. *Hormonotus modestus*, Ghana. By P. Naskrecki

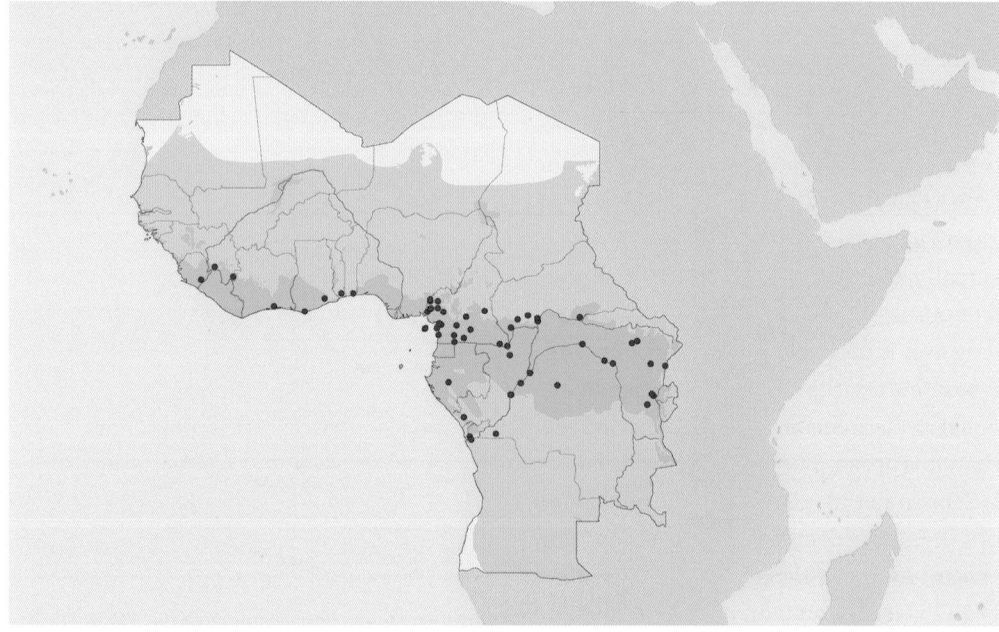

FIGURE 10.48. *Hormonotus modestus* (blue). By K. Jackson

bral row is not enlarged relative to the other dorsal scale rows. There are 152–193 ventrals and 29–71 divided subcaudals. The cloacal scale is single. The maxillary dentition is aglyph, with 18–25 teeth and no diastema. The hemipenes are unforked (or distally bilobed). The sulcus spermaticus is divided. Madsen and Osterkamp (1982) found *Lycodonomorphus bicolor* to be easily captured by hand and present in enormous numbers at the southern end of Lake Tanganyika (they estimated 38,000 snakes/km^2 based on the results of their mark-recapture study). Water Snakes were found to feed entirely on cichlid fishes, which were abundantly available. Snakes were active only at night and were more active on dark nights than on moonlit ones. Females laid clutches of from 4 to 8 eggs, and reproduction did not appear to be restricted to a particular time of year as in most snake species, possibly because the lake provides a more stable habitat year-round in terms of temperature and prey availability than the habitats of terrestrial snakes.

Tanganyika Water Snake: *Lycodonomorphus bicolor* (Günther 1893a)

The Tanganyika Water Snake is restricted to Lake Tanganyika and its shores. These include southeastern Democratic Republic of Congo, western Tanzania, Burundi, and northeastern Zambia. The type locality is Lake Tanganyika (country not specified).

The body is grayish brown above and pale yellow underneath. The pupil is round or slightly vertically elliptical. The nasal is divided or occasionally single. The internasals are shorter and narrower than the prefrontals. The prefrontals are longer than broad. The temporal formula is 1+2. There is 1 preocular, the depth of which is less than or equal to the diameter of the eye. There are 2 postoculars of roughly equal size. There are usually no suboculars, but sometimes there may be 1 or 2. There are usually 8(4) upper labials, occasionally 7(4) or 9(5), or 7–9(4,5). If suboculars are present, there may be no contact between the eye and the upper labials. There are 9(4), sometimes 8(4), and occasionally 7(4) lower labials. The anterior chin shields are longer than the posterior pair. The dorsal scales are smooth, without apical pits, and arranged in 23 (occasionally 25) straight rows. There are 152–166 ventrals (fewer than 165 in males, more than 151 in females) and 50–71 divided subcaudals (more than 58 in males, fewer than 60 in females). The maximum length recorded for this species is 725 mm (Witte 1952).

Key to the Central and Western African Species of the Genus *Lycodonomorphus*

1	7 lower labials	*L. leleupi*
1'	8 or 9 lower labials	2
2(1)	Fewer than 167 ventrals	*L. bicolor*
2'	More than 171 ventrals	*L. subtaeniatus*

Mulanje Water Snake: *Lycodonomorphus leleupi* (Laurent 1950a)

The Mulanje Water Snake has a limited and discontinuous distribution along the shores of Lake Tanganyika, from south-eastern Democratic Republic of Congo to Mozambique and northern Zimbabwe. The type locality is Kundelungu, northwest Lake Tanganyika, Democratic Republic of Congo.

The dorsum is uniformly dark olive green above. The venter is whitish. The nasal is divided. The internasals are shorter and narrower than the prefrontals. The prefrontals are equal in length and breadth. The temporal formula is 1+2. There is 1 preocular, the depth of which is less than or equal to the diameter of the eye. The pupil is round or slightly vertically elliptical. There are 2 postoculars, the superior larger than the inferior. There are no suboculars. There are usually 7(4,5) upper labials. There are 7(3) lower labials. The anterior chin shields are longer than the posterior pair. The dorsal scales are smooth, without apical pits, and arranged in 19–21 straight rows. There are 164–174 ventrals (fewer than 172 in

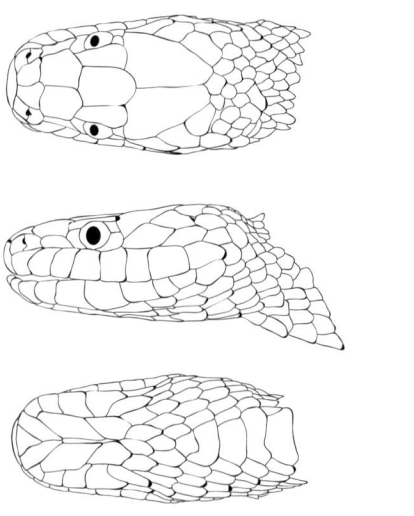

FIGURE 10.49. *Lycodonomorphus bicolor* RMCA 80-32-R-72. By T. Giri

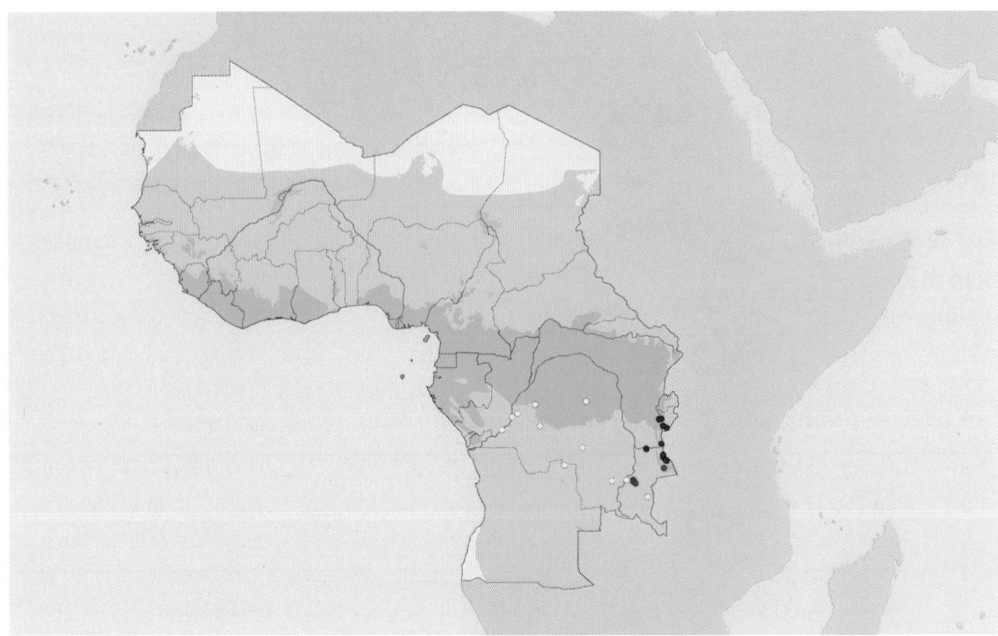

FIGURE 10.50. *Lycodonomorphus bicolor* (blue), *Lycodonomorphus leleupi* (red), *Lycodonomorphus subtaeniatus* (yellow). By K. Jackson

males, more than 163 in females) and 52–67 divided subcaudals (more than 64 in males, fewer than 62 in females). The maximum length recorded for this species is 851 mm (Loveridge 1958).

Lined Water Snake: *Lycodonomorphus subtaeniatus* Laurent 1954

Two subspecies of Lined Water Snake are recognized, differing in distribution and in ventral and subcaudal counts. The range of *Lycodonomorphus subtaeniatus subtaeniatus* Laurent 1954 extends from northern Angola to Democratic Republic of Congo. The type locality is Keseki, near Kwamouth, Democratic Republic of Congo. *L. s. upembae* Laurent 1954 is restricted to the vicinity of Lake Upemba, Democratic Republic of Congo. The type locality is Nyonga, Katanga, Democratic Republic of Congo.

The body is olive green, paler on the sides of the body with a lateral stripe of olive green. The pupil is vertically elliptical. The venter is pale. The nasal is semi-divided. The internasals are shorter and narrower than the prefrontals. The prefrontals are longer than broad. The temporal formula is 1+2. There is 1 preocular, the depth of which is less than or equal to the diameter of the eye, or occasionally 2, the superior larger than the inferior. There are 2 postoculars, the inferior larger than the superior. There are no suboculars. There are usually 8(4,5) upper labials and 8(4) lower labials. The anterior and posterior pairs of chin shields are roughly equal in length. The dorsal scales are smooth, with 2 apical pits, and arranged in 21–23 straight rows. There are 172–193 ventrals (175–193 for *L. s. subtaeniatus* and 172–188 for *L. s. upembae*) and 29–58 divided subcaudals (41–58 for *L. s. subtaeniatus* and 29–40 for *L. s. upembae*). The max-

FIGURE 10.51. *Lycodonomorphus subtaeniatus upembae*, Democratic Republic of Congo. By E. Greenbaum

imum length recorded for this species is 1,009 mm (Laurent 1954).

Wolf Snakes: Genus *Lycophidion* Fitzinger 1843

The Wolf Snakes, genus *Lycophidion*, are a genus of 19 species of small, nocturnal, terrestrial snakes, with an African distribution. Twelve species occur within our zone.

Lycophidion feed primarily on skinks and possess a suite of maxillary dentitional specializations associated with feeding on hard-bodied, cylindrical prey, including an arched shape to the maxilla and enlarged anterior maxillary teeth separated by a diastema from a series of small posterior maxillary teeth. Savitzky (1983) described this type of dentition in *Lycophidion* and in other snakes (including the File Snakes, *Gonionotophis*) and discovered that the posterior teeth are hinged, allowing them to fold backward against the maxillary bone, thereby avoiding being broken off, for example, when gripping between the hard, overlapping scales of the skinks on which Wolf Snakes feed. Little is known about the natural history of this genus. *Lycophidion capense* in east Africa feed primarily on small lizards (especially skinks), sometimes other small snakes and amphisbaenians, and females lay clutches of 3–8 eggs (Pitman 1974; Spawls et al. 2004). The natural history of other species in the genus is probably similar.

Wolf Snakes are typically small, the maximum total length ranging from 350 to 750 mm depending on the species. The head is small and somewhat flat, with a poorly defined neck. The tail is short. The eyes are small with a vertically elliptical pupil. The nasal is divided in all species within our zone, and the posterior nasal may be either separated from or in contact with the first upper labial. The loreal is present. The internasals and prefrontals are paired. There are 1–2 preoculars and 1–3 postoculars. There is a supraocular but no suboculars. There are 7–8 upper labials and 8–10 lower labials. There is 1 anterior temporal. There are 2 pairs of chin shields, separated by a submandibular groove that is usually shallow but is pronounced in some species. The dorsal scales are smooth and arranged in 15–17 straight rows midbody. The number of dorsal scale rows may or may not decrease toward the posterior end of the body, depending on the species. The dorsal scales each have 1 or more apical pits (2–3 in *Lycophidion irroratum* and *L. nigromaculatum*, 4 or more in *L. laterale*). There are 144–214 ventrals and 22–58 divided subcaudals. The cloacal scale is single. Scale characters vary relatively little between species, but relative scale proportions are often helpful for identification, as is coloration. The hemipenes are bilobed, and the sulcus spermaticus is forked.

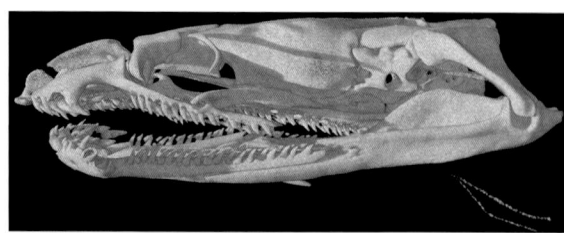

FIGURE 10.52. Computed tomography scan of the skull of the Cape Wolf Snake, *Lycophidion capense*, showing enlarged anterior maxillary tooth followed by diastema on the maxilla (enlarged anterior tooth on the mandible is broken off in this specimen). Courtesy of Digimorph.org

White-spotted Wolf Snake: *Lycophidion albomaculatum* Steindachner 1870

The range of *Lycophidion albomaculatum* extends from Senegal to Mali and to Guinea. The type locality is near Dakar, Senegal.

The overall coloration of the body is black with extensive white flecks on the black scales. The dorsal pattern is a series of pinkish-red rectangular blotches along the dorsal midline (in specimens from coastal populations), or sometimes uniformly black or dark brown (in specimens from inland populations). The ventral side is darker than the dorsal side. The nasal is divided, and the posterior nasal is separated from the first upper labial. The prefrontals are longer than they are broad. The length of the loreal is two times its depth. There is 1 preocular in broad contact with the frontal. There are 2 postoculars, the inferior larger than the superior. The frontal is broader than it is long. There are 8(3,4,5) upper labials. The anterior and posterior chin shields are approximately equal in length. There are 9(4) upper labials. There are 17 dorsal scale rows, with a reduction in number toward the posterior end of the body. There is 1 apical pit. There are 180–210 ventrals (fewer than 199 in males, more than 194

Key to Central and Western African Species of the Genus *Lycophidion*

1	15 dorsal scale rows	*L. meleagre*
1'	17 dorsal scale rows	2
2(1)	2 upper labials in contact with the eye	3
2'	3 upper labials in contact with the eye	4
3(2)	9 lower labials; first upper labial not in contact with the posterior nasal	*L. laterale*
3'	7 lower labials; first upper labial in contact with the posterior nasal	*L. depressirostre*
4(2)	First upper labial not in contact with the posterior nasal	5
4'	First upper labial in contact with the posterior nasal	8
5(4)	17 dorsal scale rows the length of the body, 17 rows just before the cloacal scale	6
5'	17 dorsal scale rows anteriorly, diminishing to 15 rows just before the cloacal scale	12
6(5)	Dorsum pale with staggered dark paravertebral blotches	*L. nigromaculatum*
6'	Dorsum dark, lightly stippled with white	7
7(6)	Only 1 apical pit on each dorsal scale	*L. ornatum*
7'	At least 2 apical pits on each dorsal scale	*L. irroratum*
8(4)	Anterior chin shields distinctly longer than the posterior pair	9
8'	Anterior chin shields equal in length to or shorter than the posterior pair	11
9(8)	Dorsum with black transverse blotches; individual dorsal scales uniform in color	*L. multimaculatum*
9'	Dorsum uniformly dark; individual scales lightly stippled with white	10
10(9)	Frontal distinctly broader than long; snout dark	*L. taylori*
10'	Frontal as long as or longer than broad; tip of the snout pale	*L. capense*
11(8)	Dorsum pale brown, lightly stippled with white	*L. hellmichi*
11'	Dorsum dark, usually with pale rings or blotches	12
12(11)	Black with 30 or so silvery rings	*L. semicinctum*
12'	Dorsum uniformly black or patterned with 40 or so red rectangles	*L. albomaculatum*

in females) and 33–53 subcaudals (more than 41 in males, fewer than 39 in females). The maximum recorded length is 620 mm (Trape and Mané 2004).

Cape Wolf Snake: *Lycophidion capense* Smith 1831

The distribution of the Cape Wolf Snake covers a large part of east and central Africa from Egypt in the north and Cameroon in the west, to Namibia in the southwest and Mozambique in the southeast. Of five subspecies recognized for *Lycophidion capense*, one subspecies, *L. c. jacksoni* Boulenger 1893, occurs within our zone. The range of this subspecies extends from northeastern Democratic Republic of Congo in the west to Ethiopia in the north Tanzania in the south. The type locality is Kilimanjaro, Tanzania.

The body is gray or brown above, with

FIGURE 10.53. *Lycophidion albomaculatum*, Senegal. By W. Wüster

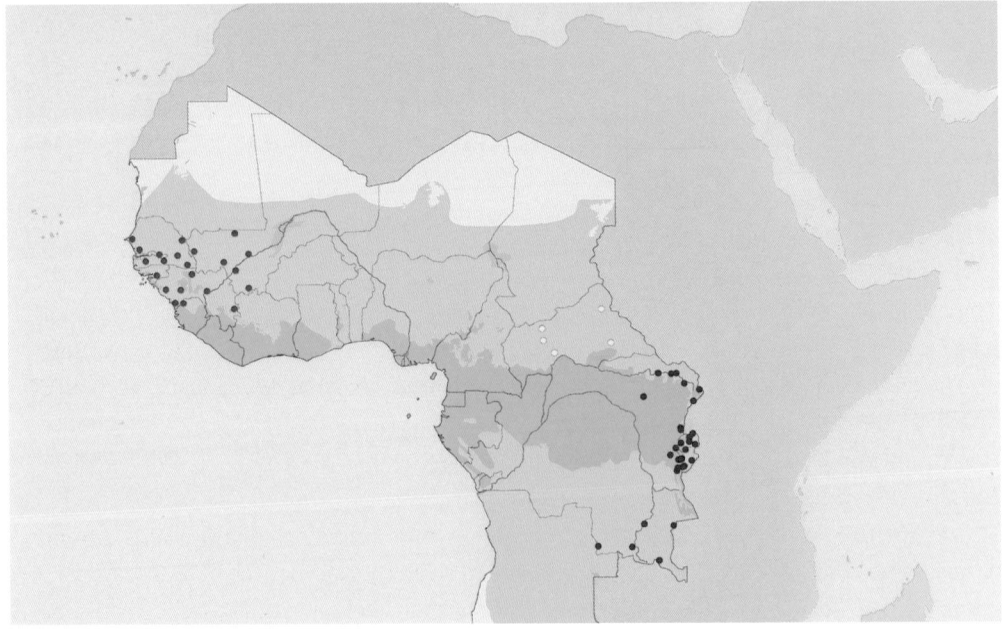

FIGURE 10.54. *Lycophidion albomaculatum* (red), *Lycophidion capense* (blue), *Lycophidion depressirostre* (yellow). By K. Jackson

pale tips (and sometimes edges) to each dorsal scale. The tip of the snout is pale. The venter is pale in juveniles and dark in adults. The nasal is divided, and the posterior nasal is in contact with the first upper labial. The prefrontals are as long as or longer than they are broad. The length of the loreal is two times its depth. There is 1 preocular, in broad contact with the frontal. There are 2 postoculars, the inferior longer than the superior. The frontal is equal in length and breadth. There are 8(3,4,5) upper labials. The anterior chin shields are longer than the posterior pair. There are 8(4) lower labials. There are 17 dorsal scale rows midbody, with a reduction posteriorly. There is 1 apical pit on each dorsal scale. There are 167–193 ventrals (fewer than 181 in males, more than 168 in females) and 24–45 subcaudals (more than 33 in males, fewer than 40 in females). The maximum recorded length is 580 mm (Broadley and Hughes 1993).

Flat-snouted Wolf Snake: *Lycophidion depressirostre* Laurent 1968

The range of *Lycophidion depressirostre* extends from Central African Republic to Chad to Sudan. The type locality is Torit, Sudan.

The body is uniformly dark brown, with a white fleck at the tip of each dorsal scale. The nasal is divided, and the posterior nasal is in contact with the first upper labial. The length of the loreal is two times its depth. There is 1 preocular in contact with the frontal and 2 postoculars of equal size. The frontal is equal in length and breadth. There are 8(4,5) upper labials. The posterior chin shields are equal to or longer than the anterior pair. There are 7(5) lower labials. There are 17 dorsal scale rows midbody, with a reduction in number of rows posteriorly.

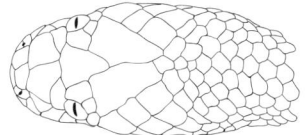
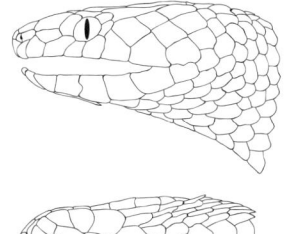
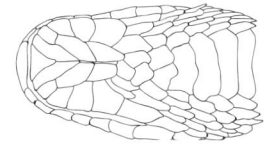

FIGURE 10.55. *Lycophidion capense* RMCA 86-01-R-51. By T. Giri

FIGURE 10.56. *Lycophidion capense*, Kenya. By S. Spawls

There are 153–178 ventrals (fewer than 175 in males, more than 160 in females) and 22–40 subcaudals (more than 31 in males, fewer than 32 in females). The maximum recorded length is 492 mm (Laurent 1968).

Hellmich's Wolf Snake: *Lycophidion hellmichi* Laurent 1964

The range of *Lycophidion hellmichi* extends from Namibia to southern Angola and southern Democratic Republic of the Congo. The type locality is Kapolopopo in the Mocamedes Desert of Angola.

The body is dark brown above, each dorsal scale with a white tip or border. There is sometimes a paler vertebral stripe. The first 1–3 lateral dorsal scale rows and the lateral ends of the ventrals are orange or white. The ventrals are dark brown with pale edges. The nasal is divided, and the posterior nasal is in contact with the first upper labial. The prefrontals are longer than they are broad. The length of the loreal is two times its depth. There is 1 preocular, in broad contact with the frontal. There are 1–3 postoculars. The frontal is equal in length and breadth. There are 8(3,4,5) or 9(3,4,5) upper labials. The anterior and posterior pairs of chin shields are of roughly equal length. There are 8–10(4) or (5) lower labials. There are 17 dorsal scale rows, with a reduction posteriorly. There is 1 apical pit on each dorsal scale. There are 195–214 ventrals and 24–58 subcaudals (more than 30 in males,

fewer than 45 in females). The maximum recorded length is 471 mm (Broadley 1991a).

Pale Wolf Snake: *Lycophidion irroratum* (Leach 1819)

The range of *Lycophidion irroratum* extends from Guinea-Bissau to Central African Republic. The type locality is Fanti (close to what is now Cape Coast), Ghana.

The body is dark gray above, with a tiny white dot on each dorsal scale. In juveniles, there is a dorsal pattern of small back blotches alternating side to side along the paravertebral line. The snout is paler than the body, with a white lateral line. The ventral surface of the body is blackish. The nasal is divided, and the posterior nasal is separated from the first upper labial. The prefrontals are longer than they are broad. The length of the loreal is more than two times its depth. There is usually 1 preocular,

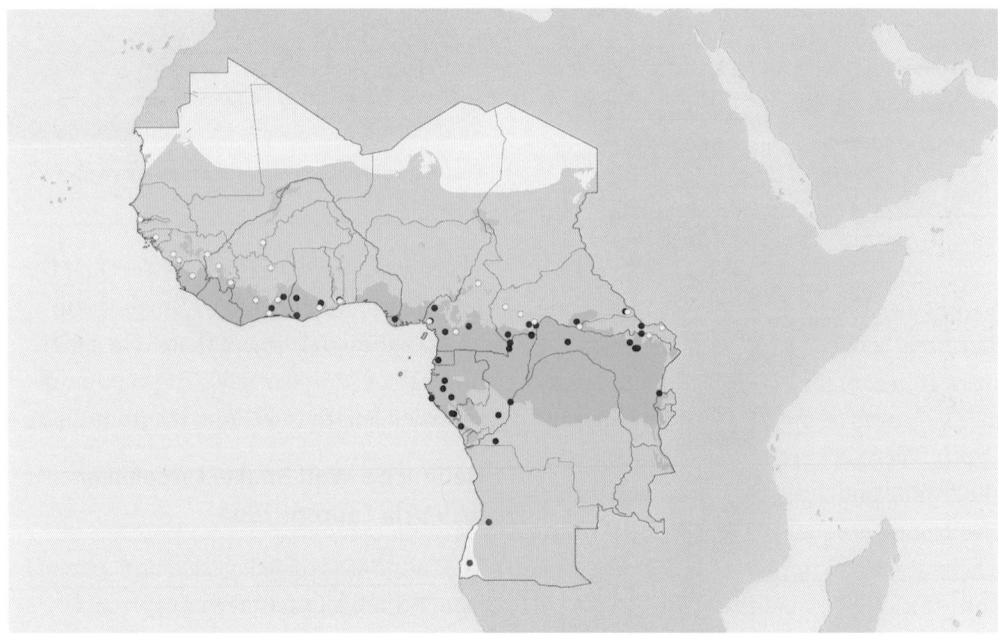

FIGURE 10.57. *Lycophdion hellmichi* (red), *Lycophidion irroratum* (yellow), *Lycophidion laterale* (blue). By K. Jackson

FIGURE 10.58. *Lycophidion irroratum*, Ghana. By M. Fujita

in broad contact with the frontal. Occasionally there may be 2 preoculars, the superior larger than the inferior. There are 2 postoculars, the inferior larger than the superior. The frontal is longer than broad. There are usually 8(3,4,5) and sometimes 7(3,4,5,) upper labials. The anterior chin shields are much longer than the posterior pair. There are 8(4) or 9(5) lower labials. There are 17 dorsal scale rows with no reduction in number of rows toward the posterior end of the body. There are 2–4 apical pits. There are 158-193 ventrals (fewer than 188 in males, more than 163 in females) and 27–54 subcaudals (more than 31 in males, fewer than 53 in females). The maximum recorded length for *L. irroratum* is 450 mm (Villiers 1975).

Flat Wolf Snake: *Lycophidion laterale* Hallowell 1857

The range of *Lycophidion laterale* extends from Liberia to Angola to the Democratic Republic of the Congo. The type locality is Gabon.

The body is uniformly brown above, with two lighter brown lateral stripes on the snout. The ventral side is dark with pale edges to each scale. The nasal is divided, and the posterior nasal is separated from the first upper labial. The prefrontals are broader than they are long. The length of the loreal is more than two times its depth. There is 1 preocular in broad contact with the frontal. There are 2 postoculars, the inferior larger than the superior. The fron-

FIGURE 10.59. *Lycophidion laterale*, Republic of Congo. By K. Jackson

tal is equal in length and breadth. There are 7(4,5) or 8(4,5) upper labials. The anterior chin shields are longer than the posterior pair. There are usually 9(4) and occasionally 8(4) upper labials. There are 17 dorsal scale rows, with no reduction posteriorly. There are 4 or more apical pits. There are 170–212 ventrals. There are 27–45 subcaudals, with a trend toward higher numbers of subcaudals in the eastern part of its range. The maximum recorded length is 500 mm (Villiers 1975).

Speckled Wolf Snake: *Lycophidion meleagre* Boulenger 1893

The range of *Lycophidion meleagre* extends from Angola to Kenya. The type localities are Ambriz and Ambrizette, Angola.

The body is gray with a small white dot on each dorsal scale. The snout is pale. The tail is uniformly dark. The nasal is divided, and the posterior nasal is in contact with the first upper labial. The prefrontals are equal in length and breadth. The length of the loreal is two times its depth. There is 1 preocular in broad contact with the frontal. There are 2 postoculars, the inferior larger than the superior. The frontal is broader than it is long. There are 8(3,4,5) upper labials. The anterior chin shields are longer than the posterior pair. There are 8(4) lower labials. There are 15 dorsal scale rows midbody, with no reduction posteriorly. There are 144–174 ventrals and 22–34 subcaudals. The maximum recorded length is 350 mm (Spawls et al. 2004).

Many-spotted Wolf Snake: *Lycophidion multimaculatum* Boettger 1888

The range of *Lycophidion multimaculatum* extends from Central African Republic to Angola and Tanzania. The type localities

are Povo Nemlao and Povo Netonna, near Banana, Democratic Republic of the Congo.

The body is reddish brown above, with pale stippling, and a dorsal pattern of a series of black blotches or bands. The venter is white, sometimes with a dark mid-ventral stripe or line of blotches. The nasal is divided, and the posterior nasal is in contact with the first upper labial. The length of the loreal is two times its depth. There is 1 preocular, in broad contact with the frontal. There are 1 or 2 postoculars, the inferior larger than the superior. The frontal is broader than it is long. There are 8(3,4,5) upper labials. The anterior chin shields are longer than the posterior pair. There are 8(4) or 8(5) lower labials. There are 17 dorsal scale rows, decreasing toward the posterior end of the body. There is 1 apical pit on each dorsal scale. There are 159–187 ventrals (fewer than 181 in males, more than 164 in females) and 22–42 subcaudals (more than 27 in males, fewer than 32 in females). The maximum recorded length is 392 mm (Broadley 1983).

Black-spotted Wolf Snake: *Lycophidion nigromaculatum* (Peters 1863)

The Range of *Lycophidion nigromaculatum* extends from Guinea to Ghana. The type locality is Dabocron, Ghana.

The body is light brown to reddish brown above, with a pattern of black rectangular blotches alternating from side to side along the paravertebral lines. The head is the same color as the body, with a pale lateral stripe outlined in black. The ventral side of the body is darker than the dorsal side. The nasal is divided, and the posterior nasal is separated from the first upper labial. The length of the prefrontals is equal to their breadth. The length of the loreal is 1.5 times

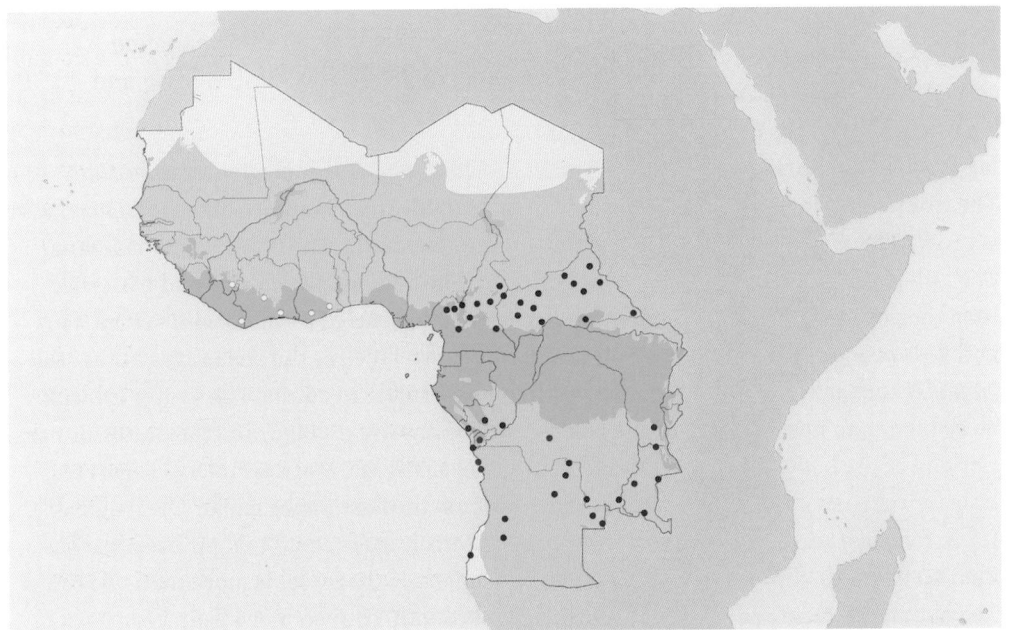

FIGURE 10.60. *Lycophidion meleagre* (red), *Lycophidion multimaculatum* (blue), *Lycophidion nigromaculatum* (yellow). By K. Jackson

its depth. There is 1 preocular, in broad contact with the frontal. There are 2 postoculars, the inferior larger than the superior. The frontal is as long as or longer than it is broad. There are 8(3,4,5) upper labials. The anterior chin shields are longer than the posterior pair. There are 8(4) or 9(5) upper labials. There are 17 dorsal scale rows, with no reduction toward the posterior end of the body. There are 2–3 apical pits. There are 179–212 ventrals (fewer than 186 in males, more than 178 in females) and 42–53 subcaudals (more than 49 in males, fewer than 48 in females). The maximum recorded length is 367 mm (Leston and Hughes 1968).

Forest Wolf Snake: *Lycophidion ornatum* Parker 1936

The range of *Lycophidion ornatum* extends from Nigeria to Sudan to Tanzania. The type locality is Congulu, Angola.

FIGURE 10.61. *Lycophidion nigromaculatum*, Ghana. By A. Leaché

The body is uniformly dark brown or gray above, with white stippling on each scale. The head is marked with a pale lateral stripe extending to the neck. The venter is dark. The nasal is divided, and the posterior nasal is not in contact with the first upper labial. The length of the loreal is 1.5 times its depth. There is 1 preocular in contact with the frontal. There are 2 postoculars, the inferior larger than the superior. The frontal is as long as or longer than it is broad. There are 8(3,4,5) upper labials. The anterior chin shields are much longer than the posterior pair. The submandibular groove is pronounced. There are 8(4) lower labials. There are 17 dorsal scale rows, with no reduction in number toward the posterior end of the body. There are 175–212 ventrals (fewer than 207 in males, more than 187 in females) and 32–53 subcaudals (more than 40 in males, fewer than 47 in females). The maximum recorded length is 590 mm (Broadley and Hughes 1993).

Ringed Wolf Snake: *Lycophidion semicinctum* Duméril, Bibron and Duméril 1854b

The range of *Lycophidion semicinctum* extends from Mali to Guinea to Central African Republic. The type locality is Ghana.

The body is black above and patterned with a series of 30 or so white rings. This pattern fades as the snake ages but is still discernible in adults. The venter is black. The nasal is divided, and the posterior nasal is usually (80% of specimens) separated from the first upper labial. The length of the prefrontals is equal to their breadth. The length of the loreal is more than two times its depth. In specimens from West Africa, there are usually 2 preoculars, the superior larger than the inferior and in broad contact with the frontal. In specimens from central Africa, there is usually just 1 preocular.

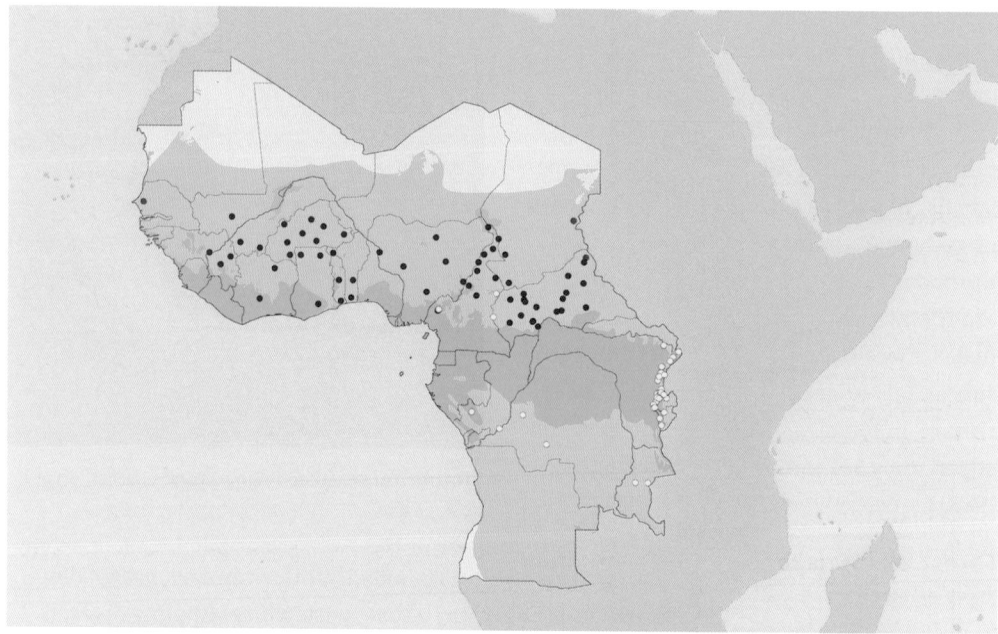

FIGURE 10.62. *Lycophidion ornatum* (yellow), *Lycophidion semicinctum* (blue), *Lycophidion taylori* (red). By K. Jackson

FIGURE 10.63. *Lycophidion semicinctum*, Ghana. By S. Spawls

There are usually 2 postoculars, the inferior larger than the superior, but sometimes there are 3 postoculars of equal size. The frontal is broader than it is long. There are 8(3,4,5) upper labials. The anterior chin shields are shorter than or equal to the posterior pair. There are 8(4) to 9(4) lower labials. There are 17 dorsal scale rows, decreasing posteriorly. There is 1 apical pit. There are 182–208 ventrals (fewer than 200 in males, more than 198 in females) and 35–57 subcaudals (more than 46 in males, fewer than 47 in females). The maximum recorded length for *L. semicinctum* is 750 mm (Villiers 1975).

Taylor's Wolf Snake: *Lycophidion taylori* Broadley and Hughes 1993

The range of *Lycophidion taylori* extends from Senegal to Somalia. The type locality is Borama, Somalia.

The body is gray or brownish gray, with a white tip to each dorsal scale. The head is dark with white flecks. Many specimens have a white nuchal band. The venter is dark, with white edges to the ventral scales. The nasal is divided, and the posterior nasal is in contact with the first upper labial. The prefrontals are longer than they are broad. The length of the loreal is two times its depth. There is 1 preocular in contact with the frontal and 2 small postoculars. The frontal is broader than it is long. There are 8(3,4,5) upper labials. The anterior chin shields are longer than the posterior pair. There are 8(4) or 8(5) lower labials. There are 17 dorsal scale rows, with a reduction posteriorly. There is 1 apical pit on each dorsal scale. There are 158–184 ventrals (fewer than 180 in males, more than 164 in females) and 26–36 subcaudals (more than 29 in males, fewer than 31 in females). The maximum recorded length is 496 mm (Broadley and Hughes 1993).

Subfamily Pseudoxyrhophiinae

The Pseudoxyrhophiinae are a group of about 80 species that represent the major radiation of snakes in Madagascar. Apart from a few species of boids (*Sanzinia* and *Acrantophis*), some Blind Snakes (Typhlopidae and Xenotyphlopidae), and the psammophiine genus *Mimophis*, all the snakes in Madagascar are pseudoxyrhophiines. Pseudoxyrhophiines are restricted to Madagascar, with a few possible exceptions occurring on the African mainland. The taxonomic affiliations of the African genera *Duberria* and *Buhoma* are not certain, but they may be pseudoxyrhophiines and are included here as such.

The genus *Duberria* was grouped by Bogert (1940) in Group 5 with the Mole Snakes, *Pseudaspis* (Family Lamprophiidae; subfamily Pseudaspidinae), and the African Water Snakes, *Grayia* (Family Colubridae; subfamily Grayiinae), on the basis of absence of vertebral hypapophyses on the posterior part of the body, forked sulcus spermaticus, and aglyph dentition. But subsequent work has not supported Bogert's Group 5 taxa as being closely related. Recent molecular phylogenetic evidence (Vidal et al. 2008; Kelly et al. 2009; Pyron et al. 2011, 2013) finds *Duberria* nested within the Madagascan Pseudoxyrhophiinae.

We tentatively include Forest Snakes of the genus *Buhoma* among the Pseudoxyrhophiinae here. Species now assigned to *Buhoma* formerly belonged to the genus *Geodipsas*, then a genus of snakes found both in Madagascar and in Africa. Ziegler et al. (1997) moved all the African species of *Geodipsas* to their own genus, *Buhoma*, on the basis of their hemipenal morphology. The remaining Madagascar *Geodipsas* have since been subsumed into *Compsophis*, a genus of Malagasy pseudoxyrhophiine snakes (Glaw et al. 2007). Recent molecular studies (Vidal et al. 2008; Kelly et al. 2009; Pyron et al. 2011), however, found *Buhoma* not to be closely related the Madagascan pseudoxyrhophiines, but without reaching any consensus on the actual phylogenetic affiliations of the genus. Pyron et al. (2013) found *Buhoma* to be most closely related to species included among the pseudaspidine lamprophiids, such as the Mole Snake, *Pseudaspis cana*.

Forest Snakes: Genus *Buhoma* Ziegler, Vences, Glaw, and Böhme 1997

The Forest Snakes, genus *Buhoma*, are four species of small, terrestrial, nocturnal Forest Snakes in central and eastern Africa. Two species occur within our zone. Little is known of their natural history. Knoepffler (1966) collected a female *Buhoma depressiceps* who contained 6 eggs, each 8 mm in length. Werner (1899) reports that *B. depressiceps* feed on amphibians.

The head is small, the neck indistinct. The tail is short. The eye is medium sized with a round or slightly vertically elliptical pupil. The nasal may be single or divided. The internasals and prefrontals are paired. The loreal is present. There are 1–3 preoculars and no suboculars. There are 1–3 postoculars. The frontal is longer than it is broad. The temporal formula is 1+(2 or 3). There are 7 or 8 upper labials, 2 of which are in contact with the eye, and 8–11 lower labials. There are 2 pairs of chin shields. The submandibular groove is pronounced. The dorsal scales are keeled, without apical pits, and arranged in 17 or 19 straight rows at midbody. The cloacal scale is single. The subcaudals may be single or divided. The vertebral row is not enlarged relative to the other dorsal scale rows. The maxillary dentition is opisthoglyph with 15–19 ungrooved maxillary teeth followed by 2 slightly larger, grooved teeth at the posterior end of the maxilla. The hemipenes are unforked, and the sulcus spermaticus is forked.

Pale-headed Forest Snake: *Buhoma depressiceps* (Werner 1897)

The range of the Pale-headed Forest Snake extends from Cameroon to Uganda. The type locality is Barombi, Cameroon.
The dorsum is brown or gray, with a distinct dark vertebral stripe on the posterior part of the body, replaced anteriorly with a double row of spots, sometimes coming together to form a zig-zag pattern. The head is darker than the body. There is usually a pale nuchal band. The venter is pale, with a series of lateral spots that sometimes merge to form a pair of continuous ventrolateral stripes along the middle third of the body.

Key to Central and Western African Species of the Genus *Buhoma*

1 19 dorsal scale rows at midbody .... *B. depressiceps*
1′ 17 dorsal scale rows at midbody .... *B. marlieri*

The nasal is single. The loreal is equal in depth and length and is smaller than the eye. There are 2 or 3 preoculars, of which the superior is the largest. There are 2 postoculars, the superior larger than the infe-

FIGURE 10.64. *Buhoma depressiceps*, Cameroon. By V. Gvoždík

rior. The temporal formula is 1+2. There are 7(3,4) upper labials. There are usually 8(4) lower labials, of which the fourth lower labial is the largest. The anterior chin shields are longer than the posterior pair. The dorsal scales are keeled, without apical pits, and arranged in 19 straight rows midbody. There are 143–151 ventrals and 27–36 divided subcaudals. The cloacal scale is single. The maximum length recorded for this species is 298 mm (Knoepffler 1966).

Marlier's Forest Snake: *Buhoma marlieri* (Laurent 1956)

Marlier's Forest Snake is endemic to the Albertine Rift. The type locality is Mwana, Kivu, Democratic Republic of Congo.

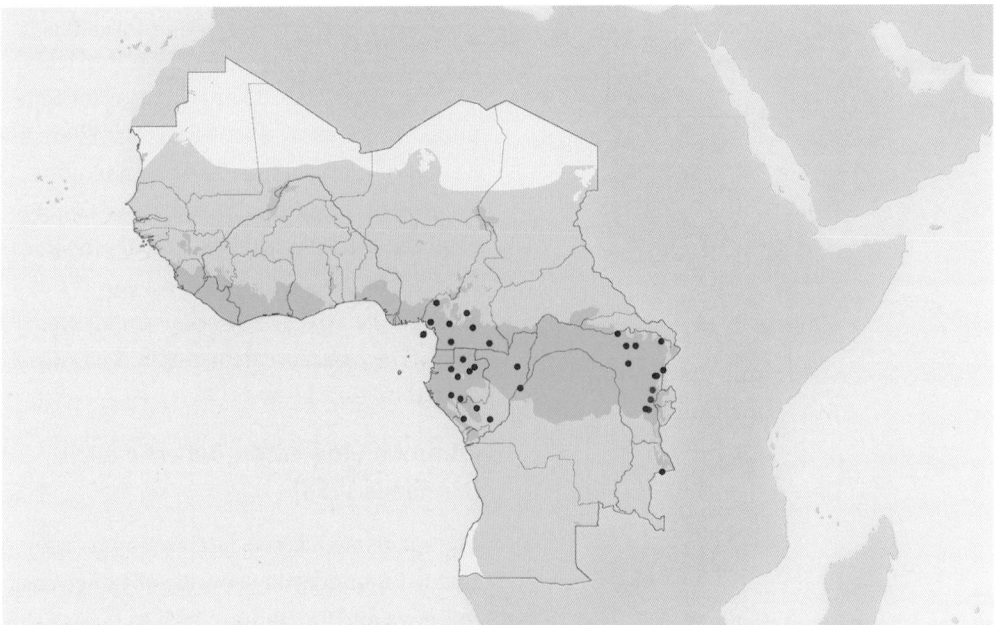

FIGURE 10.65. *Buhoma depressiceps* (blue), *Buhoma marlieri* (red). By K. Jackson

The coloration of *Buhoma marlieri* is similar to that of *B. depressiceps*. The dorsal lateral spots are larger in *B. marlieri* than in *B. depressiceps*, more often merging to form bands. The head is paler in *B. marlieri* than in *B. depressiceps*. The venter is pale, with a pair of ventrolateral stripes made up of small dark spots.

The nasal is semi-divided. The loreal is equal in depth and length, and in some individuals it may be broken up into 2 or 3 smaller scales. There are 3 preoculars, of which the superior is the largest. There are 2 postoculars, the superior larger than the inferior. The temporal formula is 1+2. There are 7(3,4) upper labials. There are usually 8(3) lower labials, of which the third lower labial is the largest. The anterior chin shields are longer than the posterior pair. The dorsal scales are keeled, without apical pits, and arranged in 17 straight rows midbody. There are 146–163 ventrals and 35–43 divided subcaudals. The cloacal scale is single. The maximum length recorded for this species is 440 mm (Spawls et al. 2004).

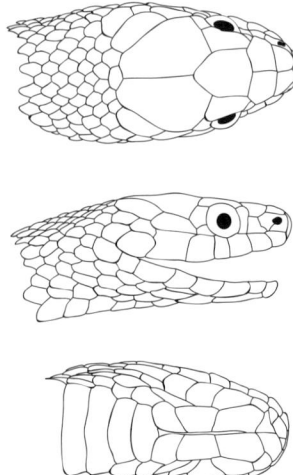

FIGURE 10.66. *Buhoma marlieri* RMCA 18091. By T. Giri

The Slug-eaters: Genus *Duberria* Fitzinger 1826

Duberria is a genus of two species found in southern and eastern Africa. One of these species occurs within our zone. *Duberria* are dietary specialists, feeding on slugs and snails, which they find by following the slimy trails they leave. Slug-eaters spend their time in tufts of grass or under rocks. They are small, harmless to humans, and rarely encountered. They give birth to litters of 4–13 live young, which Loveridge (1933) reports that they sometimes eat.

The maxillary dentition is aglyph, with 10–12 maxillary teeth decreasing in size posteriorly.

The head is small and the neck not distinct from the stout, cylindrical body. The tail is short and ends in a sharp, pointed tip. The eye is small, with a round pupil. The nasal is undivided. The internasals and prefrontals are paired. The loreal scale may be present or absent. There are 1–2 preoculars and 1–2 postoculars. The frontal is longer than it is broad. There are 6–7 upper labials and no suboculars. There is 1 anterior temporal. There are 6–8 lower labials. There are 2 pairs of chin shields. The submandibular groove is shallow. The dorsal scales are smooth, with 2 apical pits, and arranged in 15 straight rows. The vertebral row is not enlarged relative to the other dorsal scale rows. The cloacal scale is single. The subcaudals are divided.

Common Slug-eater: *Duberria lutrix* (Linnaeus 1758)

The range of *Duberria lutrix* extends from eastern Democratic Republic of Congo east to Kenya and Tanzania, north as far as Ethiopia, and south as far as South Africa. The type locality for the species is "Indiis" (in error).

The dorsum is olive brown above, usually uniformly so but sometimes with a darker vertebral stripe. The venter is yellow in the middle, darker at the sides. There is 1 preocular and 1 postocular. The temporal formula is 1+2. There are 6(3,4) upper labials. There are 6(3) or sometimes 7(3) lower labials. There are 116–148 ventrals (fewer than 133 in males, more than 123 in females) and 17–46 subcaudals (more than 27 in males, fewer than 39 in females). The maximum length recorded for this species is 450 mm (Spawls et al. 2004).

Several subspecies of *D. lutrix* exist, including three that occur within our zone: *D. l. shirana* (Boulenger 1894), *D. l. atriventris* (Sternfeld 1912), *D. l. currylindhali* Laurent 1956. The type localities for the subspecies are as follows. For *D. l. shirana*, the type locality is Shire Highlands, Malawi. For *D. l. atriventris*, the type locality is Kisenyi, Rwanda. For *D. l. currylindhali*, the type locality is Upper Lubitshako River, Fizi Territory, Kivu Province, Democratic Republic of Congo. These subspecies are distinguished from one another on the basis of ventral and subcaudal counts and the condition of the loreal scale. *D. l. shirana* has 124–148 ventrals (124–132 in males, 141–148 in females) and 24–46 subcaudals (27–46 in males, 24–38 in females). *D. l. atriventris* has 120–144 ventrals (120–130 in males, 131–144 in females) and 17–37 subcaudals (28–37 in males, 17–24 in females). *D. l. currylindahli* has 116–135 ventrals (116–125 in males, 124–135 in females) and 24–39 subcaudals (35–39 in males, 24–29 in females). The condition of the loreal scale is quite variable in this species. Usually in *D. l. atriventris* there is a small loreal scale. The loreal scale is

FIGURE 10.67. *Duberria lutrix*, South Africa. By A. Barlow

FIGURE 10.68. *Duberria lutrix* feeding on a snail, South Africa. By A. Barlow

usually absent in *D. l. shirana* and in *D. l. currylindahli*. Broadley (1983) found the loreal scale to be absent in 7% of specimens of *D. l. lutrix*.

Subfamily Pseudaspidinae

The subfamily Pseudaspidinae includes two monotypic genera with distributions in sub-Saharan Africa: *Pseudaspis*, which occurs within our zone, and *Pythonodipsas*, which occurs farther south. The taxonomic affiliations of the Pseudaspidinae remain uncertain, however. Bogert included *Pseudaspis* in his Group 5, along with the Slug-eaters, *Duberria* (Family Lamprophiidae; subfamily Pseudoxyrhophiinae), and

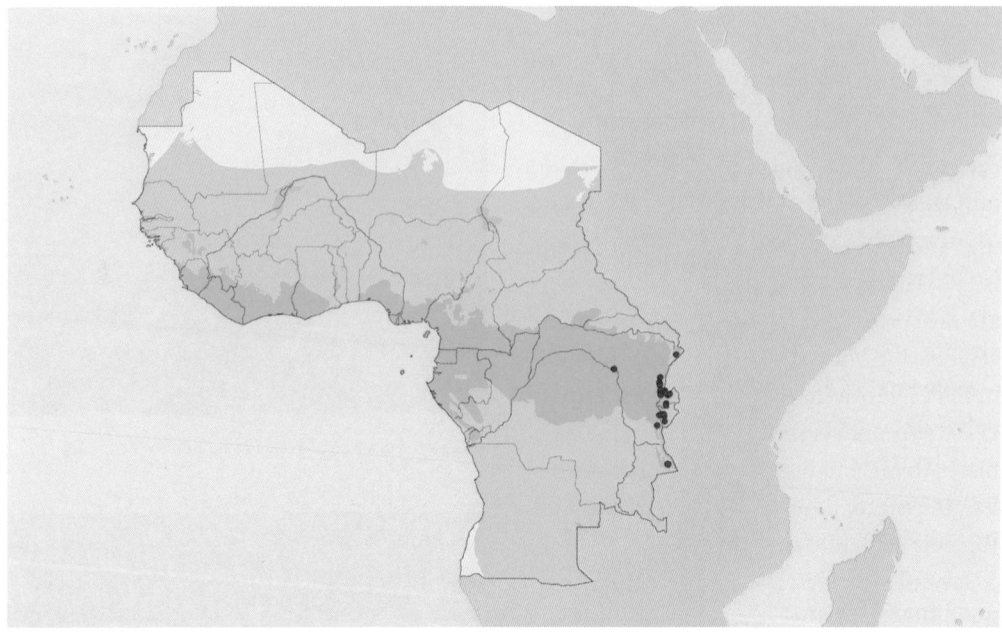

FIGURE 10.69. *Duberria lutrix* (red). By K. Jackson

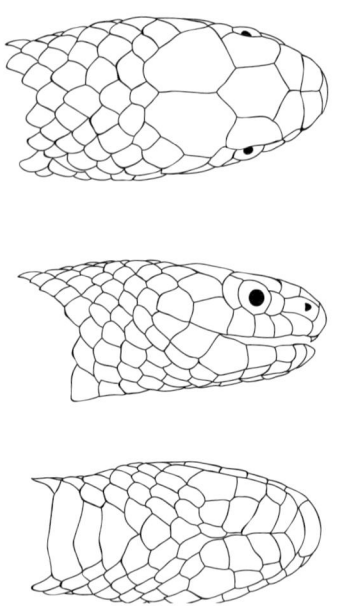

FIGURE 10.70. *Duberria lutrix* RMCA 21634. By T. Giri

the African Water Snakes, *Grayia* (Family Colubridae; subfamily Grayiinae), on the basis of absence of vertebral hypapophyses on the posterior part of the body, forked sulcus spermaticus, and aglyph dentition. But subsequent work has not supported Bogert's Group 5 taxa as being closely related. Recent molecular evidence consistently supports *Pseudaspis* and *Pythonodipsas* as being each other's closest relatives (Vidal et al. 2008; Kelly et al. 2009; Pyron et al. 2013) but differs on the affiliations of these two genera relative to other taxa. Vidal et al. (2008) find the Pseudaspidinae (*Pseudaspis* and *Pythonodipsas*) to be most closely related to the Shovel-snouts, genus *Prosymna* (family Lamprophiidae; subfamily Prosymninae). Pyron et al. (2013) find the Pseudaspidinae (*Pseudaspis* and *Pythonodipsas*) to be most closely related to the Asian genus *Psammodynastes* and to the African Forest Snakes of the genus *Buhoma* (included here with the Pseudoxyrhophiinae).

Genus *Pseudaspis* Fitzinger 1843

Mole Snake: *Pseudaspis cana* (Linnaeus 1758)

Pseudaspis is a monotypic genus. The range of *Pseudaspis cana* extends from Namibia to Swaziland in the east and to southern Democratic Republic of Congo and southern Kenya in the north. The type locality is "Indiis" (in error).

Mole Snakes are large, terrestrial snakes, active during the day in savannah and grassland habitats, where they are quite noticeable and frequently encountered by humans. When threatened, they raise the front part of the body off the ground and lunge at the perceived threat with the mouth open, a behavior that often leads to their being killed by people who mistake them for venomous snakes. The maxillary dentition is aglyph, with 12–14 unspecialized maxillary teeth increasing in size posteriorly. Although they are not dangerously venomous, bites to humans by *P. cana* have been known to sometimes cause localized edema and mild pain (Rippey et al. 1976). The diet of adults consists of small mammals, especially mole rats (rodents of the family Bathyergidae). Juveniles feed primarily on lizards. Mole Snakes give birth to live young, usually 30–50 at a time, though Broadley et al. (2003) report as many as 95 in a single litter. The hemipenes are bilobed with a forked sulcus spermaticus and are remarkably long. Bogert (1940) describes the hemipenes of *P. cana*, dissected while retracted inside the base of the tail, as extending to the thirty-second subcaudal, bifurcating at the eighth, and with the sulcus dividing at the fourth.

Pseudaspis is a stout-bodied snake with a short slender tail. The head is small with a prominent snout. The neck is not distinct from the body. The eye is small with a round pupil. The dorsum is light or dark brown, often with darker spots or blotches. These blotches may merge so as to produce a banded pattern. The pattern is prominent in juveniles and usually disappears in adults. The venter is uniformly yellowish to grayish. The rostral scale is pointed. The nasal is divided. The internasals and prefrontals are paired. The loreal scale is present. There is 1 preocular and 2–3 postoculars. The temporal formula is 2+3 or 3+4. There are usually 7(4) upper labials. There are 10–13(5 or 6) lower labials, usually 11(6) or 12(6). There may be 1 or 2 pairs of chin shields, depending on whether the posterior pair are partially separated by gulars (in which case they still count as a posterior pair of chin shields) or completely separated by gulars (in which case they are no longer considered chin shields). The submandibular groove is not pronounced. The dorsal scales

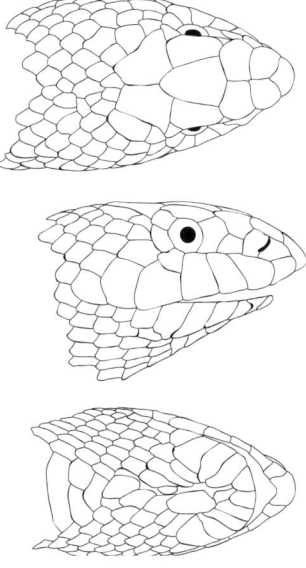

FIGURE 10.71. *Pseudaspis cana* RMCA 18208. By T. Giri

may be smooth or weakly keeled, with 1–2 apical pits, and arranged in 25–31 straight rows. The vertebral row is not enlarged relative to the other dorsal scale rows. There are 175–218 ventrals (more than 190 in females, fewer than 195 in males) and 43–70 divided subcaudals (more than 54 in males, fewer than 58 in females). The cloacal scale is divided. The maximum length recorded for this species is 2,100 mm (Branch 1994). Two subspecies of *P. cana* have been pro-

posed (Laurent 1956)—*P. c. cana* (Linneaus 1758) in the southern part of the range of the species and *P. c. anchietae* (Bocage 1882) in the northern part of the range—on the basis of lower subcaudal and dorsal scale row counts in *P. c. anchietae*. But these differences appear to represent a morphocline in scale counts, decreasing from south to north across the range of the species, rather than a division between two subspecies (FitzSimons 1974).

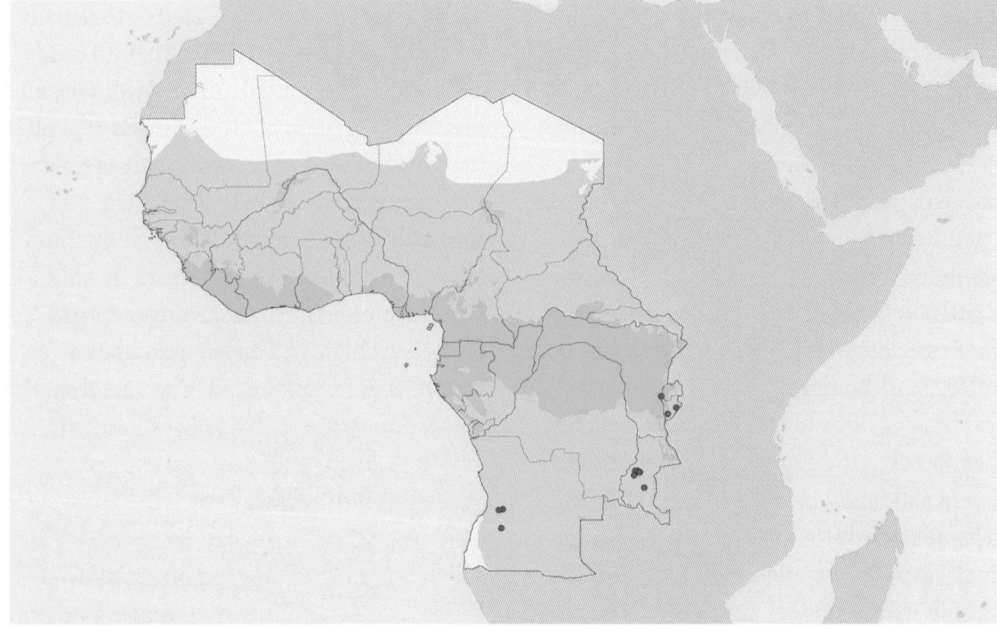

FIGURE 10.72. *Pseudaspis cana* (red). By K. Jackson

FIGURE 10.73. *Pseudaspis cana* large juvenile, Botswana. By S. Spawls

FIGURE 10.74. *Pseudaspis cana*, Tanzania. By S. Spawls

11

Family Lamprophiidae

Subfamilies Psammophiinae and Prosymninae

The Psammophiinae is a subfamily of the family Lamprophiidae, made up of 50 species in 7 genera. The Psammophiinae are found throughout Africa, including one species in Madagascar (*Mimophis mahfalensis*), and are presumed to have had an African origin. Their distribution also extends into southern Europe, the Middle East, and central Asia. Within our zone, the Psammophi-inae are represented by 26 species in 5 genera: *Hemirhagerrhis* (Bark Snakes), *Malpolon* (Montpellier Snakes), *Psammophis* (Sand Snakes), *Psammophylax* (Skaapstekers), and *Rhamphiophis* (Beaked Snakes). We include the small lamprophiid subfamily Prosymninae (the Shovel-snouts) at the end of this chapter, following the Psammophiinae.

Subfamily Psammophiinae: Sand Snakes to Skaapstekers

The subfamily Psammophiinae corresponds to Bogert's (1940) Group 16, grouped on the basis of the absence of vertebral hypapophyses in the posterior part of the body, an unforked sulcus spermaticus, and opisthoglyph dentition. Bourgeois (1965) assigned these same genera (except for *Malpolon*, which was not included among her specimens) to the subfamily Psammophiinae on the basis of her study of skulls of African snakes. More recently, all molecular phylogenetic studies that have included these taxa (Cadle 1994; Nagy et al. 2005; Vidal et al. 2007, 2008; Kelly et al. 2008; Pyron et al. 2011, 2013; Zheng and Wiens 2016) have without exception supported the psammophiines as a monophyletic group.

Psammophiines are typically long, slender, fast-moving terrestrial snakes that are active by day and prefer open habitats, where they chase down a varied assortment of prey. Characteristic of the face of most

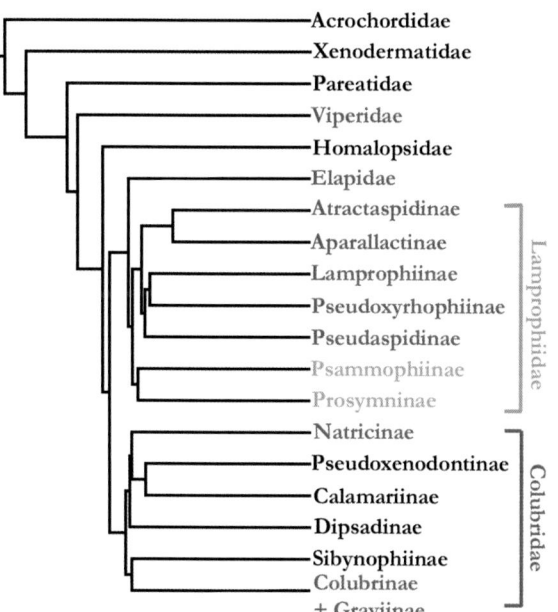

FIGURE 11.1. Higher-level phylogeny of caenophidian families and subfamilies, with families and subfamilies occurring within our zone highlighted (red or orange). The subfamilies Psammophiinae and Prosymninae, the topic of this chapter, are highlighted in orange. Modified by K. Jackson from Zheng and Wiens (2016)

psammophiines is a pronounced *canthus rostralis*—a sort of brow ridge extending from above the eye to the snout, giving the eye and the scales in front of it a sunken appearance. Other distinctive characteristics of psammophiines include unusual hemipenes and a social system that relies on pheromone-rich secretions from glands in the skin. Psammophiine hemipenes are short and slender, lacking any ornamentation of spines or calyces characteristic of snake hemipenes. Each hemipenis is unforked, with an unforked sulcus spermaticus extending the full length of the hemipenis (Fig. 11.2). Bogert (1940) described the hemipenes of psammophiines as "extremely small and difficult to examine satisfactorily" and added that the sex of an individual can only be determined by means of careful dissection. The possible selective advantage often proposed to explain the smallness of psammophiine hemipenes is ease in disengaging during copulation. Whereas the larger hemipenes, ornamented with spines and calyces, of most snakes help to prevent the female from disengaging from the male during copulation, perhaps in the open habitats preferred by most psammophiines, snakes are at greater than average risk

from predators, so that disengaging easily in order to make a quick getaway is advantageous. Shine et al. (2006b) offer another possibility, that reduction of hemipenis size and associated reduction in tail length have some locomotor advantage in these fast-moving snakes. In other snakes (e.g., Tree Snakes, genus *Dipsadoboa*, Vine Snakes, genus *Thelotornis*, and others), a slender tail (e.g., associated with arboreality) is often associated with smaller-than-average hemipenes. Clutch size is typically low among psammophiines, perhaps associated with their slender body shape. Based on dissection of 700 museum specimens, Shine et al. (2006b) found that larger species tended to lay more eggs than smaller species, but that within species there was no correlation between maternal body size and clutch size.

Compounding the difficulty of determining sex of psammophiines (because of the small size of the hemipenes) is the general lack of sexual dimorphism within the group. It is unfortunate that males and females are so hard to tell apart except by dissection, since psammophiines are notable for their often complex reproductive behavior, which is thus difficult to study on live individuals in the field. Much of what is known of the natural history of psammophiines is therefore based on studies involving dissection of museum specimens (e.g., Butler 1993; Shine et al. 2006b).

Psammophiines engage in pheromone-mediated social and reproductive behavior to a degree that is unusual in snakes. All psammophiines possess an extranarial valve bearing a secretion outlet of a special nasal gland that secretes a pheromone-rich substance from the nostril. Snakes engage in "self-rubbing" behavior, rubbing the secretions from their nose on their venter and

sulcus spermaticus

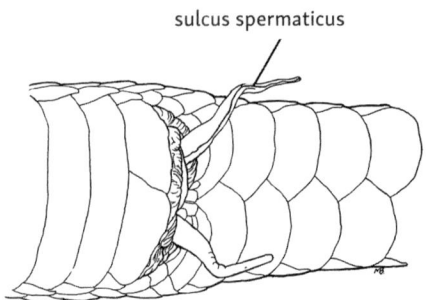

FIGURE 11.2. Cloacal region of the Hissing Sand Snake, *Psammophis* cf. *sibilans*, showing rudimentary hemipenes in everted position. Adapted by K. Jackson from Bogert (1940)

tail so as to leave a pheromone trail detectable by others of their species (Fig. 11.3). Pitman (1974) described these scent trails as "usually evil smelling." Some psammophiines are territorial (highly unusual among snakes), males of some species engage in combat, and some, the Sand Snakes (genus *Psammophis*), have an additional pheromone-secreting gland that they use in a complex system involving chemically marking other individuals of their species (Haan 2003a, 2003b), a system we describe in greater detail in the introduction to the genus *Psammophis*, below. Relatively little is known about these glands or the system of social behaviors associated with them. Perhaps these pheromone secretions help with conspecific recognition in a way that compensates for the lack of sexual dimorphism seen in psammophiine snakes.

The effects on humans of bites by psammophiine snakes are known only from a few isolated reports. Such case studies as exist describe no examples of clinically significant symptoms resulting from the bites of psammophiines. Bites to humans by *Psammophis phillipsii* and *P. sibilans* may sometimes cause localized edema and mild pain (Rippey et al. 1976; Chippaux pers. obs.).

Bark Snakes: Genus *Hemirhagerrhis* Boettger 1893

Bark Snakes are a genus of four species, two of which are found within our zone. Unusual among psammophiines, *Hemirhagerrhis* are specialized for living under bark in trees (as their common name suggests) and in

FIGURE 11.3. A Moila Snake, *Malpolon moilensis*, in Mali, self-rubbing, with the right side valvular nostril zigzagging, pressed onto and along the belly. By M. Aymerich

cracks in rocks. Brown or gray, with a darker pattern, Bark Snakes are well camouflaged, looking a lot like the bark that provides their home. Their heads are flattened relative to those of other psammophiines, perhaps an adaptation for prizing bark from trees in order to get underneath it. They feed on lizards, primarily geckos and their eggs (Trape and Mané 2006a). Like other psammophiines, they engage in self-rubbing behavior even though an arboreal habitat must reduce the effectiveness of leaving a pheromone trail, as their terrestrial relatives do.

Bark Snakes resemble other psammophiines in having rudimentary hemipenes and a grooved posterior fang and are said to be notably docile when handled (Pitman 1974). The nasal is single or divided. There are 2 internasals and 2 prefrontals. The loreal is present. There are 1 or 2 preoculars and 2 or 3 postoculars. The temporal formula is variable. There are 8 or 9 upper labials, 2 of which are in contact with the eye. There are 2 pairs of chin shields, and the sublingual groove is pronounced between the posterior pair. There are 9–12 lower labials. The dorsal scales are smooth with 1 apical pit and are arranged in 17 straight rows. The cloacal scale and the subcaudals are divided.

Common Bark Snake: *Hemirhagerrhis nototaenia* (Günther 1864)

The Common Bark Snake has a broad distribution, ranging from Burkina Faso to Somalia to South Africa. The type locality is Rio de Sena, Mozambique. *Hemirhagerrhis nototaenia* is an arboreal species commonly found under bark in dry savanna (Broadley 1997).

The body is light brown or gray. The top of the head is dark, a marking that extends into a broad dark mid-dorsal stripe. Dark blotches along the edge of the stripe run into it, creating a zig-zag pattern. The underside is pale with dark spots. The nasal may be single or divided. There are 1 preocular and 2 postoculars. There are 8(4,5) upper labials. The temporal formula is 1+2 or occasionally 2+3. There are usually 9(4) lower labials, but these can range from 8 to

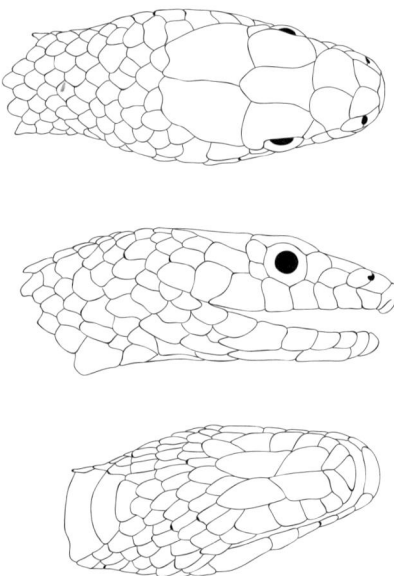

FIGURE 11.4. *Hemirhagerrhis nototaenia.* Royal Museum of Central Africa (RMCA) catalog number 1901. By T. Giri

Key to Central and Western African Species of the Genus *Hemirhagerrhis*

1   At least 65 subcaudals; usually 1 anterior temporal .... *H. nototaenia*
1' No more than 66 subcaudals; usually 2 anterior temporals .... *H. viperina*

FIGURE 11.5. *Hemirhagerrhis nototaenia*, Kenya. By W. Wüster

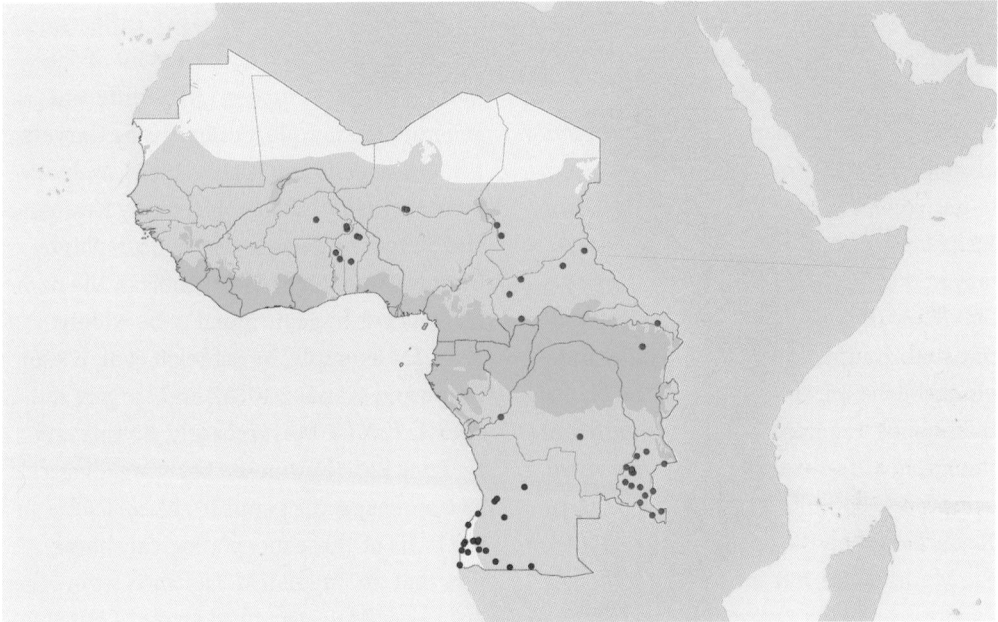

FIGURE 11.6. *Hemirhagerrhis nototaenia* (red), *Hemirhagerrhis viperina* (blue). By K. Jackson

11 and (4) or (5). There are 151–179 ventrals and 65–90 divided subcaudals. The maximum length recorded for this species is 419 mm (Broadley and Cock 1975).

Western Bark Snake: *Hemirhagerrhis viperina* (Bocage 1873)

The Western Bark Snake, formerly a subspecies of the Common Bark Snake, is restricted to Namibia and southern Angola. The type locality is Dombe Grande, Angola.

In contrast to the arboreal *Hemirhagerrhis nototaenia*, *H. viperina* lives primarily in cracks between rocks.

The body is light brown or gray with a series of rounded blotches along the mid-dorsal line. The underside is pale with dark flecks. The nasal is semi-divided. There is 1 preocular and 2 or occasionally 3 postoculars. There are 8(4,5) upper labials. The temporal formula is usually 2+3 but sometimes 1+3. There are usually 10(5) lower labials, but these can range from 9(4) to 11(5). There are 152–183 ventrals and 52–66 divided subcaudals. The maximum length recorded for this species is 492 mm (Broadley 1997).

Montpellier Snakes: Genus *Malpolon* Fitzinger 1826

The genus *Malpolon* comprises between two and four species (depending on the author), one of which is found within our zone. The Montpellier Snake, *Malpolon monspessulanus*, of southern Europe and northern Africa, is the best studied *Malpolon* species and is renowned for its complex social behavior. *M. moilensis*, the Moila Snake, occurs within our zone. Like other psammophiines, *Malpolon* are characterized by rudimentary hemipenes and a grooved posterior maxillary fang. An account of a bite, under unusual circumstances, from a large individual of the European species, the *M. monspessulanus*, reported clinically significant, though not life-threatening, neurological symptoms (Pommier and de Haro 2007). No data exist on the effect of a bite from the Moila Snake, but there is no reason to expect it to pose a danger to humans.

Malpolon engage in self-rubbing behavior like other psammophiines. The overall body shape is also characteristic of psammophiines, with a prominent canthus rostralis, a long body, and a slender tail. The nasal may be single or divided. The internasals and prefrontals are paired. The loreal is present and may be divided. There are 1 or 2 preoculars and 2 or occasionally 3 postoculars. The temporal formula is generally 2+3. There are typically 8 or 9 upper labials, 2 of which are in contact with the eye. There are 2 pairs of chin shields and from 8 to 12 lower labials. The dorsal scales are smooth, usually with a single apical pit, and arranged in 17 or 19 rows midbody. The cloacal scale and subcaudals are divided.

Moila Snake: *Malpolon moilensis* (Reuss 1834)

Malpolon moilensis is a Sahara Desert snake that gets its name from Moila, Arabia, the type locality of the species. The taxonomic position of this species remains controversial (see also the related discussion regarding *Rhamphiophis maradiensis*). Different authors have proposed placement of the Moila Snake in a succession of different genera; for example, *Coelopeltis* by Gervais (1857), *Rhagerhis* by Peters (1862), *Malpolon* by Parker (1931), *Rhamphiophis* by Kramer and Schnurrenberger (1963), *Scutophis* by Brandstätter (1995). Nevertheless, the name *M. moilensis* has continued to be widely used, for example, by Schleich et al. (1996), Trape and Mané (2006a), and Largen and Spawls (2010). Most recently, Böhme and De Pury (2011) proposed reviving *Rhagerhis* as a monospecific genus for *M. moilensis* on the basis of three morphological characters that distinguish *M. moilensis* from other species of *Malpolon*: head shape, elongate cervical ribs (associated with hooding), and micro-ornamentation of the dorsal scales. In all other morphological characters examined, *M. moilensis* is consistent with other species of *Malpolon*. In a molecular phylogenetic study of psammophiine snakes, Kelly et al. (2008) found *M. moilensis* and *M. monspessulanus* (the two members of the genus *Malpolon* included in the study) to be each other's closest relatives. But the divergence between them was deep, and little separated them from *Rhamphiophis rostratus*, sister taxon to the two *Malpolon* species.

The argument for moving *M. moilensis* to its own genus rests entirely on the perceived distinctiveness of the species and on its long separation from other members of the genus. The genus *Malpolon* is monophyletic with *M. moilensis* included in it, however, so there is no problem (such as a paraphyletic genus) being solved to justify the creation of the monospecific genus, *Rhagerhis*. We therefore maintain the use of *M. moilensis* here.

The Moila Snake is gray to brown above, scattered with small darker spots. There are 2 dark markings on each side of the head where it meets the neck. The underside is whitish, speckled with brown. The nasal is divided. There is 1 preocular and 2 or 3 postoculars. There are 8(4,5) or sometimes 7(4,5) upper labials. The temporal formula is 2+3 but in rare instances may be 1+2 or 2+2. There are usually 8 or 9 lower labials, but these may range from 8(4) to 10(5). The smooth dorsal scales are typically arranged in 17, but sometimes 19, straight rows. Although apical pits are present according to the taxonomic literature (Boulenger 1896), we have not been able to confirm their presence despite thorough examination. There are 139–176 ventrals and 48–73 divided subcaudals. Smaller than *M. monspessulanus*, the maximum length recorded for a Moila Snake is 1,300 mm (Le Berre 1989).

The Moila Snake is most active at night during the hot season but is active by day when temperatures are cooler. It feeds on a wide variety of prey, ranging from lizards to rodents to birds, and has an undeserved reputation for aggressiveness, attributable to its impressive threat display: when threatened, a Moila Snake will hood like a cobra, hiss, and attempt to strike if harassed further.

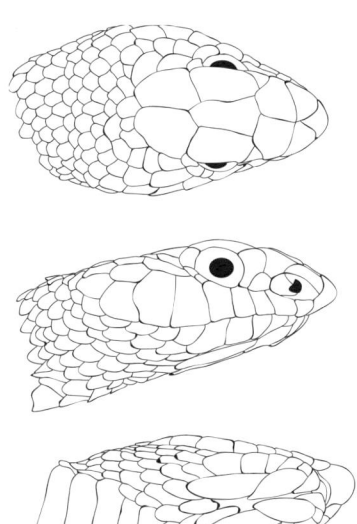

FIGURE 11.7. *Malpolon moilensis* RMCA 28589. By T. Giri

FIGURE 11.8. *Malpolon moilensis* defensive display, Mali. By M. Aymerich

Sand Snakes: Genus *Psammophis* Boie 1826

There are some 32 species of Sand Snakes (also sometimes called Whip Snakes or Sand Racers), making this by far the largest psammophiine genus. Although most species in the genus are found in Africa, the range of the genus extends into the Middle East and Asia. Seventeen species occur

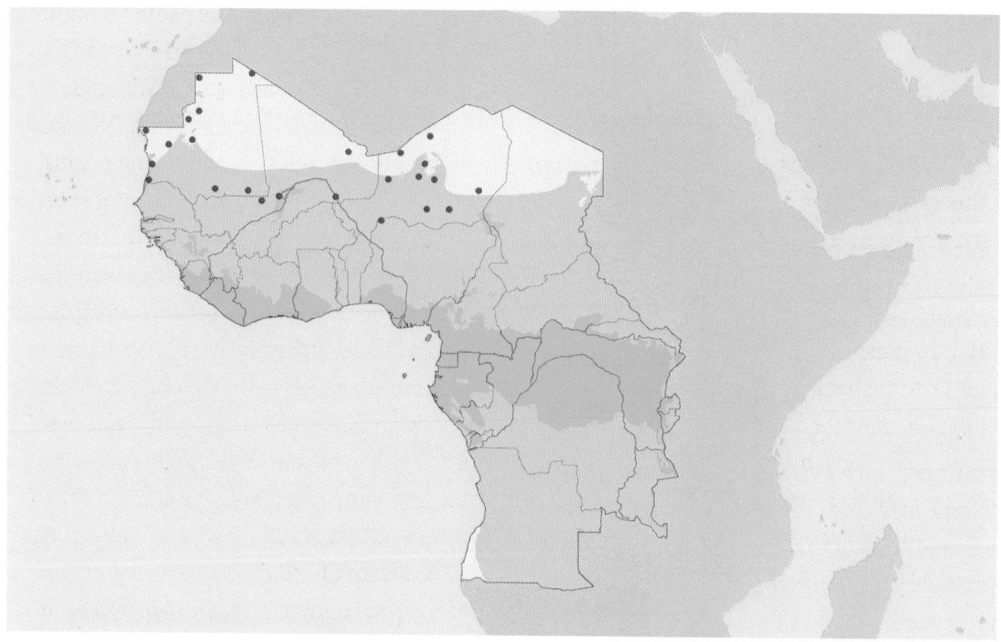

FIGURE 11.9. *Malpolon moilensis* (red). By K. Jackson

within our zone, seven of which are really southern African species whose range extends into the southern edges of the area covered by this book.

Sand Snakes are slender, elongate snakes, fast moving, diurnal, and convergent on North American Coachwhips (genus *Masticophis*). "Generally handsome snakes" was Pitman's (1974) assessment. The head is elongate, with the canthus rostralis characteristic of the psammophiinae giving the eye and side of the snout a sunken appearance. The pupil is round. The nasal is always divided, sometimes in fact into three sections, these divisions perhaps serving some function in helping to conduct secretions from the extranarial valve by capillary action. The internasals are paired, as are the prefrontals. The loreal is present. There are 1 or 2 preoculars, often a single large preocular that extends upward over the ridge of the canthus rostralis so that it is visible anterior to the supraocular in

dorsal view. Important for telling species apart is whether the preocular is in contact with the frontal or separated from it by the contact between the supraocular and prefrontal. There are 1–3 postoculars. There are 7–10 upper labials and no suboculars. The temporal formula also helps to define species, and there may be from 1 to 3 anterior temporals. There are 2 pairs of chin shields and a pronounced submandibular groove. There are 9–13 lower labials. The dorsal scales have 2 apical pits and are arranged in 11–19 smooth, oblique rows. The ventrals and the divided subcaudals are rarely sexually dimorphic. The cloacal scale is almost always divided.

Subcaudal scale counts in *Psammophis* should be accepted with caution, as snakes of this genus are capable of tail breakage as a mechanism to escape a predator. Sand Snakes, when grasped by the tail, will spin their bodies until the tail breaks off where it is held (Broadley 1987; Akani et al. 2002b).

While common in lizards, tail breakage is relatively rare in snakes. In contrast to the highly developed caudal autotomy seen in lizards, tail breakage in snakes occurs between two vertebrae (as opposed to across a special fracture plane within an individual vertebra, as in lizards), and the broken-off piece of tail does not continue to move, as it does in lizards (Hoogmoed and Avila-Pires 2011). In most species in which tail breakage does occur, a broken tail is easily recognized by the blunt end of the regenerated tail. *Psammophis* tails, however, regenerate with a sharp point like that of the original.

Though the Psammophiinae are a clearly well-defined group, defining species within the genus *Psammophis* remains one of the most intransigent problems in African snake systematics. Ongoing efforts to sort out the taxonomy of the Sand Snakes have not yet reached a consensus, so the species as we present them here should be considered a work in progress rather than the last word. Hughes (1999) sensibly recommends dealing with the Sand Snakes in terms of "working names" rather than prematurely assigning poorly understood snakes to new species. There has been an unfortunate tendency for morphological taxonomists to underestimate the possible extent of individual variation and thus to describe too many new species on the basis of minor differences in scale characters. By contrast, molecular systematics, more recently arrived on the scene (e.g., Kelly et al. 2008), divides the genus into a few large "complexes" that have the potential to shed light on the evolutionary history of the group but are of little help in identifying species in the field.

Kelly et al.'s (2008) and Vidal et al.'s (2008) molecular phylogenetic studies of the Psammophiinae provide some insights into the relationships within *Psammophis*. The status of the genus *Dromophis* (the Olympic Snakes) is one such case. Externally, *Dromophis* much resemble *Psammophis*. They were traditionally accorded their own genus based on their maxillary dentition. While *Psammophis* have a distinctive maxilla, bearing not only a grooved posterior fang but also 1 or 2 greatly enlarged, ungrooved teeth about halfway along the maxilla, *Dromophis* have no such enlarged teeth, and, with the exception of their posterior fang, have unremarkable maxillary teeth all of similar size (Fig. 11.4). But Kelly et al.'s (2008) study found the two species of *Dromophis*, *Dromophis lineatus* and *D. praeornatus*, to be nested within the genus *Psammophis* and not each other's nearest relatives. This is consistent with the findings of Vidal et al. (2008). We therefore follow their synonymizing of *Dromophis* with *Psammophis* here (see *P. lineatus* and *P. praeornatus*).

Like other psammophiines, Sand Snakes show little sexual dimorphism, and they display the most complex sexual behavior of all psammophiine genera. *Psammophis* differ from other psammophiines in having an additional outlet for pheromone secretions under the chin (Haan 2003a, 2003b). These infralabial secretion outlets (ILOs), located on the lower labial scales, are visible only periodically, being especially evident after

FIGURE 11.10. Maxillae of *Psammophis sibilans* (*left*) and *Psammophis* (formerly "*Dromophis*") *lineatus* (*right*). Note the enlarged mid-maxillary teeth in *Psammophis* and their absence in "*Dromophis*." Adapted by K. Jackson from Phisalix (1922) and Boulenger (1896), respectively

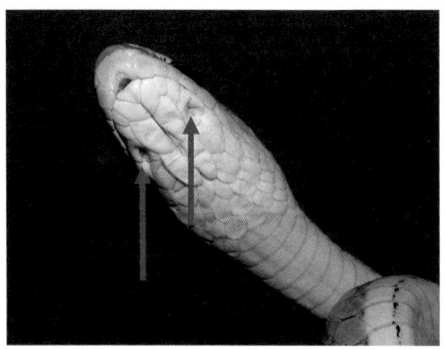

FIGURE 11.11. A captive *Psammophis* cf. *phillipsii* with red arrows indicating infralabial outlets visible on the fourth lower labial scale. By K. Jackson using a photo by C. de Haan

shedding, when they are thought to be at their most active (Fig. 11.5).

In contrast to self-rubbing using the extranarial valve (see description in introduction to the Psammophiinae), the function of the ILOs is conspecific rubbing, in which one snake rubs its chin against the back and neck of another (Haan 2003a, 2003b). This conspecific rubbing takes many forms, suggesting that it plays a complex role in communication associated with

reproduction. On the basis of many years of observation of captive specimens, Haan (2003a, 2003b) described "automatic" conspecific rubbing (i.e., every time one snake crawls over another) and "purposeful" rubbing. For example, female *Psammophis* cf. *phillipsii* perform a zig-zag rubbing motion on the neck of the male. All Sand Snakes perform a more generalized back-length stroke, but primarily by males on females.

Interestingly, the ILOs are present on roughly the fourth lower labial scale, thus matching up with that enlarged mid- maxillary tooth of Sand Snakes, suggesting perhaps that the enlarged tooth acts to channel pheromones from a gland in the upper lip into the ILOs. Because ILOs are present in Olympic Snakes (*Psammophis lineatus*), which lack the enlarged tooth, this may perhaps be explained as evidence of a secondary loss of the enlarged tooth.

While Sand Snakes, like other psammophiines, show little sexual dimorphism, in a few species of *Psammophis* the male is slightly larger than the female. Although, among

FIGURE 11.12. Sand Snakes are only rarely sexually dimorphic. These two captive *Psammophis phillipsii* are indistinguishable, though the one on top happens to be the female and the one underneath the male. By C. de Haan

FIGURE 11.13. Four captive adult *Psammophis phillipsii* (two brothers, two sisters, same generation), forming a clan with the help of their visual landmark recognition and their vomerolingually detectable chemical markings via various "rubbing" behaviors, among which self-rubbing is the general starting point. By C. de Haan

Key to Central and Western African Species of the Genus *Psammophis*

1	Cloacal scale single	2
1'	Cloacal scale divided	3

2(1)	1 preocular; supraocular in broad contact with the prefrontal	*P. cf. phillipsii*
2'	2 preoculars; supraocular not in contact with the prefrontal	*P. notostictus*

3(1)	11 to 15 dorsal scale rows	4
3'	17 or 19 dorsal scale rows	6

4(3)	11 dorsal scale rows	*P. angolensis*
4'	15 dorsal scale rows	5

5(4)	7 upper labials; preocular in broad contact with the frontal	*P. jallae*
5'	8–10 upper labials; preocular without or with only slight contact with the frontal	*P. praeornatus*

6(3)	More than 130 subcaudals	7
6'	Fewer than 131 subcaudals	8

7(6)	Usually 1 preocular and 2 anterior temporals; preocular not in contact with the frontal	*P. elegans*
7'	Usually 2 preoculars and 1 anterior temporal; preocular in contact with the frontal	*P. trigrammus*

8(6)	Preocular in contact with the frontal	9
8'	Preocular not in contact with the frontal	11

9(8)	Venter patterned with a dark mid-ventral stripe	*P. schokari*
9'	Venter without dark mid-ventral stripe	10

10(9)	Length of frontal is more than three times its breadth; more than 184 ventrals	*P. aegyptius*
10'	Length of frontal less than three times its breadth; fewer than 189 ventrals	*P. namibensis*

11(8)	3 upper labials in contact with the eye	*P. subtaeniatus*
11'	2 upper labials in contact with the eye	12

12(11)	First upper labial in contact with the loreal	13
12'	First upper labial not in contact with the loreal	14

13(12)	Frontal more than three times as long as broad	*P. lineatus*
13'	Frontal less than three times as long as broad	*P. leopardinus*

14(12)	Length of prefrontals is less than twice the length of the internasals	*P. sibilans*
14'	Length of prefrontals is almost three times the length of the internasals	15

15(14)	Dorsum of body gray or olive green without markings or only weakly patterned	*P. phillipsii*
15'	Dorsum of body dark brown with a pair of pale lateral stripes	*P. sudanensis*

snakes in general, females are usually larger, the reverse is often true of species in which males engage in combat, which at least some species of *Psammophis* are known to do, though it has been observed only rarely (Pitman 1974; Shine et al. 2006b).

Egyptian Sand Snake: *Psammophis aegyptius* Marx 1958

Psammophis aegyptius is a Saharan species. Its type locality is Siwa, Egypt.

This species is uniformly light brown or olive green above. The labials are cream. On either side of the head a dark stripe, often indistinct, extends from the nostril to the neck. The underside is uniformly cream, occasionally with spots but never with lateral stripes. The nasal is divided (in two). There is 1 preocular, which is in contact with the frontal, and 2 postoculars. There are 9(5,6) upper labials. The temporal formula is 2+2 or 2+3. The lower labials are 12(5) or 13(6). There are 17 or 19 dorsal scale rows. The ventrals range from 185 to 199. There are 111–123 subcaudals. The cloacal scale is divided. The maximum length recorded for this species is 1,500 mm (Trape and Mané 2006a).

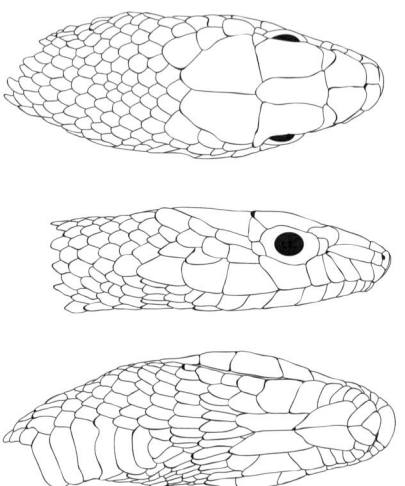

FIGURE 11.14. *Psammophis aegyptius* FMNH 205936. By T. Giri

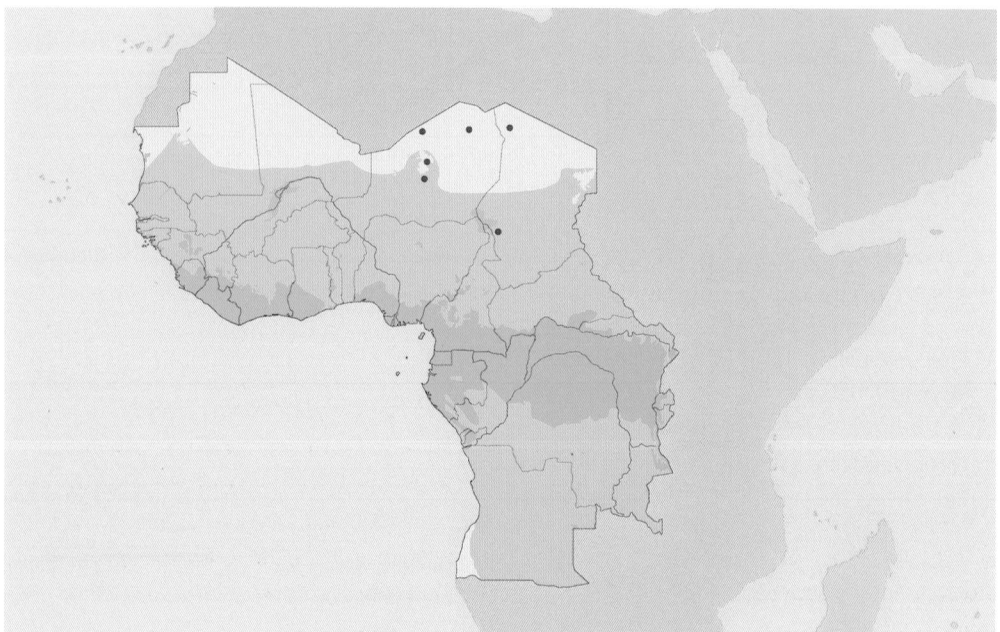

FIGURE 11.15. *Psammophis aegyptius* (red). By K. Jackson

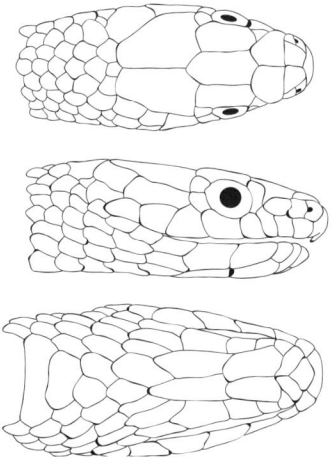

FIGURE 11.16. *Psammophis angolensis* FMNH 78251. By T. Giri

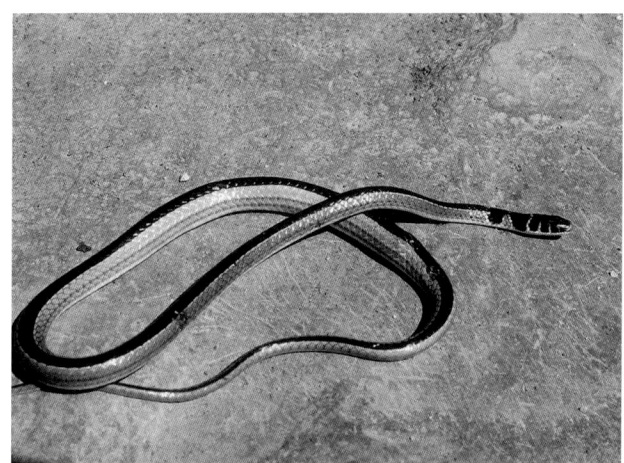

FIGURE 11.17. *Psammophis angolensis*, Botswana. By S. Spawls

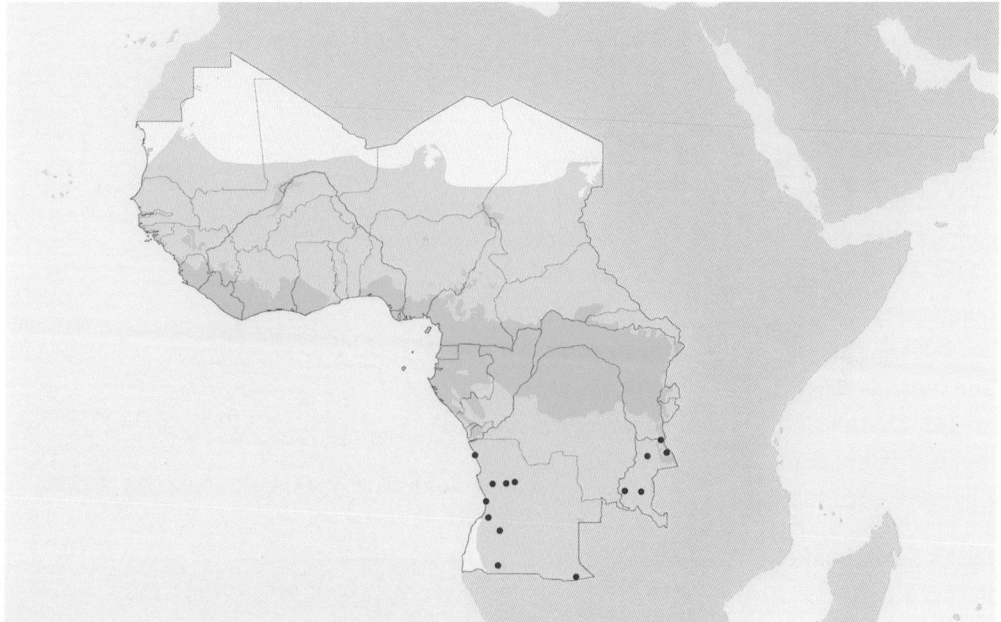

FIGURE 11.18. *Psammophis angolensis* (blue). By K. Jackson

Dwarf Sand Snake: *Psammophis angolensis* (Bocage 1872)

The range of the Dwarf Sand Snake extends from South Africa to Tanzania in the east and Angola and southern Democratic Republic of Congo in the north. The type locality is Dondo, Angola.

The body is light brown above with a broad black vertebral stripe. The head and neck are black with a pattern consisting of a few vertical markings in light brown. The underside is white or cream. The nasal is divided (in two). There is 1 preocular, which is not in contact with the frontal, and 2 or 3 postoculars. There are 8(4,5) upper labials. The temporal formula is usually 1+2, occasionally 2+2. The lower labials are 8(4). There are just 11 dorsal scale rows. The ventrals range from 135 to 160. There are 58–82 subcaudals. The cloacal scale is divided. The maximum length recorded for this species is just 500 mm (Broadley and Cock 1975).

Elegant Sand Snake: *Psammophis elegans* (Shaw 1802)

The range of the Elegant Sand Snake extends from Mauritania to Cameroon to Central African Republic. The type locality is "South America" (in error).

Psammophis elegans is light brown above with three broad, light brown stripes outlined in black. The head scales are brown mottled with black, and the upper labials are white. The underside is white with dark flecks or stripes. The nasal is divided (in two). There is 1 preocular, which is not in contact with the frontal, and 2 postoculars. There are usually 9(5,6) upper labials. The temporal formula is 2+2. The lower labials are 9–11(5). There are 17 dorsal scale rows. The ventrals range from 186 to 211. There are 142–172 subcaudals. The cloacal scale is divided. The maximum length recorded for this species is 1,750 mm (Angel 1933).

Jalla's Sand Snake: *Psammophis jallae* Peracca 1896

The range of Jalla's Sand Snake extends from South Africa to Zambia, Namibia, and into southeastern Angola. The type locality is Kazungula to Bulawayo, Zimbabwe.

Psammophis jallae has a dark brown vertebral stripe bounded on either side by a pair of broad, light brown lateral stripes, followed by a pair of thin white paraventral stripes. The underside is pale with black

flecks on the chin extending as far as the neck. The nasal is divided (in three). There is 1 preocular, which is in contact with the frontal, and 2 postoculars. There are 7(3,4) upper labials. The temporal formula is 2+2 or occasionally 1+2. The lower labials are 9(4). There are 15 dorsal scale rows. The ventrals range from 152 to 178. There are 84–112 subcaudals. The cloacal scale is divided. The maximum length recorded for this species is 1,135 mm (Broadley and Cock 1975). Although sexual dimorphism is rare

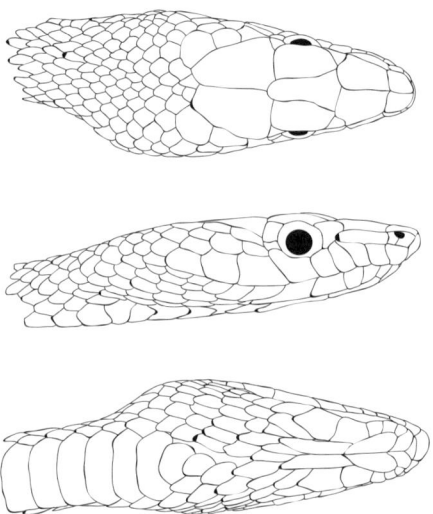

FIGURE 11.19. *Psammophis elegans* RMCA 29572. By T. Giri

FIGURE 11.20. *Psammophis elegans*, Ghana. By S. Spawls

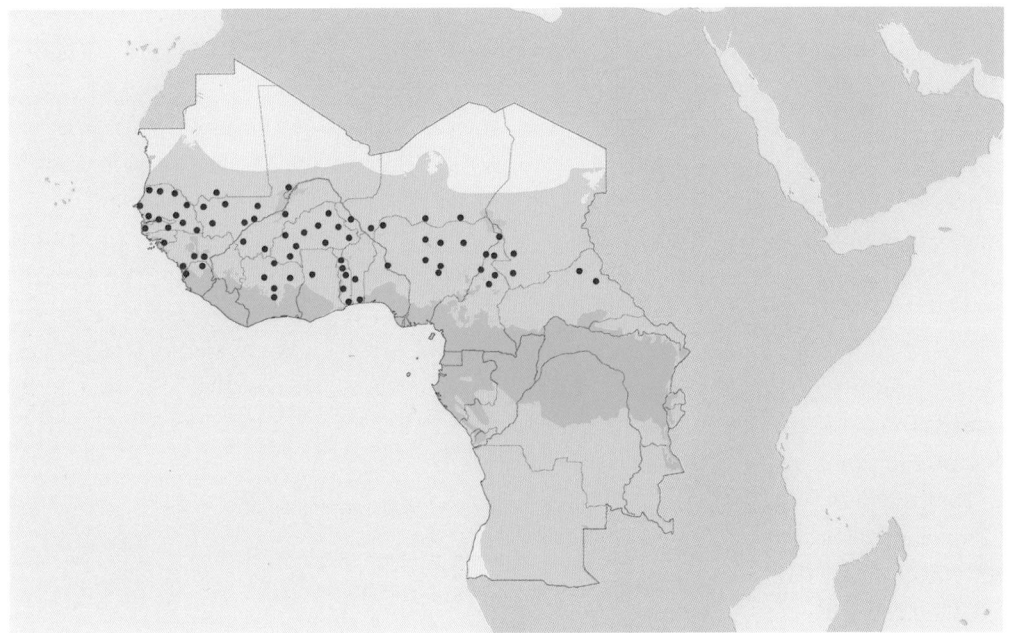

FIGURE 11.21. *Psammophis elegans* (blue). By K. Jackson

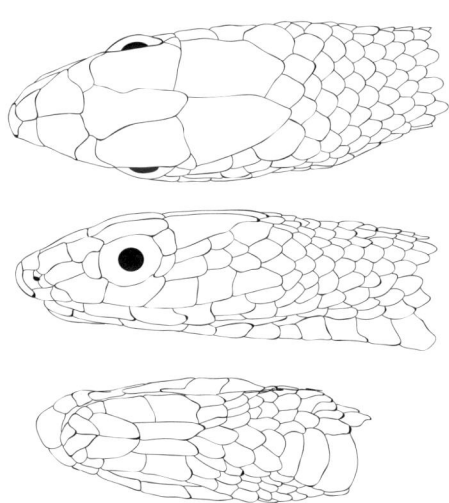

FIGURE 11.22. *Psammophis jallae* USNM 163398. By T. Giri

FIGURE 11.23. *Psammophis jallae*, Botswana. By S. Spawls

in *Psammophis*, Shine et al. (2006b) found male Jalla's Sand Snakes to be on average 19% longer than females.

Leopard Sand Snake: *Psammophis leopardinus* Bocage 1887c

The range of the Leopard Sand Snake is limited to northwestern Namibia and southwestern Angola. The type locality is Catumbela, Angola.

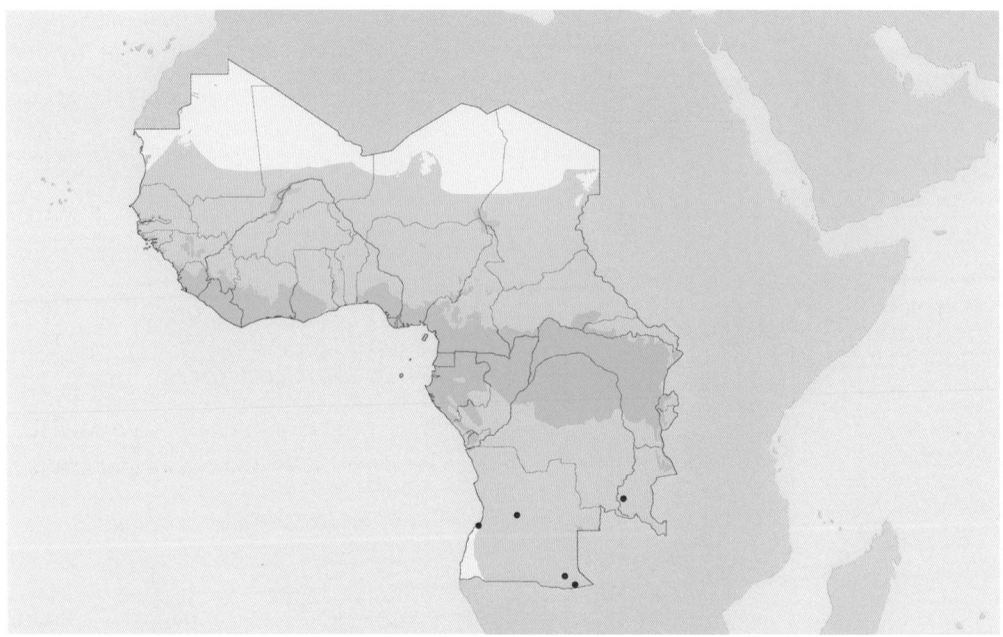

FIGURE 11.24. *Psammophis jallae* (blue). By K. Jackson

Psammophis leopardinus is reddish to grayish brown above with a pair of pale, black-edged lateral stripes originating on the head. In addition, there is a pale but dark-edged median vertebral stripe and a pair of pale dorsolateral stripes. A series of pale bars cross the back of the head, and there may be faint crossbands on the neck. The nasal is divided (usually in two but occasionally in three). There is 1 preocular, which is not in contact with the frontal, and 2 postoculars. There are 8(4,5) upper labials. The temporal formula is quite variable, with 1 or 2 anterior and 1–3 posterior temporals. The lower labials are 11(4). There are 17 dorsal scale rows. The ventrals range from 151 to 174. There are 79–105 subcaudals. The cloacal scale is divided. The maximum length recorded for this species is 1,345 mm (Broadley 1983). Shine et al. (2006b) found females to have longer tails than males relative to their snout–vent length.

Lined Olympic Snake: *Psammophis lineatus* (Duméril, Bibron and Duméril 1854a)

Though often mistaken for *Psammophis sibilans*, the Lined Olympic Snake was considered to belong to a separate genus (*Dromophis*) until recently. (The relationship between the genera *Psammophis* and *Dromophis* was explained in the generic description above.) Like *P. sibilans*, the Lined Olympic Snake has an extensive range, extending from Senegal to the Sudan to Zambia. The type locality is White Nile, Sudan. *P. lineatus* is a savanna species, often found near water. It feeds on amphibians, lizards, and sometimes other snakes (Manaças 1955; Menzies 1966). Pitman (1974) describes this species as being "of amiable disposition, easily handled." Regenerated tails are extremely common in *P. lineatus*, he adds, probably indicating attempts at predation by aquatic birds and perhaps even fishes.

P. lineatus is patterned with alternating light and dark stripes above, the light stripes narrower than the dark. The pattern becomes indistinct toward the head, but there is no band across the neck, which distinguishes it from *P. praeornatus*. The

underside and the sides of the head are cream colored. The nasal is divided (in two). There is 1 preocular, which is not in contact with the frontal, and 2 postoculars. There are usually 8(4,5) upper labials, of which the fourth or fifth is the longest. The temporal formula is usually 1+2 but occasionally 2+2 or 2+3. The lower labials are usually 9(4). There are 17 dorsal scale rows. The ventrals range from 138 to 159. There are 78–107 subcaudals. The cloacal scale is divided. The maximum length recorded for this species is 1,090 mm (Boulenger 1896).

Namib Sand Snake: *Psammophis namibensis* Broadley 1975

The range of the Namib Sand Snake extends from South Africa north into southern Angola. The type locality is Harus, Namibia.

The body is light brown above with a white vertebral stripe and a pair of narrow, white lateral stripes. The underside is white. The nasal is divided (in three). There

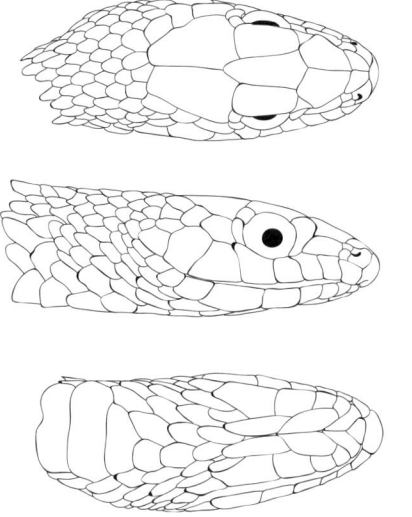

FIGURE 11.25. *Psammophis leopardinus* RMCA A7-028-R-0088. By T. Giri

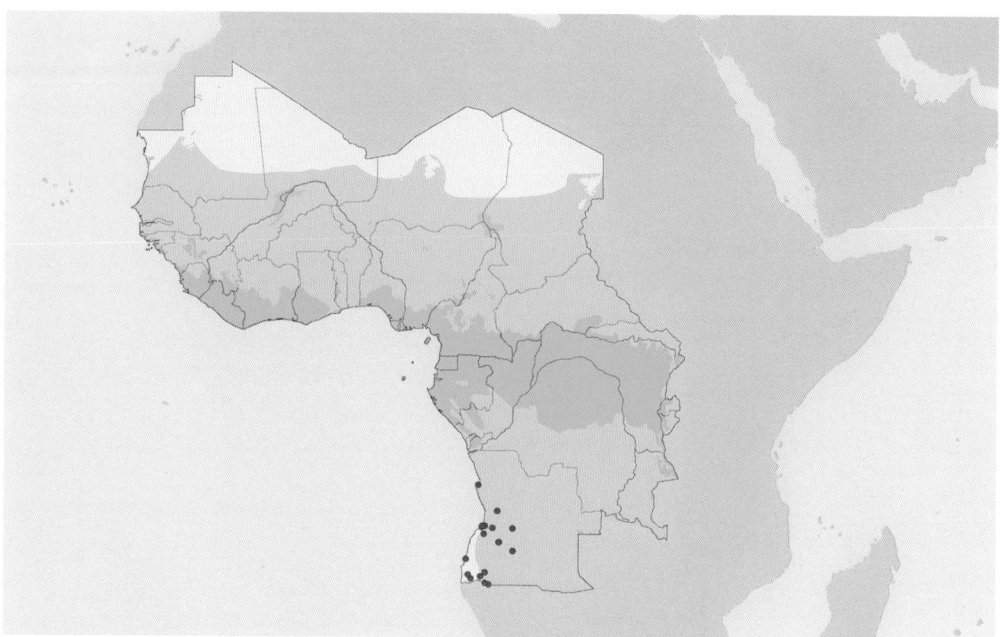

FIGURE 11.26. *Psammophis leopardinus* (red). By K. Jackson

is 1 preocular, which is in contact with the frontal, and 2 postoculars. There are 8(4,5) upper labials. The temporal formula is 2+2. The lower labials are 10(5). There are 17 dorsal scale rows. The ventrals range from 167 to 188. There are 90–116 subcaudals. The cloacal scale is divided. The maximum

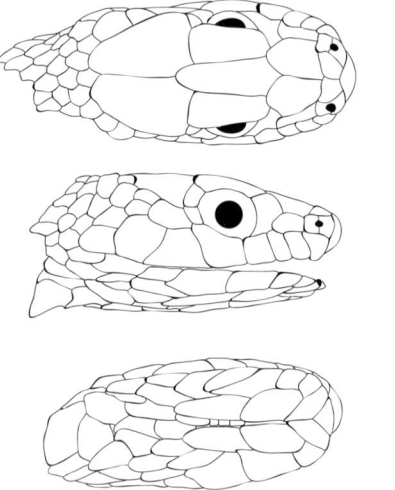

FIGURE 11.27. *Psammophis lineatus* RMCA 14861. By T. Giri

length recorded for this species is 1,360 mm (Branch 1994).

Karoo Sand Snake: *Psammophis notostictus* Peters 1867

The range of the Karoo Sand Snake extends from Lesotho to southwestern Angola. The type locality is Otjimbingue, Namibia.

The body is uniformly pale olive green above, sometimes with a pair of thin, paler stripes along the sides. The beige underside is speckled with gray at the sides. The nasal is divided (in three). There are 2 preoculars, one of which is in contact with the frontal, and 2, or occasionally 3, postoculars. There are 8(4,5), or occasionally 9(4,5,6), upper labials. The temporal formula is 2+2, or occasionally 1+2. The lower labials are 10(5). There are 17 dorsal scale rows. The ventrals range from 155 to 184. There are 76–107 subcaudals. Notably, the cloacal scale is single. The maximum length recorded for this species is 1,000 mm (Branch 1994).

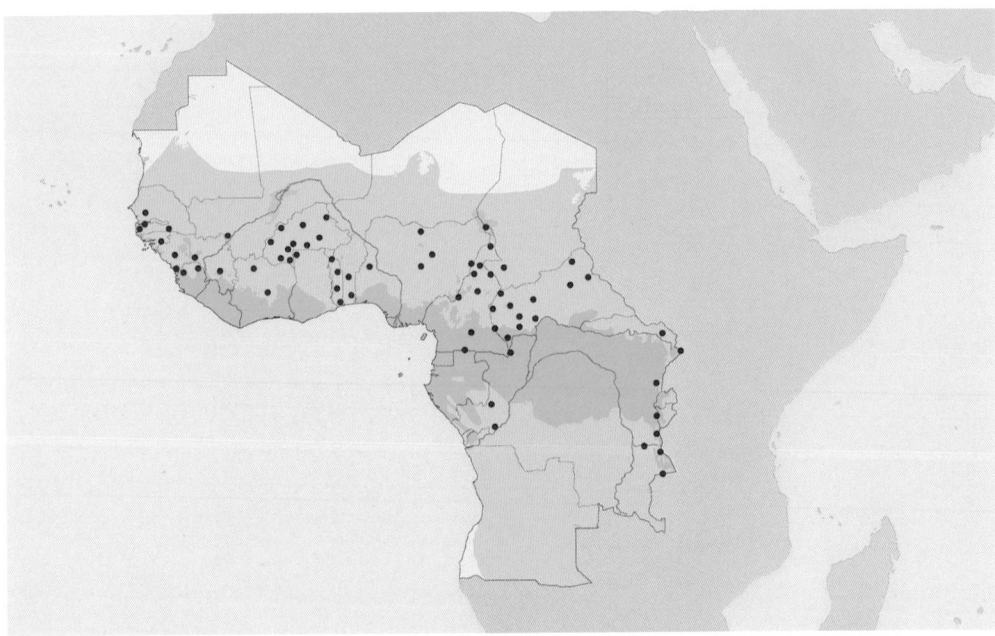

FIGURE 11.28. *Psammophis lineatus* (blue). By K. Jackson

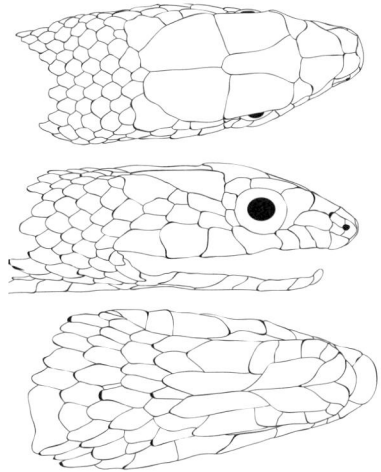

FIGURE 11.29. *Psammophis namibensis* BMNH 1988.578. By T. Giri

FIGURE 11.30. *Psammophis namibensis*, South Africa. By W. Wüster

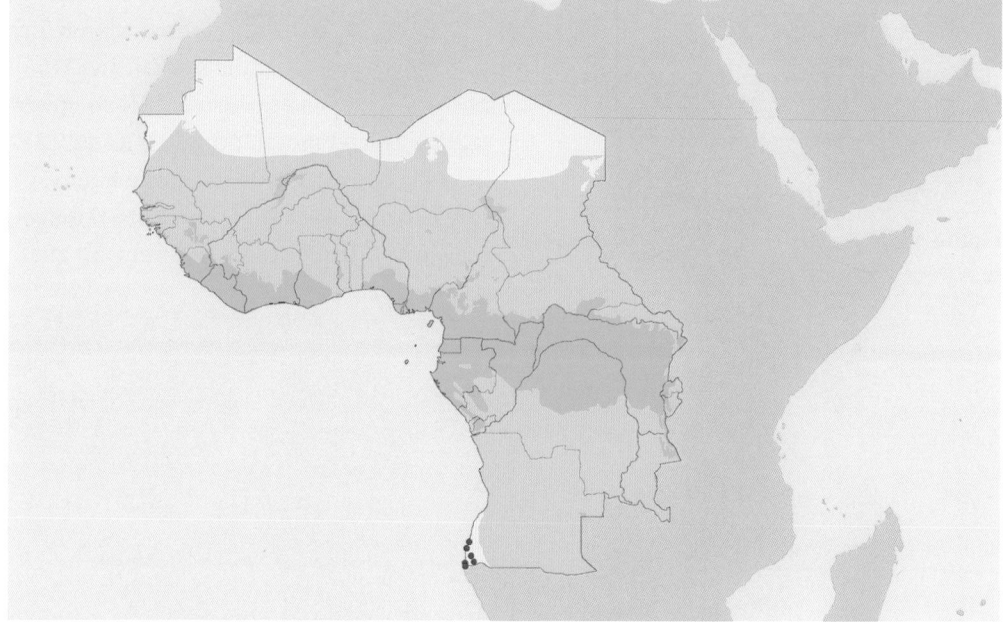

FIGURE 11.31. *Psammophis namibensis* (red). By K. Jackson

Olive Sand Snake: *Psammophis phillipsii* (Hallowell 1844)

The status of *Psammophis phillipsii* as a species is disputed. Here we use the "working" species name, *P. phillipsii*, in its broadest sense, with our rationale explained briefly below. With a sub-Saharan distribution, extending from the Gambia in west Africa to Botswana in the south, *P. phillipsi* has one of the most extensive ranges of all species in the genus, inhabiting both forested and open habitats (see Fig. 11.38). The type locality is "Liberia." Like the type specimen, most specimens from West Africa have a single cloacal scale, while in other parts of its range, most individuals have a divided cloacal scale. Hughes (1999) considers the two forms to represent different species, but Luiselli et al. (2004) could find no ecological difference between the two forms in Nigeria, an area of overlap, suggesting in fact a single species with the condition of the cloacal scale variable. A molecular genetic study is needed to settle the matter, but for now we will treat both forms as a single species. Broadley (2002) found central African *P. phillipsii* to be indistinguishable from *P. mossambicus*, a result supported by Kelly et al.'s (2008) molecular findings. We therefore treat *P. mossambicus* here as synonymous with *P. phillipsii*.

The Olive Sand Snake is uniformly brown to olive brown above. The dorsal scales sometimes have dark edges. The underside is pale and may have a pair of thin dark stripes along the lateral sides of the ventrals, either in the form of continuous lines or broken up into dotted lines. The nasal is divided (in two). There is 1 preocular, which is not in contact with the frontal, and 2 postoculars. There are usually 8(4,5) upper labials. The temporal formula is usually 2+2 but may be 1+3 or 2+3. The lower labials range from 9(4) to 11(5). There are 17 dorsal scale rows. There are 151–185 ventrals and

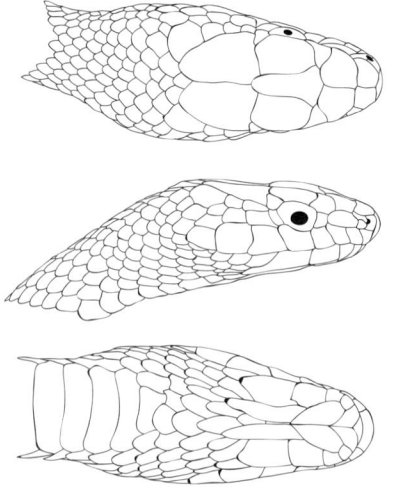

FIGURE 11.32. *Psammophis notostictus* USNM 56449. By T. Giri

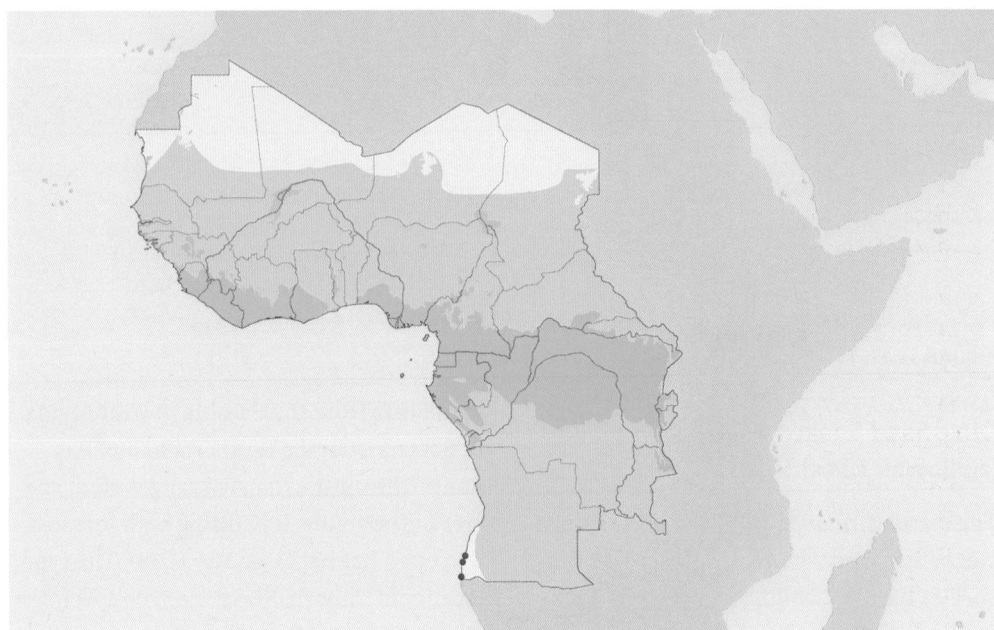

FIGURE 11.33. *Psammophis notostictus* (red). By K. Jackson

FIGURE 11.34. A captive female Olive Sand Snake (*Psammophis phillipsii*) with her eggs. By C. de Haan

82–121 subcaudals. The cloacal scale may be single or divided. The maximum length recorded for this species is 1,800 mm (Villiers 1975).

The natural history of the Olive Sand Snake is the best studied of any species within the genus. Akani et al. (2003) carried out a field study of the diet of this species in Nigeria, over a period of five years, including sampling in both wet and dry seasons. The methodology consisted of catching snakes and causing them to regurgitate or defecate. Like all field studies of psammophiines, this one was constrained by the impossibility of sexing live snakes (because of the small size of the hemipenes), except in the case of obviously gravid females. Akani's team therefore recognized three groups: (1) gravid females, (2) males and non-gravid females, and (3) juveniles. They found the average prey/predator mass to be roughly 10%, with no difference between groups. Perhaps the most surprising result was the discovery that females continue to feed while gravid, which is unusual in snakes. In all groups, lizards were the primary prey, followed by small mammals and by miscellaneous items ranging from insects to smaller snakes (including conspecifics). Lizards made up 45% to 50% of prey items in adults, but a higher proportion (67%) in juveniles. A possible explanation for this difference is that *P. phillipsii* juveniles hatch at the onset of the rainy season, as do juvenile lizards (*Agama agama*), so that the little snakes are provided with an abundant supply of small lizards to eat. Like other large species of Sand Snake, large individuals of *P. phillipsii* have been known to eat other snakes.

Butler (1993) investigated reproduction in *P. phillipsii*, a project that could only have been carried out by measuring and dissecting museum specimens. In all, 49 adult males, 30 adult females, and 46 unsexed juveniles were examined. Males were found to have longer bodies (snout–vent length) than females, but tail length did not differ between the sexes. *P. phillipsii* were found to lay their eggs during the dry season, with hatching occurring at the onset of the rainy season. The mean clutch size was 10, and no correlation was found between the size of the mother and the number of eggs laid (Butler 1993; Akani et al. 2003).

African Swamp Snake: *Psammophis praeornatus* (Schlegel 1837)

The African Swamp Snake is the other *Psammophis* species (along with *Psammophis lineatus*) of the two within our zone that were formerly assigned to the genus *Dromophis*. Two subspecies are recognized by some authors: *P. p. praeornatus* (Schlegel 1837); *P. p. gribinguiensis* (Angel 1921). These are distinguished from one another by the number of upper labials, ventrals, and subcaudals. The type locality of *P. p. praeornatus* is Walo, Senegal, and its range extends from

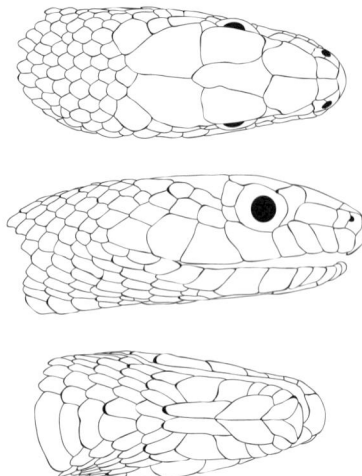

FIGURE 11.35. *Psammophis phillipsii* RMCA 30816. By T. Giri

FIGURE 11.36. *Psammophis phillipsii*, Republic of Congo. By K. Jackson

FIGURE 11.37. *Psammophis phillipsii* juvenile, Republic of Congo. By K. Jackson

Senegal to Nigeria. The type locality of *P. p. gribinguiensis* is Gribingui, Chad, and its range extends from Nigeria to Central African Republic. Like the Lined Olympic Snake, the African Swamp Snake is a savanna species, but, as its name suggests, it is more dependent on wetlands than the former.

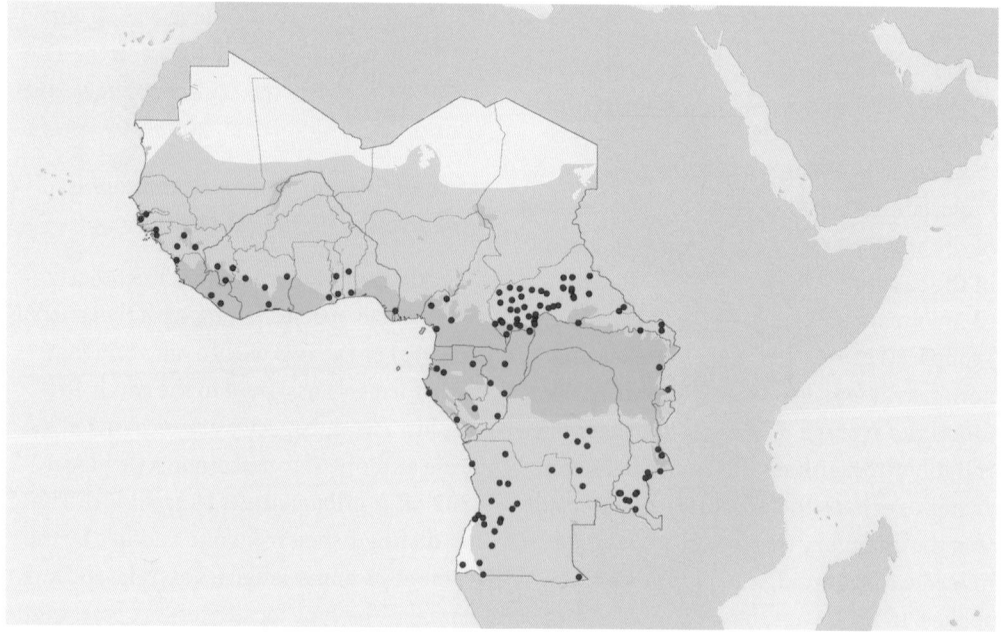

FIGURE 11.38. *Psammophis phillipsii* (blue). By K. Jackson

Brown or pale olive above, the front part of the body is patterned with a series of black bands. At midbody a vertebral stripe appears, extending to the tail and joined for the last third of the snake's length by a pair of dark paraventral stripes. The underside is pale, sometimes with black spots. The nasal is divided (in two). There is 1 preocular and 2 postoculars. There are 8(4,5), 9(5,6), or 10(6,7) upper labials. The temporal formula is usually 1+2 but sometimes 2+2 or 2+3. The lower labials are usually 10(5), the sixth lower labial being the longest. There are 15 dorsal scale rows. The ventrals range from 161 to 190. There are 107–133 subcau-

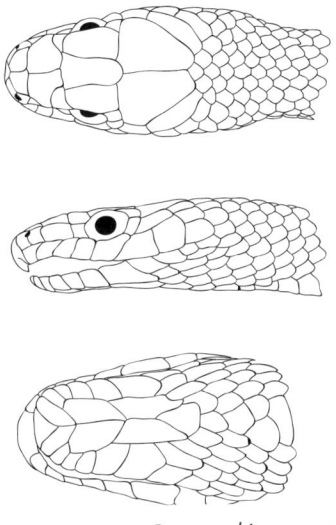

FIGURE 11.39. *Psammophis praeornatus* RMCA 29464. By T. Giri

FIGURE 11.40. *Psammophis praeornatus*, Ghana. By S. Spawls

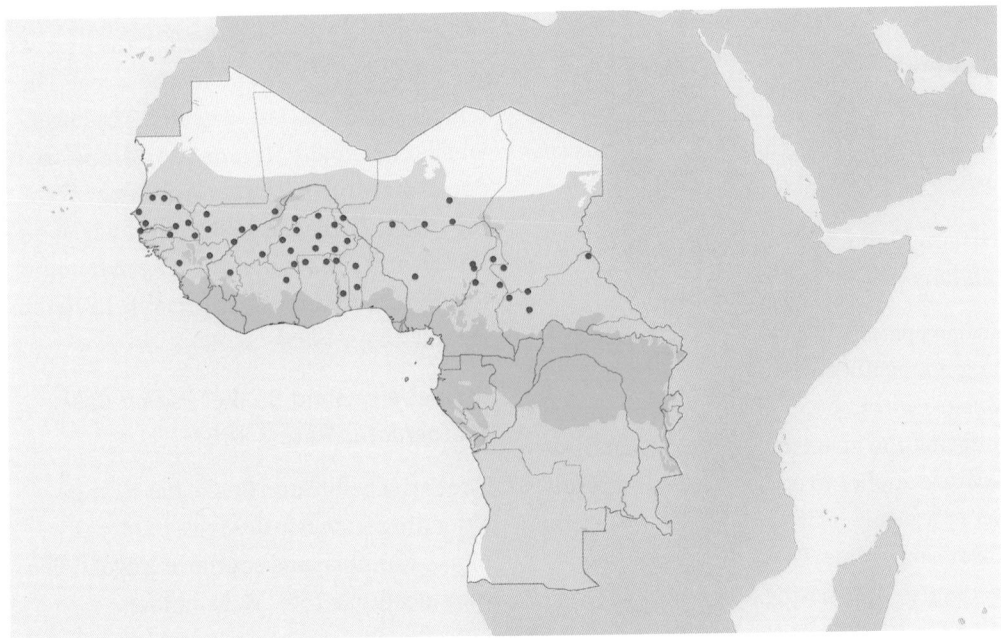

FIGURE 11.41. *Psammophis praeornatus* (red). By K. Jackson

dals. The cloacal scale is divided. The maximum length recorded for this species is 665 mm (Chippaux unpub. data).

Schokari Sand Snake: *Psammophis schokari* (Forskål 1775)

The range of the Schokari Sand Snake includes all of sub-Saharan Africa north of the equator and extends into the Middle East. The type locality is Yemen.

Psammophis schokari is light brown above with two broad lateral stripes of darker brown outlined in black and a vertebral stripe. The head scales are brown mottled with black, and there is a black stripe along each side of the head. The underside has two narrow white stripes separated by a broad brown stripe mottled with black. The nasal is divided (in two). There is usually 1 but sometimes 2 preoculars, one of which is in contact with the frontal, and 2 postoculars. There are 9(5,6) upper labials. The temporal formula is 2+2 or 2+3. The lower labials are usually 11(5). There are 17, or occasionally 19, dorsal scale rows. The ventrals range from 156 to 210. There are 104–121 subcaudals. The cloacal scale is divided. The maximum length recorded for this species is 1,480 mm (Anderson 1898).

Hissing Sand Snake: *Psammophis sibilans* (Linnaeus 1758)

Psammophis sibilans is perhaps the most taxonomically problematic of all the Sand Snakes within our zone. As a result, much information given for this species in the literature applies to other species as a result of errors in identification. Here we use the "working" name, *P. sibilans*, in its broadest sense. The Hissing Sand Snake has an almost pan-African distribution, and while it may in fact represent more than one species, at this point there is not yet agreement about how the species should be divided up. The Hissing Sand Snake's range extends from western to southern Africa. The type locality is "Asia" (in error). *P. sibilans* is a fast-moving snake that hunts during the day on the ground and in vegetation. It is reported to feed on rodents and amphibians (Pauwels et al. 2004) but probably takes a broad range of prey.

The Hissing Sand Snake is dark brown to olive green above, with a pair of pale lateral stripes outlined in black. There is also a pale vertebral stripe that is sometimes speckled with black. This pattern becomes indistinct as the snake ages. In juveniles, the head scales are mottled and outlined in pale yellow. The labials are cream colored with light brown spots faintly visible on the upper labials. The underside is uniformly cream, occasionally with spots, but never with lateral stripes. The nasal is divided (in two). There is 1 preocular, which is not in contact with the frontal, and 2 postoculars. There are usually 8(4,5) upper labials, the first of which is not in contact with the loreal. The temporal formula is 2+2 or 2+3. The lower labials are usually 11(4) or 11(5). There are 17 dorsal scale rows. The ventrals range from 155 to 186. There are 95–121 subcaudals. The cloacal scale is divided. The maximum length recorded for this species is 1,870 mm (Brandstätter 1995).

Stripebelly Sand Snake: *Psammophis subtaeniatus* Peters 1867

The Stripebelly Sand Snake has a range extending from South Africa to Mozambique, Namibia, and southern Angola. The type locality is Tete, Mozambique.

Psammophis subtaeniatus has a broad brown mid-dorsal stripe, separated from

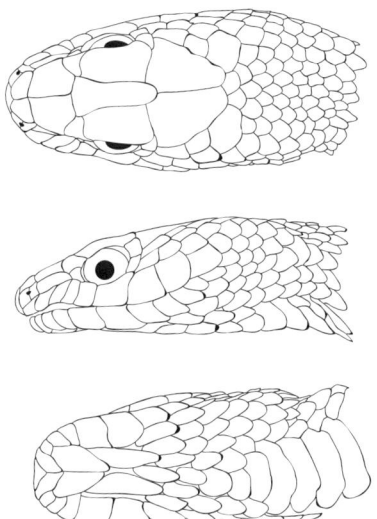

FIGURE 11.43. *Psammophis schokari*, United Arab Emirates. By W. Wüster

FIGURE 11.42. *Psammophis schokari* RMCA 2203. By T. Giri

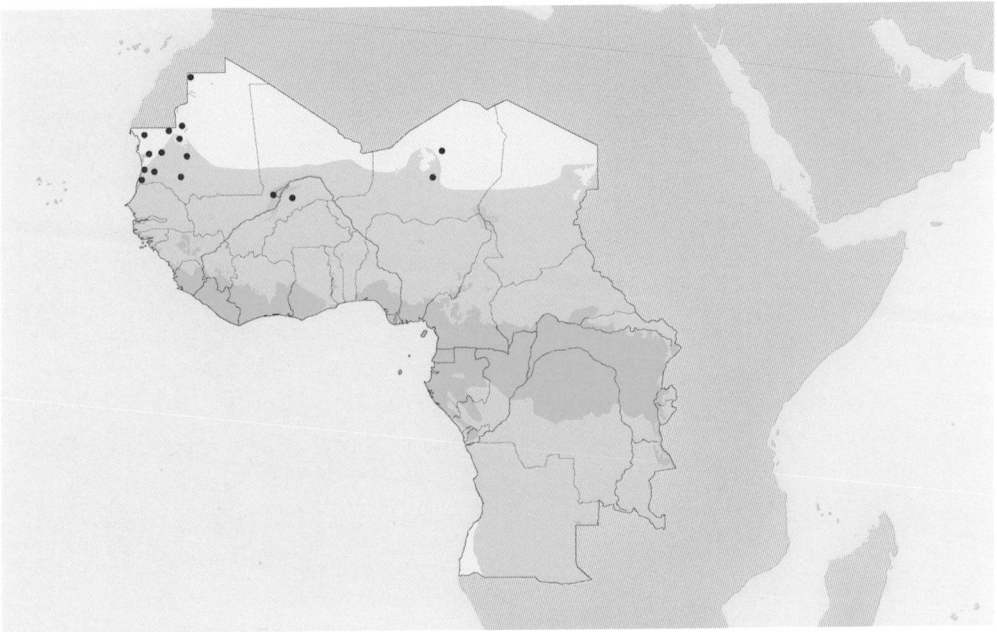

FIGURE 11.44. *Psammophis schokari* (blue). By K. Jackson

a broad brown lateral stripe by a narrower yellow or white stripe. The sides of the ventrals are white, but what gives the Stripe-belly Sand Snake gets its name is a broad yellow stripe outlined in black along its underside. The head is brown above, often with gray markings that may continue onto the neck as a series of faint crossbars. The supralabials and underside of the head may be white, yellow, or reddish, usually speckled with black. The nasal is divided. There is 1 preocular, which is not in contact with the frontal, and 2 postoculars. There are 9(4,5,6), rarely 8(4,5), 8(3,4,5), 9(5,6),

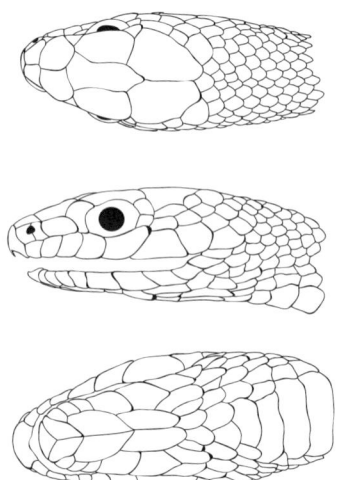

FIGURE 11.46. *Psammophis sibilans*, Senegal. By W. Wüster

FIGURE 11.45. *Psammophis sibilans* RMCA 21200. By T. Giri

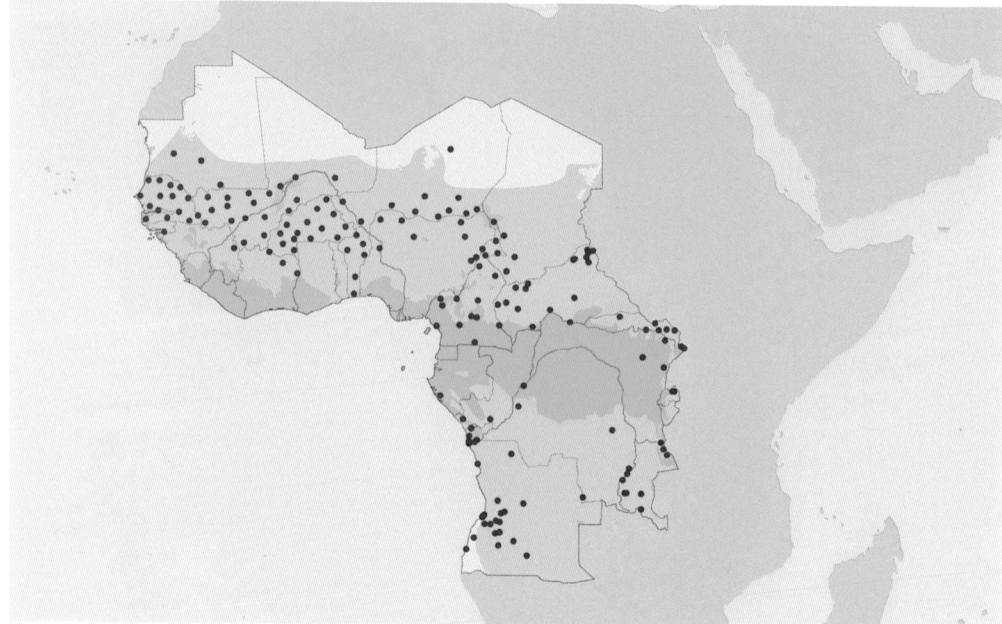

FIGURE 11.47. *Psammophis sibilans* (blue). By K. Jackson

10(5,6,7), or 10(4,5,6,7). The temporal formula is 2+2. The lower labials are 10(4). There are 17 dorsal scale rows. The ventrals range from 155 to 181. There are 101–130 subcaudals. The cloacal scale is divided. The maximum length recorded for this species is 1,370 mm (Broadley and Cock 1975).

Sudan Sand Snake: *Psammophis sudanensis* Werner 1919

The range of the Sudan Sand Snake extends from Senegal to the Sudan. The type locality is Kadugli, Sudan.

The body is dark brown or olive above,

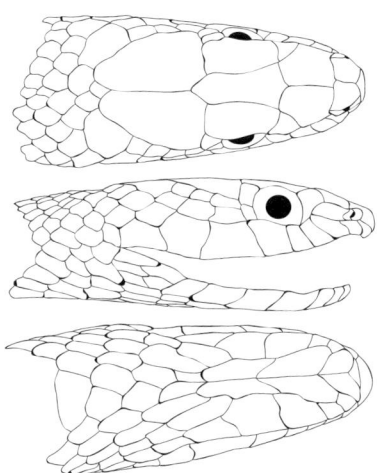

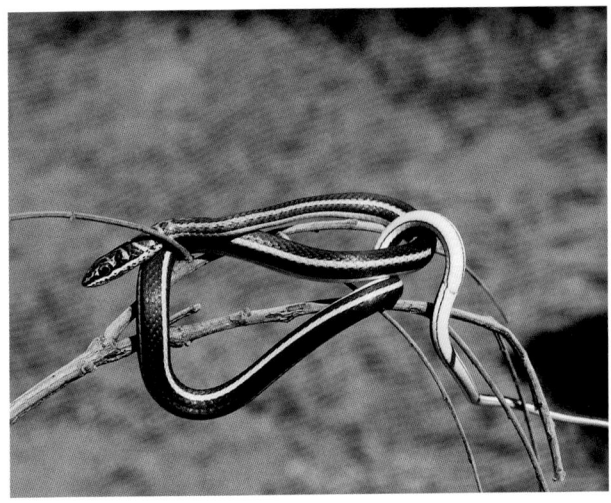

FIGURE 11.48. *Psammophis subtaeniatus* BMNH 1988.664.003. By T. Giri

FIGURE 11.49. *Psammophis subtaeniatus*, Botswana. By S. Spawls

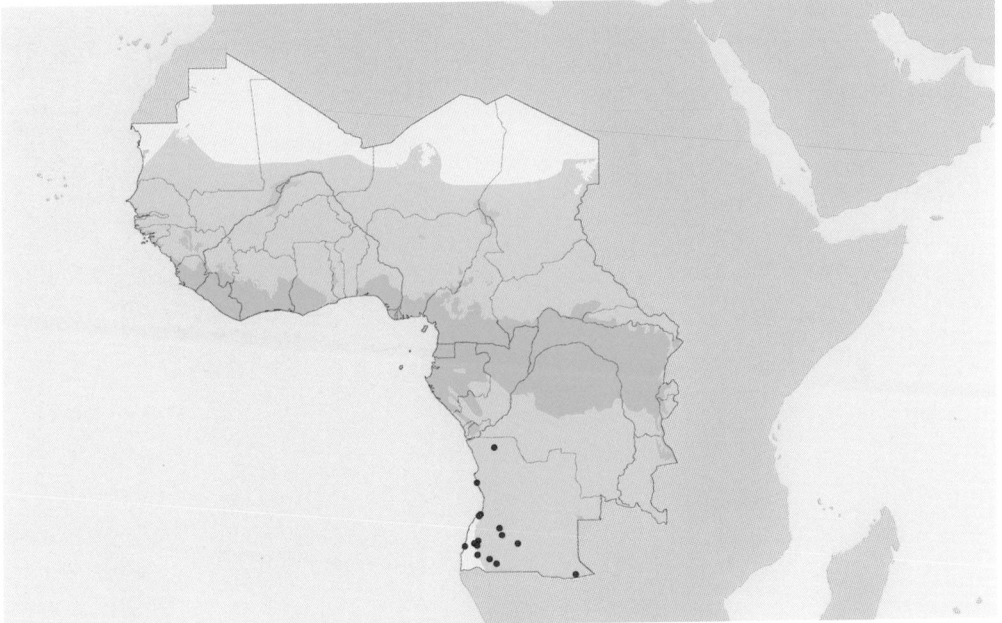

FIGURE 11.50. *Psammophis subtaeniatus* (blue). By K. Jackson

becoming paler posteriorly. There is a pair of broad light brown or yellowish dorsolateral stripes that are separated by a dark vertebral stripe. Each scale in the vertebral row has a white spot at its base, giving the vertebral stripe a speckled appearance. The head is uniformly light brown. The labials, chin, and throat are cream colored. The underside is uniformly yellowish white with a dark stripe along either side. The nasal is divided (in two). There is 1 preocular, which is not in contact with the frontal, and 2 postoculars. There are 8(4,5) upper labials. The temporal formula is 2+2. The lower labials

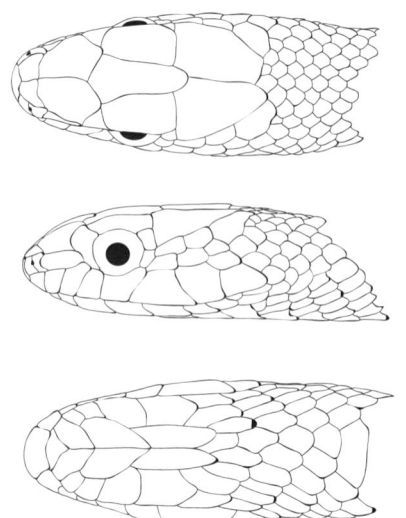

FIGURE 11.52. *Psammophis sudanensis*, Kenya. By S. Spawls

FIGURE 11.51. *Psammophis sudanensis* USNM 533142. By T. Giri

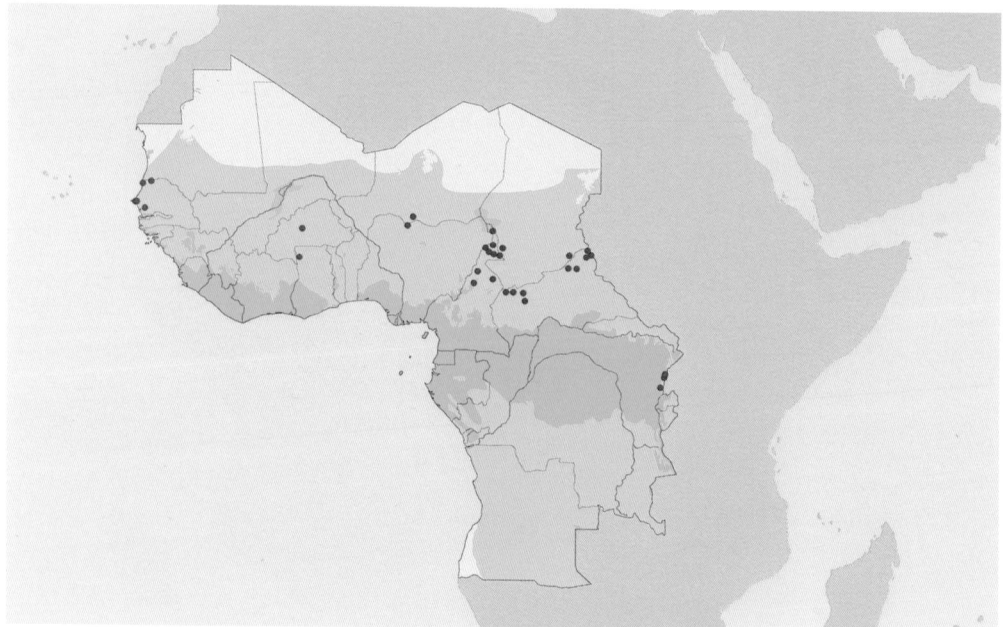

FIGURE 11.53. *Psammophis sudanensis* (red). By K. Jackson

are 9 or 10(4) or occasionally 11(5). There are 17 dorsal scale rows. The ventrals range from 148 to 183. There are 83–129 subcaudals. The cloacal scale is divided. The maximum length recorded for this species is 1,665 mm (Brandstätter 1995).

Western Sand Snake: *Psammophis trigrammus* Günther 1865

The range of the Western Sand Snake extends from South Africa to Namibia and southern Angola. The type locality is Moçâmedes, Angola.

Psammophis trigrammus is uniformly light brown to olive green above. The underside is cream colored. The nasal is divided (in three). There are 2 preoculars, one of which is usually in contact with the frontal, and 2 postoculars. There are 9(5,6) upper labials. There may be 1 or 2 anterior and 1 or 2 posterior temporals. The lower labials are 10(5). There are 17 dorsal scale rows. The ventrals range from 182 to 201. There are 132–156 subcaudals. The cloacal scale is divided. The maximum length recorded for this species is 1,380 mm (Broadley 2002).

Skaapstekers: Genus *Psammophylax* Fitzinger 1843

Skaapstekers (*Psammophylax*) are a genus composed of four species with a southern and eastern African distribution. Although none could truly be considered a central or western African species, the edges of the ranges of all species in the genus extend into our zone.

"Skaapsteker" is an Afrikaans word meaning "sheep-stabber," which originated among the early Boers or Afrikaans farmers as they herded their stock over the mountains in search of grazing. Although they held *Psammophylax* to blame for occasion-

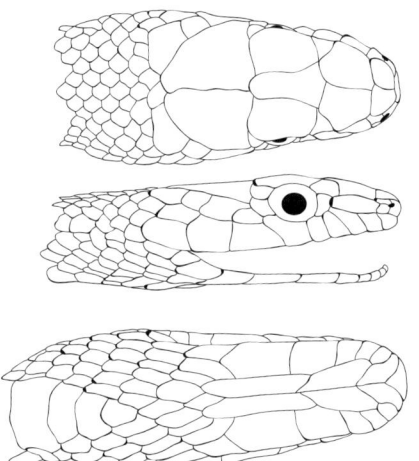

FIGURE 11.54. *Psammophis trigrammus* BMNH TYPE 1946.1.8.12. By T. Giri

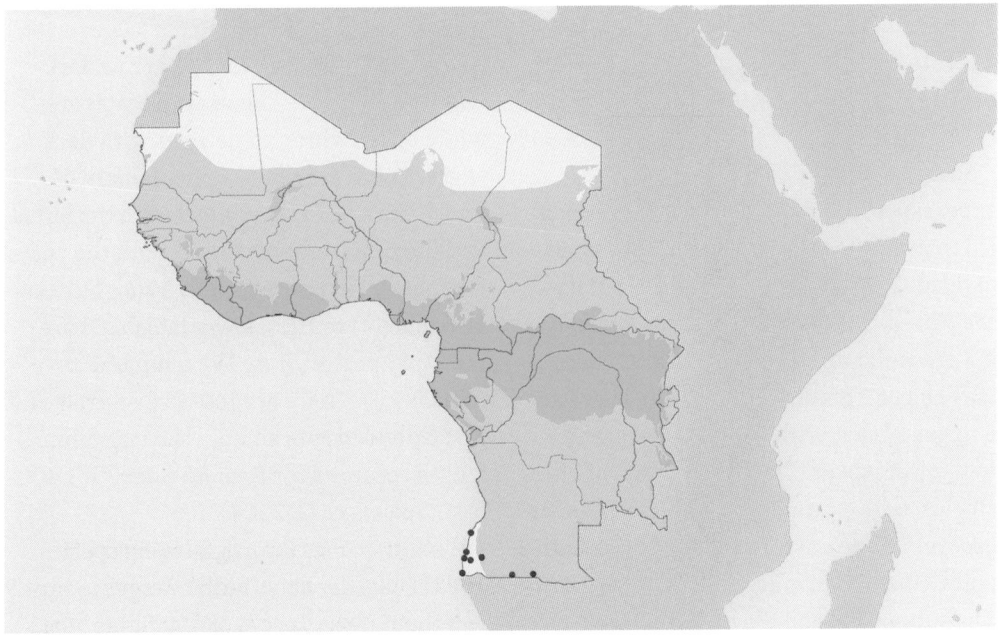

FIGURE 11.55. *Psammophis trigrammus* (red). By K. Jackson

ally biting and killing their sheep, the Puff Adder, *Bitis arietans*, is a more likely culprit (C. Elstob pers. comm.).

Compared with *Psammophis*, *Psammophylax* are relatively heavier bodied, which may account for their larger clutch size and diet of mostly mammalian prey reported by Shine et al. (2006b). Skaapstekers do not engage in male combat as some Sand Snakes do, although, like other psammophiines, Skaapstekers show relatively little sexual dimorphism. Shine et al. (2006b) found that female *P. tritaeniatus* (the only species of Skaapsteker they examined) were 27% longer and had larger heads than males, suggesting that in species that exhibit sexual size dimorphism, it is the males who tend to be larger among *Psammophis*, and the females larger among *Psammophylax*. But Broadley (pers. comm.) found that male Skaapstekers reached larger maximum lengths than females. These results are not necessarily mutually exclusive. Shine et al. (2006b) found many mature males to be small, bringing down the average size of males to below that of females, although the largest individuals in both studies were males.

Skaapstekers are similar in overall appearance, though generally heavier bodied than Sand Snakes. The nasal is always divided. The internasals and prefrontals are paired. The loreal is present. There is 1 preocular and 2–3 postoculars. There may be 1 or 2 anterior temporals. There are 8 upper labials, with no suboculars, and 9–11 lower labials. There are 2 pairs of chin shields, with a pronounced submandibular groove. The dorsal scales are smooth, with 2 apical pits, and arranged in 17 straight rows. The subcaudals and the cloacal scale are divided. Although all Skaapstekers share

the same formula for upper labials, 8(4,5), useful scale characters for distinguishing the species are contact between particular upper labials with other scales such as the loreal and posterior nasal.

Like all psammophiines, Skaapstekers possess an extranarial valve and engage in self-rubbing behavior but are not known to possess the more complex system of pheromones and behavior seen in Sand Snakes. Skaapstekers are unusual among snakes in engaging in parental care, which the different species do to different degrees (Shine et al. 2006b).

Northern Greybelly Skaapsteker: *Psammophylax multisquamis* (Loveridge 1932)

Psammophylax multisquamis is an east African species with a range extending from Tanzania north to Ethiopia. The western edge of its range extends into Rwanda. The type locality is Nairobi, Kenya.

P. multisquamis is gray or olive green above, with a pair of broad dark lateral stripes and usually also a narrow dark mid-dorsal stripe. The ventral scales are white or off-white, sometimes with dark edges. There are 8(4,5) upper labials, of which the third is in contact with the loreal and the first is not in contact with the posterior nasal. There are usually 10(5), but may be from 9(4) to 11(5), lower labials. There are 2 or 3 postoculars. The temporal formula is 2+3. There are 160–184 ventrals and 51–66 divided subcaudals. The maximum length recorded for *P. multisquamis* is 1,400 mm (Spawls et al. 2004).

P. multisquamis is an egg-laying species, and the mother stays with the eggs to protect them (Spawls et al. 2004; Shine et al. 2006b).

Key to Central and Western African Species of the Genus *Psammophylax*

1 First upper labial not in contact with the posterior nasal 2
1' First upper labial in contact with the posterior 3

2(1) Supraocular broader than the frontal; venter patterned with dark blotches ....
.... *P. rhombeatus*
2' Supraocular narrower or equal in breadth to the frontal; venter uniformly
pale, sometimes with a dark distal edge to the ventral scales .... *P. multisquamis*

3(1) Dark markings on the upper labials; dark mid-vertebral stripe either absent
or more than 2 scales in breadth .... *P. variabilis*
3' Upper labials uniformly pale, with no markings; dark mid-vertebral stripe
less than 2 scales in breadth .... *P. tritaeniatus*

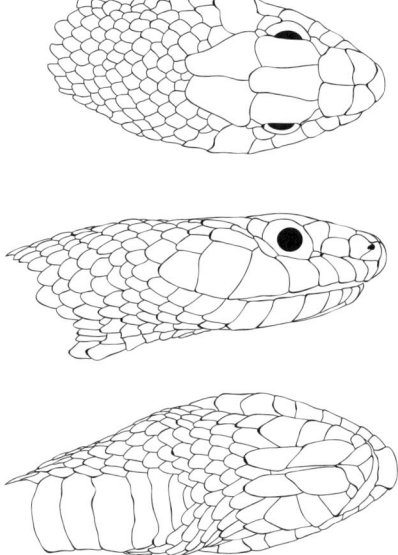

FIGURE 11.56. *Psammophylax multisquamis* USNM 161993. By T. Giri

FIGURE 11.57. *Psammophylax multisquamis*, Kenya. By S. Spawls

FIGURE 11.58. *Psammophylax multisquamis*, Ethiopia. By S. Spawls

Rhombic Skaapsteker: *Psammophylax rhombeatus* (Linnaeus 1758)

The Rhombic Skaapsteker is a southern African species whose range extends northward into southern Angola. The type locality is "Indiis" (meaning South Africa).

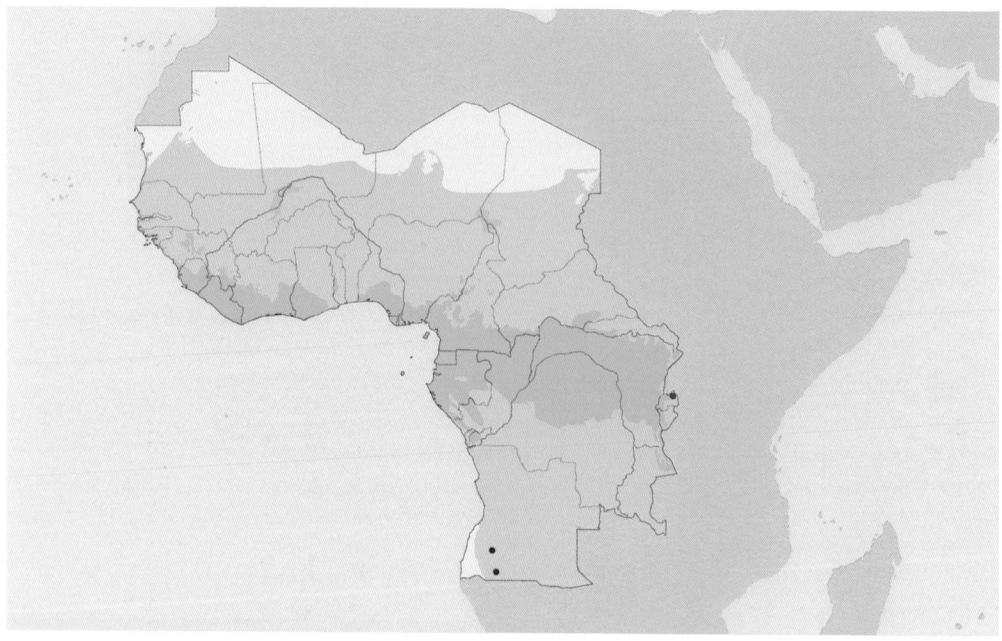

FIGURE 11.59. *Psammophylax multisquamis* (red), *Psammophylax rhombeatus* (blue). By K. Jackson

Sometimes called the "Spotted Skaap-steker," *Psammophylax rhombeatus* is olive green above, either with a pattern of dark brown lozenges or with two pale lateral stripes. The underside is light green with a pair of thin dark stripes. There are 8(4,5) upper labials, of which the first is not in contact with the posterior nasal and the second is in contact with the loreal. There are 10(4) or 10(5) lower labials. There are 2 postoculars. The temporal formula is 2+3 or sometimes 2+2. There are 140–183 ventrals and 60–84 divided subcaudals. The maximum length recorded for *P. rhombeatus* is 1,500 mm (FitzSimons 1974).

The Rhombic Skaapsteker is an egg-laying species, and the mother stays with the eggs to protect them (Broadley 1983).

Two subspecies are recognized: *P. r. rhombeatus* (Linnaeus 1758); *P. r. ocellatus* (Bocage 1873). They are distinguished on the basis of the size and shape of the rostral.

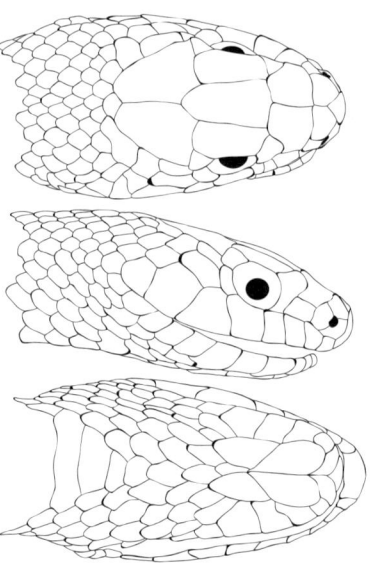

FIGURE 11.60. *Psammophylax rhombeatus* BMNH 59.5.30.6. By T. Giri

FIGURE 11.61. *Psammophylax rhombeatus* (spotted phase), South Africa. By W. Wüster

FIGURE 11.62. *Psammophylax rhombeatus* (striped phase), South Africa. By A. Barlow

Striped Skaapsteker: *Psammophylax tritaeniatus* (Günther 1868)

The Striped Skaapsteker is a southern African species whose range extends north into southern Angola and southern Democratic Republic of Congo. The type locality is no more specific than "southeast Africa."

Similar in coloration to *Psammophylax variabilis*, *P. tritaeniatus* is gray or olive green above, with a pair of broad dark lateral stripes and usually also a narrow dark mid-dorsal stripe. The underside is uniformly pale. There are 8(4,5) upper labials, of which the first is in contact with the posterior nasal and the second is in contact with the loreal. There are 10(5) or 11(5) lower labials. There are 2 postoculars. The temporal formula is 1+2 or 2+3. There are 139–176 ventrals and 49–69 divided subcaudals. The maximum length recorded for *P. tritaeniatus* is 900 mm (Spawls et al. 2004).

The Striped Skaapsteker is an egg-laying species. The mother abandons the eggs once they are laid, but it can be said in her defense that the embryos are at an advanced

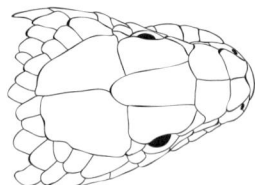

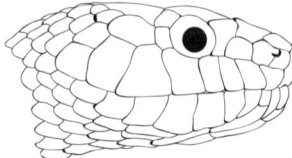

FIGURE 11.64. *Psammophylax tritaeniatus*, Namibia. By S. Spawls

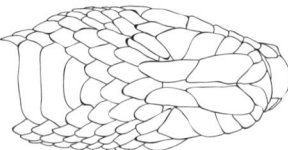

FIGURE 11.63. *Psammophylax tritaeniatus* RMCA 11180. By T. Giri

FIGURE 11.65. *Psammophylax tritaeniatus*, South Africa. By C. Elstob

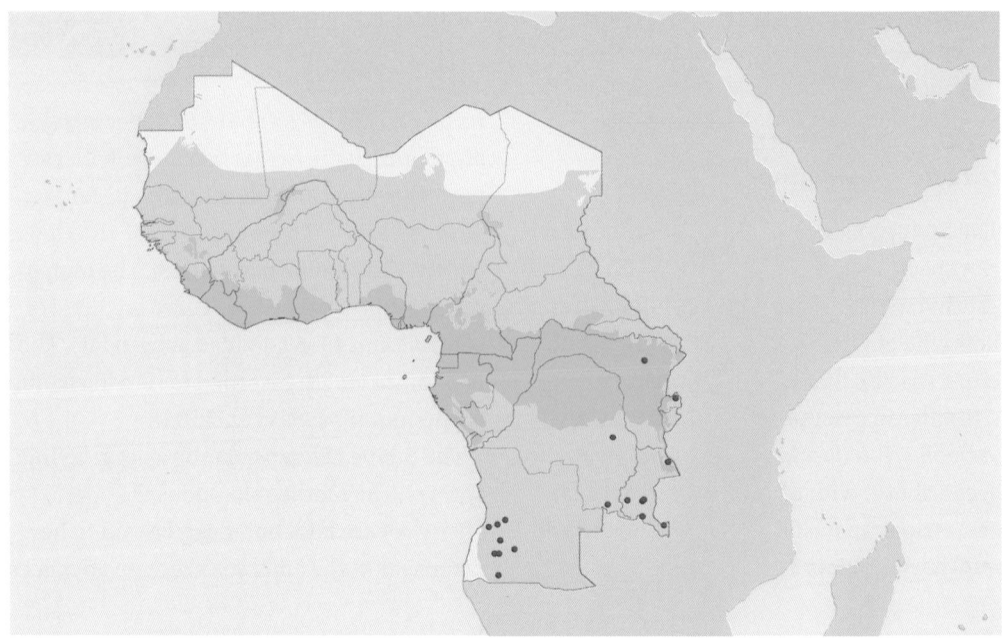

FIGURE 11.66. *Psammophylax tritaeniatus* (red). By K. Jackson

stage of development when the eggs are laid, so the incubation period is relatively short compared to *P. rhombeatus* and *P. multisquamis*, in which the mother stays with the eggs to protect them (Shine et al. 2006b).

Greybelly Skaapsteker: *Psammophylax variabilis* Günther 1893b

The Greybelly Skaapsteker is a southeastern African species whose range extends from Botswana and Mozambique to the northwestern edge of its range in Rwanda and eastern Democratic Republic of Congo. The type locality is Shire Highlands, Malawi.

Psammophylax variabilis is gray or olive green above, with a pair of broad dark lateral stripes and usually also a narrow dark mid-dorsal stripe. The underside is uniformly pale gray. There are 8(4,5) upper labials, the first of which may or may not be in contact with the posterior nasal and the second is in contact with the loreal. There are 10(5) lower labials. There are 2 postoculars. The temporal formula is 1+2 or 2+3. There are 149–167 ventrals and 49–61 divided subcaudals. The maximum length recorded for *P. variabilis* is 1,000 mm (Spawls et al. 2004).

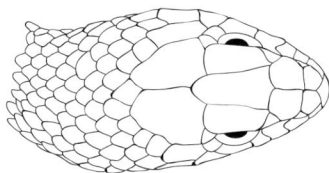
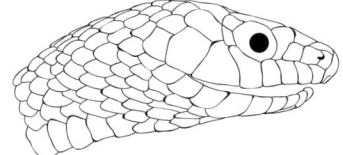
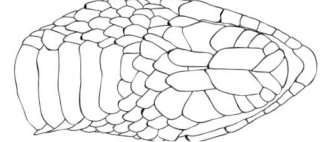

FIGURE 11.67. *Psammophylax variabilis* RMCA 11267. By T. Giri

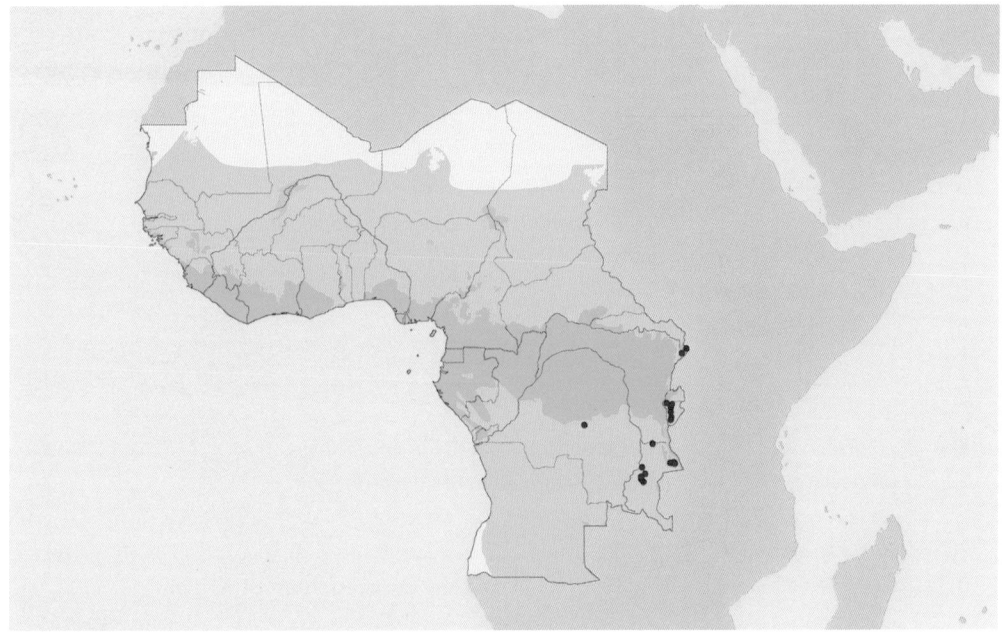

FIGURE 11.68. *Psammophylax variabilis* (blue). By K. Jackson

The Greybelly Skaapsteker is the only species in the genus that gives birth to live young rather than laying eggs (Broadley 1983; Blackburn 1985; Shine 1985).

Laurent (1956) described three subspecies of *P. variabilis*: *P. v. festivus*, *P. v. subniger*, *P. v. vanoyei*. The first and second are no longer recognized, while the third probably represents the subspecies native to our zone.

Beaked Snakes: Genus *Rhamphiophis* Peters 1854

Rhamphiophis is a genus of two to six species (depending on the author), four of which occur within our zone. As their name suggests, Beaked Snakes have an enlarged, hooked, or downward pointing rostral scale and a reinforced skull, adapted for digging. Though diurnal, they are encountered only rarely, as most of their time is spent underground in burrows excavated by small mammals. The latter make up part of their varied diet, which also includes lizards, amphibians, snakes, and sometimes small birds.

Like other psammophiines, they have rudimentary hemipenes and possess a grooved posterior maxillary fang. They have an extranarial valve and engage in self-rubbing behavior. The nasal is divided. There are 2 internasals and 2 prefrontals. The loreal is present. There are 1 or 2 preoculars and 2 or 3 postoculars. There are 2 anterior temporals. There are 8(4,5) upper labials and 9–11 lower labials. There are 2 pairs of chin shields, with the submandibular groove pronounced between the posterior pair. The smooth dorsal scales have 2 apical pits and are arranged in 17 or 19 straight rows. The cloacal scale and the subcaudals are divided. The number of ventrals is sexually dimorphic (females having more), but the number of subcaudals does not usually differ between the sexes.

Striped Beaked Snake: *Rhamphiophis acutus* (Günther 1888)

The Striped Beaked Snake inhabits wet savanna and rainforest. Its range extends from Burundi, Tanzania, and southeast Democratic Republic of Congo to Zambia and Angola. The type locality is Pungo Andogo, Angola.

Key to Central and Western African Species of the Genus *Rhamphiophis*

1 More than 75 subcaudals; 5 lower labials in contact with the anterior chin shields .... *R. oxyrhynchus*

1' Fewer than 75 subcaudals; 4 lower labials in contact with the anterior chin shields 2

2(1) 1 preocular; anterior chin shields shorter than the posterior pair .... *R. maradiensis*

2' 2 preoculars; anterior chin shields longer than the posterior pair 3

3(2) Underside uniformly white .... *R. acutus*

3' Underside pale with a black stripe extending along the ventrals on either side .... *R. togoensis*

The body is light brown above with a pair of broad dark lateral stripes outlined in black, extending from head to tail. The top of the head and the back of the neck are black. The underside is white. There is usually 1 preocular (occasionally 2) and 2 (occasionally 3) postoculars. The temporal formula is 2+3 (sometimes 2+2 or 2+4). There are 9(4) to 10(5) lower labials. There are 155–190 ventrals (168–190 in males, 155–184 in females) and 53–67 divided subcaudals (58–67 in males, 53–65 in females). The maximum length recorded for *Rhamphiophis acutus* is 1,077 mm (Broadley 1971c).

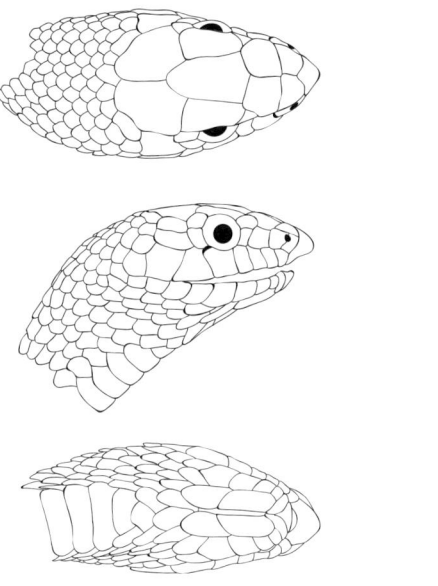

FIGURE 11.69. *Rhamphiophis acutus* RMCA 30787. By T. Giri

In their molecular phylogenetic study of psammophiines, Kelly et al. (2008) found *R. acutus* to be more closely related to the Skaapstekers (genus *Psammophylax*) than to other species of *Rhamphiophis*. This result led Kelly et al. (2008) to conclude that the Striped Beaked Snake belongs in the genus *Psammophylax* and that its hooked rostral and reinforced snout represented adaptations for digging derived independently of those of *Rhamphiophis*. A detailed study of cranial anatomy is needed to evaluate this hypothesis. Unfortunately, Bourgeois (1965), who carried out the greatest study

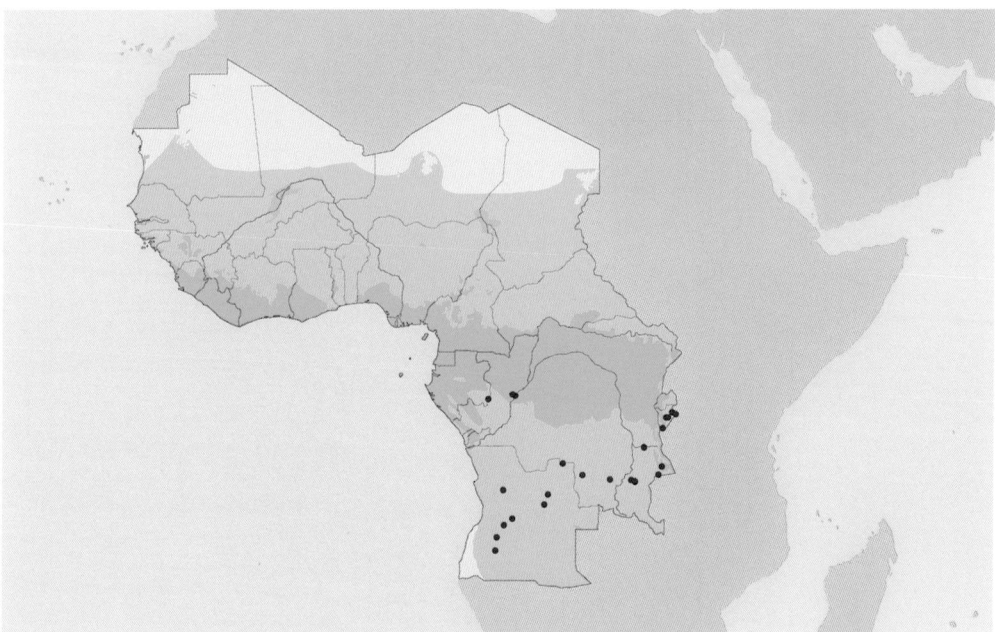

FIGURE 11.70. *Rhamphiophis acutus* (blue). By K. Jackson

ever of the cranial anatomy of African snakes, studied *R. acutus* in detail, but no other *Rhamphiophis* species for comparison.

Maradi Beaked Snake: *Rhamphiophis maradiensis* Chirio and Ineich 1991

The Maradi Beaked Snake is known from just four specimens from southern Niger but probably also exists in northern Nigeria. The type locality is Gari'n Bakwai, Niger.

The body is rusty beige above with dark spots arranged in a pair of paravertebral stripes. The sides are yellow, scattered with reddish flecks. The snout is whitish, and the top of the head and the back of the neck are black. The underside is white with a broad pale-yellow midline stripe. There is usually 1 preocular and 2 postoculars. The temporal formula is 2+3. There are 9(5) lower labials. There are 178–188 ventrals and 58–64 divided subcaudals. The maximum length recorded for *R. maradiensis* is 573 mm (Chirio and Ineich 1991).

Trape and Mané (2006a) expressed the view that *R. maradiensis* in fact represents simply a misidentified Moila Snake (*Malpolon moilensis*), but there is sufficient evidence to make a good case for its status as a separate species in a separate genus. The relative sizes and proportions of many head scales of *R. maradiensis* differ from those of *M. moilensis*, though these could arguably be dismissed as resulting from the small sample size (n = 4) of the former species. Differences more difficult to ignore include its lack of the black markings on the sides of the head characteristic of the Moila Snake, as well as the fact that there is absolutely no overlap in number of ventrals between the two species: 178–188 for *R. maradiensis* (Chirio and Ineich 1991) and just 139–176 for *M. moilensis*. Chirio (unpub. data) kept Maradi Beaked Snakes in captivity for several months and never observed the trademark threat display of the Moila Snake. This observation may simply reflect

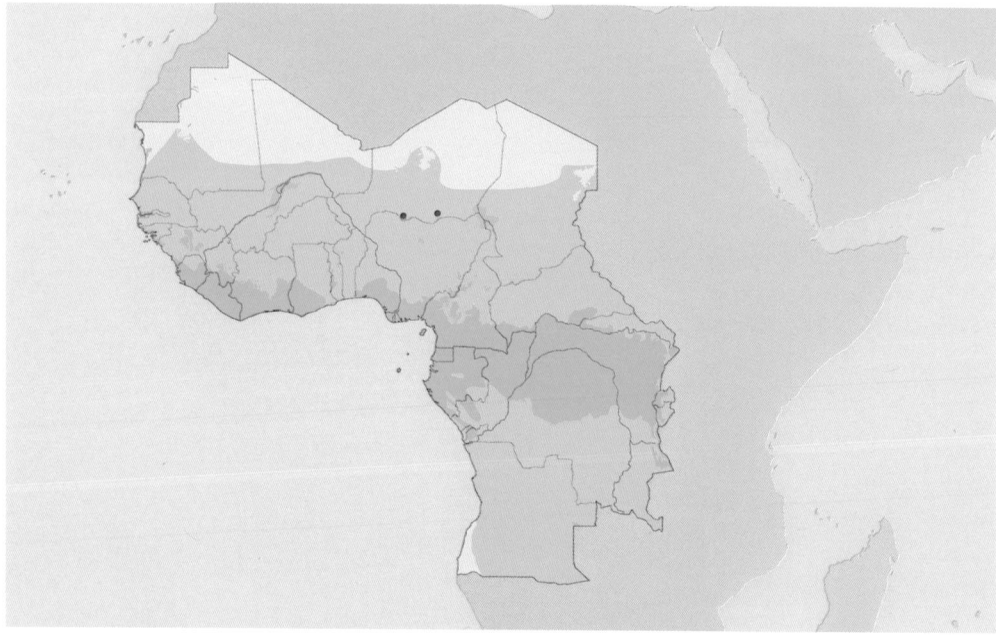

FIGURE 11.71. *Rhamphiophis maradiensis* (red). By K. Jackson

atypical behavior by tame captive animals, or it may represent a significant behavioral difference between species.

Western Rufous Beaked Snake: *Rhamphiophis oxyrhynchus* (Reinhardt 1843)

The Western Rufous Beaked Snake has a broad distribution in west Africa, extending from Senegal to Somalia to South Africa. The type locality is "Guinea," but presumably meaning "Guinea Coast" rather than the country, Guinea. It is likely that the true type locality is somewhere in Ghana. Two subspecies (sometimes considered full species) are recognized, both of which occur within our zone. *Rhamphiophis o. oxyrhynchus* (Reinhardt 1843) is the west African subspecies, with a range extending from Senegal to Chad, and the same type locality as the species. The eastern and southern African subspecies, *R. o. rostratus* (Peters 1854), has a range extending from Sudan to Somalia to South Africa. Its type locality is Tete, Mozambique. The two subspecies differ in ventral and subcaudal counts and in coloration. There appears to be a gap between the ranges of the two subspecies (Chirio and Ineich 1991).

 R. o. oxyrhynchus is light brown above with white upper labials and venter. *R. o. rostratus* is uniformly pale, with a pair of broad dark lateral stripes extending from head to tail. There is usually 1 preocular and 2 postoculars. The temporal formula is 2+3 or 2+4. There are 10(5) or occasionally 11(5) lower labials. *R. o. oxyrhynchus* has 162–199 ventrals (162–179 in males, 167–199 in females) and 80–104 divided subcaudals (80–104 in males, 81–101 in females). *R. o. rostratus* has 148–194 ventrals (148–186 in males, 154–194 in females) and 87–124 divid-ed subcaudals (117–124 in males, including some individuals with truncated tails, 87–121 in females). The maximum length recorded for *R. oxyrhynchus* is 1,600 mm (Spawls et al. 2004).

Togo Beaked Snake: *Rhamphiophis togoensis* (Matschie 1893b)

The Togo Beaked Snake, formerly a subspecies of the Striped Beaked Snake, is also a species of rainforests and wet savannas.

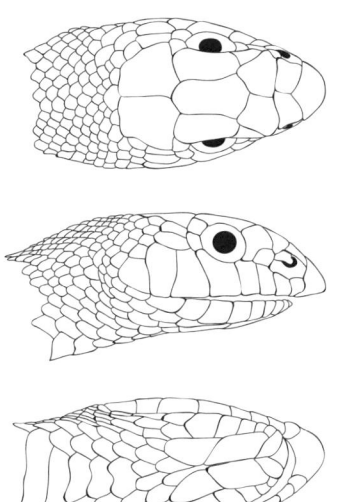

FIGURE 11.72. *Rhamphiophis oxyrhynchus* RMCA 29650. By T. Giri

FIGURE 11.73. *Rhamphiophis oxyrhynchus*, Ghana. By S. Spawls

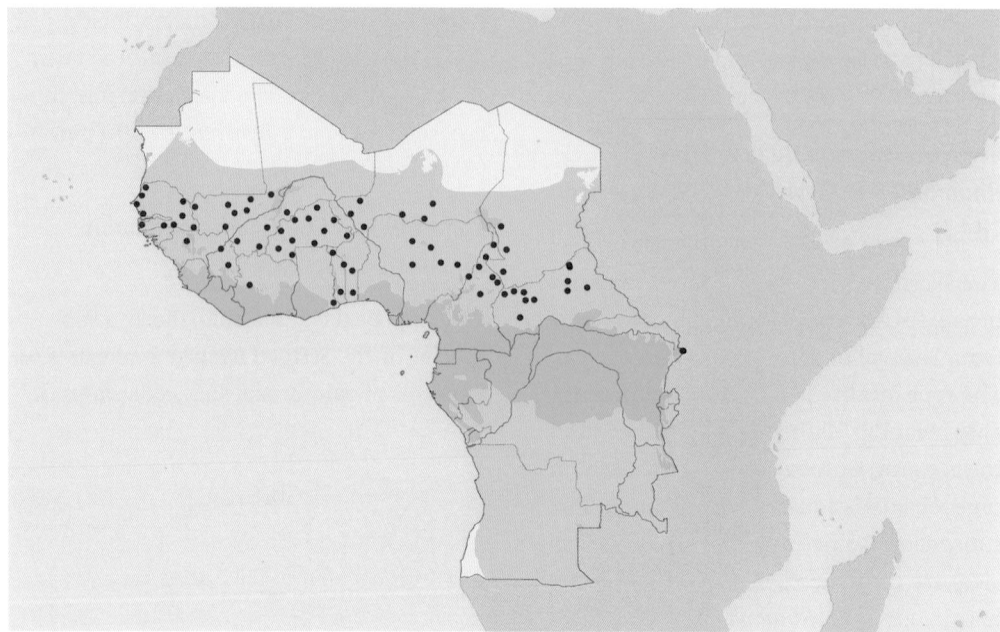

FIGURE 11.74. *Rhamphiophis oxyrhynchus* (blue). By K. Jackson

Its range extends from Guinea to northern Democratic Republic of Congo to Uganda. The type locality is Togo.

The body is brown above, with a pair of broad black lateral stripes extending from snout to tail. The underside is pale with a pair of thin black stripes. There are usually 2 preoculars and 2 postoculars. The temporal formula is 2+3. There are 9(4) lower labials. There are 165–188 ventrals (171–188 in males, 165–179 in females) and 57–76 divided subcaudals (60–76 in males, 57–72 in females). The maximum length recorded for *Rhamphiophis togoensis* is 830 mm (Chirio and Ineich 1991).

In terms of scale counts, *R. togoensis* is very close to *R. acutus*. It can be distinguished from the latter by the presence of the ventral stripes and by the higher average number of ventrals. Some authors (Broadley 1971c; Pitman 1974; Chirio and Ineich 1991) consider *R. togoensis* to be a sub-species of *R. acutus*, but without explaining their reasons for doing so.

Subfamily Prosymninae

The subfamily Prosymninae contains only the Shovel-snout Snakes, genus *Prosymna*, a genus with a sub-Saharan African distribution, containing 16 species. The Shovel-snouts are a distinctive genus, morphologically specialized for a semi-fossorial lifestyle and for a diet of reptile eggs.

Bogert (1940) assigned the genus *Prosymna* alone to Group 12 on the basis of the absence of hypapophyses on the posterior vertebrae, an unforked sulcus spermaticus, and an aglyph dentition. It is perhaps worth noting that only the aglyph dentition (the last of the three criteria for Bogert's groupings) set *Prosymna* apart from his Group 16, the Psammophiinae (hypapophyses absent, unforked sulcus spermaticus, opisthoglyph dentition).

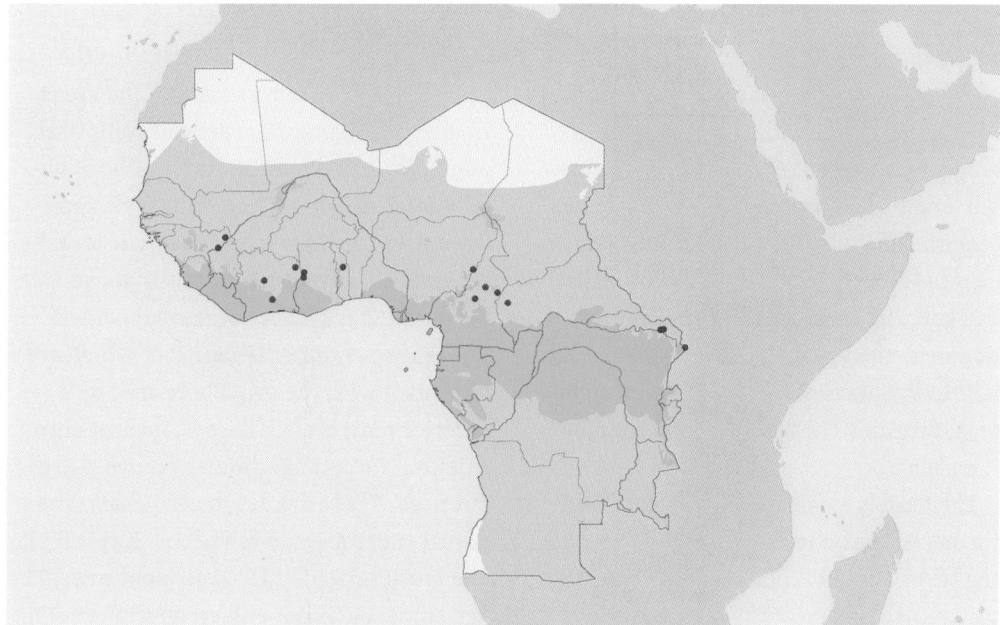

FIGURE 11.75. *Rhamphiophis togoensis* (red). By K. Jackson

Molecular studies find the genus *Prosymna* to be monophyletic and belonging somewhere within the Lamprophiidae. Vidal et al. (2008) and Pyron et al. (2011) found *Prosymna* to be most closely related to the Pseudaspidinae (*Pseudaspis* and *Pythonodipsas*). Kelly et al. (2009) proposed assigning *Prosymna* by itself to its own family, Prosymnidae. Pyron et al. (2013) and Zheng and Wiens (2016) find the Prosymninae to be most closely related to the Psammophiinae.

Shovel-snout Snakes: Genus *Prosymna* Gray 1849

Prosymna is an African genus of sixteen species, five of which occur within our zone. Shovel-snouts are burrowing snakes whose diet consists almost entirely of reptile eggs (Broadley 1979). This dietary specialization is reflected in the maxillary dentition, which is aglyph, with reduction of the anterior maxillary teeth, and enlargement of the posteriormost teeth into a blade, presumably for slicing open the shells. Boulenger (1894) describes the maxilla as "short, with seven or eight teeth increasing in size posteriorly, the first tooth minute, falling below the centre of the eye, the hindermost teeth very large, strongly compressed, blade-like" (see Fig. 11.76). The hemipenes of Shovel-snouts are remarkable in being exceptionally long—longer, in fact, than the tail of these short burrowing snakes. Schmidt (1923) described the hemipenes of a specimen of *Prosymna ambigua*, collected by the American Museum of Natural History expedition to the Democratic Republic of Congo, 1909–1915. The snake had a tail length of 48 mm, and the hemipenes exceeded this length by at least 10 mm. Unforked, and with an unforked sulcus spermaticus, the organ is "telescoped" when withdrawn, which was why Schmidt's estimate of the length of the hemipenes could not be more precise (see Fig. 11.77). Schmidt (1923) speculated that the telescoped hemipenes reflected a reduction in tail length associated with adaptation to a fossorial lifestyle, and that modern *Prosymna* must be derived from an ancestor that had had a longer tail—one more proportionate to the length of the hemipenes held within it. Shovel-snouts lay small clutches (e.g., Broadley 1979a reports 3-6 eggs for *P. ambigua*).

The head is small and not distinct from the neck. The body and tail are short, and the tip of the tail is pointed. The snout is

FIGURE 11.76. Maxilla of *Prosymna sundevalli.* Modified by K. Jackson from Boulenger (1894)

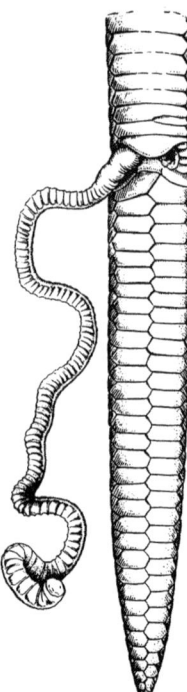

FIGURE 11.77. Everted right hemipenis of *Prosymna ambigua.* Modified K. Jackson from Schmidt (1923)

rounded, and the rostral scale flattened into a shovel-like edge, giving this genus the common name "Shovel-snout." The eye is small, with a round or vertically elliptical pupil. The nasal may be single or horizontally semi-divided. There is a single prefrontal and a single internasal. The loreal is present. There are 1-2 preoculars and 1-3 postoculars. There are no suboculars. There are 5-7 upper labials, 2 of which are in contact with the eye. There are 1 or 2 anterior temporals. There is 1 pair of chin shields. The submandibular groove is pronounced. There are 7-9 lower labials. The dorsal scales are smooth with 1-2 apical pits and are arranged in 15-21 straight rows. The vertebral row is not enlarged relative to the other dorsal scale rows. The cloacal scale is single.

East African Shovel-snout: *Prosymna ambigua* Bocage 1873

Two subspecies are recognized: *Prosymna ambigua ambigua* Bocage 1873; *P. a. bocagii* Boulenger 1897a. The range of *P. a. ambigua* extends from the Republic of Congo in the north to Angola in the south, and east to Lake Tanganyika. The type locality is Duque de Bragança, Angola. The range of *P. a. bocagii* extends from southern Cameroon to South Sudan, including northern Democratic Republic of Congo and Uganda. The type locality is Zongo, Democratic Republic of Congo. The dorsum is dark brown or gray, with a white spot on the tip of each dorsal scale. The ventrals are black proximally, with brown and white spots, and with pale distal edges. The nasal is single or semi-divided horizontally. The loreal is longer than deep. There is 1 small preocular, deeper than long. There are usually 2 postoculars, the inferior larger than the superior.

Key to Central and Western African Species of the Genus *Prosymna*

1   5 upper labials 2
1'   Usually 6 upper labials (occasionally 5 or 7) 3

2(1) 2 apical pits on each dorsal scale; 136–151 ventrals in males, 153–168 ventrals in females .... *P. meleagris*
2'   1 apical pit on each dorsal scale; 149–165 ventrals in males, 166–187 ventrals in females .... *P. greigerti*

3(1) Length of frontal equal to length of parietals; 153–174 ventrals in males, 169–199 ventrals in females .... *P. frontalis*
3'   Frontal longer than parietals; 121–155 ventrals in males, 130–171 ventrals in females 4

4(3) Loreal longer than deep; coloration uniformly dark except for a pale dot on each dorsal scale.. *P. ambigua*
4'   Depth of loreal greater than or equal to its length; coloration pale with darker bands or blotches.. *P. angolensis*

FIGURE 11.78. *Prosymna ambigua*, Republic of Congo. By K. Jackson

The frontal is triangular, broader than long, and longer than the parietals. The supraocular is narrow. There are usually 6(3,4), sometimes 5 or 7(3,4), upper labials and 8(3) lower labials. The temporal formula is usually 1+2, occasionally 2+3. The dorsal scales are smooth, with 1 apical pit, and arranged in 15 (occasionally 17) straight rows. There are 124–171 ventrals (fewer than 155 in males, more than 160 in females) and 15–34 divided subcaudals (more than 24 in males, fewer than 25 in females). The maximum length recorded for this species is 398 mm (Broadley 1980).

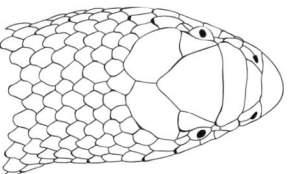

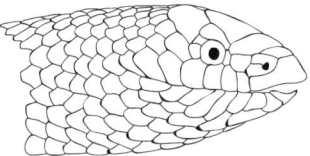

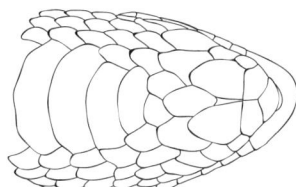

FIGURE 11.79. *Prosymna ambigua* RMCA 8245. By T. Giri

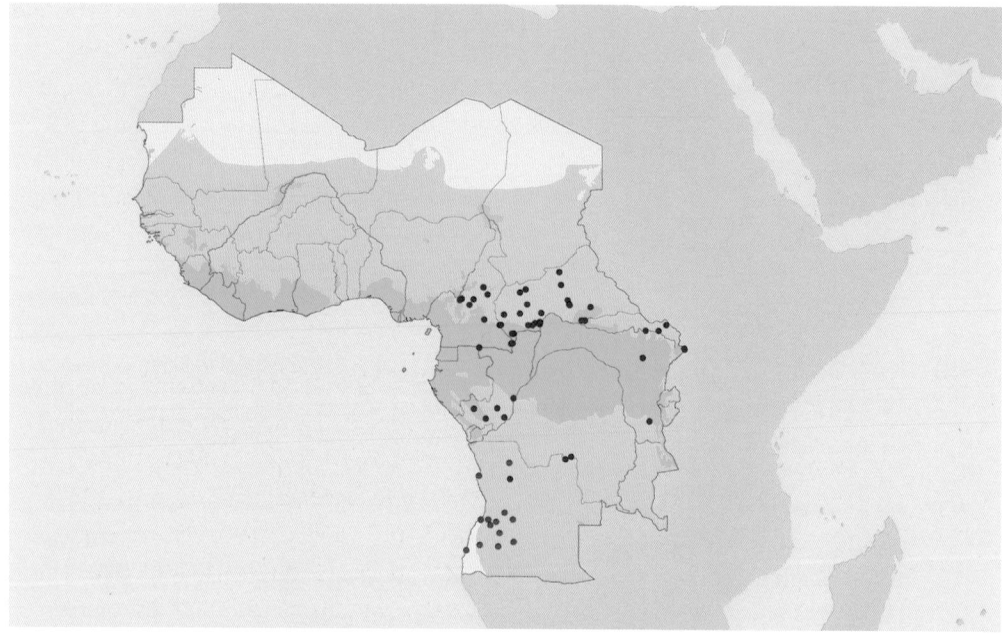

FIGURE 11.80. *Prosymna ambigua* (blue), *Prosymna angolensis* (red). By K. Jackson

Angola Shovel-snout: *Prosymna angolensis* Boulenger 1915

The range of *Prosymna angolensis* extends from Angola and Namibia in the west to Zambia and Zimbabwe in the east. The type locality is Caconda, Angola. Coloration is highly variable for this species. The dorsum is light brown and may be patterned with large black blotches merging together or with two rows of small black spots. The venter is uniformly white. The nasal is single or semi-divided horizontally. The loreal is equal in length and depth. There is usually 1 preocular, the depth of which is less than the diameter of the eye. There is usually 1 postocular, the depth of which is greater than the diameter of the eye. The frontal is equal in length and breadth and longer than the parietals. The supraocular is narrow. There are usually 6(3,4), occasionally 5, or 7(3,4) upper labials and 7(3) or sometimes 8(3) lower labials. The temporal formula is 1+2, occasionally 2+2 or 2+3. The dorsal scales are smooth, with 1 apical pit, and arranged in 15 straight rows. There are 121–163 ventrals (fewer than 156 in males, more than 129 in females) and 16–28 divided subcaudals (more than 22 in males, fewer than 21 in females). The maximum length recorded for this species is 360 mm (Bocage 1895a).

Southwestern African Shovel-snout: *Prosymna frontalis* (Peters 1867)

The range of *Prosymna frontalis* extends from northwestern South Africa to southwestern Angola. The type locality is Otjimbingue, Namibia.

The dorsum is yellowish brown, each dorsal scale edged with black. The venter is uniformly white. The nasal is semi-divided horizontally. The loreal is large and longer than it is deep. There is usually 1 preocular, the depth of which is less than the diameter of the eye. There are 2 postoculars, the inferior larger than the superior. The frontal is triangular, equal in length and breadth, and longer than the parietals. The supraocular is narrow. There are 6(3,4) upper labials and 8(3) or 9(3) lower labials. The temporal formula is 1+2. The dorsal scales are smooth, with 1 apical pit, and arranged in 15 straight rows. There are 153–199 ventrals (fewer than 175 in males, more than 168 in females) and 32–54 divided subcaudals (more than 40 for males, fewer than 44 for females). The maximum length recorded for this species is 440 mm (Broadley 1980).

Greigert's Shovel-snout: *Prosymna greigerti* Mocquard 1906

Greigert's Shovel-snout is found in savanna from Senegal to Ethiopia. Central African Republic, South Sudan, and Uganda represent the southern limit of its range. Once considered a subspecies of *Prosymna meleagris*, *P. greigerti* is now accepted as a distinct species, two subspecies of which are recognized: *P. g. greigerti* Mocquard 1906; *P. g. collaris* Sternfeld 1908b. The type localities are Lobi Region (Ivory Coast) for *P. g. greigerti* and Misahöhe, Togo, for *P. g. collaris*.

The dorsum is reddish brown or gray, with a pale apical spot on each dorsal scale.

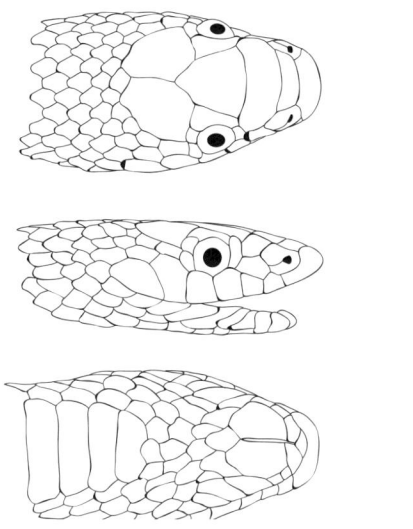

FIGURE 11.81. *Prosymna frontalis* BMNH 1976.1770. By T. Giri

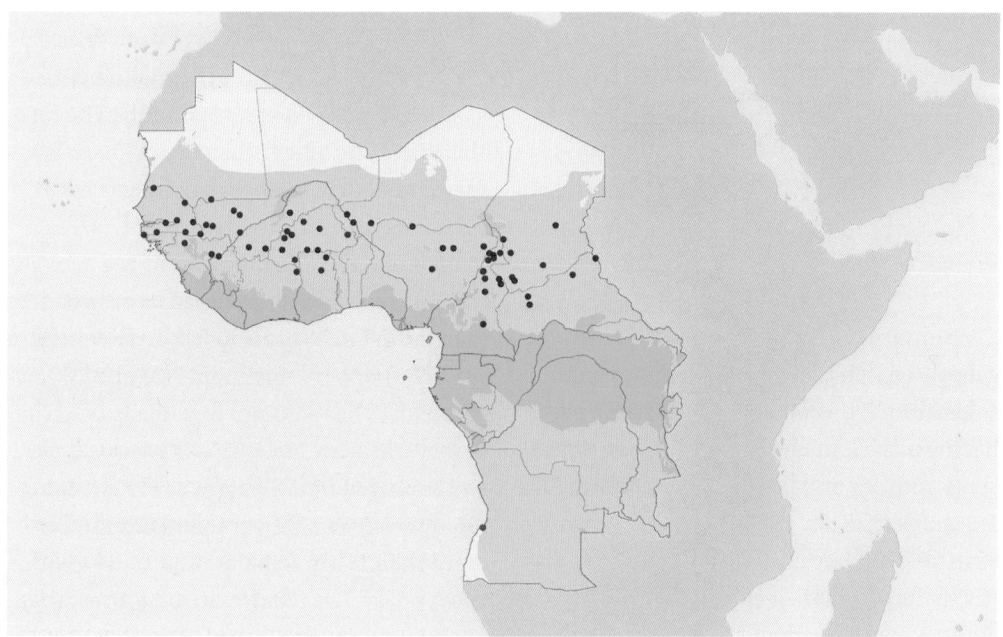

FIGURE 11.82. *Prosymna frontalis* (red), *Prosymna greigerti* (blue). By K. Jackson

FIGURE 11.83. *Prosymna greigerti*, Ghana. By W. Branch

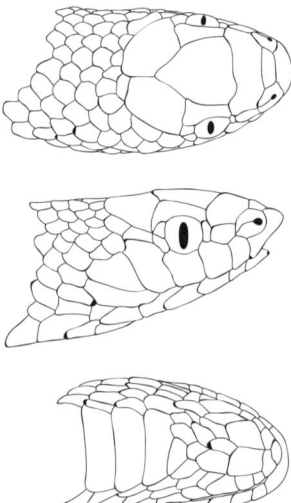

FIGURE 11.84. *Prosymna greigerti* USNM 161993. By T. Giri

The venter is uniformly white. The two subspecies can be distinguished by their coloration. *P. g. collaris* earns its name by having a black nuchal band with two white spots anterior and posterior to it, which *P. g. greigerti* lacks. The nasal is single or semi-divided horizontally. The loreal is distinctly longer than deep. There is 1 small preocular, deeper than long. There is 1 pos-

tocular, the depth of which is less than or equal to the diameter of the eye. The frontal is triangular, as broad as or broader than long. The supraocular is narrow. There are 5(2,3) upper labials and 7–8(3) lower labials. The temporal formula is 1+2. The dorsal scales are smooth, with 1 apical pit, and arranged in 15 straight rows. There are 149–187 ventrals (fewer than 166 in males, more than 165 in females) and 19–39 divided subcaudals (more than 31 in males, fewer than 28 in females). The maximum length recorded for this species is 360 mm (Trape and Mané 2006a).

Speckled Shovel-snout: *Prosymna meleagris* (Reinhardt 1843)

The range of *Prosymna meleagris* extends from Senegal to Ethiopia. The type locality is "Guinea," but presumably meaning "Guinea Coast" rather than the country, Guinea. It is likely that the true type locality is somewhere in Ghana.

The dorsum is reddish brown or gray, with a white spot at the tip of each dorsal scale. The venter is uniformly white. The nasal is semi-divided horizontally. The loreal is distinctly longer than deep. There is 1 small preocular, deeper than long. There is 1 postocular, the depth of which is less than or equal to the diameter of the eye. The frontal is triangular, as broad as or broader than long. The supraocular is narrow. There are 5(2,3) upper labials and 7(3) or 8(3) lower labials. The temporal formula is 1+2. The dorsal scales are smooth, with 2 apical pits, and arranged in 15 straight rows. There are 136–168 ventrals (fewer than 152 in males, more than 152 in females) and 17–39 divided subcaudals. The maximum length recorded for this species is 372 mm (Broadley 1980).

FIGURE 11.85. *Prosymna mealeagris*, Ghana. By M. Fujita

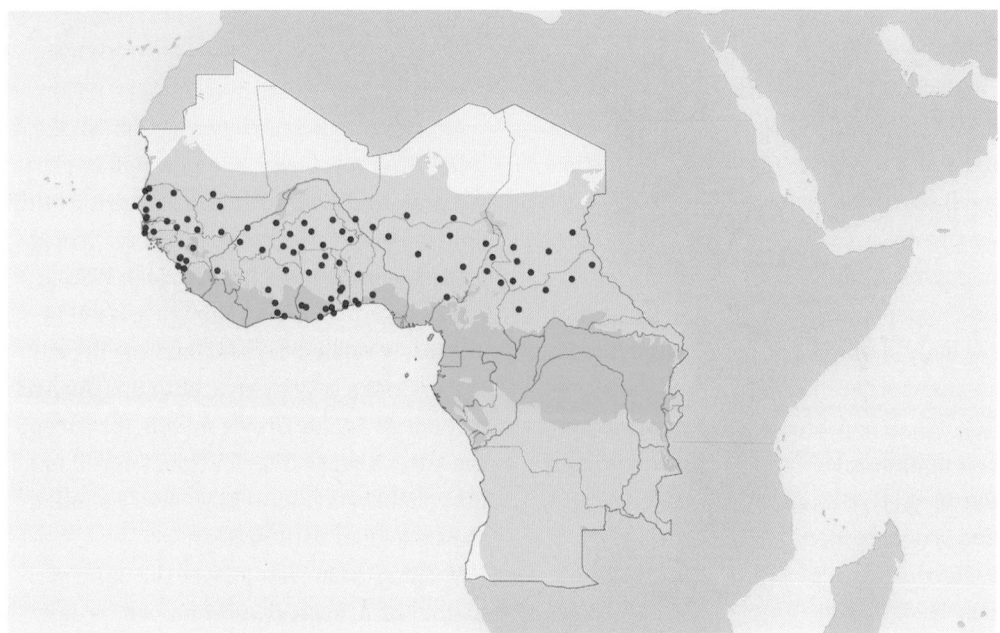

FIGURE 11.86. *Prosymna meleagris* (blue). By K. Jackson

12

Family Colubridae

Subfamily Natricinae

Snakes of the colubrid subfamily Natricinae are semiaquatic or terrestrial snakes. The entire subfamily includes about 225 species with a broad distribution globally, including Africa, Eurasia, and North America, as well as one genus in Australia. They are generally thought to have originated in Asia (e.g., Nagy et al. 2005). Nine species in five genera (*Afronatrix, Helophis, Hydraethiops, Limnophis, Natriciteres*) occur within our zone. These correspond to Bogert's Group 4 (except for *Helophis*, which was not included in Bogert's analysis). These genera are unique among African colubrids in possessing the combination of vertebral hypapophyses in the anterior part of the body, hemipenes with an undivided sulcus spermaticus, and aglyph dentition. Bogert's Group 4 has been supported by recent molecular systematic studies (e.g., Pyron et al. 2013; Zheng and Wiens 2016) and is now recognized as the subfamily Natricinae. Globally, natricine species may be egg-layers or may give birth to live young. All natricines occurring within our zone are egg-layers. Many of the natricines within our zone have dorsally oriented eyes and nostrils, reflecting their aquatic habitat. They have round pupils and tend to be active during the day, feeding on fishes and amphibians. Their bite poses absolutely no danger to humans.

Genus *Afronatrix* Rossman and Eberle 1977

The Brown Water Snake: *Afronatrix anoscopus* (Cope 1861)

Afronatrix is a monotypic genus. The Brown Water Snake, *Afronatrix anoscopus*, is a west African species with a range extending from Senegal to Cameroon. The type locality is Liberia.

Afronatrix is primarily an aquatic snake, inhabiting rainforest swamps, ponds, streams, and rivers but also venturing onto land close to these bodies of water. Ecologically it closely resembles African Water Snakes (genus *Grayia*), though it is not at all closely related to them, and the Brown Water Snake is sympatric with Smith's African Water Snake, *Grayia smithii*, for part of the latter's broad distribution. Luiselli and his collaborators have extensively studied the ecology of both in Nigeria.

Much of what is known of the natural history of *A. anoscopus* centers on its diet. Of 323 individuals collected in all seasons and over several years, the pooled results showed the Brown Water Snake's top prey item to be tadpoles, followed closely by the aquatic frog, *Silurana tropicalis*, and then by terrestrial amphibians and fishes. Juveniles consumed more tadpoles and adults more fishes, presumably because of their sizes (Luiselli et al. 2003). In the dry season, when prey is harder to come by, *A. anoscopus* does not aestivate to conserve energy. Instead, it moves from ponds containing an abundance of *S. tropicalis* to rivers, where they spend the dry season eating fish while the preferred ponds are dried up (Luiselli et al. 2005). And as for coexisting with *Grayia smithii*, the two species eat the same diet during the rainy season, when prey is so abundant that they are not in competition with one another. During the dry season, however, they occupy different dietary niches. *Grayia smithii*, which is a larger snake, feeds on larger prey items, while *A. anoscopus* feeds on smaller ones (Luiselli 2006c). In both species, females are significantly larger than males.

Though prehensile tails are widespread in arboreal snakes, which use them to hang from branches, Senter's (2000) observation that Brown Water Snakes have strongly prehensile tails requires some explaining. The explanation lies in *A. anoscopus*'s method of predation. The snake uses its tail to anchor itself to a submerged stick or rock, waiting for a potential prey item to come within striking range. When it does and the snake manages to catch it, it then continues to use its tail grip as an anchor as it swallows its meal.

The body is reddish brown above, with dark spots along the sides, a pattern that becomes indistinct in older individuals. The underside is yellow or orange with a few small black spots. The pupil is round. The maxillary dentition is aglyph and without diastemata. There are approximately 26 maxillary teeth, increasing in size posteriorly. The nasal is semi-divided. The internasals and prefrontals are paired, and the loreal is present. There are usually 2 preoculars, but sometimes just 1, and 2 postoculars. There are from 1 to 4 (most often 2) suboculars, making *Afronatrix* unusual in having no upper labials in contact with the eye. Slight differences in position may determine whether a scale counts, say, as a preocular or a subocular, but the total number of scales in contact with the eye (perioculars) ranges from 6 to 8 but is most often 7. The temporal formula is 1+2 or 1+3. There are 9(0) or less often 10(0) upper labials and usually 9(5) or 10(6) lower labials, though their total number may range from 8 to 11. There is a pronounced submandibular groove. The dorsal scales are keeled, without apical pits, and arranged in 21–27 (usually 23 or 25) straight rows. There are 134–159 ventrals and 56–75 divided subcaudals. The cloacal scale is divided. The hemipenes are forked distally. The sulcus spermaticus

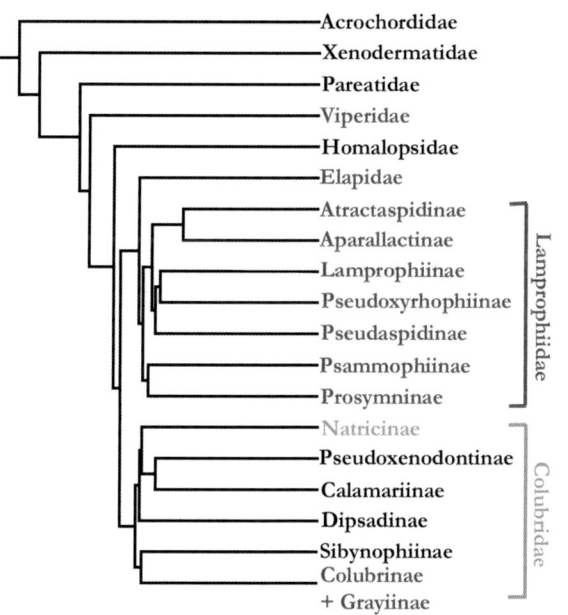

FIGURE 12.1. Higher-level phylogeny of caenophidian families and subfamilies, with families and subfamilies occurring within our zone highlighted (red or orange). The subfamily Natricinae, the topic of this chapter, is highlighted in orange. Modified by K. Jackson from Zheng and Wiens (2016)

FIGURE 12.2. *Afronatrix anascopus*, Ghana. By M. Fujita

FIGURE 12.3. *Afronatrix anoscopus*, Guinea. By P. Naskrecki

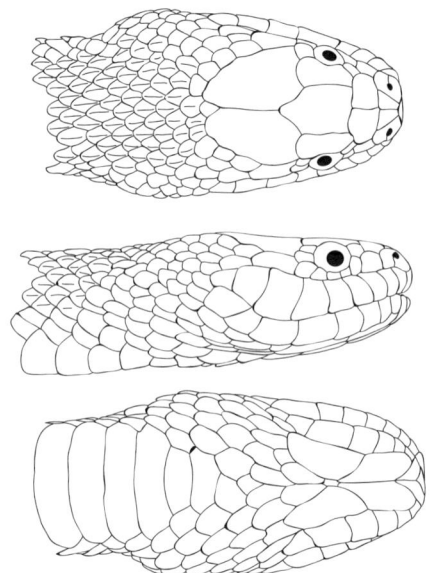

FIGURE 12.4. *Afronatrix anoscopus* RMCA 29566. By T. Giri

is not forked. The maximum length recorded for the Brown Water Snake is 750 mm (Villiers 1975), but the average adult length is closer to 400 mm.

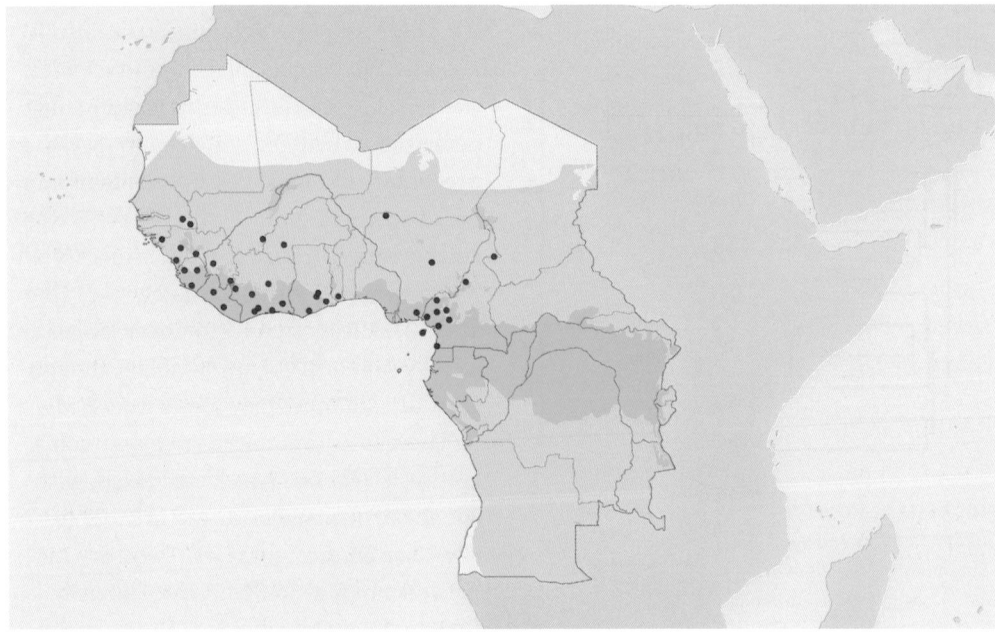

FIGURE 12.5. *Afronatrix anoscopus* (blue). By K. Jackson

Genus *Helophis* Witte and Laurent 1942

Schouteden's Mud Snake: *Helophis schoutedeni* (Witte 1922)

Helophis is a monotypic genus. *Helophis schoutedeni* is a central African forest species, known from the Republic of Congo and from the Democratic Republic of Congo. The type localities are Kwamouth, Democratic Republic of Congo, and Tondu, on Lake Tumba, Democratic Republic of Congo. The common name most often given for this species is "Schouteden's Sun Snake," but this may not have been the intent of the original description. Witte (1922) originally described the species as "*Pelophis schoutedeni*." The genus name *Pelophis* was presumably from the Greek *pelos* (mud) and *ophis* (snake), and Schouteden was the name of the collector. So "Schouteden's Mud Snake" would have been the common name that Witte intended. Subsequently, Witte and Laurent (1942) changed the genus name to "*Helophis*," because *Pelophis* had turned out already to be in use. The common name Schouteden's Sun Snake, perpetuated since (e.g., Frank and Ramus 1996), is perhaps derived from interpreting *Helophis* as *helios* ("sun") and *ophis* ("snake"). But Witte and Laurent chose the name *Helophis* because of the semiaquatic habitat of this species (*helos* = "swamp" and *ophis* = "snake"). We therefore suggest here the common name Schouteden's Mud Snake.

 H. schoutedeni is a stout-bodied aquatic snake with dorsally oriented nostrils and dorsolaterally oriented eyes. The head is small with a poorly defined neck. The eye is small, with a round pupil. In life, the iris is orange. The maxillary dentition is aglyph with 16–17 small teeth, without diastemata.

The body is black above and below. The dorsum is patterned with narrow light brown or orange bands, each the thickness of 1–2 dorsal scales, alternating with slightly thicker black spaces. Posteriorly, the pattern of bands breaks up into small spots. The sides of the body are patterned with small white flecks. Scales of the dorsum of the head are brown or orange speckled with black. The labial, temporal, and anterior gular scales are white with black edges. The head is short, with a rounded snout. The nasal is divided. The prefrontals are paired. The internasals are paired but small and triangular in shape (associated with the dorsal orientation of the nostrils). The loreal is present. There is 1 preocular. There are usually 2 postoculars, the superior larger than the inferior, but sometimes just 1. The frontal is equal in length and breadth. There are no suboculars. There are usually 10(6,7,8) or 11(6,7,8) upper labials, sometimes with 2 rather than 3 upper labials in contact with the eye. The temporal formula is 1 or 2+2 or 3. There are 2 pairs of chin shields, the anterior pair shorter than the posterior pair. The submandibular groove is shallow. There are 9(4) to 11(5) lower labials. The dorsal scales are weakly keeled, without apical pits, and arranged in 23 straight rows. The vertebral row is not enlarged relative to the other dorsal scale rows. There are 153–180 ventrals and 36–59 divided subcaudals. The cloacal scale is divided. The maximum length recorded for this species is 655 mm (Nagy et al. 2014).

 Witte (1922) considered *Helophis* to be a distinctive genus but to resemble the Blackbelly Snakes, genus *Hydraethiops*, more than any other. The genus *Helophis* differs from *Hydraethiops* in having fewer maxillary teeth and paired internasals. Nagy et

FIGURE 12.6. *Helophis schoutedeni*, Democratic Republic of Congo. By E. Greenbaum

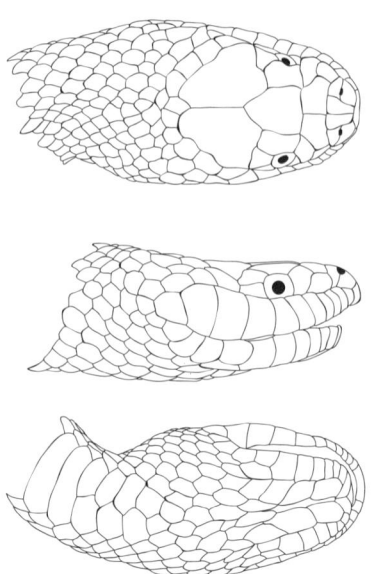

FIGURE 12.7. *Helophis schoutedeni* RMCA 85-1-R-20. By T. Giri

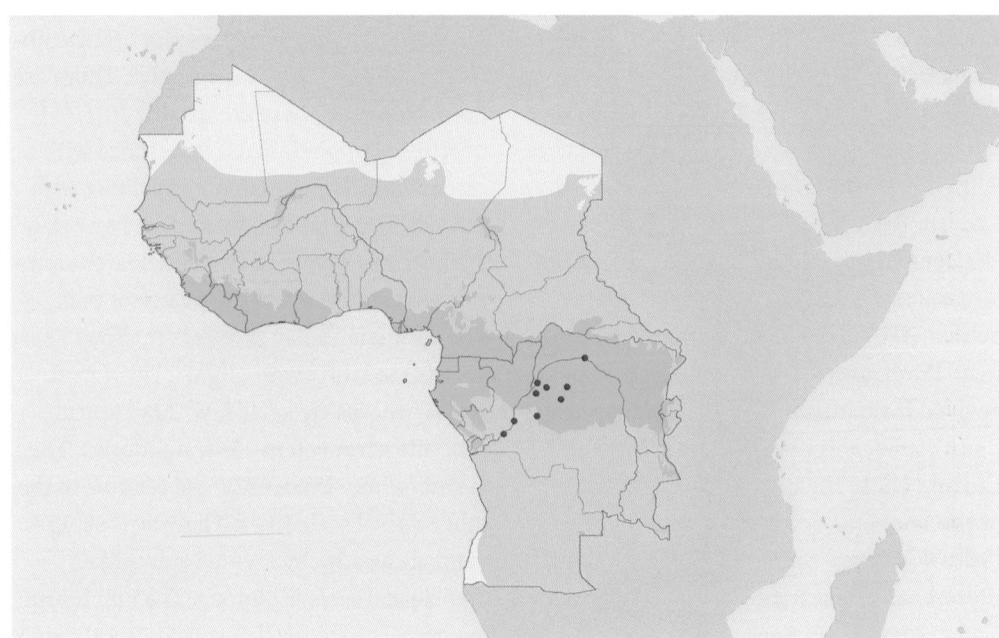

FIGURE 12.8. *Helophis schoutedeni* (red). By K. Jackson

al. (2014) point out that the condition of the internasals is variable in *Hydraethiops* (usually single but sometimes partially divided or paired). This leaves maxillary tooth count as the only difference between *Helophis* and *Hydraethiops*, which they suggest is perhaps not sufficient to justify separate genera.

Blackbelly Snakes: Genus *Hydraethiops* Günther 1872

Hydraethiops is a genus comprising two species, both of which occur within our zone. These semiaquatic snakes are found in the rainforests of central Africa. Little is known of their natural history, but of the two species, more is known about the habits of *Hydraethiops melanogaster* than of *H. laevis*, the latter species being known from only a few specimens. Schmidt (1923) examined stomach contents from three individuals of *H. melanogaster* collected in the Democratic Republic of Congo. One contained fish remains (including one recognizable as a catfish), one contained a tadpole, and a third contained a mass of mud with some vegetation matter, presumably the stomach contents of a prey item ingested by the snake. Schmidt also reports a female *H. melanogaster* collected in December 1913 with 13 eggs in the oviducts. *Hydraethiops* are striking in appearance, with a black venter and lighter dorsum, an unusual color pattern in snakes. Blackbelly Snakes have broad, almost triangular heads. The eyes are small and dorsolaterally oriented, with round pupils. The maxillary dentition is aglyph with 20–22 teeth. The nostrils are dorsally oriented. The prefrontals are paired, but the internasals are usually fused into a single, triangular median scale (associated with the dorsal orientation of the nostrils). In occasional cases, however, the internasal may be partially or even completely divided. The nasal is semi-divided. The loreal is present. There is 1 preocular, 1 or 2 postoculars, and no suboculars. There are from 9 to 12 upper labials and from 0 to 2 anterior temporals. There may be either 1 or 2 pairs of chin shields, and the submandibular groove is not pronounced. There are from 11 to 13 lower labials. The dorsal scales have 1 apical pit, may be smooth or keeled, depending on the species, and are arranged in 21–25 straight rows. There are 143–163 ventrals and 39–55 divided subcaudals. The cloacal scale is divided. The hemipenes are forked distally, with an unforked sulcus spermaticus.

Key to the Identification of Central and Western African Species of the Genus *Hydraethiops*

1 Dorsal scales keeled .... *H. melanogaster*
1' Dorsal scales smooth .... *H. laevis*

Smooth-scaled Blackbelly Snake: *Hydraethiops laevis* Boulenger 1904b

This species is known from only three specimens, two collected at the type locality of Efulen, Cameroon, the other in Moudouma, Gabon. It is probably also present in primary forest in Cameroon and the Republic of Congo.

The body is reddish brown or yellowish above with darker blotches forming a zig-zag pattern. The underside is dark except for small pale spots in the middle part of the body. There is usually 1, sometimes 2, postoculars. There are usually 9(4,5) upper labials, of which the sixth and seventh are in contact with the parietal. The few known specimens all have 2 pairs of chin shields, but the possibility remains that this character may be variable, as it is in *Hydraethiops melanogaster*. The temporal formula is 0+1 or 0+2. There are 11–13(4) or 12(5) lower labials. The dorsal scales are smooth and arranged in 21 straight rows. There are 154–163 ventrals and 51–52 divided subcaudals. The maximum length recorded for *H. laevis* is 570 mm (Boulenger 1904a).

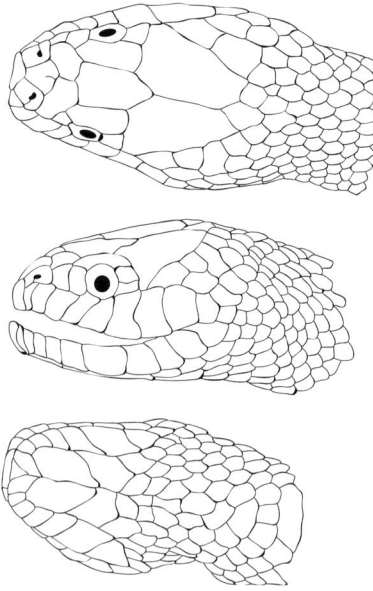

FIGURE 12.9. *Hydraethiops laevis* BMNH 1946.1.15.36. By T. Giri

Rough-scaled Blackbelly Snake: *Hydraethiops melanogaster* Günther 1872

The range of the Rough-scaled Blackbelly Snake extends from Cameroon to the Democratic Republic of Congo. The type locality is simply "Gabon."

The body is reddish brown or dark olive brown above, the dorsal scales finely outlined in yellow. There are dark spots along the sides, a pattern that becomes indistinct in older individuals. The underside is uniformly dark with the exceptions of the underside of the head and the upper labials, which are mottled with white. There is usually 1, sometimes 2, postoculars. There are usually 10(5,6) but sometimes 9(5,6) or 11(5,6) upper labials. There may be either 1 or 2 pairs of chin shields. The temporal formula is usually 1+3 but sometimes 1+2 or 2+4. There are 11(5) or less often 12(5) lower labials. The dorsal scales are keeled and

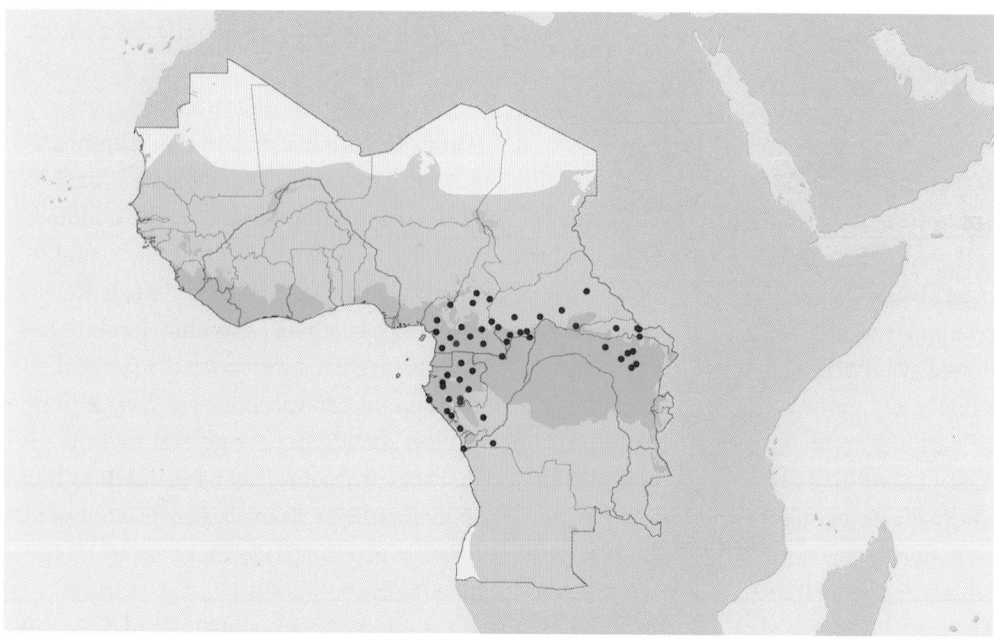

FIGURE 12.10. *Hydraethiops melanogaster* (blue), *Hydraethiops laevis* (red). By K. Jackson

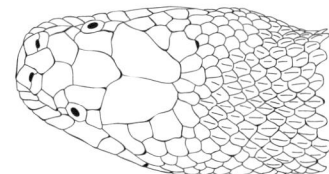

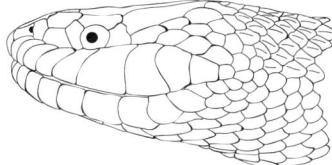

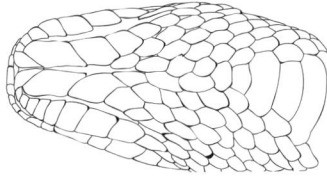

FIGURE 12.11. *Hydraethiops melanogaster* RMCA 76-14-R-121-2.786. By T. Giri

FIGURE 12.12. *Hydraethiops melanogaster*, Democratic Republic of Congo. By E. Greenbaum

arranged in 23, or sometimes 25, straight rows. There are 143–156 ventrals and 39–55 divided subcaudals. The maximum length recorded for *H. melanogaster* is 945 mm (Villiers 1966).

Swamp Snakes: Genus *Limnophis* Günther 1865

Striped Swamp Snake: *Limnophis bicolor* Günther 1865

Limnophis bicolor is a stout-bodied semi-aquatic snake found in and around water. Its range is thought to extend from Zimbabwe north to southwestern Angola and the Democratic Republic of Congo. The type locality is Duque de Bragança, Angola.

The Striped Swamp Snake is olive green to black above, often with a broad pale stripe along either side of the body. The underside is uniformly yellowish to reddish. The labials, chin shields, gulars, and subcaudals are edged with black. The head is relatively small and the distinction between head and neck not well defined. The nasal is usually divided or semi-divided but occasionally single. The prefrontals are paired, but the internasal is a single scale. The loreal is present. There is 1 preocular, 2 postoculars, and 8(3,4) upper labials. The temporal formula is usually 1+2 or 1+3, but the anterior temporal may be lost as a result of fusion to the parietal, so that the parietal is in contact with the sixth upper labial. There are 2 pairs of chin shields, and the submandibular groove is not pronounced. There are typically 9(4) lower labials. The dorsal scales are smooth and arranged in 19 straight rows without apical pits. There are 132–148 ventrals and 45–68 divided subcaudals. The cloacal scale may be single or divided. The maximum length recorded for *L. bicolor* is 760 mm, which was for a female specimen (Bocage 1895a). Bogert (1940) noted that females of this species are usually larger than males.

Two subspecies are recognized: *L. b. bicolor* (Günther 1865); *L. b. bangweolicus* (Mertens 1936). Some authors consider these to represent two separate species (Mertens 1936; Laurent 1964). Mertens (1936) described the new species *L. bangweolicus* on the basis of three morphological characters that ostensibly distinguish it from *L. bicolor*: absence of an anterior temporal, lower maxillary tooth count (22–27 compared with 26–31 in *L. bicolor*), and a black midline stripe on the ventral surface of the tail in *L. bangweolicus*. Other arguments in support of *L. bangweolicus* as a separate species include geographical distribution (*L. b. bicolor* is possibly endemic to Angola, whereas *L. b. bangweolicus* is found farther east in Botswana, the Democratic Republic of Congo, and Zimbabwe) (Laurent 1964; Broadley 1974, 1991b) and diet (*L. b. bangweolicus* possibly feeds exclusively on fishes, while *L. b. bicolor* feeds on amphibians) (Broadley 1974). But all these observations are based on a small number of specimens, and even so there is variability in all three supposedly diagnostic morphological characters (Bogert 1940; Witte 1953). We therefore treat *L. bicolor* as a single species here until there is further morphological or molecular evidence to support the validity of *L. bangweolicus* as a species.

Marsh Snakes: Genus *Natriciteres* Loveridge 1953

Natriciteres is a genus of three species of small, semiaquatic, largely nocturnal snakes, all of which have broad distributions within our zone. Marsh Snake adults average just 30 cm in total length. Vulnerable to many potential predators, including waterfowl, fish, freshwater crabs, and turtles, as well as some larger snakes, they discourage predators and herpetologists

by producing an offensive cloacal discharge when handled (Pitman 1974) and, more significantly, by means of caudal "pseudoautotomy"—having tails that will break off when grasped as a defense mechanism. (See the discussion of tail breakage in snakes in the genus *Psammophis* introduction in chap. 11.) Broadley (1987) reported truncated tails, generally remodeled to form a blunt cone after breaking, in a large percentage of wild-caught *Natriciteres*, evidence of attacks by predators, and was surprised by how easily Marsh Snakes will drop their tails. When changing the water dish of a tame, captive *Natriciteres olivacea*, his hand brushed the snake's back, causing the tail to break off. Counts of subcaudals for Marsh Snakes should therefore be accepted with caution. *Natriciteres* are sexually dimorphic, the females typically being larger than the males.

In all members of the genus, the pupil is round. The maxillary dentition is aglyph, without diastemata. There are 23–25 maxillary teeth, increasing in size posteriorly. The nasal is divided. There are 2 internasals and 2 prefrontals, and the loreal is present. There are 1 or 2 preoculars and usually 3 but sometimes 2 postoculars. The temporal formula is 1+2. There are 8(4,5) upper labials. There are 2 pairs of chin shields, with a pronounced submandibular groove. There are 8–10 lower labials. The dorsal scales lack apical pits and are arranged in 15–19 straight rows. These can be difficult to count accurately at midbody, as the number of rows decreases posteriorly. The cloacal scale may be single or divided. The subcaudals are divided. The hemipenes are unforked, and the sulcus spermaticus is unforked.

Much of what is known of the ecology of Marsh Snakes comes from the field studies

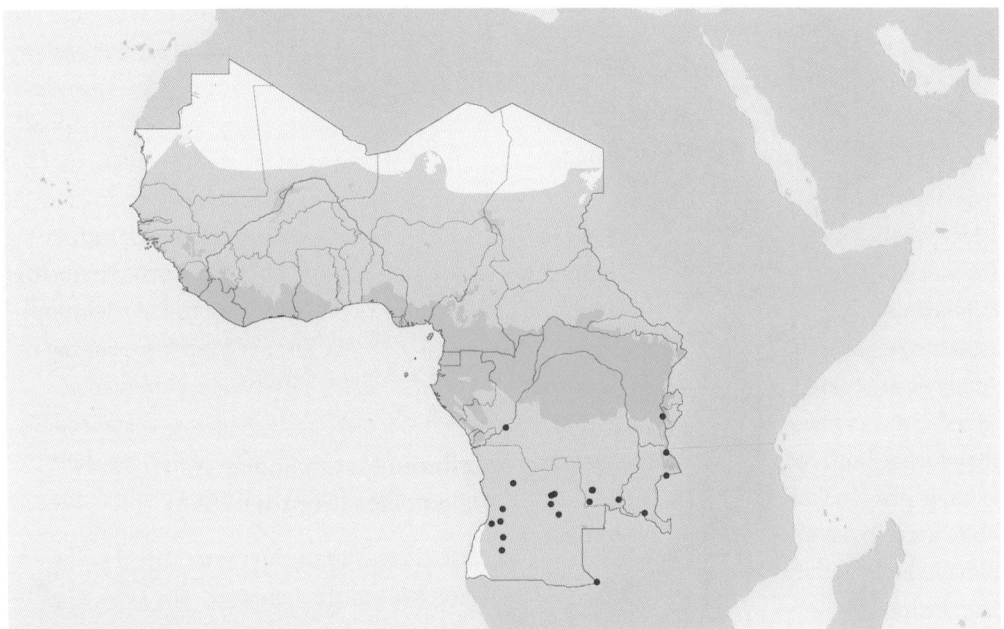

FIGURE 12.13. *Limnophis bicolor* (blue). By K. Jackson

FIGURE 12.14. *Limnophis bicolor bangweolicus*, Angola. By W. Conradie

FIGURE 12.15. *Limnophis bicolor bicolor*, Angola. By W. Conradie

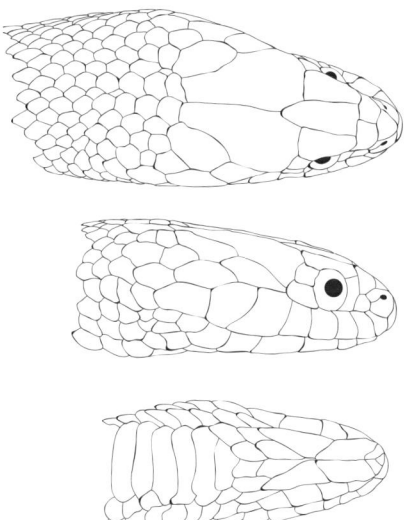

FIGURE 12.16. *Limnophis bicolor* USNM 320724. By T. Giri

of Akani and Luiselli, carried out in Nigeria where all three species occur. Their data are almost exclusively from *N. variegata* and *N. fuliginoides* because these were much more abundant than *N. olivacea*, which proved too rare to collect enough specimens to make statistically significant conclusions. Akani and Luiselli (2001a) found no seasonal difference in activity in the species studied. Unlike many aquatic snakes that aestivate during the dry season, *N. fuliginoides* lays 2–5 eggs during this period (December–March in Nigeria) but none in the wet season. Non–statistically significant evidence suggested that in both species, clutch size increases with maternal size. Analysis of stomach contents indicated that Marsh Snakes have a varied diet and that they forage both on land and in the water, as their prey included both terrestrial (e.g., terrestrial invertebrates, amphibian metamorphs) and aquatic (e.g., tadpoles, small fish) animals.

Because *N. fuliginoides* and *N. variegata* seemed to occupy identical ecological niches, Luiselli (2003) compared the diet of both species from (1) sites where only *N. fuliginoides* was present and (2) sites where both species were present. *N. fuliginoides* had a varied diet where *N. fuliginoides* was absent. In the presence of *N. variegata*, the diet of *N. fuliginoides* shifted, creating separate dietary niches for the two species. Specifically, *N. fuliginoides* shifted from feeding on small vertebrates to primarily invertebrate prey, from a mixture of aquatic and terrestrial to primarily terrestrial organisms, and from larger prey to smaller prey. Akani and Luiselli (2001a) found *N. fuliginoides* and *N. variegata* to be widespread in moist forest and swamp forest, though possibly penetrating forest savanna mosaic. However, they found *N. olivacea* to be primarily a savanna species that rarely makes its way into rainforest. Of the few *N. olivacea* captured, one was in a highly polluted site outside Lagos, suggesting a capacity for tolerating extremely disturbed habitats.

Key to Central and Western African Species of the Genus *Natriciteres*

1   Cloacal scale single .... *N. fuliginoides*
1'   Cloacal scale divided 2

2(1) 19 dorsal scale rows counted halfway between midbody and neck .... *N. olivacea*
2'   15 dorsal scale rows counted halfway between midbody and cloacal scale .... *N. variegata*

Collared Marsh Snake: *Natriciteres fuliginoides* (Günther 1858)

Natriciteres fuliginoides is a central and western African forest species. The type locality is "West Africa."

The body is brownish or blackish above often with pale spots arranged in two stripes. In many parts of its range, this species has a distinctive pale band across the neck, giving the Collared Marsh Snake its

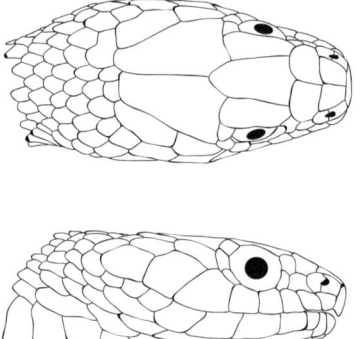

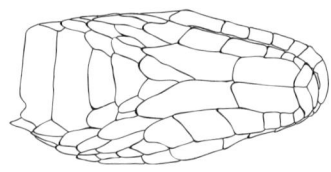

FIGURE 12.17. *Natriciteres fuliginoides* RMCA 89-20-R-99. By T. Giri

FIGURE 12.18. *Natriciteres fuliginoi-des*, Republic of Congo. By K. Jackson

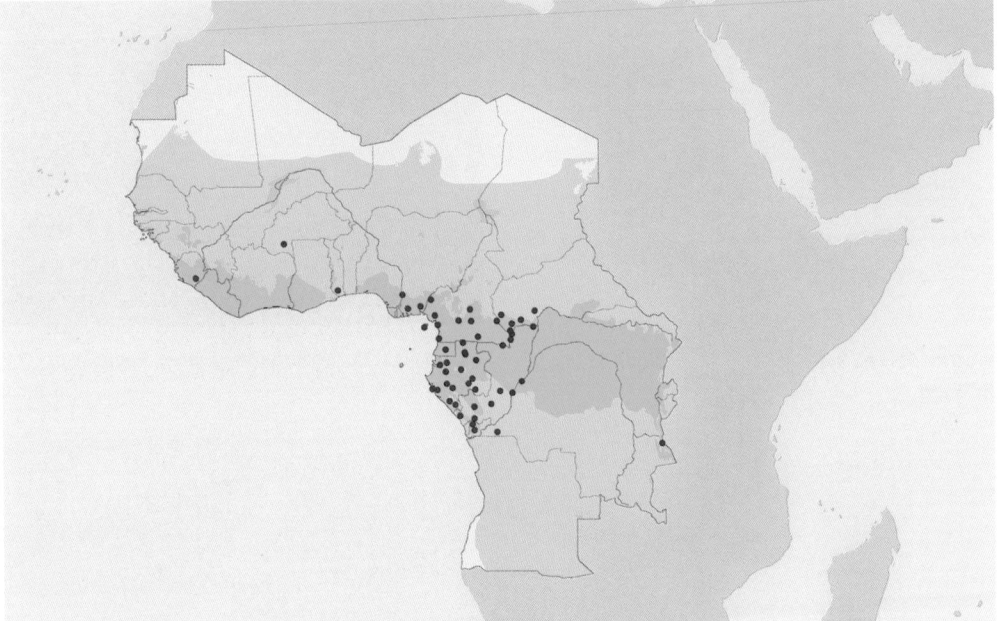

FIGURE 12.19. *Natriciteres fuliginoides* (blue). By K. Jackson

common name. The ventral and labial scales are usually whitish with black edges. There are 2 or occasionally 3 preoculars. There are 9(4) or sometimes 10(5) lower labials. There are 17 dorsal scale rows the length of the body. There are 115–138 ventrals and 69–97 subcaudals. The cloacal scale is single. The maximum length recorded for this species is 492 mm (Roux-Estève 1965).

Olive Marsh Snake: *Natriciteres olivacea* (Peters 1854)

The Olive Marsh Snake has a broad distribution in tropical Africa. The type locality is Tete, Mozambique.

The body is either uniformly olive brown to reddish above or with a broad dark mid-dorsal stripe, often with pale edges. The underside is uniformly whitish, usually with dark edges to the ventrals. There is usually 1 preocular. There are 10(4) or 10(5) lower labials. There are 19 dorsal scale rows anteriorly, decreasing to 17 posteriorly. There are 128–160 ventrals (fewer than 151 in males, more than 134 in females) and 52–95 subcaudals (more than 59 in males, fewer than 71 in females). The cloacal scale is divided. The maximum length recorded for this species is 600 mm (Villiers 1975).

Variegated Marsh Snake: *Natriciteres variegata* (Peters 1861)

The Variegated Marsh Snake has a distribution that extends from Guinea to Central African Republic. The type locality is Ghana.

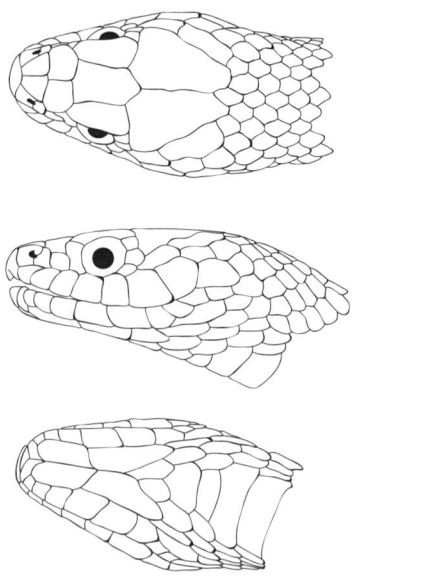

FIGURE 12.20. *Natriciteres olivacea* RMCA 6576. By T. Giri

FIGURE 12.21. *Natriciteres olivacea*, Republic of Congo. By K. Jackson

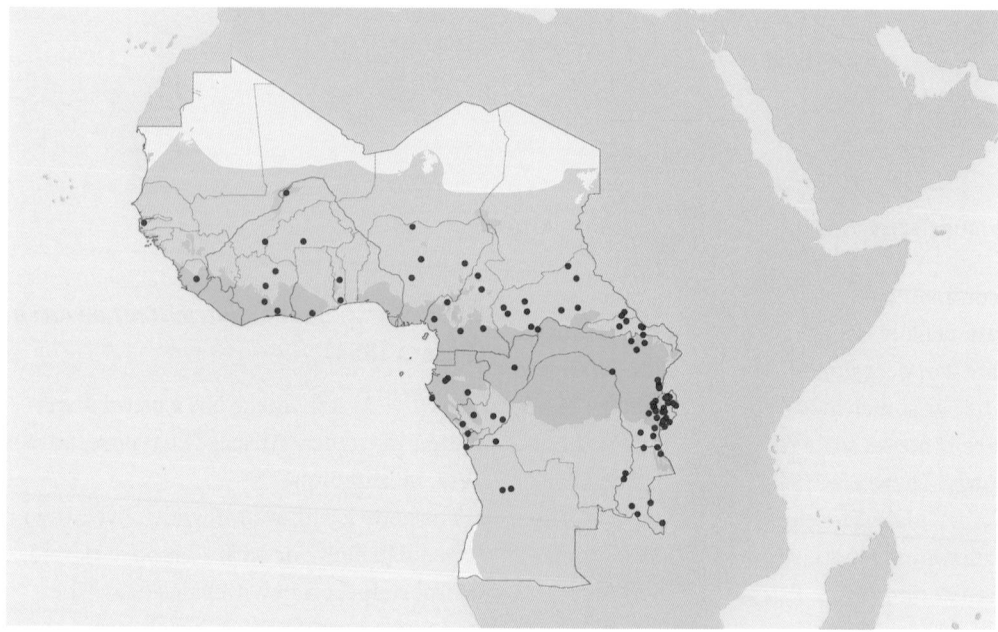

FIGURE 12.22. *Natriciteres olivacea* (blue). By K. Jackson

The body is brownish or reddish above, sometimes with dark stripes flecked with white scales. There is a pale band across the back of the neck in most parts of its range, and the center of the parietals is pale. The upper labials are pale with black edges, and the underside is whitish, sometimes with dark flecks along the sides. There are 2 or occasionally 3 preoculars. There are 9(4) or sometimes 10(5) lower labials. There are 15 dorsal scale rows for most of the body length, but there may be 17 anteriorly. There are 121–143 ventrals and 60–80 subcaudals. The cloacal scale is divided. The maximum length recorded for this species is 400 mm (Loveridge 1958).

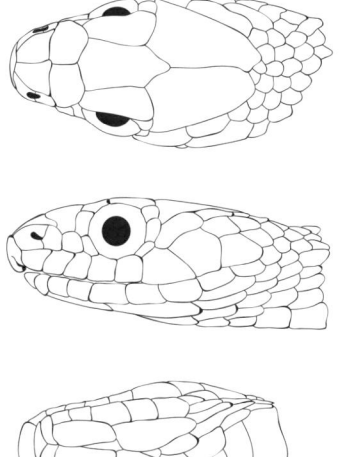

FIGURE 12.23. *Natriciteres variegata* RMCA 29881. By T. Giri

FIGURE 12.24. A juvenile *Natriciteres variegata*, Liberia. By W. Branch

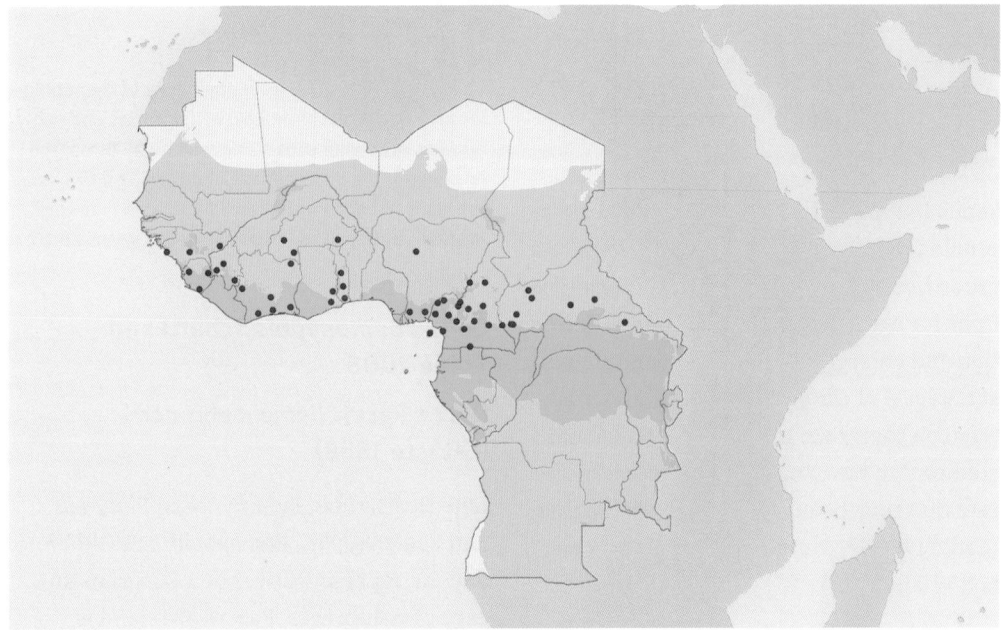

FIGURE 12.25. *Natriciteres variegata* (blue). By K. Jackson

Family Colubridae

Subfamilies Colubrinae and Grayiinae

Subfamily Colubrinae

The Colubrinae is an enormous subfamily (almost 700 species in more than 100 genera) with an almost worldwide distribution. The Colubrinae are arboreal, semi-arboreal, and terrestrial snakes with opisthoglyph or aglyph dentition. Within our zone, they are represented by 63 species in 17 genera. The Colubrinae as understood here includes genera included in Bogert's (1940) Groups 8, 9, 10, 11, 13, 14, and 15, assigned to those groups on the basis of the absence of vertebral hypapophyses in the posterior part of the body, an unforked sulcus spermaticus, and an aglyph or opisthoglyph dentition. The Colubrinae also corresponds to Bourgeois's (1965) subfamilies Philothamninae, Dispholidinae, and Boiginae, based on skull morphology and dentition. The genus *Dasypeltis* (Egg-eating Snakes) proved problematic for Bogert (1940) to assign to a group because of morphological specializations for egg eating, such as reduced dentition and enlarged vertebral hypapophyses at the level of the esophagus. He assigned them on their own to Group 18. Molecular studies that have included *Dasypeltis* indicate that they belong among the Colubrinae (Cadle 1994; Pyron et al. 2011, 2013; Zheng and Wiens 2016).

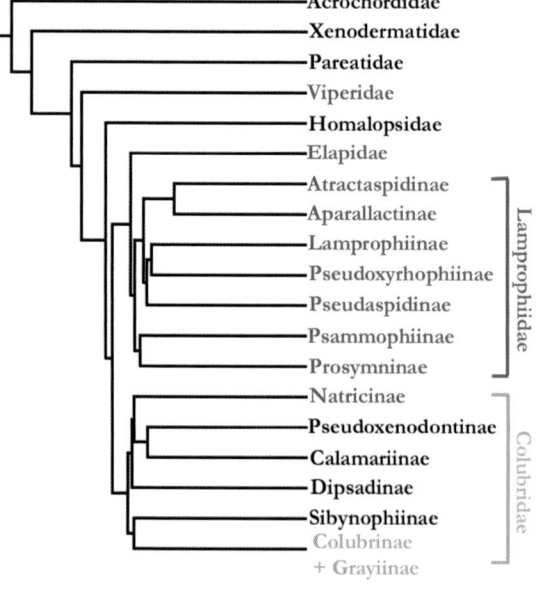

FIGURE 13.1. Higher-level phylogeny of caenophidian families and subfamilies, with families and subfamilies occurring within our zone highlighted (red or orange). The subfamilies Colubrinae and Grayiinae, the topic of this chapter, are highlighted in orange. Modified by K. Jackson from Zheng and Wiens (2016)

Genus *Bamanophis* Schätti and Trape 2008

Dorr's Racer: *Bamanophis dorri* (Lataste 1888)

Like *Hemorrhois*, *Lytorhynchus*, *Platyceps*, and *Spalerosophis*, *Bamanophis* are Old World Racers, representatives of a Eurasian lineage of colubrines. Racers are slender, fast-moving snakes, active by day in desert, sahel, and dry savanna habitats. *Bamanophis* is a monotypic genus. The range of Dorr's Racer, *Bamanophis dorri*, extends from Senegal to Niger and Benin. The type locality is Bakel, Senegal.

The dorsum is brown or reddish brown with a darker pattern of X's along the vertebral midline and matching dark lateral blotches. The venter is pinkish white or straw yellow, sometimes with dark spots. The head is broad with a rounded snout and well-defined neck. The tail is relatively short compared to other racers. The eye is small with a round or slightly vertically elliptical pupil. The nasal is divided or semi-divided. The internasals and prefrontals are paired. The loreal is present and longer than deep. There is 1 preocular. The supraocular is narrower than the frontal. There are 2 or 3 postoculars, of which the inferior is the largest and sometimes occupies the position of a subocular. There are usually 9(4,5) upper labials, sometimes 10(4,5), and occasionally 11(4,5). The temporal formula is usually 2+2, but sometimes 2+3 or 3+3. The lower labials range from 9(4) to 12(5). The anterior chin shields are shorter than or equal in length to the posterior pair. The submandibular groove is pronounced. The dorsal scales are smooth, with 2 apical pits, and arranged in 29–31 (occasionally 33) straight rows (most often 29 in males, 31 in females). The vertebral row is not enlarged relative to the other dorsal scale rows. There are 229–265 ventrals (fewer than 249 for males, more than 241 for females). There are 75–95 divided subcaudals. The cloacal scale is divided. The maximum length recorded for this species is 995 mm (Schätti and Trape 2008). The maxillary dentition is aglyph, with 15–19 anterior maxillary teeth decreasing is size posteriorly and followed by 2 larger ungrooved posterior teeth. The hemipenes are unforked, and the sulcus spermaticus is not forked.

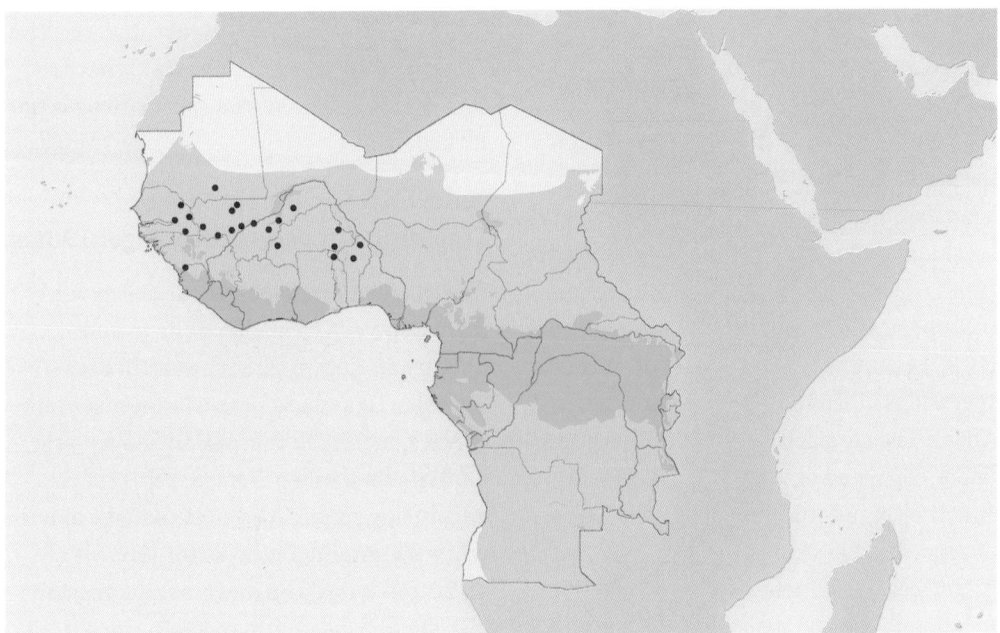

FIGURE 13.2. *Bamanophis dorri* (blue). By K. Jackson

Originally described as *Periops dorri* Lataste 1888, this species has been moved to several different genera (most recently *Hemorrhois*) before being assigned to its own genus, *Bamanophis*, by Schätti and Trape (2008). On the basis of several morphological features, Schätti and Trape argued that (*Hemorrhois*) *dorri* differed from all other *Hemorrhois* and did not clearly belong in any other existing genus. Recent molecular studies (Nagy et al. 2004; Pyron et al. 2011) support the removal of *B. dorri* from *Hemorrhois*, finding it to be most closely related to the genus *Macroprotodon*.

Herald Snakes: Genus *Crotaphopeltis* Fitzinger 1843

The Herald Snakes, *Crotaphopeltis*, are a genus of six species found in sub-Saharan Africa. Three of these occur within our zone. Herald Snakes are nocturnal and terrestrial or semiaquatic. They feed primarily on amphibians, especially toads (Bufonidae) that are poisonous to many other snakes. The White-lipped Herald Snake, *Crotaphopeltis hotamboeia*, is the most widely distributed species and the one about which the most is known. When threatened, *C. hotamboeia* flattens its head into a triangular shape so that it resembles a Night Adder (genus *Causus*), another nocturnal, terrestrial species that feeds on toads. Keogh et al. (2000) examined a large series of museum specimens of this species from southern Africa. The stomach contents of these specimens consisted of 97% anurans, 39% of which were bufonids (toads). Examination of reproductive structures indicated that *C. hotamboeia* lay clutches of 4–12 eggs. Bites to humans by *C. hotamboeia* may sometimes cause localized edema and mild pain (Chip-

paux pers. obs.). The maxillary dentition is opisthoglyph, with 12–20 anterior maxillary teeth separated by a diastema from 1–3 grooved posterior fangs. The head is medium sized, broad, and fairly flat. The neck is well defined. The snout is rounded. The eye is medium or large with a vertically elliptical pupil. The nasal is usually divided. The internasals and prefrontals are paired. The loreal is present. There is 1 preocular, or sometimes 2, but no suboculars. The frontal is longer than broad and is separated from the eye by a supraocular. There are 2 postoculars of equal size. There are 7–8 upper labials, 3 or less often 2 of which are in contact with the eye. There is 1 anterior temporal. There are 8–12 lower labials. There are 2 pairs of chin shields, the anterior pair longer than the posterior pair. The submandibular groove is pronounced. The dorsal scales may be smooth or keeled, with 1–2 apical pits. They are arranged in 19 straight or oblique rows at midbody. The vertebral row is not enlarged relative to the other dorsal scale rows. The subcaudals are divided. The cloacal scale is single. The hemipenes are unforked, and the sulcus spermaticus is not forked.

Yellow-flanked Herald Snake: *Crotaphopeltis degeni* (Boulenger 1906b)

The range of *Crotaphopeltis degeni* extends from Central African Republic to South Sudan and Kenya. The type locality is Entebbe, Uganda. *C. degeni* is a semiaquatic species, more tied to riparian and aquatic habitats than other *Crotaphopeltis* species within our zone. *C. degeni* is thought to hunt while swimming in the water (Spawls et al. 2004). It feeds on anurans and possibly also small fishes (Pitman 1974), since it is a good swimmer and diver. This species lays clutches of 5–11 eggs (Rasmussen 1997a).

The dorsum is iridescent dark brown. The upper labials and venter are light yellow. There is a dark stripe along the midline of the underside of the tail. The temporal formula is 1+2. There are usually 8(3,4,5) upper labials. There are usually 10(5) lower labials. The dorsal scales are smooth, with 1–2 apical pits, and are arranged in 19 straight rows. There are 160–175 ventrals and 30–40 subcaudals (both without sexual dimorphism). The cloacal scale is single. The maximum length recorded for this species is 690 mm (Rasmussen 1997a).

Western Herald Snake: *Crotaphopeltis hippocrepis* (Reinhardt 1843)

The range of *Crotaphopeltis hippocrepis* extends from Mali to Central African Republic. The type locality is "Guinea," but presumably meaning "Guinea Coast" rather than the country, Guinea. It is likely that the true type locality is somewhere in Ghana. The body is brown above, with each dorsal scale paler at the center than at the edges. The sides of the head have the same dark markings as *C. hotamboeia*. The venter is uniformly white. The temporal formula is usually 1+2, occasionally 1+1 or 1+3. There are usually 8(3,4,5), sometimes just 2, upper labials in contact with the eye. Occasional-

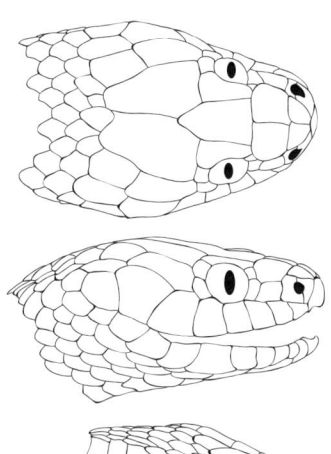
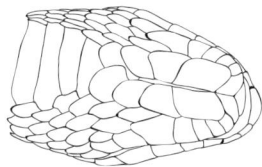

FIGURE 13.3. *Crotaphopeltis degeni* RMCA 76-003-R-0146. By T. Giri

FIGURE 13.4. *Crotaphopeltis degeni*, Kenya. By S. Spawls

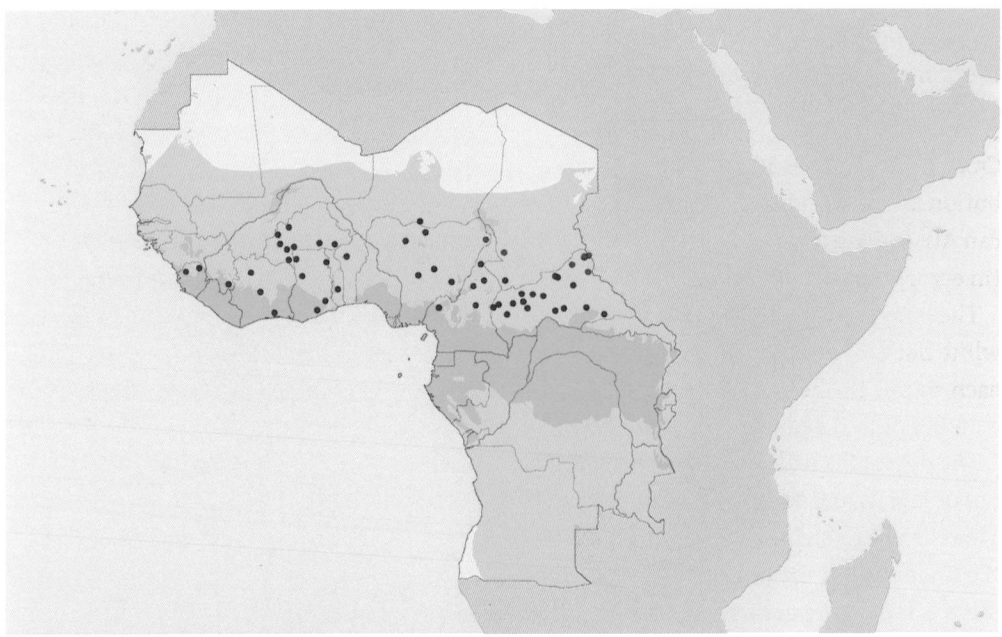

FIGURE 13.5. *Crotaphopeltis degeni* (red), *Crotaphopeltis hippocrepis* (blue). By K. Jackson

FIGURE 13.6. *Crotaphopeltis hippocrepis*, Ghana. By S. Spawls

ly, there may be 9 upper labials. There are usually 10(4) or 10(5) lower labials, sometimes 9(4) or 11(5). The dorsal scales are smooth, with 1 or occasionally 2 apical pits, and arranged in 19 straight or occasionally slightly oblique rows. There are 160–188 ventrals (fewer than 182 in males, more than 165 in females) and 38–60 divided subcaudals (more than 42 in males, fewer than 56 in females). The cloacal scale is single. The maximum length recorded for this species is 750 mm (Rasmussen et al. 2000).

White-lipped Herald Snake: *Crotaphopeltis hotamboeia* (Laurenti 1768)

Crotaphopeltis hotamboeia has a vast distribution and is found throughout sub-Saharan Africa. The type locality is East Indies (in error).

The body is dark olive above, with small white flecks. There is a dark marking on each side of the head, posterior to the eye. The venter is uniformly whitish. The temporal formula is usually 1+2, occasionally 1+1 or 1+3. There are usually 8(3,4,5) upper labials, sometimes 8(3,4) or 8(4,5). There are usually 9(5) or 10(5) lower labials, but this may range from 8(4) to 11(6). The dorsal scales are weakly keeled, with 1-2 apical pits, and are arranged in 19 straight or occasionally slightly oblique rows. There are 155–182 ventrals (without sexual dimorphism) and 27–49 divided subcaudals (more

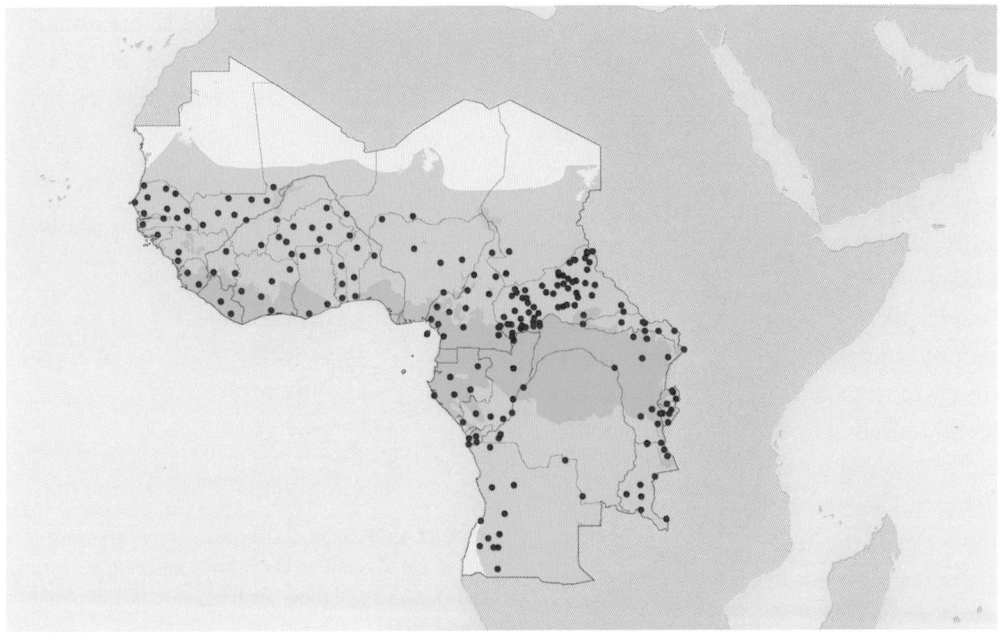

FIGURE 13.7. *Crotaphopeltis hotamboeia* (blue). By K. Jackson

FIGURE 13.8. *Crotaphopeltis hotamboeia*, Democratic Republic of Congo. By E. Greenbaum

than 35 in males, fewer than 42 in females). The cloacal scale is single. The maximum length recorded for this species is 810 mm (Broadley and Cock 1975).

Egg-eating Snakes: Genus *Dasypeltis* Wagler 1830

Dasypeltis is an African genus of 11 species, 9 of which occur within our zone. *Dasypeltis* are nocturnal, terrestrial, and semi-arboreal. They specialize in feeding on bird eggs and possess a suite of morphological specializations to allow them to do this, including modifications of dentition, scutellation, and vertebral hypapophyses (Gans 1952). In contrast to reptile eggs, which have leathery shells that can be sliced open with a bladelike tooth, bird eggs have rigid shells. To feed on them, *Dasypeltis* must first engulf the egg. This is facilitated by a reduced aglyph dentition, characterized by the small size and number (5–9 maxillary teeth) as well as by the absence of teeth on the anterior part of the maxilla and the dentary. There is also a specialization of the scales of the head and neck, which are anchored to the underlying skin in such a way as to allow the skin in this area to stretch dramatically as the unbroken bird egg is engulfed. As the egg is swallowed, it is finally broken by specialized vertebral hypapophyses. The contents of the egg are then swallowed, and the snake regurgitates the shell. *Dasypeltis* are unusual among snakes, and indeed unique among African snakes, in possessing this suite of morphological adaptations for feeding on bird eggs. Gartner and Greene (2008) note that more birds lay eggs of a readily ingestible size in Africa than in a comparable area of the United States, and that the largest radiation

of African ground-nesting birds existed in Africa before the diversification of colubrine snakes in the Miocene. These conditions may have set the stage for the evolution of the unique egg-eating morphology seen in *Dasypeltis*. *Dasypeltis* appear to also be morphologically specialized to mimic small vipers. When threatened, Egg-eating Snakes carry out an impressive warning display (Gans and Richmond 1957). This display involves rubbing modified paraventral scales against one another (like *Echis*, *Cerastes*) so as to produce a hissing sound, puffing up the body so that it appears stouter and more viper-like, and spreading the

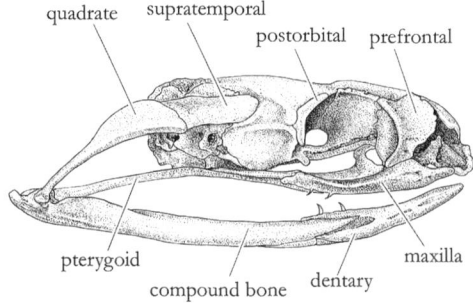

FIGURE 13.9. Skull of *Dasypeltis scabra* showing reduction of teeth on the maxilla and dentary associated with egg eating. Modified by K. Jackson from Gans (1974)

FIGURE 13.10. *Dasypeltis atra* eating an egg, Kenya. By S. Spawls

FIGURE 13.11.
Causus rhombeatus (*left*) with *Dasypeltis scabra* (*right*). *Dasypeltis* often mimic sympatric small vipers. By S. Spawls

quadrates apart, making the head shape appear more triangular and viper-like. Gans (1961) found regional differences in color pattern in *Dasypeltis* consistent with the hypothesis that *Dasypeltis* is mimicking small vipers, and different vipers in different localities (e.g., *Echis* in Egypt, *Causus* and *Atheris* in central and western African forest).

The head is small, with a rounded snout, and the neck not well defined. The eye is medium sized with a vertically elliptical pupil. The body is slender and cylindrical, and the tail relatively short. The hemipenes are unforked, and the sulcus spermaticus is undivided. The nasal is single or semi-divided. There are 2 internasals and 2 prefrontals. The loreal is absent, a notable feature since *Dasypeltis* is the only African colubrine genus to lack a loreal scale. The frontal is as long as or slightly longer than it is broad. The supraoculars are narrow. There is 1 preocular and 2–3 postoculars. There are usually 2 anterior temporals, but sometimes just 1. There are 6–7 upper labials, 2 of which are in contact with the eye.

There are 2 pairs of chin shields, the anterior pair longer than the posterior pair. The submandibular groove is not pronounced. There are 6–9 lower labials, 3 of which are in contact with the anterior pair of chin shields. The dorsal scales are keeled and arranged in 19–27 oblique rows at midbody. There may be 0–2 apical pits on the dorsal scales. The vertebral row is not enlarged relative to the other dorsal scale rows. The ventrals and subcaudals follow the usual trend in sexual dimorphism, with females having a generally higher ventral count and males a higher subcaudal count. The cloacal scale is undivided.

Gans (1959) undertook a major taxonomic revision of the genus *Dasypeltis*. Recently, Trape and Mané (2006b) and Trape et al. (2012) have greatly clarified the taxonomy of West African species of *Dasypeltis*, like Gans, relying heavily on coloration and pattern, characters that snake taxonomists rarely use but turn out to be taxonomically useful for this genus. Further work is needed to clarify the taxonomy of central and east African *Dasypeltis*. *Dasypeltis fasciata*,

Key to Central and Western African Species of the Genus *Dasypeltis*

1   Dorsal scales of the paraventral row more than 3 times as broad as the other
     dorsal scales .... *D. atra*
1'   Dorsal scales of the paraventral row less than 2 times as broad as the other
     dorsal scales 2

2(1) Nasal scale single 3
2'   Nasal scale divided 4

3(2) Broad dark stripe on the frontal scale .... *D. sahelensis*
3'   Dark blotches or crossbands on the frontal scale .... *D. parascabra*

4(2) Venter uniformly pale 5
4'   Venter pale with dark markings 6

5(4) Dorsal pattern clear and well defined, consisting of a series of dark diamond-shaped
     spots along the vertebral midline and of dark vertical lines at the level of the spots;
     frontal shorter than the parietals .... *D. confusa*
5'   Dorsal pattern indistinct, or with outlines of markings that are not well defined;
     frontal equal in length to the parietals .... *D. fasciata*

6(4) 2 apical pits clearly visible on the dorsal scales 7
6'   Apical pits indistinct or absent 8

7(6) Vertebral spots each covering fewer than 5 scales; frontal equal in length and
     breadth .... *D. scabra*
7'   Vertebral spots each covering more than 5 scales; frontal longer than broad .... *D. palmarum*

8(6) Vertebral spots each covering more than 3 scales; spots much longer than the interspaces
     between the spots .... *D. gansi*
8'   Vertebral spots each covering fewer than 3 dorsal scales; length of spots approximately
     equal to the length of the interspaces between the spots .... *D. latericia*

especially, is likely to see taxonomic changes as the species is better understood.

Montane Egg-eater: *Dasypeltis atra* Sternfeld 1912

Dasypeltis atra is a montane species found in high-altitude savannah habitats. Its range extends from eastern Democratic Republic of Congo to Rwanda and Burundi, and into east Africa. The type locality is virgin forest behind the Boundary Mountains on the northwest shore of Lake Tanganyika (Democratic Republic of Congo).

The dorsum is usually uniformly black or reddish or dark brown. Sometimes there may be a pattern of dark ovals along the vertebral midline outlined against the background color of the dorsum by lighter-colored scales. The venter is uniformly pale. The buccal mucosa is black.

The nasal is semi-divided. There are usually 2 postoculars, the inferior equal to or larger than the superior. The temporal formula is 2+3. There are 7(3,4) upper labials and 7(3) lower labials. The dorsal scales are keeled, with 2 apical pits, and are arranged in 19–27 oblique rows at midbody.

The scales of the paravertebral row are greatly enlarged relative to the other dorsal scales, a feature unique to *D. atra*. There are 202–237 ventrals (fewer than 219 in males, more than 213 in females) and 49–72 divided subcaudals (more than 57 in males, fewer than 63 in females). The maximum length reported for this species is 1,100 mm (Spawls et al. 2004).

FIGURE 13.12. *Dasypeltis atra*, Democratic Republic of Congo. By E. Greenbaum

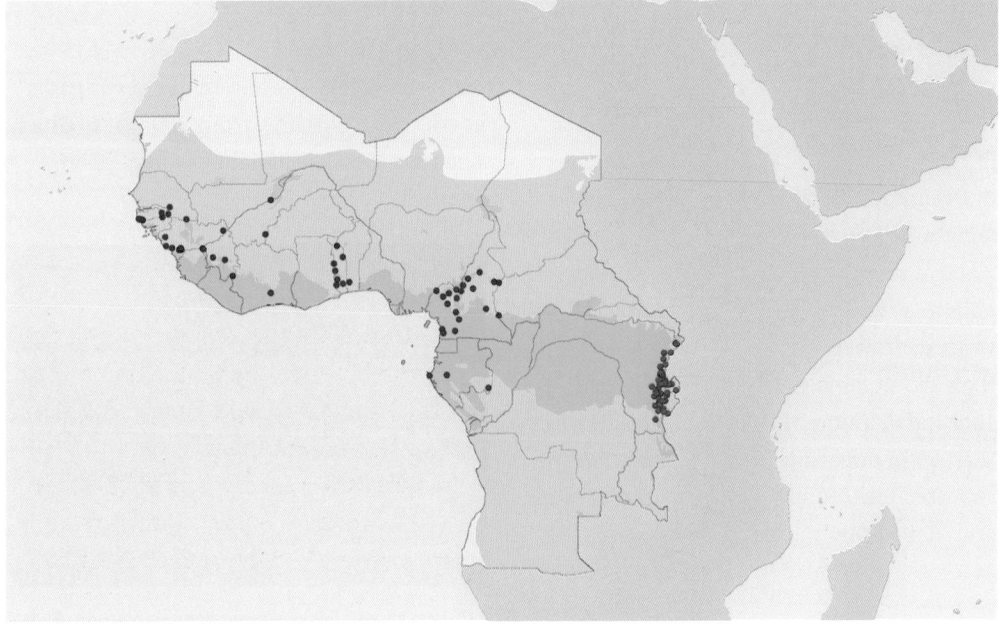

FIGURE 13.13. *Dasypeltis atra* (red), *Dasypeltis confusa* (blue). By K. Jackson

Confusing Egg-eater *Dasypeltis confusa* Trape and Mané 2006b

Dasypeltis confusa is a species found in wet savannah habitats. The type locality is Ibel, Senegal. The species was described based on examination of specimens from West Africa, but the range of the species is thought to extend not only through West Africa from Senegal to Cameroon and Gabon, but also into central Africa east as far as South Sudan and south as far as Angola and western Zambia (Trape et al. 2012).

The body is light brown above, with a series of dark brown or black blotches along the vertebral midline, alternating with light-colored areas. Along the sides of the body is a series of dark brown or black bands, narrower than the mid-vertebral blotches but lining up with them. This dorsal pattern extends the length of the body but becomes indistinct toward the end of the tail. On the dorsum of the head are two dark brown chevrons with the point directed anteriorly. The venter is light beige. The nasal is semi-divided. There are usually 2 postoculars, the inferior equal to or larger than the superior. The temporal formula is 2+2 or 2+3. There are 7(3,4), sometimes 6(3,4), upper labials. There are 6(3) lower labials. The dorsal scales are keeled, with 2 apical pits, and are arranged in 23–26 oblique rows at midbody. There are 213–242 ventrals (fewer than 228 in males, more than 223 in females) and 53–73 divided subcaudals (more than 65 in males, fewer than 68 in females). The maximum length reported for this species is 970 mm (Trape and Mané 2006b).

Central African Egg-eater: *Dasypeltis fasciata* Smith 1849

Dasypeltis fasciata is a forest species, with a range encompassing forested areas of west, central, and east Africa. The type locality is Sierra Leone.

The dorsum is uniformly light brown or greenish with faintly defined, slightly darker markings. The venter is uniformly gray. The nasal is semi-divided. There are usually 2 postoculars, the inferior smaller than the superior. The temporal formula is 2+3. There are 7(3,4) upper labials and 7(3) lower labials. The dorsal scales are keeled, without apical pits, and are arranged in 19–25 oblique rows at midbody. There are 221–262 ventrals (fewer than 250 in males, more than 232 in females) and 61–91 divided subcaudals (more than 67 in males, fewer than 84 in females). The maximum length reported for this species is 1,021 mm (Roux-Estève 1965).

Gans's Egg-eater: *Dasypeltis gansi* Trape and Mané 2006b

Dasypeltis gansi is found in dry savannas, north of the equator, from Senegal to Chad. The type locality is Mahamouda Chérif, Senegal.

FIGURE 13.14. *Dasypelits fasciata*, Cameroon. By V. Gvoždík

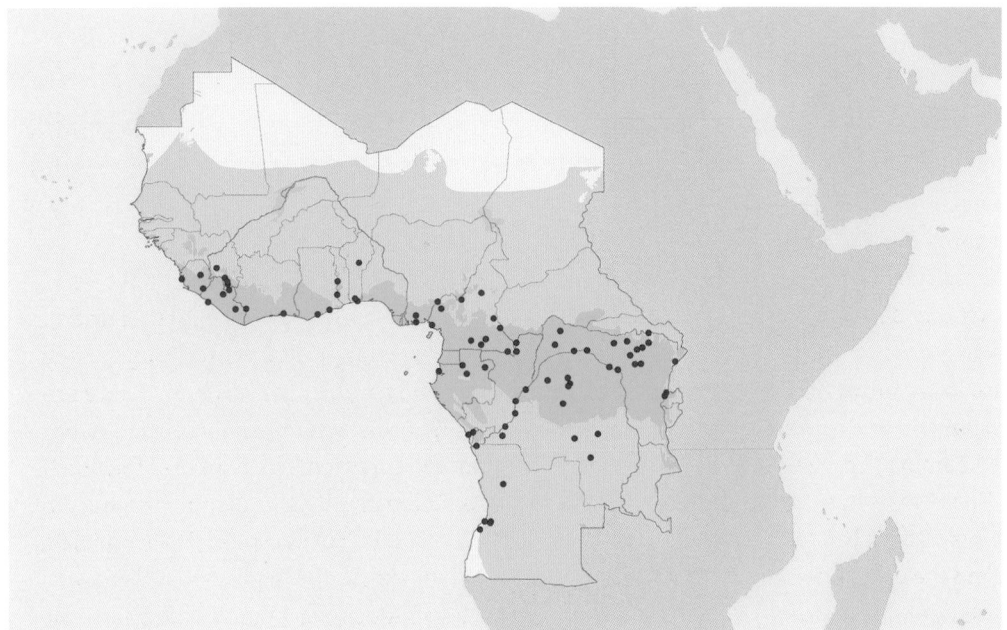

FIGURE 13.15. *Dasypeltis fasciata* (blue). By K. Jackson

The dorsum is either uniformly light beige or with a pattern of darker transverse bands along the anterior part of the body, disappearing posteriorly and replaced by a dark vertebral stripe. The transverse bands include a few whitish scales at the vertebral midline. The venter is uniformly yellowish. The nasal is semi-divided. There are usually 2 postoculars, the inferior equal to or larger than the superior. The temporal formula is usually 2+3 but sometimes 2+2 or 2+4. There are 7(3,4) upper labials and 6(3) lower labials. The dorsal scales are keeled, without apical pits, and are arranged in 21–25 oblique rows at midbody. There are 221–252 ventrals (fewer than 240 in males, more than 237 in females) and 59–80 divided subcaudals (more than 67 in males, fewer than 74 in females). The maximum length reported for this species is 1,020 mm (Trape and Mané 2006b).

FIGURE 13.16. *Dasypeltis gansi*, Ghana. By S. Spawls

Laterite Egg-eater: *Dasypeltis latericia* Trape and Mané 2006b

Dasypeltis latericia is found in dry savanna habitat in Guinea, Mali, and Senegal. The type locality is Boundoukondi, Senegal.

The species name refers to laterite soil, a rusty red being characteristic of the iron-ore-rich soil where this species is found, and also the color of this species of

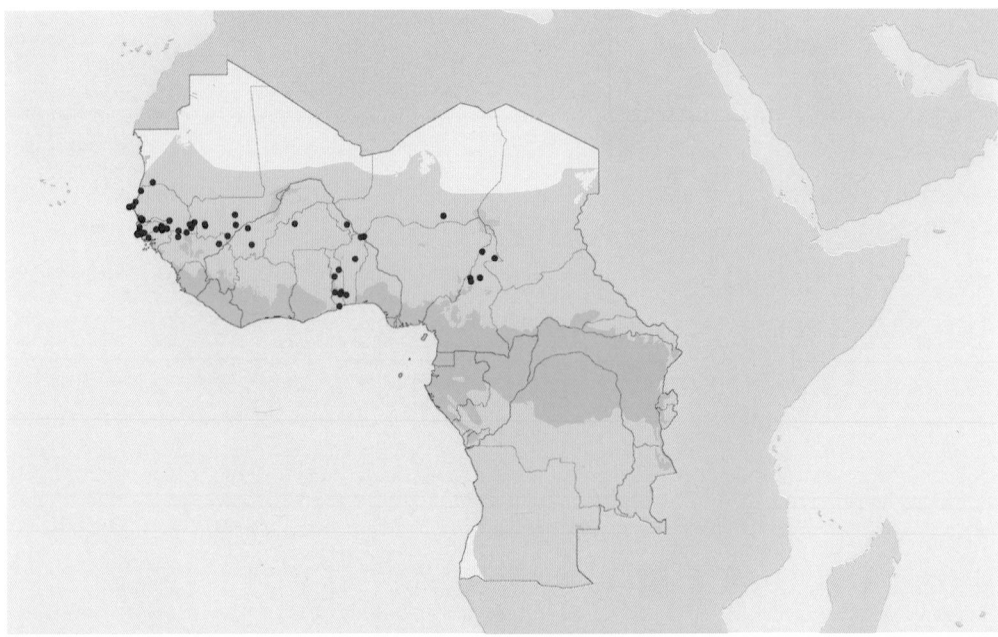

FIGURE 13.17. *Dasypeltis gansi* (blue). By K. Jackson

Egg-eater. The dorsum is either uniformly rusty or pinkish or with dark bands on the anterior part of the body, giving way posteriorly to a dark vertebral stripe. The dark bands include a few whitish scales at the vertebral midline. The venter is light brown with dark lateral flecks. The nasal is semi-divided. There are usually 2 postoculars, the inferior equal to or larger than the superior. The temporal formula is usually 2+3, occasionally 2+2 or 2+4. There are 7(3,4) upper labials and 6(3) lower labials. The dorsal scales are keeled, without apical pits, and are arranged in 21–23 oblique rows at midbody. There are 224–262 ventrals (fewer than 232 in males, more than 235 in females) and 63–75 divided subcaudals (more than 67 in males, fewer than 69 in females). The maximum length reported for this species is 800 mm (Trape and Mané 2006b).

Palm Egg-eater: *Dasypeltis palmarum* (Leach 1818)

Dasypeltis palmarum is known from Atlantic coastal forests from Gabon to Angola. The type locality is Boma, Democratic Republic of Congo.

The coloration of *D. palmarum* is similar to that of *D. scabra*. But the markings on the head and venter of *D. palmarum* are more prominent than those of *D. scabra*. The rhombic vertebral spots of *D. palmarum* are smaller than those of *D. scabra*, and the spaces between the spots are as large as the length of one of the spots, whereas in *D. scabra* the spots are larger and the spaces between the spots are much less than the length of one of the spots. The nasal is semi-divided. There are usually 2 or 3 postoculars of approximately equal size. The temporal formula is 2+3. There are 7(3,4) upper labials and 7(3) lower labials. The dorsal scales are keeled, with 2 apical pits,

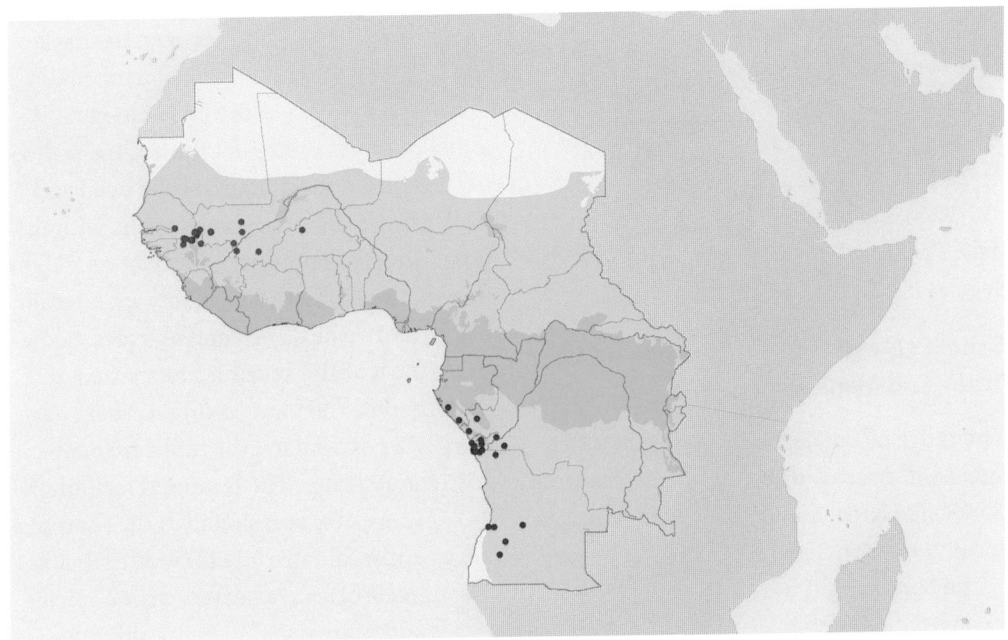

FIGURE 13.18. *Dasypeltis latericia* (red), *Dasypeltis palmarum* (blue). By K. Jackson

and are arranged in 21–27 oblique rows at midbody. There are 213–248 ventrals (fewer than 242 in males, more than 219 in females) and 62–86 divided subcaudals (more than 73 in males, fewer than 74 in females). The maximum length reported for this species is 822 mm (Villiers 1966).

Pararhombic Egg-eater: *Dasypeltis parascabra* Trape, Mediannikov, and Trape 2012

Dasypeltis parascabra is found in habitats along the border between forest and woodland savanna, with a distribution thought to extend from Guinea to Nigeria. The type locality is Dalakan, Guinea.

The coloration of *D. parascabra* is similar to that of *D. scabra*. The dorsum is gray, patterned with a series of dark circles or ovals along the vertebral midline, the length of each spot being more than 5 times the length of the interspace that separates one dark circle from the next. Along the sides

FIGURE 13.19. *Dasypeltis palmarum*, Republic of Congo. By M. Burger

of the body, elongate crossband-like spots extend from the interspaces to the ventrals. The top of the head is patterned with dark crossbands. The venter is gray, patterned with a pair of irregular lateral ventral lines made up of small dark spots on the lateral edges of the ventral scales. The nasal is single. There are usually 2 postoculars, the inferior equal to or larger than the superior. The temporal formula is 2+2, 2+3, or 2+4. There are 7(3,4) upper labials and 6(3) or 7(3) lower labials. The dorsal scales are keeled, without apical pits, and arranged in 21–23 oblique rows at midbody. There are 216–224 ventrals and 64–69 divided subcaudals. The maximum length reported for this species is 605 mm (Trape et al. 2012).

Sahel Egg-eater: *Dasypeltis sahelensis* Trape and Mané 2006b

The range of *Dasypeltis sahelensis* includes sahel and dry savannas of West Africa from Senegal to Niger and Nigeria. The type locality is Tialé, Senegal.

The body is light brown above with a series of dark-brown or black blotches along the vertebral line alternating with areas of light brown or white. Along the sides of the body there are irregular bands of the same color as the dark blotches but narrower. These dark bands usually line up with the pale spaces between the dark vertebral blotches. This dorsal pattern extends the length of the body but generally becomes indistinct toward the end of the tail. On the dorsum of the head and neck are 2–3 dark brown chevrons with the point directed anteriorly. The venter is light brown, usually with a small brown lateral dot every two or three ventral scales. Sometimes most of the ventrals are spotted with dark brown. The nasal is single. There are usually 2 postoculars, the inferior smaller than the superior. The temporal formula is 2+2, 2+3, or 2+4, occasionally 1+2. There are 7(3,4) upper labials and 6(3) lower labials. The dorsal scales are keeled, with 2 apical pits, and are arranged in 21–23 oblique rows at midbody. There are 207–237 ventrals (fewer than 222 in males, more than 213 in females) and 45–67 divided subcaudals (more than 58 in males, fewer than 58 in females). The maximum length reported

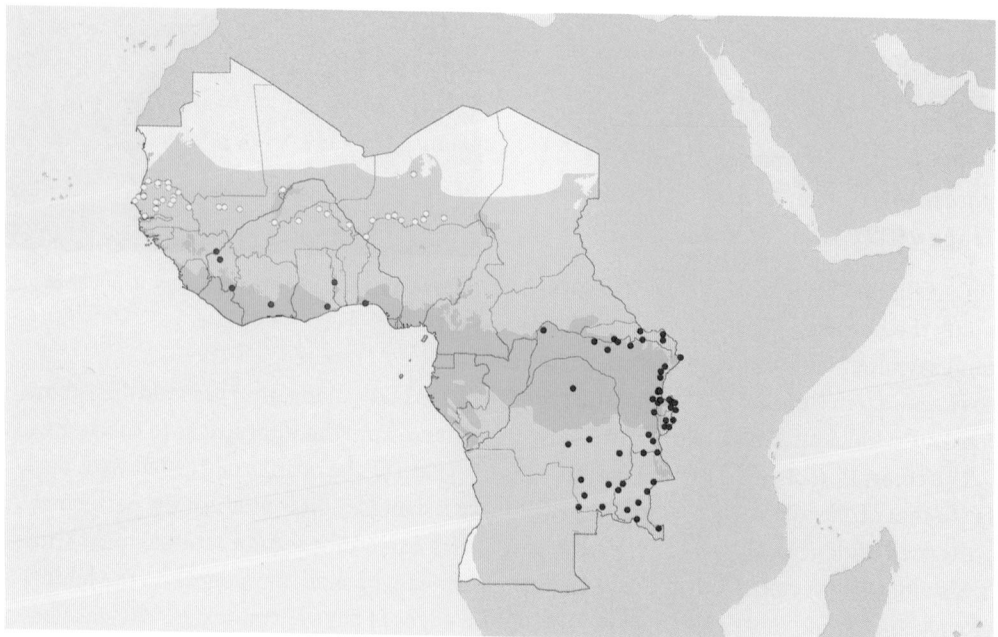

FIGURE 13.20. *Dasypeltis parascabra* (red), *Dasypeltis sahelensis* (yellow), *Dasypeltis scabra* (blue). By K. Jackson

for this species is 620 mm (Trape and Mané 2006b).

Rhombic Egg-eater: *Dasypeltis scabra* (Linnaeus 1758)

Dasypeltis scabra is found in forests in southern and eastern Africa, from South Sudan to South Africa, including eastern Democratic Republic of Congo. The type locality is "Indiis" (in error). Specimens from West Africa identified in the past as *D. scabra* probably belong to one of the three new species from West Africa described by Trape and Mané (2006b): *D. gansi*, *D. confusa*, *D. sahelensis*.

The dorsum is light brown or gray, with a pattern along the vertebral midline of a series of dark diamond shapes (or rhombus shapes) separated by a pale spot. Along the sides of the body, a dark vertical band connects each pale spot to the ventrals. At the nape, the diamond shape is replaced by a thick "V" shape, with the point directed anteriorly. The venter is gray with black spots outlining the ventral scales. The nasal is semi-divided. There are usually 2 postoculars, the inferior equal to or larger than the superior. The temporal formula is 2+3. There are 7(3,4) upper labials and 7(3) lower labials. The dorsal scales are keeled, with 2 apical pits, and are arranged in 20–27 (most often 23 or 25) oblique rows at midbody. There are 179–241 ventrals (fewer than 230 in males, more than 197 in females) and 43–79 divided subcaudals (more than 48 in males, fewer than 65 in females). The maximum length reported for this species is 1,000 mm (Villiers 1975).

Tree Snakes: Genus *Dipsadoboa* Günther 1858

Dipsadoboa is a genus of 10 species of nocturnal arboreal snakes, occurring in the rainforests and open woodlands of sub-Saharan Africa. Seven species occur within our zone. The taxonomy of this genus has been extensively reviewed by Rasmussen (1986, 1989, 1993, 1994).

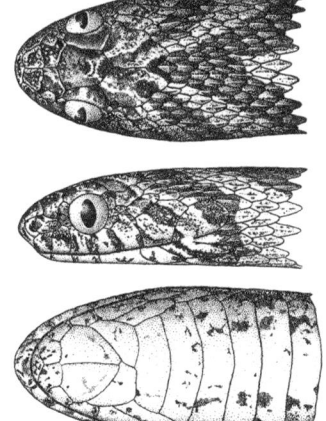

FIGURE 13.21. *Dasypeltis scabra*. Modified by K. Jackson from Witte (1953)

FIGURE 13.22. *Dasypeltis scabra*, Kenya. By S. Spawls

The head is broad and flat, almost triangular, with a thin neck. The body and tail are long and slender. The snout is rounded. The size of the eye may range from small to large, with a vertically elliptical pupil.

The maxillary dentition is opisthoglyph, with 11–24 unspecialized maxillary teeth, followed by 2–3 grooved fangs at the posterior end of the maxilla. Rasmussen (1996) found that the number of maxillary teeth in *Dipsadoboa* species is correlated with prey type. Species with higher maxillary tooth counts (e.g., *D. unicolor*, *D. viridis*, *D. duchesnii*) fed exclusively on slippery anurans, while those with low maxillary tooth counts (e.g., *D. shrevei*) fed on non-slippery chameleons. Two species with intermediate maxillary tooth counts, *D. aulica* and *D. flavida* (not present in our zone), had diets that included both lizards and anurans. The hemipenes are unforked with an undivided sulcus spermaticus.

The nasal is single. The internasals and prefrontals are paired. The loreal is present. There are 1–2 preoculars and 2 postoculars. There are no suboculars. The frontal is as long as or longer than broad and is separated from the eye by a supraocular. The temporal formula is 1 or 2+1+2. There are usually 8 upper labials, 2 or 3 of which are in contact with the eye. There are 2 pairs of chin shields. The submandibular groove is pronounced. In some species, a pair of gulars resembles a third pair of chin shields, but these are considered false chin shields because they are not in contact with the lower labials. The anterior chin shields are separated from the mental scale by the first pair of lower labials. There are 8–12 lower labials. The dorsal scales are smooth, with or without apical pits, and arranged in 17–19 oblique rows at midbody. The vertebral row is enlarged relative to the other dorsal scale rows. The ventrals are angulate or weakly keeled. The cloacal scale is single. The subcaudals may be single or divided.

Dipsadoboa are unusual in showing a reversal of the usual trend of sexual dimorphism in numbers of ventrals. (In most snake species, females tend to have more ventrals than males. In *Dipsadoboa*, males have as many or more ventrals than females.) Myers (1982) observed the same phenomenon in Neotropical Vine Snakes (genus *Imantodes*) and speculated that the increased number of ventrals correlates with an increased number of trunk vertebrae, which confers an advantage in stretching across gaps between branches while traveling through the forest canopy, especially for males, which are never constrained by the added weight of carrying eggs when gravid. *Dipsadoboa* in Africa occupy a similar ecological niche to *Imantodes* in the Neotropics, and it seems likely that the trend toward higher ventral counts in males represents a case of convergence associated with use of a similar arboreal habitat. This reverse trend in number of ventrals is more pronounced in some species than in others. Rasmussen (1993) noted that the only species of *Dipsadoboa* in which females have more ventrals than males is *D. weileri*, which is also the least arboreal species in the genus, consistent with the idea that the higher ventral count in males represents an adaptation for arboreality.

Short-snouted Tree Snake: *Dipsadoboa brevirostris* Sternfeld 1908c

The range of *Dipsadoboa brevirostris* extends from Guinea to Cameroon. The type locality is Yabassi, Cameroon.

The dorsum is uniformly light brown.

Key to Central and Western African Species of the Genus *Dipsadoboa*

1   Subcaudals single 2
1'   Subcaudals divided 5

2(1) Underside of the body a different color from underside of tail .... *D. weileri*
2'   Underside of body and tail uniform in color 3

3(2) Usually 3 upper labials in contact with the eye; underside paler than dorsum ....
        .... *D. underwoodi*
3'   Usually 2 upper labials in contact with the eye; underside the same color as, or only
    slightly paler than, the dorsum 4

4(3) Diameter of eye less than the distance between eye and nostril; frontal equal in
       length and breadth .... *D. unicolor*
4'   Diameter of eye equal to the distance between eye and nostril; frontal longer
       than broad .... *D. viridis*

5(1) Frontal approximately equal in length and breadth .... *D. shrevei*
5'   Frontal distinctly longer than broad 6

6(5) Dorsal scales pale with dark edges; not more than 220 ventrals .... *D. duchesnii*
6'   Dorsal scales uniformly pale; at least 217 ventrals .... *D. brevirostris*

The venter is slightly paler. There are 19–21 maxillary teeth, excluding the grooved posterior fangs. The diet consists of anurans.

The internasals are shorter and narrower than the prefrontals. The prefrontals are longer than broad. The loreal is usually deeper than long. There is 1 preocular, which is not in contact with the frontal. The supraocular is narrower than the frontal. There are 2 postoculars of equal size. There are 8(3,4,5) upper labials and 9(4) lower labials. The temporal formula is 1+1, sometimes 2+1. The anterior chin shields are longer than the posterior pair. The dorsal scales are smooth, without apical pits, and arranged in 17 oblique rows at midbody. The vertebral row is enlarged relative to the other dorsal scale rows. There 217–229 angulate ventrals, with a slightly higher ventral count in males than in females (217–227 in females, 222–229 in males). There are 91–111 divided subcaudals (91–95 in females, 109–111 in males). The cloacal scale is single. The maximum length recorded for this species is 800 mm (Rasmussen 1989).

Blue-tailed Tree Snake: *Dipsadoboa duchesnii* (Boulenger 1901)

The range of *Dipsadoboa duschesnii* extends from Nigeria east to the Democratic Republic of Congo. The type locality is Mandungu, Ituri, Democratic Republic of Congo.

The dorsum is uniformly light brown, with dark edges to the dorsal scales. The venter is whitish or yellowish anteriorly, becoming darker posteriorly so that

FIGURE 13.23. *Dipsadoboa brevirostris*, Ghana. By M. Fujita

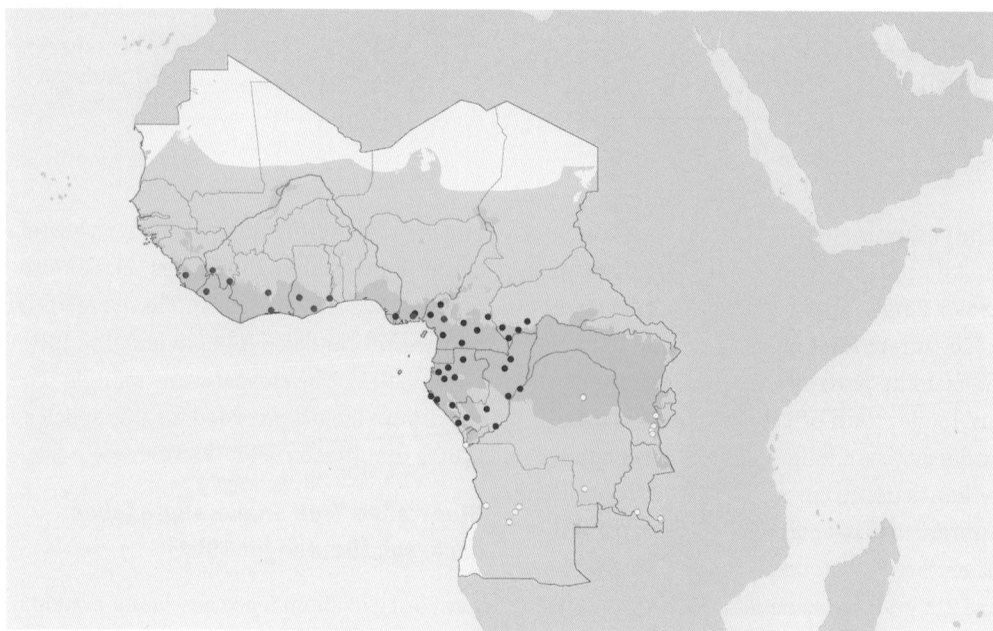

FIGURE 13.24. *Dipsadoboa brevirostris* (red), *Dipsadoboa duchesnii* (blue), *Dipsadoboa shrevei* (yellow). By K. Jackson

the underside of the tail is as dark as the dorsum. There are 17–23 maxillary teeth, excluding the grooved posterior fangs. The diet consists of forest frogs and toads.

The internasals are shorter and narrower than the prefrontals. The prefrontals are equal in length and breadth. The loreal is deeper than long and often fused to the inferior preocular. There may be 2 preoculars, but often the inferior is fused to the loreal so that there is just 1 preocular. The preocular is usually not in contact with the frontal. The supraocular is narrower than the frontal. There are 2 postoculars, the superior larger than the inferior. There are 8(3,4,5) upper labials and 10(5), sometimes 10(4), lower labials. The temporal formula is 1+1, occasionally 2+1. The anterior chin shields are longer than the posterior pair. The dorsal scales are smooth, without apical pits, and arranged in 17 oblique rows at midbody. The vertebral row is enlarged relative to the other dorsal scale rows. There are 198–220 angulate ventrals, with a higher ventral count in males than in females (198–215 in females, 201–220 in males). There are 92–121 single subcaudals (96–121 in males, 92– 114 in females). The cloacal scale is usually single, but in rare cases it may be divided. The maximum length recorded for this species is 1,305 mm (Villiers 1966).

Shreve's Tree Snake: *Dipsadoboa shrevei* (Loveridge 1932)

Shreve's Tree Snake is a moist savanna species with a range extending from Angola to Tanzania, through Zambia and the Democratic Republic of Congo. The type locality is Missao de Dondi, Bela Vista, Angola.

The dorsum is uniformly dark brown or dark bluish gray. The venter is paler than the dorsum, cream or beige anteriorly. Juveniles are paler than adults, and their tails are speckled with brown. There are 11–15 maxillary teeth, excluding the grooved posterior fangs. The diet consists of chameleons.

The internasals are much shorter and narrower than the prefrontals. The prefrontals are broader than long. The loreal is deeper

than long. There is 1 preocular, which is not in contact with the frontal. The supraocular is distinctly narrower than the frontal. There are 2 postoculars, the superior larger than the inferior. There are 8(3,4,5) upper labials and 10(5) lower labials. The temporal formula is 1+2 or 1+1. The anterior chin shields are longer than the posterior pair. The dorsal scales are smooth, with 1–2 apical pits, and arranged in 19 oblique rows at midbody. The vertebral row is enlarged relative to the other dorsal scale rows. There are 199–219 angulate ventrals, with a higher ventral count in males than in females (199–212 in females, 203–219 in males). There are 74–96 divided subcaudals, with little or no sexual dimorphism (74–96 in females, 74–91 in males). The cloacal scale is usually single, but in rare cases it may be divided. The maximum length recorded for this species is 1,110 mm (Rasmussen 1986).

FIGURE 13.25. *Dipsadoboa shrevei* RMCA 8257. By T. Giri

Underwood's Tree Snake: *Dipsadoboa underwoodi* Rasmussen 1993

The range of *Dipsadoboa underwoodi* extends from Guinea to Congo. The type locality is Mundame, Cameroon.

In adults, the dorsum is dark brown or black. The venter is whitish with dark edges to the subcaudals. In juveniles, the dorsum is pale and patterned with black spots that merge as the snake ages. The venter is greenish or yellowish, and palest at the throat. The ventral scales are edged with black.

There are 18–22 maxillary teeth, excluding the grooved posterior fangs. The diet consists of slippery anurans. The internasals are much shorter and narrower than the prefrontals. The prefrontals are broader than long. The loreal is equal in length and depth. There is 1 preocular, which is in contact with the frontal. The supraocular is distinctly narrower than the frontal. There are 2 postoculars, the superior larger than

FIGURE 13.26. *Dipsadoboa underwoodi*, Democratic Republic of Congo. By J. Kielgast

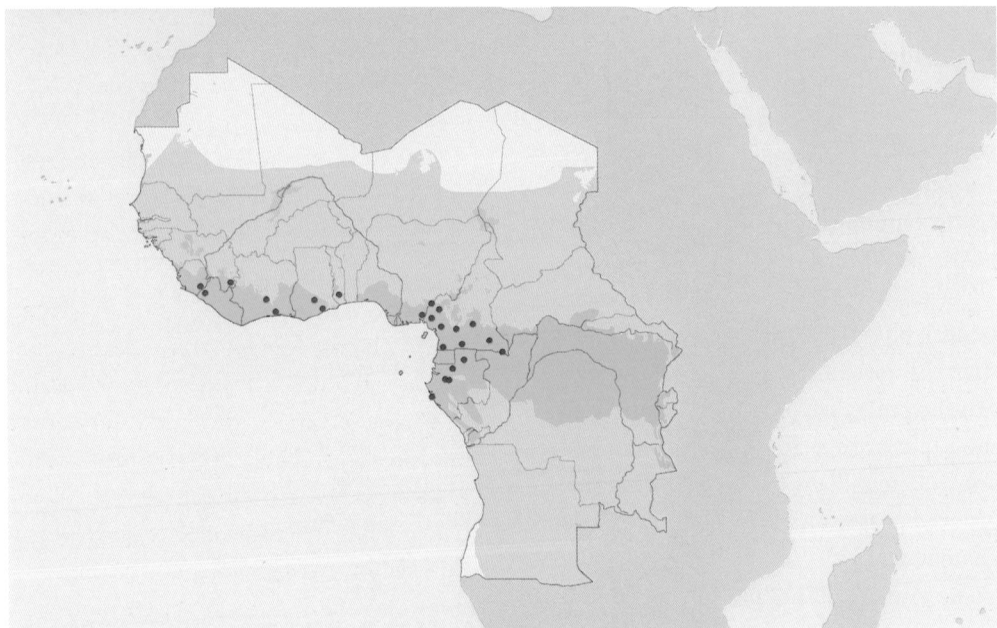

FIGURE 13.27. *Dipsadoboa underwoodi* (red). By K. Jackson

the inferior. There are 8(3,4,5) or sometimes 8(4,5) upper labials and 9(4) or 10(5) lower labials. The temporal formula is 1+1. The anterior chin shields are longer than the posterior pair. The dorsal scales are smooth, without apical pits, and arranged in 17 oblique rows at midbody. The vertebral row is enlarged relative to the other dorsal scale rows. There are 177–202 keeled ventrals, with a higher ventral count in males than in females (177–199 in females, 178–202 in males). There are 71–87 single subcaudals, with little or no sexual dimorphism (72–84 in females, 71–87 in males). The cloacal scale is single. The maximum length recorded for this species is 610 mm (Rasmussen 1993).

Günther's Tree Snake: *Dipsadoboa unicolor* Günther 1858

Günther's Tree Snake is a montane species with a disjunct distribution extending from Guinea-Bissau to Uganda. This species occurs at high altitudes (800–3,000 m) and is active at night at temperatures of 7°C (Böhme 1975). This is a lower temperature than is tolerated by other species and presumably represents an adaptation by *Dipsadoboa unicolor* to its montane habitat. The type locality is West Africa.

The dorsum is uniformly green or olive brown, each scale edged with black. The venter is greenish or yellowish. The throat is paler than the rest of the underside.

There are 16–22 maxillary teeth, excluding the grooved posterior fangs. The diet consists of slippery anurans.

The internasals are shorter and narrower than the prefrontals. The prefrontals are longer than broad. The loreal is equal in depth and length. There is 1 preocular, usually in contact with the frontal. The supraocular is distinctly narrower than the fron-

tal. There are 2 postoculars, the superior larger than the inferior. There are usually 8(4,5), sometimes 9(4,5,6), upper labials and 9(5) to 12(7) lower labials. The temporal formula is 1+1, occasionally 1+2. The anterior chin shields are longer than the posterior pair. The dorsal scales are smooth, without apical pits, and arranged in 17 oblique rows at midbody. The vertebral row is enlarged relative to the other dorsal scale rows. There are 181–220 angulate ventrals, with a higher ventral count in males than in females (183–211 in females, 181–220 in males). There are 52–78 single subcaudals, with little or no sexual dimorphism (54–78 in males, 52–77 in females). The cloacal scale is single. The maximum length recorded for this species is 1,280 mm (Rasmussen 1993).

Green Tree Snake: *Dipsadoboa viridis* (Peters 1869)

The range of *Dipsadoboa viridis* extends from Guinea and Liberia in the west to the Democratic Republic of Congo, Rwanda, and Burundi in the east. The type locality is New Caledonia (in error). Two subspecies are recognized: *D. v. viridis* (Peters 1869); *D. v. gracilis* Laurent 1956. *D. v. viridis* is the western subspecies, with a distribution extending from Liberia to southeastern Cameroon and Congo. *D. v. gracilis* is the eastern subspecies, with a distribution extending from Central African Republic to the Democratic Republic of Congo to Rwanda and Burundi. The Ubangi River (northern tributary of the Congo River) probably represents the border between the two subspecies.

The dorsum is uniformly olive green. The venter is also olive green, and the ventral coloration differs between the two subspecies as follows. In *D. v. viridis*, the upper labials, underside of the body, and underside of the tail are all the same shade of green, whereas in *D. v. gracilis*, the upper labials and underside of the body are a lighter shade of green than the underside of the tail. There are 16–23 maxillary teeth, excluding the grooved posterior fangs. The diet consists of slippery anurans. When threatened, this species rolls itself into a ball with the head protected inside the coils of the body.

The internasals are much shorter and narrower than the prefrontals. The prefrontals are broader than long. The loreal is deeper than long. There is 1 preocular, in contact with the frontal. The supraocular is distinctly narrower than the frontal. There

FIGURE 13.28. *Dipsadoboa unicolor*, Cameroon. By V. Gvoždík

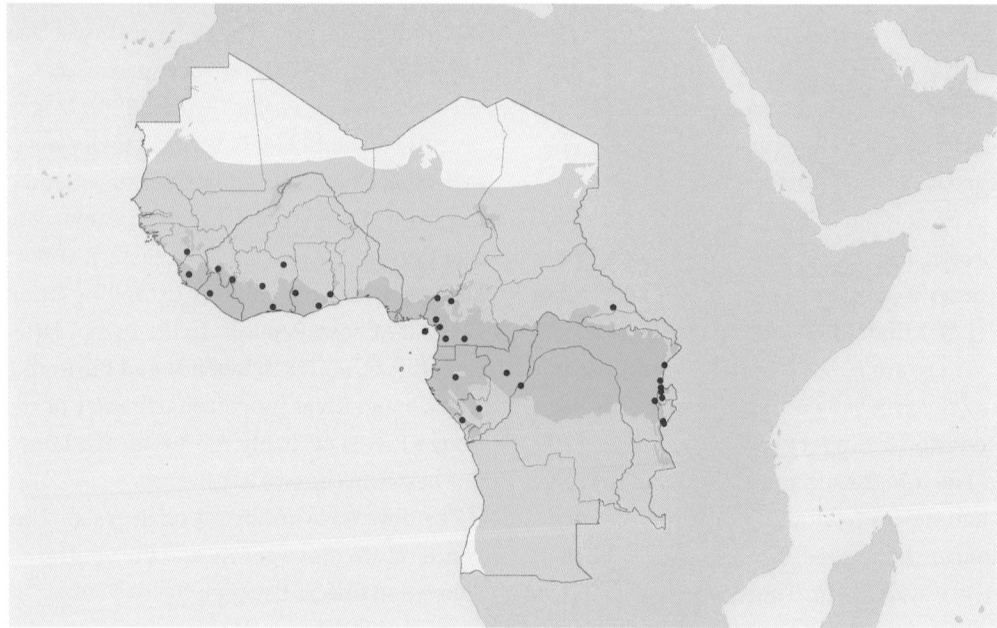

FIGURE 13.29. *Dipsadoboa unicolor* (blue). By K. Jackson

FIGURE 13.30. *Dipsadoboa viridis*, Cameroon. By V. Gvoždík

FIGURE 13.31. *Dipsadoboa viridis* defensive posture, Republic of Congo. By K. Jackson

FIGURE 13.32. *Dipsadoboa viridis* defensive posture with head hidden, Republic of Congo. By K. Jackson

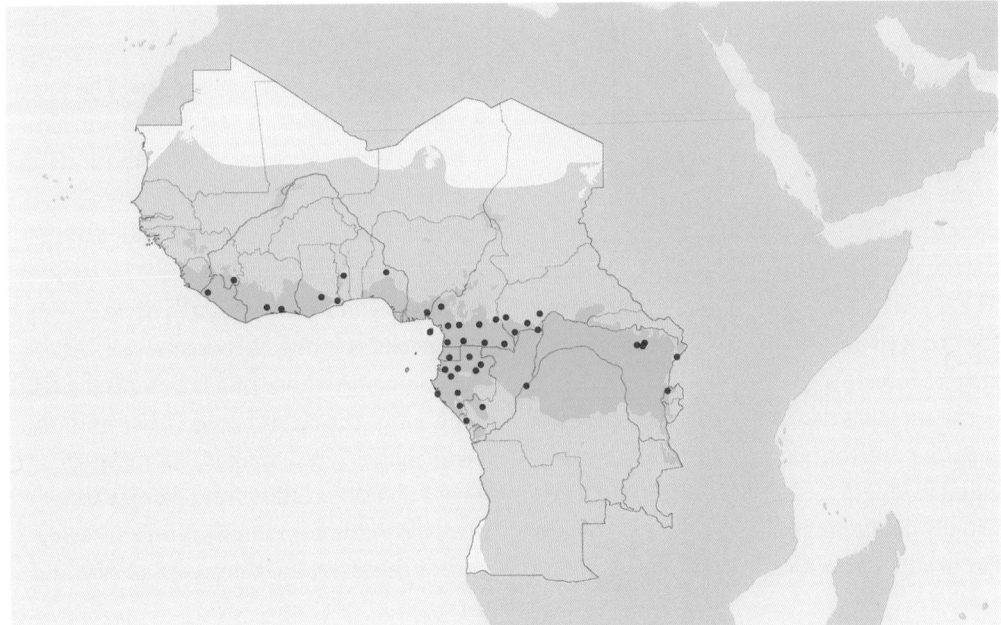

FIGURE 13.33. *Dipsadoboa viridis* (blue). By K. Jackson

are 2 postoculars, the superior larger than the inferior. There are usually 8(4,5) upper labials, sometimes 9(4,5,6), and 9(5) to 11(6) lower labials. The temporal formula is 1+1. The anterior chin shields are longer than the posterior pair. The dorsal scales are smooth, without apical pits, and arranged in 17 oblique rows at midbody. The vertebral row is enlarged relative to the other dorsal scale rows. There are 193–238 angulate ventrals: 193–232 in *D. v. viridis* (193–232 in females, 195–228 in males); 208–238 in *D. v. gracilis* (213–231 in females, 208–238 in males). Thus *D. v. gracilis*, but not *D. v. viridis*, has a higher ventral count in males than in females. There are 71–112 single subcaudals: 71–103 in *D. v. viridis* (76–103 in males, 71–100 in females) and 87–112 in *D. v. gracilis* (93–112 in males, 87–106 in females), with a higher subcaudal count and a higher degree of sexual dimorphism in *D. v. gracilis* than in *D. v. viridis*. The cloacal scale is single. The maximum length recorded for this species is 1,240 mm (Rasmussen 1993).

Weiler's Tree Snake: *Dipsadoboa weileri* (Lindholm 1905)

The range of *Dipsadoboa weileri* extends from Guinea to the Sudan and the Democratic Republic of Congo. The type locality is Cameroon.

The dorsum is uniformly olive green or dark gray. The venter is yellowish except for the underside of the tail, which is dark. There are 15–22 maxillary teeth, excluding the grooved posterior fangs. The diet consists of forest frogs and toads.

The internasals are shorter and narrower than the prefrontals. The prefrontals are equal in length and breadth. The loreal is usually deeper than long. There is 1 preocular, usually in contact with the frontal. The

supraocular is narrower than the frontal. There are 2 postoculars, the superior larger than the inferior. There are usually 8(4,5), sometimes 9(4,5,6) or 8(3,4,5), upper labials and 10(5) to 12(7) lower labials. The temporal formula is 1+1, sometimes 1+2. The anterior chin shields are much longer than the posterior pair. The dorsal scales are smooth, without apical pits, and arranged in 17 or 19 oblique rows at midbody. The vertebral row is enlarged relative to the other dorsal scale rows. There are 181–205 angulate ventrals, and *D. weileri* is the only species in the genus with a higher ventral count in females than in males (181–205 in females, 182–203 in males), a characteristic thought to be correlated with a less arboreal lifestyle than other species. There are 56–73 single subcaudals, with little or no sexual dimorphism (56–73 in males, 56–71 in females). The cloacal scale is single. The maximum length recorded for this species is 960 mm (Rasmussen 1993).

Genus *Dispholidus* Duvernoy 1832

Boomslang: *Dispholidus typus* (Smith 1829)

Dispholidus is a monotypic genus. The Boomslang, *Dispholidus typus*, is a diurnal, arboreal snake with a vast distribution in sub-Saharan Africa, extending from Senegal in the west to the coast of Kenya in the east and south to South Africa. The type locality is Old Latakoo (South Africa).

Dispholidus is one of two genera of African colubrine whose bite is potentially fatal to humans, the other being *Thelotornis*, the Twig Snakes/Bird Snakes. Boomslangs have a characteristic threat display that involves puffing up the neck and anterior part of the laterally compressed body, and making striking gestures with the mouth open. Spawls (1985) reports *D. typus*, when harassed by a herpetologist with a grab-stick, alternately carrying out this threat display, then fleeing and lying motionless on an area of open ground with its body kinked in such a way as to resemble a stick. Bites to humans by Boomslangs are relatively rare. The diet consists of a variety of mostly arboreal prey, including chameleons and other arboreal lizards, frogs, small mammals, birds (especially baby birds), and bird eggs. *Dispholidus* lay clutches of 8–23 eggs (FitzSimons 1962).

The head is small and short with a rounded snout and a well-defined neck. The eye is large with a characteristic "tear-drop"-shaped pupil, which is round but with the anterior edge drawn out to a narrower point (Fig. 13.37). The body is long

FIGURE 13.34. *Dipsadoboa weileri*, Cameroon. By V. Gvoždík

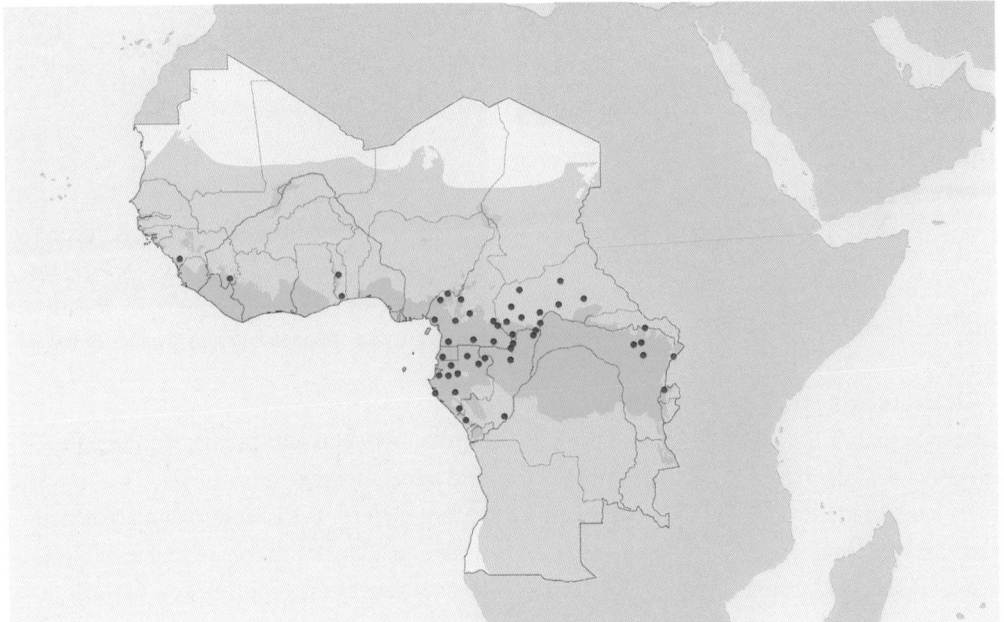

FIGURE 13.35. *Dipsadoboa weileri* (red). By K. Jackson

FIGURE 13.36. *Dispholidus typus* threat display, Botswana. By S. Spawls

FIGURE 13.37. *Dispholidus typus*, Kenya. By S. Spawls

FIGURE 13.38. *Dispholidus typus* juvenile, Botswana. By S. Spawls

and somewhat laterally compressed. The tail is long and slender. The hemipenes are unforked, and the sulcus spermaticus is not forked.

This species has a vast distribution throughout sub-Saharan Africa and is enormously variable in coloration. Attempts have been made to describe subspecies on the basis of dramatic differences in coloration supported by small differences in ventral and subcaudal counts. But coloration in this species varies between juveniles and adults, between adult males and females, and in other ways. It is not yet clear to what extent differences in coloration are attributable to adaptation to particular habitats or may represent geographical variation. Further study is needed to make sense of these variations in coloration and the extent to which they may be phylogenetically informative or useful to taxonomists. For example, Perret (1961) found that, in Cameroon, the males were green with black striations while females were reddish brown. In both sexes, there was a black oval spot on each side of the neck. Laurent (1955) proposed the subspecies *D. t. kivuensis* for specimens from eastern Democratic Republic of Congo, in which the adult males were green (the females were brown or gray). The term "kivuensis phase" is often used to refer to specimens of *D. typus* that are green with thick black edges to the scales. Spawls et al. (2004) report that this type of coloration is seen in Boomslangs from woodland areas in east Africa. Laurent (1955) describes adult male *D. typus* from Angola that were brown or black and patterned with yellow or orange spots. The coloration of juveniles is different from that of adults, often having a pale venter and upper labials, sharply delineated from a darker dorsal coloration. This juvenile coloration of *D. typus* is not unlike the adult coloration of the Forest Vine Snake / Bird Snake, *Thelotornis kirtlandii*.

The nasal scale is undivided. The internasals and prefrontals are paired. The frontal is longer than it is broad. There is 1 loreal scale and 1 preocular. There are 2–3 postoculars (usually 3 of similar size). Occasionally, the inferiormost postocular occupies the position of a subocular. There are 7–8 upper labials, usually 7(3,4). The temporal formula is 1+2. There are 8–10 lower labials, usually 8(4) or 9(5). There are 2 pairs of chin shields, the anterior pair equal to or longer than the posterior pair. The submandibular groove is pronounced. The dorsal scales are keeled, without apical pits, and arranged in 17–21 (usually 19, sometimes 21) oblique rows. The vertebral row is some-what enlarged relative to the other dorsal scale rows. There are 164–201 keeled ventrals (fewer than 192 in males, more than 172 in females) and 87–131 divided subcaudals (more than 104 in males, fewer than 127 in females). The cloacal scale is divided (but see comments below regarding the death of Karl P. Schmidt). The maximum length recorded for this species is 1,820 mm (Broadley and Cock 1975).

The maxillary dentition is opisthoglyph. The maxilla is short and bears 5-6 small similar-sized anterior maxillary teeth, followed after a diastema by 3 strongly enlarged and deeply grooved posterior fangs. Boomslang envenomation is characterized by rapid appearance of localized pain at the site of the bite, headache, vomiting, and bleeding (Broadley and Cock 1975; Haagner and Smith 1987). Envenomation by *D. typus* causes severe hemorrhaging (Simbotwe 1982). The venom of *D. typus* is rich in enzymes, especially proteolytic ones (Robertson and Delpierre 1969). It contains a prothrombin activator (Guillin et al. 1978; Rosing et al. 1988) and a thrombin-like enzyme (Marsh 1994). According to Kress (1988), it may also contain a metalloprotease. A specific antivenom against the venom of *D. typus* exists, but it is difficult to obtain outside South Africa, where it is manufactured.

This species was responsible for the death in 1957 of herpetologist Karl P. Schmidt, whose monograph on the snakes collected by the American Museum Congo Expedition of 1909-15 remains to this day one of the most useful studies ever of central African snakes (Schmidt 1923). The fatal bite occurred when a juvenile Boomslang was brought to the herpetology department of the Chicago Natural History Museum from the local zoo for identification. The snake had proven difficult to identify because it was an aberrant specimen, having an undivided cloacal scale, so the identification key the zoo staff had been using did not work.

FIGURE 13.39. *Dispholidus typus* RMCA 73-17-R-149. By T. Giri

Schmidt wrote an account of the incident: "[we] were discussing the possibility of its being a boomslang when I took it from Dr. Inger without thinking of any precaution and it promptly bit me on the fleshy lateral aspect of the first joint of the left thumb. The mouth was widely opened and the bite was made with the rear fangs only, only the right fang entering to its full length of about 3 mm. Only one other tooth mark, from the penultimate tooth, appeared on the thumb when the snake was disengaged." Schmidt died 24 hours later. The account of his death, including Schmidt's own observations, was later published (Pope 1958). This incident occurred at a time when little was known about the venoms of colubrid snakes, and all the herpetologists involved underestimated the potential danger of the juvenile Boomslang.

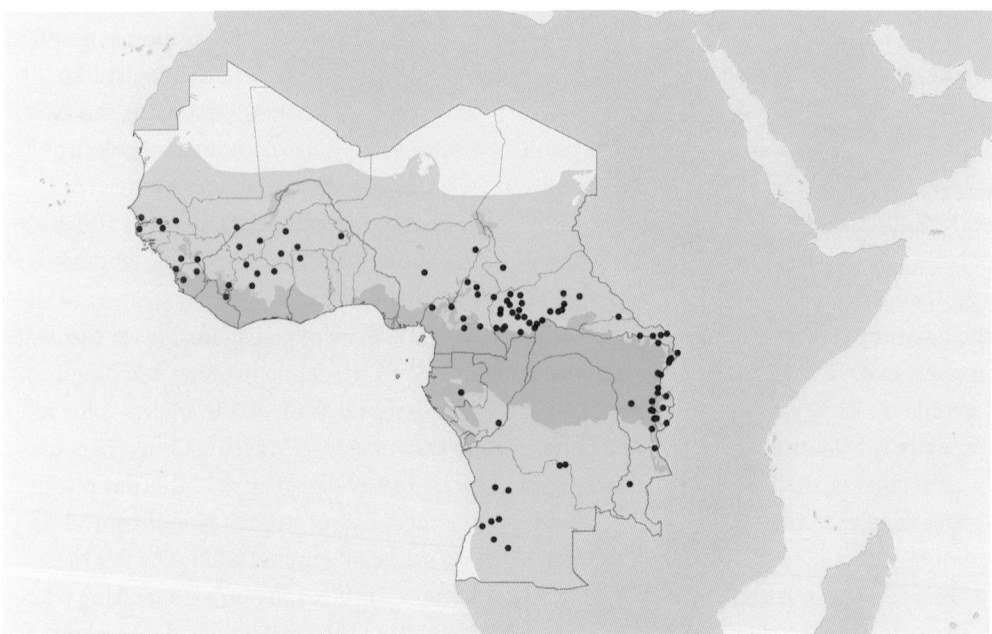

FIGURE 13.40. *Dispholidus typus* (blue). By K. Jackson

Emerald Snakes: Genus *Hapsidophrys* Fischer 1856

Hapsidophrys is a genus of three species of forest snakes, all of which are present in our zone. Emerald Snakes much resemble the Bush Snakes (genus *Philothamnus*), from which they differ in having keeled, rather than smooth, dorsal scales. Like *Philothamnus*, *Hapsidophrys* are diurnal and arboreal. Despite the broad distribution of *Hapsidophrys* in west and central Africa, their natural history has not been extensively studied. Luiselli et al. (2001b) examined stomach contents from 20 individuals of *Hapsidophrys lineatus* captured in southern Nigeria. Diet was found to consist primarily of forest frogs and lizards, though some shrews (*Crocidura* sp.) were also eaten.

The head is small and the neck slender and well defined. The snout is long and rounded. The eye is medium sized with a round pupil. The body is long and cylindrical, with a long, slender tail. The maxillary dentition is aglyph, with 19–33 teeth, increasing in size posteriorly and without diastemata. The hemipenes are unforked, and the sulcus spermaticus is undivided. The rostral is rounded. The nasal is divided. The internasals and prefrontals are paired. The loreal is present. There are 1–2 preoculars but no suboculars. The frontal is longer than broad and is separated from the eye

by a supraocular. There are 2–3 postoculars. The temporal formula is 1-2+1-2. There are 8–11 upper labials, two of which are in contact with the eye. There are 9–11 lower labials. There are 2 pairs of chin shields. The submandibular groove is pronounced. The dorsal scales are keeled, without apical pits, and arranged in 15 straight or slightly oblique rows midbody. The vertebral row is not enlarged relative to the other dorsal scale rows. There are 129–176 keeled ventrals and 90–172 divided subcaudals, which may be keeled or rounded. The cloacal scale may be single or divided.

Black-lined Emerald Snake: *Hapsidophrys lineatus* Fischer 1856

The range of *Hapsidophrys lineatus* extends from Sierra Leone south to Angola and east to Uganda. The type locality is Elmina, Ghana. In contrast to *H. smaragdina*, this species is rarely found close to human habitations.

The dorsum is emerald green, the dorsal scales outlined in black. The venter is pale green. *H. lineatus* lacks the black stripe across the eye seen in *H. smaragdina*, from which it also differs in having the dorsal scales edged with black. There are 30–33 maxillary teeth. There are 2 preoculars, the superior larger than the inferior. There are usually 2 postoculars of equal size. There are usually 8(4,5), sometimes 9(4,5), upper labials and 10(5) lower labials. The tempo-

---

Key to Central and Western African Species of the Genus *Hapsidophrys*

1	Cloacal scale single	*H. lineatus*
1'	Cloacal scale divided	2
2(1)	Fewer than 175 ventrals	*H. smaragdina*
2'	More than 184 ventrals	*H. principis*

ral formula is usually 2+2, sometimes 3+2 or 3. There are 152–176 keeled ventrals and 90–158 rounded subcaudals. The cloacal scale is single. The maximum length recorded for this species is 1,225 mm (Villiers 1966).

Príncipe Emerald Snake: *Hapsidophrys principis* (Boulenger 1906a)

The Príncipe Emerald Snake is endemic to the islands of São Tomé and Príncipe. The type locality is Prince's Island, Gulf of Guinea, West Africa.

In coloration, *Hapsidophrys principis* is morphologically almost indistinguishable from *H. smaragdina*, which it resembles in most scale counts. *H. principis* differs from *H. smaragdina* in ventral count and differs slightly (overlapping) in subcaudal count. *H. principis* has 185–191 keeled ventrals and 170–177 keeled subcaudals. Jesus et al. (2009) found sufficient difference in mitochondrial gene sequences between *H. prin-*

cipis compared to *H. smaragdina* collected from Gabon to justify considering them as two separate species, concluding that *H. principis* should be considered a sister taxon of *H. smaragdina*. The maximum length recorded for this species is 1,150 mm (Boulenger 1906a).

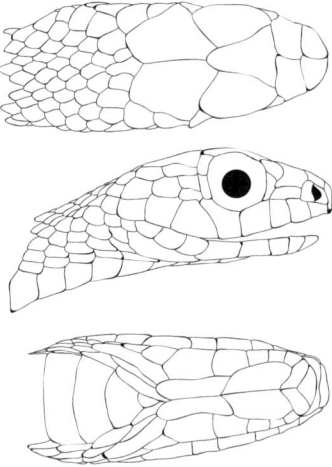

FIGURE 13.41. *Hapsidophrys lineatus* RMCA 9781. By T. Giri

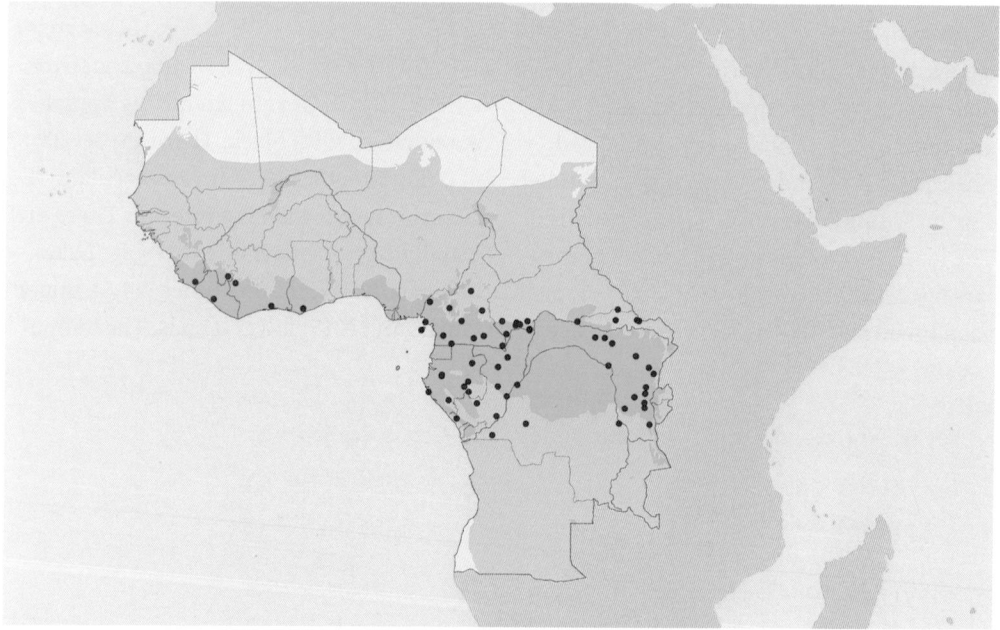

FIGURE 13.42. *Hapsidophrys lineatus* (blue). By K. Jackson

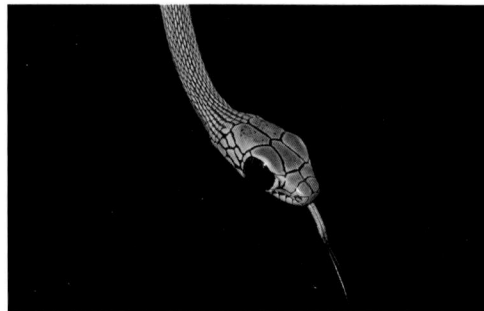

FIGURE 13.43. *Hapsidophrys lineatus*, Cameroon. By V. Gvoždík

FIGURE 13.44. *Hapsidophrys lineatus*, Democratic Republic of Congo. By E. Greenbaum

Common Emerald Snake: *Hapsidophrys smaragdina* (Schlegel 1837)

The range of *Hapsidophrys smaragdina* extends from Gambia south to Angola and east to Uganda. The type locality is the Gold Coast (meaning Ghana). This species is more often found close to human habitations than *H. lineatus*.

The dorsum is emerald green with occasional dorsal scales that are turquoise. The venter is pale green. The head is marked with a thick black lateral stripe across the eye from the nasal scale to the last upper labials. There are 19–23 maxillary teeth. There is 1 preocular, deeper than long, and with a depth that exceeds the diameter of the eye. There are 2 postoculars, the superior larger than the inferior. There are usually 9(5,6) upper labials and 10(5) lower labials. The temporal formula is usually 1+2, sometimes 1+1. There are 150–174 keeled ventrals and 129–172 keeled subcaudals.

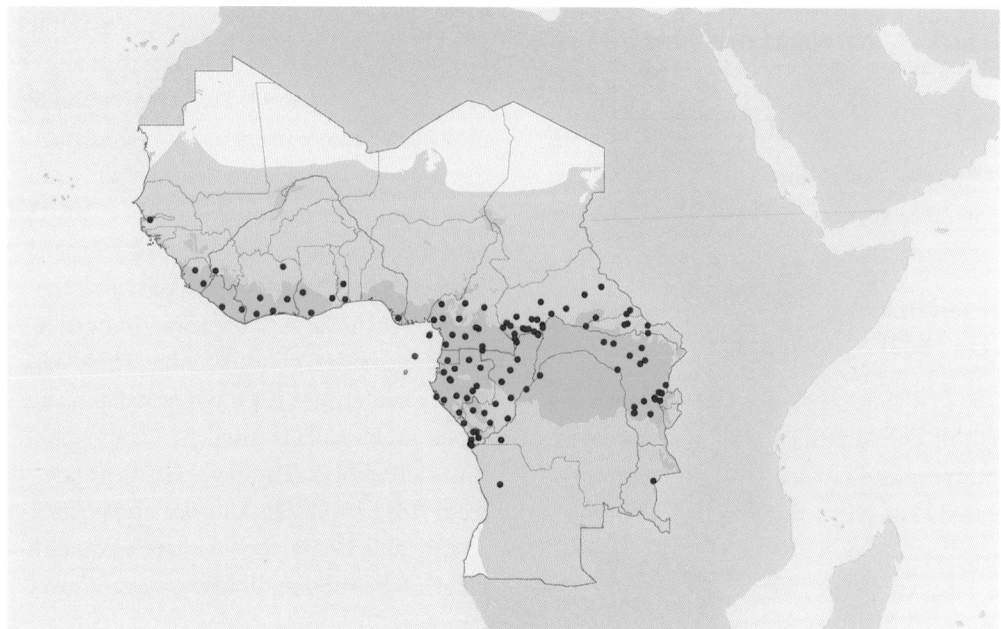

FIGURE 13.45. *Hapsidophrys principis* (red), *Hapsidophrys smaragdina* (blue). By K. Jackson

FIGURE 13.46. *Hapsidophrys sma-ragdina* juvenile, Republic of Congo. By K. Jackson

The cloacal scale is divided. The maximum length recorded for this species is 1,191 mm (Roux-Estève 1965).

Whip Snakes/Racers: Genus *Hemorrhois* Boie 1826

Like *Bamanophis*, *Lytorhynchus*, *Platyceps*, and *Spalerosophis*, *Hemorrhois* are Old World Racers, representatives of a Eurasian lineage of colubrines. Racers are slender, fast-moving snakes, active by day in desert, sahel, and dry savanna habitats. The genus *Hemorrhois* currently includes four species in desert to dry savanna habitats of north Africa, southern Europe, and the Middle East. One species, the Algerian Racer, *Hemorrhois algirus*, occurs within our zone.

Algerian Racer: *Hemorrhois algirus* (Jan 1863)

The distribution of *Hemorrhois algirus* is within the Sahara Desert, from southern Morocco and Libya in the north to Mauritania and Niger in the south. The type locality is Sfax, Tunisia. One subspecies, *H. a. intermedius* Werner 1929, occurs within our zone, in the western part of the range of the species, with a distribution extend-ing from Morocco and the western edge of Algeria to Mauritania in the south. The type locality for *H. a. intermedius* is Aïn Sefra, Algeria. The dorsum is brown, olive, or yellowish with dark bands and dark lateral spots. The upper labial scales are white with dark edges. The venter is grayish white with dark lateral spots on the ventral scales. The head is broad with a rounded snout and well-defined neck. The body and tail are long and slender. The eye is small with a round or slightly vertically elliptical pupil. The nasal is divided. The internasals and prefrontals are paired. The loreal is divided into 2 or 3 scales. There are 2 preoculars, the superior larger than the inferior, which may occupy the position of a subocular. The supraocular is narrower than the frontal. There are 2 or 3 postoculars, of which the inferior is the largest and sometimes occupies the position of a subocular. There are 9(5) or sometimes 10(5) upper labials. The temporal formula is usually 2+2 but sometimes 2+3 or 3+3. The lower labials range from 7(4) to 9(5). The anterior and posterior chin shields are approximately equal in length. The submandibular groove is pronounced. The dorsal scales are smooth, with 2 apical pits, and arranged in 23–25 straight rows. The vertebral row is not enlarged relative to the other dorsal scale rows. There are 205–240 ventrals (221–237 for *H. a. intermedius*; fewer than 232 for males, more than 226 for females). There are 83–117 divided subcaudals (96–110 for *H. a. intermedius*). The cloacal scale may be single or divided. The maximum length recorded for this species is 1,400 mm (Schleich et al. 1996). The maxillary dentition is aglyph, with 12–14 anterior maxillary teeth decreasing in size posteriorly and followed by 2 larger ungrooved posterior teeth. The hemipenes are unforked, and the sulcus spermaticus is not forked.

Genus *Lytorhynchus* Peters 1862

Like *Bamanophis, Hemorrhois, Platyceps*, and *Spalerosophis, Lytorhynchus* are Old World Racers, representatives of a Eurasian lineage of colubrines. Racers are slender, fast-moving snakes, active by day in desert,

sahel, and dry savanna habitats. *Lytorhynchus* is a genus of desert-adapted snakes with a distinctive "leaf-shaped" rostral scale. Of the six species in the genus, five are restricted to the Middle East and/or central Asia. One species, *Lytorhynchus diadema*, occurs within our zone.

Crowned Leafnose Snake: *Lytorhynchus diadema* (Duméril, Bibron, and Duméril 1854b)

The distribution of *Lytorhynchus diadema* extends through Sahara Desert habitat from Mauritania into the Middle East. The type locality is Algeria.

The body is light brown above with a pattern of dark oval blotches along the dorsal midline. The pattern on the dorsum of the head includes a transverse dark band between the eyes that is connected to a dark ring (or "diadem") at the level of the parietals. The venter is pale, without markings. The head is narrow with an enlarged

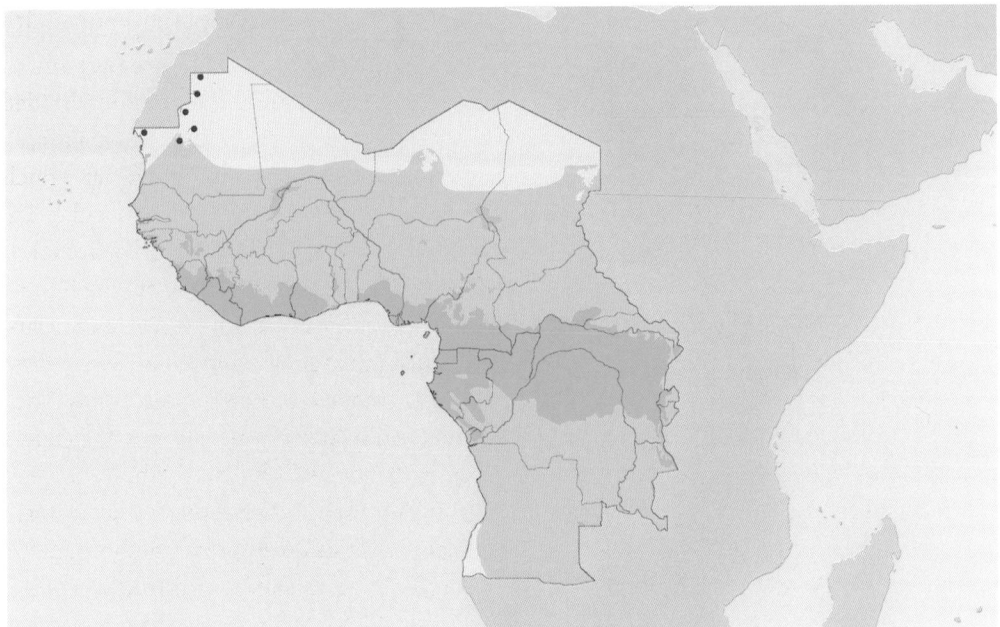

FIGURE 13.47. *Hemorrhois algirus* (red). By K. Jackson

"leaf-shaped" rostral scale. The body is long and the tail is slender. The neck is indistinct. The eye is medium sized with a round pupil. The nasal is divided. The internasals and prefrontals are paired. The loreal is small, equal in length and depth. There are usually 2 preoculars, the superior larger than the inferior. The supraocular is equal in breadth to the frontal. There are 2 postoculars, of which the superior is the larger. There are usually no suboculars but sometimes there is 1, sometimes interpreted as a preocular. There are 7(4) to 8(3,4) or 9(5,6) upper labials. The temporal formula is usually 1+2 or 2+2. There are usually 8(4) lower labials. There is 1 pair of chin shields followed by a large pair of false chin shields. The submandibular groove is pronounced. The dorsal scales are smooth, without apical pits, and arranged in 19 straight rows. The vertebral row is not enlarged relative to the other dorsal scale rows. There are 155–195 ventrals and 33–47 divided subcaudals.

The cloacal scale is divided. The maximum length recorded for this species is 450 mm (Villiers 1975). The maxillary dentition is aglyph with 6–14 teeth greatly increasing in size posteriorly. The hemipenes are unforked, and the sulcus spermaticus is not forked.

Smooth Snakes: Genus *Meizodon* Fischer 1856

The Smooth Snakes, *Meizodon*, are a genus of five species found in sub-Saharan Africa, three of which occur within our zone. *Meizodon* are terrestrial, diurnal, egg-laying snakes. Stomach contents of *Meizodon coronatus* from rainforest habitat in southern Nigeria were found to consist exclusively of small lizards (both diurnal ones such as *Trachylepis* and juvenile *Agama* and nocturnal ones such as geckoes). Feces from the same population contained arthropod parts, but these were most likely ingested by lizards

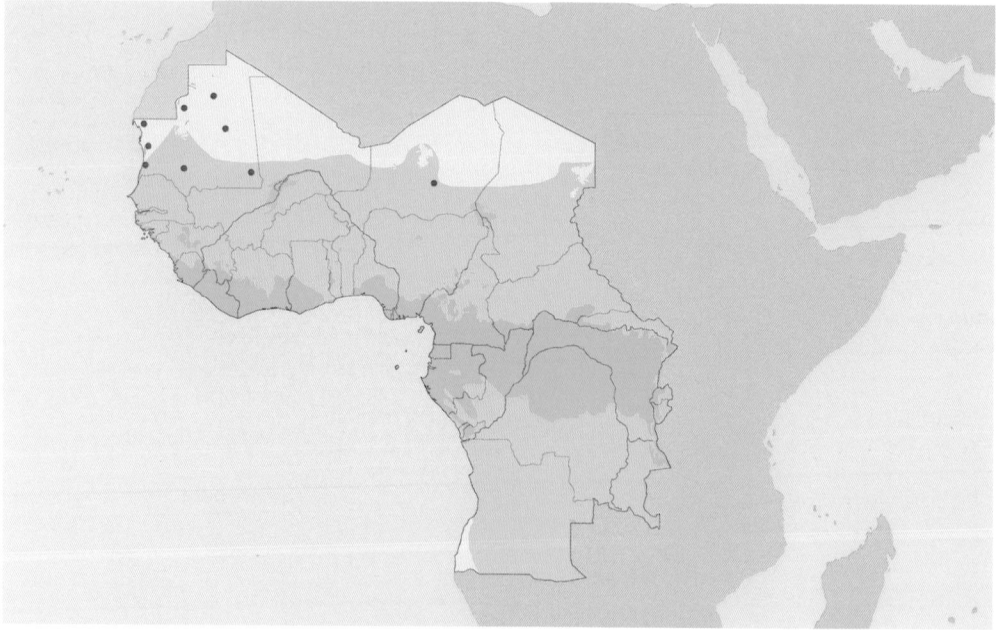

FIGURE 13.48. *Lytorhynchus diadema* (red). By K. Jackson

FIGURE 13.49. *Lytorhynchus diadema*, Egypt. By S. Spawls

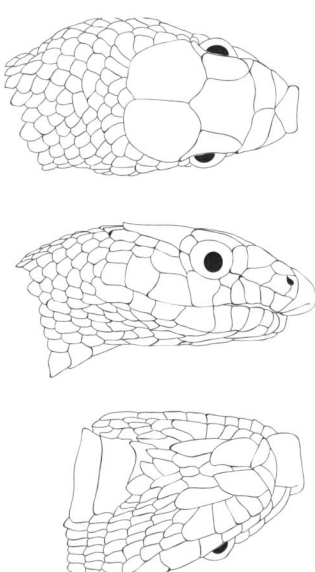

FIGURE 13.50. *Lytorhynchus diadema* RMCA 28572. By T. Giri

and then secondarily ingested by the snakes (Luiselli et al. 2001b).

The head is small and the neck not well defined. The eye is medium or small with a round pupil. The maxillary dentition is aglyph with 17–20 teeth, increasing in size posteriorly. The two posteriormost teeth are the largest and are separated from the rest by a diastema.

The rostral is rounded. The nasal may be single or divided. The internasals and prefrontals are paired. The loreal is present. There are 1–2 preoculars and 2 postoculars. The frontal is longer than broad and is separated from the eye by a supraocular. There are 1–2 anterior temporals. There are 7–9 upper labials, 2 of which are in contact with the eye. There is 1 pair of chin shields and

Key to Central and Western African Species of the Genus *Meizodon*

1   21 dorsal scale rows midbody .... *M. semiornatus*
1' 19 dorsal scale rows midbody 2

2(1) Head scales and dorsal scales have pale edges .... *M. coronatus*
2' Color of each scale uniform .... *M. regularis*

a pair of gulars acting as false chin shields. There are 8–10 lower labials. The dorsal scales are smooth, usually with 1 apical pit, and arranged in 19 or 21 straight rows. The vertebral row is not enlarged relative to the other dorsal scale rows. There are 162–225 ventrals and 60–98 divided subcaudals. The cloacal scale is divided. The hemipenes are unforked and the sulcus spermaticus is not forked.

Western Crowned Smooth Snake: *Meizodon coronatus* (Schlegel 1837)

Meizodon coronatus is a savanna species whose range extends from Senegal to Chad. The type locality is Ghana.

The dorsum is gray with whitish blotches formed by two pearly white lateral dots on each dorsal scale. There are 4 dark bands across the head, the first at the level of the border between the internasals and the prefrontals, the second between the eyes, the third across the parietals, and the fourth, a nuchal band, across the back of the neck. The nuchal band is sometimes incomplete, forming instead 2 lateral spots. The pattern fades as the snake ages.

The nasal is divided. The loreal is small, equal in depth and length. There is 1 preocular. The supraocular is narrow. There are 2 postoculars, the superior larger than the inferior. There are 8(4,5), sometimes 7(4,5), upper labials. The temporal formula is 1+2 or 3. There are usually 9(5), sometimes 9(4), lower labials. The dorsal scales are smooth, usually with 1 but sometimes without apical pits, and arranged in 19 straight rows. There are 162–190 ventrals (fewer than 184 in males, more than 174 in females) and 61–75 subcaudals. The maximum length recorded for this species is 650 mm (Villiers 1975).

Eastern Crowned Smooth Snake: *Meizodon regularis* Fischer 1856

The range of *Meizodon regularis* extends from Guinea-Bissau to Uganda. The type locality is Peki, Ghana.

The dorsum is dark gray with, in some individuals, a pattern of faint pale rings. There is an incomplete pale ring present across the throat (but absent at the nape). The upper labials are paler than the top of

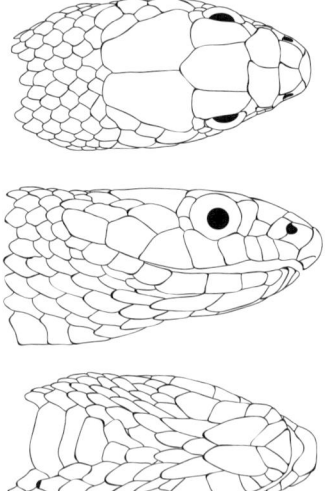

FIGURE 13.51. *Meizodon coronatus* RMCA 29718. By T. Giri

FIGURE 13.52. *Meizodon coronatus*, Cameroon. By M. LeBreton

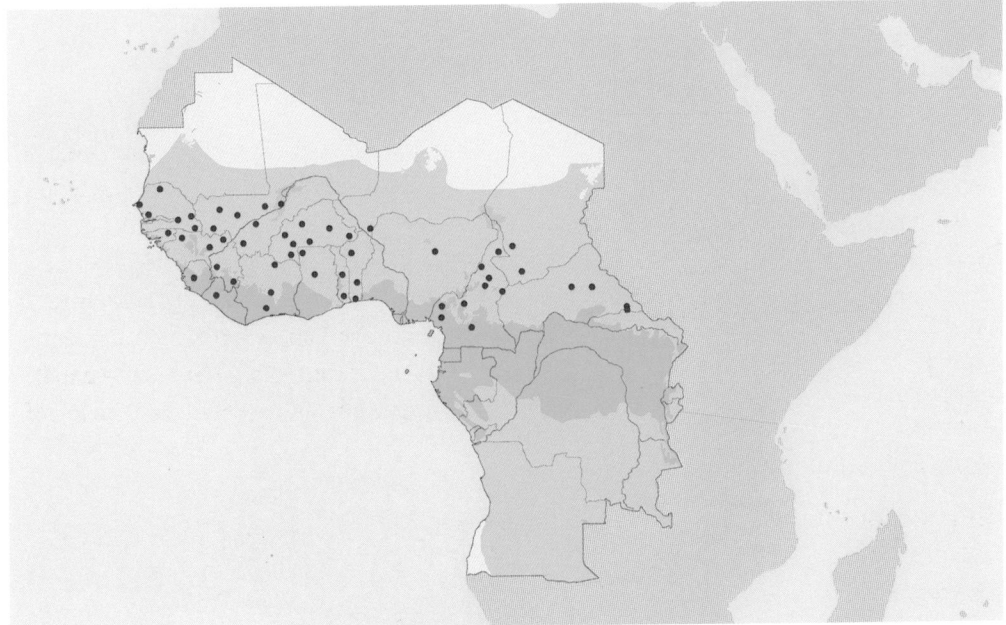

FIGURE 13.53. *Meizodon coronatus* (blue). By K. Jackson

FIGURE 13.54. *Meizodon regularis*, Cameroon. By M. LeBreton

the head, which is dark. In juveniles, there is a thin yellowish band across the head at the level of the preoculars and at the level of the postoculars. The venter is whitish. The nasal is divided. The loreal is small, equal in depth and length. There is 1 preocular. The supraocular is slightly narrower than the frontal. There are 2 postoculars, the superior larger than the inferior. There are 8(4,5) upper labials. The temporal formula is 1+2 or 3. There are usually 9(4) or 9(5) lower labials. The dorsal scales are smooth, with 1 apical pit, and arranged in 19 straight rows. There are 175–205 ventrals and 60–74 subcaudals. The maximum length recorded for this species is 706 mm (Fischer 1856).

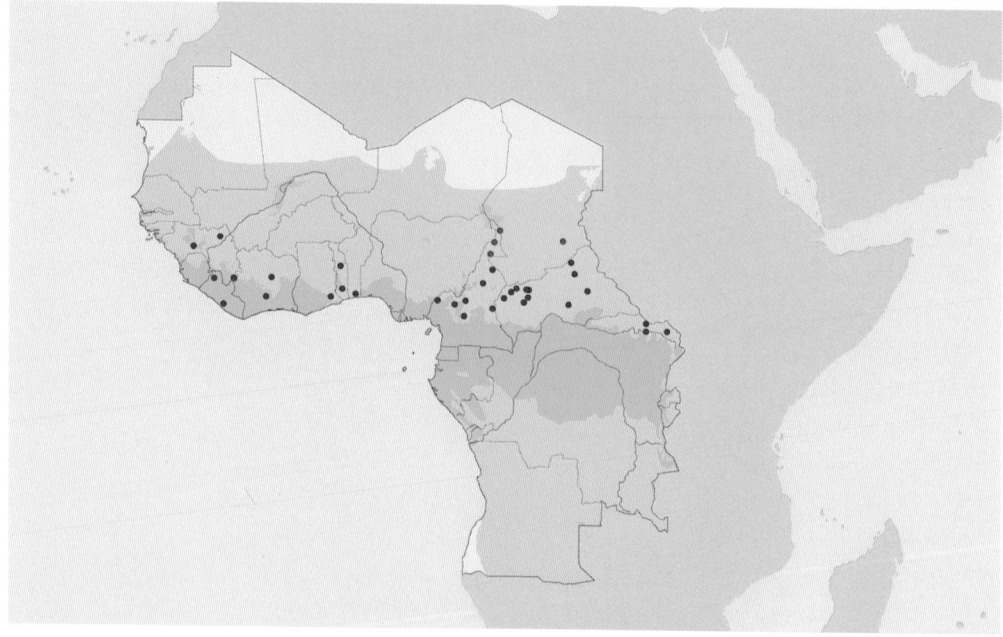

FIGURE 13.55. *Meizodon regularis* (blue), *Meizodon semiornatus* (red). By K. Jackson

Semiornate Smooth Snake: *Meizodon semiornatus* (Peters 1854)

The range of *Meizodon semiornatus* extends from Chad east to Somalia and south to South Africa. The type locality is Tete, Mozambique.

The dorsum is brown or olive green with irregularly spaced dark bands, which are most prominent at the anterior end of the body and gradually dissipate posteriorly. The top of the head is dark as far back as the posterior edge of the parietals. The occipital scales are pale. The throat is whitish, and the rest of the venter is gray. The nasal is divided. The loreal is small, equal in depth and length. There is usually 1 preocular, deeper than long, but sometimes 2, in which case the superior is always larger than the inferior. The supraocular is slightly narrower than the frontal. There are 2 postoculars, the superior larger than the inferior. There are 8(4,5) upper labials. The temporal for-mula is usually 2+3. There are usually 9(4) or 9(5) lower labials. The dorsal scales are smooth, with 1 apical pit, and arranged in 21 straight rows. There are 167–225 ven-trals and 66–98 subcaudals. The maximum length recorded for this species is 575 mm (Broadley and Cock 1975).

FIGURE 13.56. *Meizodon semiornatus*, brown phase, Tanzania. By S. Spawls

Bush Snakes: Genus *Philothamnus* Smith 1840

The Bush Snakes, *Philothamnus*, are a genus of 19 species of diurnal, mostly arboreal snake found throughout sub-Saharan Africa. Fifteen species occur within our zone. Bush Snakes are slender snakes, rarely over a meter in length, and usually green. They are completely harmless but often mistaken for green mambas (*Dendroaspis jamesoni*, *D. viridis*). *Philothamnus* lay 3–16 eggs, with some species known to lay eggs in communal nests. Most *Philothamnus* feed primarily on amphibians, but lizards, fishes, and nestling birds are also taken by some species. *Philothamnus* are found in trees and bushes, in forest and savanna habitats. They will descend to the ground and some species are most often found near water.

The body is slender with a long tail. The head is long and narrow with a well-defined neck. The eye is medium to large with a round pupil. The maxillary dentition is aglyph, consisting of 20–25 teeth increasing slightly in size toward the posterior end of the maxilla (Boulenger 1894). The genus was reviewed by Hughes (1985), who makes use of maxillary tooth position counts as a taxonomic character. Snake teeth are continuously shed and replaced, so counting the tooth position (regardless of whether tooth is present) is a more stable character than counting the number of teeth present (a number that varies) at time the snake was collected. But maxillary tooth position counts presented for the genus *Philothamnus* range from 17 to 48, possibly reflecting inconsistent methods of counting teeth used by different authors. The hemipenes are unforked, and the sulcus spermaticus is undivided.

The nasal is divided. The internasals and prefrontals are paired. The loreal is present. There are 1–2 preoculars and 2–3 postoculars. The frontal is longer than broad. The temporal formula is (1-2)+(1-2)+(0-2). There are 7–10 upper labials, of which 2 or 3 are in contact with the eye. There are 2 pairs of chin shields. The submandibular groove is pronounced. The anterior chin shields are separated from the mental scale by the first pair of lower labials. There are 8–11 lower labials. The dorsal scales are smooth, without apical pits, and arranged in 13–15 rows at midbody. The arrangement of the dorsal scale rows is oblique toward the anterior end of the body, and *Philothamnus* are usually described as having oblique dorsal scale rows, but they straighten out toward the posterior end of the body. The vertebral row is not enlarged relative to the other dorsal scale rows. The ventrals are keeled in most, but not all, species. The subcaudals are divided and are keeled in some species. The number of ventrals and subcaudals is sexually dimorphic in some species and not in others. The cloacal scale may be single or divided.

Western Bush Snake: *Philothamnus angolensis* Bocage 1882

The range of *Philothamnus angolensis* extends from Cameroon to Tanzania to Botswana. The type locality is Capangombe, Angola.

The dorsum is emerald to olive green with irregular dark blotches or bands on the anterior part of the body. There are 23–31 maxillary tooth sockets. The internasals are as long as and narrower than the prefrontals. The length of the loreal is 1.5 times its depth. There is 1 preocular, which is not in contact with the frontal. The supraocular

Key to Central and Western African Species of the Genus *Philothamnus*

1   Cloacal scale single 2
1' Cloacal scale divided 4

2(1)   More than 163 (male) or 169 (female) ventrals .... *P. ruandae*
2'   Fewer than 166 ventrals 3

3(2)   Anterior chin shields clearly longer than posterior pair; usually 13 dorsal
       scale rows .... *P. carinatus*
3'   Anterior chin shields equal to or shorter than posterior pair; usually 15
       dorsal scale rows .... *P. heterodermus*

4(1)   1 anterior temporal 5
4'   2 anterior temporals 15

5(4)   2 upper labials in contact with the eye 6
5'   3 upper labials in contact with the eye 8

6(5)   Preocular in contact with the frontal .... *P. hughesi*
6'   Preocular not in contact with the frontal 7

7(6)   Anterior and posterior chin shields of approximately equal length .... *P. battersbyi*
7'   Anterior chin shields distinctly shorter than the posterior pair .... *P. hoplogaster*

8(5)   More than 200 ventrals .... *P. thomensis*
8'   Fewer than 200 ventrals 9

9(8)   3 rows of temporals .... *P. dorsalis*
9'   2 rows of temporals 10

is slightly narrower than the frontal. There are 2 postoculars, the superior larger than the inferior. The temporal formula is 1+2. There are 9(4,5,6) upper labials and 9(5) or 10(6) lower labials. The anterior chin shields are shorter than the posterior pair. The dorsal scales are arranged in 15 rows. There are 143–175 usually rounded but sometimes keeled ventrals (fewer than 168 in males, more than 147 in females), and 90–134 rounded subcaudals (fewer than 123 in females). The cloacal scale is divided. The

maximum recorded length for this species is 980 mm (Boulenger 1894).

Battersby's Bush Snake: *Philothamnus battersbyi* Loveridge 1951b

The range of *Philothamnus battersbyi* extends from Sudan to Ethiopia to Uganda, including Rwanda and Burundi. The type locality is Sipi Forest, Mt. Elgon, Uganda. This species is often found in vegetation near water. It feeds on frogs and fishes. Individuals lay clutches of 3–11 eggs in com-

10(9)	Subcaudals keeled	*P. nitidus*
10'	Subcaudals rounded	11
11(10)	2 posterior temporals	12
11'	1 posterior temporal	13
12(11)	Preocular not in contact (or barely in contact) with the frontal	*P. angolensis*
12'	Preocular broadly in contact with the frontal	*P. irregularis*
13(11)	Ventrals rounded	*P. heterolepidotus*
13'	Ventrals keeled	14
14(13)	Dorsum uniformly green except for a brown mid-vertebral stripe; length of loreal less than 1.5 times its depth	*P. ornatus*
14'	Dorsum green speckled with paler scales, no mid-vertebral stripe; length of loreal more than 2 times its depth	*P. bequaerti*
15(4)	More than 170 ventrals	*P. semivariegatus*
15'	Fewer than 170 ventrals	*P. nitidus*

munal nests. Spawls et al. (2004) reports communal nests containing more than 100 eggs of this species in Kenya.

The dorsum is pale or dark green. Some dorsal scales have a white basal spot, and others may have black edging. Black interstitial skin is visible between the scales. The chin and throat are white, and the remainder of the venter is white, yellowish, or pale green. There are 21–30 maxillary tooth sockets. The internasals are shorter and narrower than the prefrontals. The length of the loreal is 1.5 to 2 times its depth. There is 1 preocular, which is not in contact with the frontal. The supraocular is narrower than the frontal. There are 2 postoculars of roughly equal size. The temporal formula is 1+1 or 2. There are 8(4,5) or 9(5,6) upper labials and 8–11(4) lower labials. The dorsal scales are arranged in 15 rows. There are 152–176 ventrals (fewer than 175 in males,

more than 156 in females), which may be rounded or keeled. There are 88–129 rounded subcaudals (more than 98 in males, fewer than 122 in females). The cloacal scale is divided. The maximum recorded length for this species is 900 mm (Spawls et al. 2004).

Bequaert's Bush Snake: *Philothamnus bequaerti* (Schmidt 1923)

The range of *Philothamnus bequaerti* extends from Cameroon to Uganda. The type locality is Niangara, Democratic Republic of Congo.

The dorsum is emerald green to olive green, with blue or white spots on some dorsal scales. Dark blotches or bands on the anterior part of the body. The venter is pale green or yellowish. There are 19–24 maxillary tooth sockets. The internasals are shorter and narrower than the prefrontals. The length of the loreal is twice its depth. There is 1 preocular, which is not in contact with the frontal. The supraocular is slightly narrower than the frontal. There are 2 postoculars of roughly equal size. The temporal formula is 1+1. There are 9(4,5,6) upper labials and 9(5) or 10(5) lower labials. The anterior chin shields are shorter than the posterior pair. The dorsal scales are arranged in 15 rows. There are 155–179 keeled ventrals (fewer than 174 in males, more than 159 in females) and 93–123 rounded subcaudals (more than 102 in males, fewer than 114 in females). The cloacal scale is divided. The maximum recorded length for this species is 1,000 mm (Loveridge 1958).

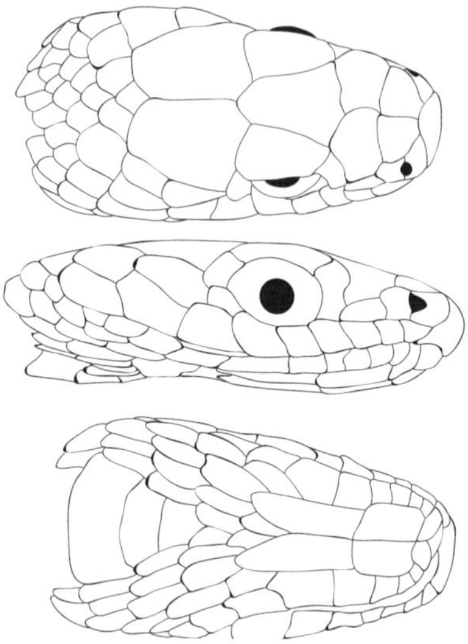

FIGURE 13.57. *Philothamnus angolensis* REDPATH 4745. By T. Giri

FIGURE 13.58. *Philothamnus angolensis*, Botswana. By S. Spawls

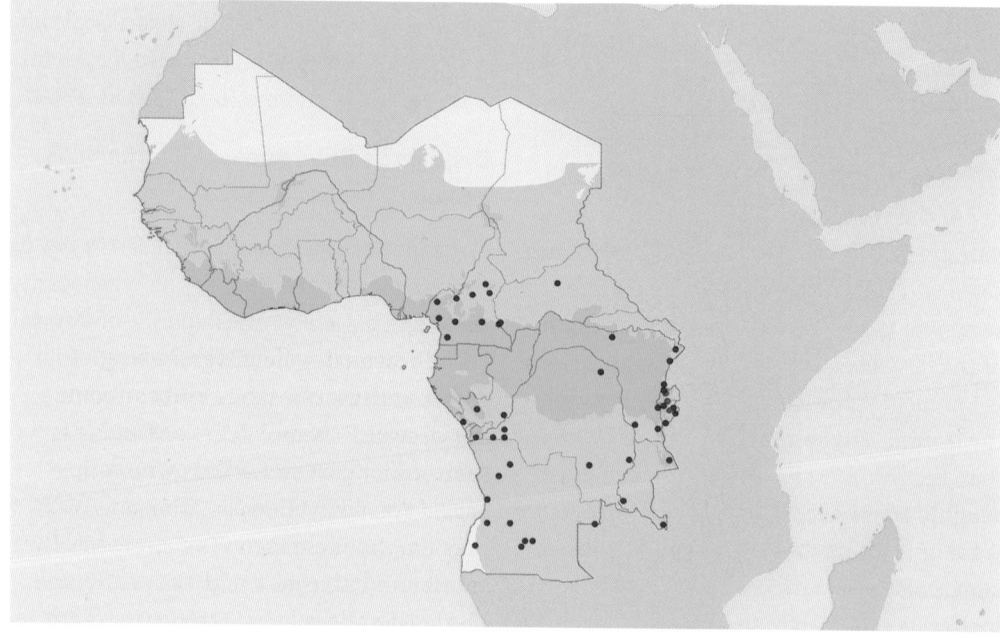

FIGURE 13.59. *Philothamnus angolensis* (blue), *Philothamnus battersbyi* (red). By K. Jackson

FIGURE 13.60. *Philotham-nus battersbyi*, Kenya. By S. Spawls

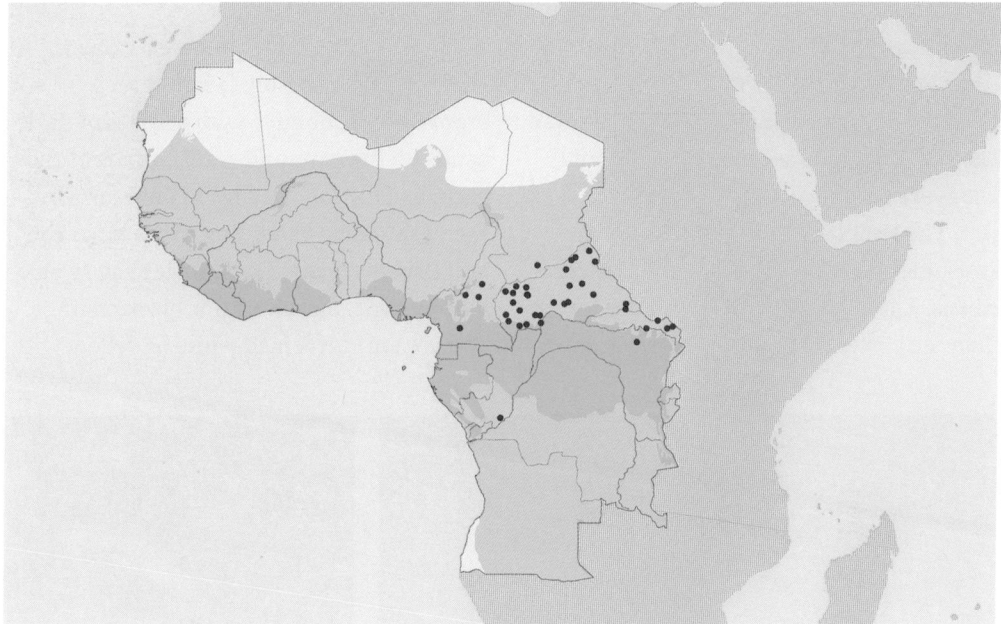

FIGURE 13.61. *Philothamnus bequaerti* (blue). By K. Jackson

Thirteen-scaled Bush Snake: *Philothamnus carinatus* (Andersson 1901)

The range of *Philothamnus carinatus* extends from Guinea east to Kenya and south to the Democratic Republic of Congo. The type locality is Mpanja, Cameroon. It is primarily a forest species.

The Thirteen-scaled Bush Snake gets its common name from the number of dorsal scale rows midbody, which is 13, in contrast to the usual 15 dorsal scale rows of other *Philothamnus*. Hughes (1985), however, noted that this character is geographically variable and thus not infallible. While the 13-dorsal-scale rule held true for snakes from the central African part of *P. carina-*

tus's range (Democratic Republic of Congo and Cameroon), with only rare exceptions, it did not hold true in the western part of the range. Out of thirty-seven specimens of *P. carinatus* collected in Ghana, thirty-two had 15 dorsal scale rows, four had 13, and one had 11.

The body is olive green above, sometimes with dark bands on the anterior part of the body. Like many *Philothamnus* species, *P. carinatus* will inflate the anterior part of the body when threatened, emphasizing the banded pattern. The venter is pale. There are 29–48 maxillary tooth sockets. The internasals are longer and narrower than the prefrontals. The length of the loreal is 1.5 to 2 times its depth. There is 1 preocular, which is not in contact with the frontal. The supraocular is as broad as the frontal. There are 2 postoculars of roughly equal size. The temporal formula is 2+2. There are 9(4,5,6) upper labials and 10(5) lower labials. The anterior chin shields are longer than the posterior pair. The dorsal scales

are arranged in 13 rows (15 in some parts of the species' range; see above). There are 138–166 keeled ventrals (fewer than 158 in males, more than 144 in females) and 67–110 rounded subcaudals (more than 75 in males, fewer than 92 in females). The cloacal scale is single. The maximum recorded length for this species is 815 mm (Loveridge 1958).

Stripe-backed Bush Snake: *Philothamnus dorsalis* (Bocage 1866b)

The range of *Philothamnus dorsalis* extends from Gabon to Angola to eastern Democratic Republic of Congo. The type locality is Molembo, Loango, Angola.

The snout is usually yellowish brown. The rest of the dorsum is light green, bronze green, or olive with a series of dark bands on the nape and anterior part of the back, usually followed by an olive-brown vertebral stripe extending to the tip of the tail. Usually, many dorsal scales have a pale blue or white streak and/or black edging. The venter is greenish white or yellow with

FIGURE 13.63. *Philothamnus carinatus*, Republic of Congo. By K. Jackson

FIGURE 13.62. *Philothamnus carinatus*, Democratic Republic of Congo. By E. Greenbaum

FIGURE 13.64. *Philothamnus carinatus,*
Republic of Congo. By M. Burger

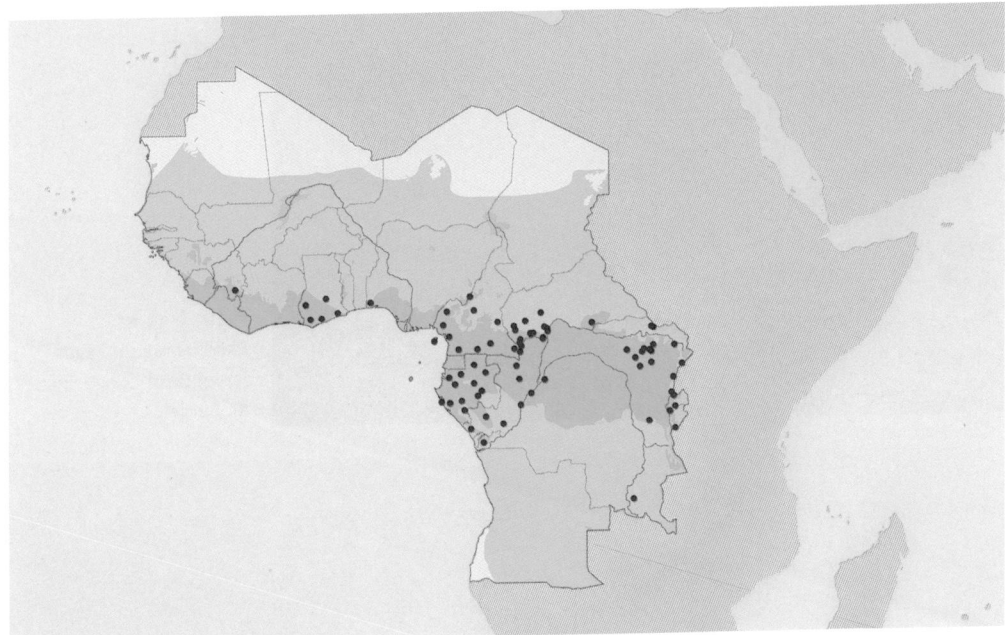

FIGURE 13.65. *Philothamnus carinatus* (blue). By K. Jackson

a brownish line along the ventral keels. The labials, chin, and throat are yellow or white. There are 20–27 maxillary tooth sockets. The internasals are longer and narrower than the prefrontals. The length of the loreal is 1.5 to 2 times its depth. There is 1 preocular, which is not in contact with the frontal. The supraocular is narrower than the frontal. There are 2 postoculars, the superior larger than or equal to the inferior. The temporal formula is 1+1+1. There are usually 9(4,5,6) upper labials and 9(5) lower labials. The anterior chin shields are shorter than the posterior pair. The dorsal scales are arranged in 15 rows. There are 167–186 keeled ventrals, without sexual dimorphism. There are 112–147 subcaudals (more than 119 in males, fewer than 136 in females), usually rounded or angulated, sometimes keeled. The cloacal scale is divided. The maximum recorded length for this species is 890 mm (Loveridge 1958).

Gabon Bush Snake: *Philothamnus heterodermus* (Hallowell 1857)

The range of *Philothamnus heterodermus* extends from Guinea-Bissau to Angola to Tanzania. The type locality is Gabon.

The dorsum is olive green, with or without dark bands or blotches. The venter is pale green. There are 26–40 maxillary tooth sockets. The internasals are slightly shorter and narrower than the prefrontals. The length of the loreal is twice its depth. There is 1 preocular, which is not in contact with the frontal. The supraocular is slightly narrower than the frontal. There are 2 postoculars, the superior larger

FIGURE 13.66. *Philothamnus dorsalis*, Republic of Congo. By M. Burger

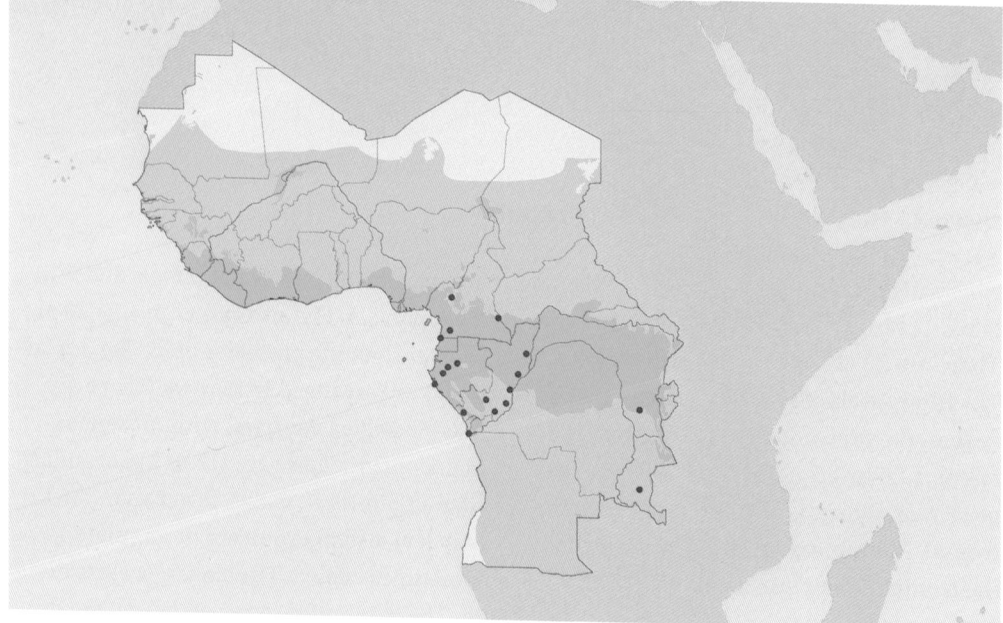

FIGURE 13.67. *Philothamnus dorsalis* (red). By K. Jackson

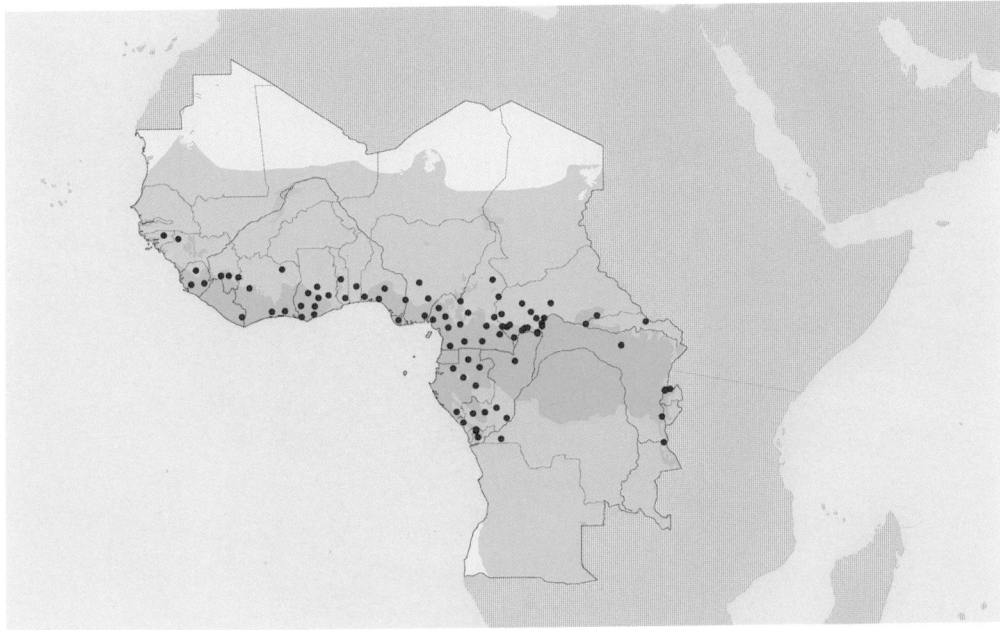

FIGURE 13.68. *Philothamnus heterodermus* (blue). By K. Jackson

than the inferior. The temporal formula is 2+2. There are usually 9(4,5,6), occasionally 8(4,5,6), upper labials and usually 9(5) lower labials. The anterior chin shields are shorter than the posterior pair. The dorsal scales are arranged in 13–15 rows. There are 141–166 keeled ventrals (fewer than 166 in males, more than 142 in females) and 71–105 rounded subcaudals (more than 78 in males, fewer than 95 in females). The cloacal scale is single. The maximum recorded length for this species is 900 mm (Villiers 1975).

FIGURE 13.69. *Philothamnus heterodermus*, Cameroon. By V. Gvoždík

Slender Bush Snake: *Philothamnus heterolepidotus* (Günther 1863b)

The range of *Philothamnus heterolepidotus* extends from Sierra Leone to Sudan and Tanzania and to Angola and Botswana. The type locality is Africa.

The dorsum is dark green or bronze green, with black interstitial skin, and no white basal dot on the dorsal scales. The chin and throat are white, and the rest of the venter is greenish white, yellowish, or pale green. There are 23–33 maxillary tooth sockets. The internasals are slightly shorter and narrower than the prefrontals. The length of the loreal is twice its depth. There is 1 preocular, which is not in contact with the frontal. The supraocular is distinctly narrower than the frontal. There are 2 postoculars, the superior larger than the inferior. The temporal formula is 1+1. There

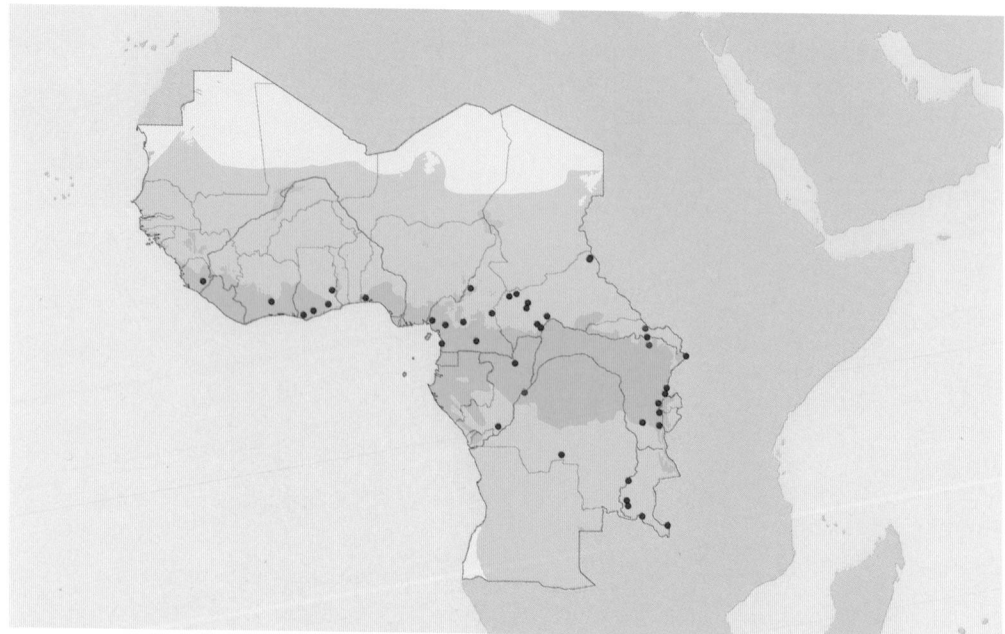

FIGURE 13.70. *Philothamnus heterolepidotus* (blue), *Philothamnus hoplogaster* (red). By K. Jackson

are usually 9(4,5,6), occasionally 8(4,5,6), upper labials and usually 9(5) lower labials. The anterior chin shields are much shorter than the posterior pair. The dorsal scales are arranged in 15 rows. There are 168–194 rounded or angulate ventrals, without sexual dimorphism, and 101–144 rounded subcaudals (more than 106 in males, fewer than 138 in females). The cloacal scale is divided. The maximum recorded length for this species is 800 mm (Villiers 1975).

Green Water Snake: *Philothamnus hoplogaster* (Günther 1863b)

The range of *Philothamnus hoplogaster* extends from southern Cameroon east to Kenya and south to South Africa. The type locality is "Port Natal," for Durban, South Africa. *P. hoplogaster* is a species often found near water. Its diet consists of frogs and fishes.

The dorsum is pale or dark green with black interstitial skin visible between the scales. Some dorsal scales may have black edging, and some may have a white basal spot. The venter is white, yellowish, or pale green, the chin and throat being white. There are 21–33 maxillary tooth sockets. The internasals are shorter and narrower than the prefrontals. The length of the loreal is 1.5 times its depth. There is 1 preocular, which is not in contact with the frontal. The supraocular is narrower than the frontal. There are 2 postoculars, the superior larg-

FIGURE 13.71. *Philothamnus hoplogaster*, Kenya. By S. Spawls

er than the inferior. The temporal formula is 1+1 or 2. There are 8(4,5) upper labials and 10(5) lower labials. The anterior chin shields are shorter than the posterior pair. The dorsal scales are arranged in 15 rows. There are 138–167 ventrals (fewer than 164 in males, more than 143 in females), usually rounded but sometimes keeled. There are 60–118 rounded subcaudals (more than 72 in males, fewer than 103 in females). The cloacal scale is divided. The maximum recorded length for this species is 960 mm (Spawls et al. 2004).

Hughes's Bush Snake: *Philothamnus hughesi* Trape and Roux-Estève 1990

The range of *Philothamnus hughesi* extends from Cameroon to the Democratic Republic of Congo. The type locality is Gangalingolo (near Brazzaville), Congo. Nothing is known of its natural history.

The dorsum is uniformly bright leaf green. The venter is pale green. Trape and Roux-Estève (1990) do not provide the number of maxillary tooth sockets in their description of the species. The internasals are shorter and narrower than the prefrontals. The length of the loreal is 1.5 times its depth. There is 1 preocular, which is in contact with the frontal. The supraocular is narrower than the frontal. There are 2 postoculars of roughly equal size. The temporal formula is 1+1. There are 8(4,5), occasionally 7(3,4), upper labials and 10(5), occasionally 9(5) or 11(5), lower labials. The anterior chin shields are shorter than the posterior pair. The dorsal scales are arranged in 13–15 rows. There are 152–165 ventrals (fewer than 163 in males, more than 154 in females), which are usually rounded but sometimes angulated or keeled. There are 93–105 rounded subcaudals. The cloacal scale is divided. The maximum recorded length for this species is 933 mm (Trape and Roux-Estève 1990).

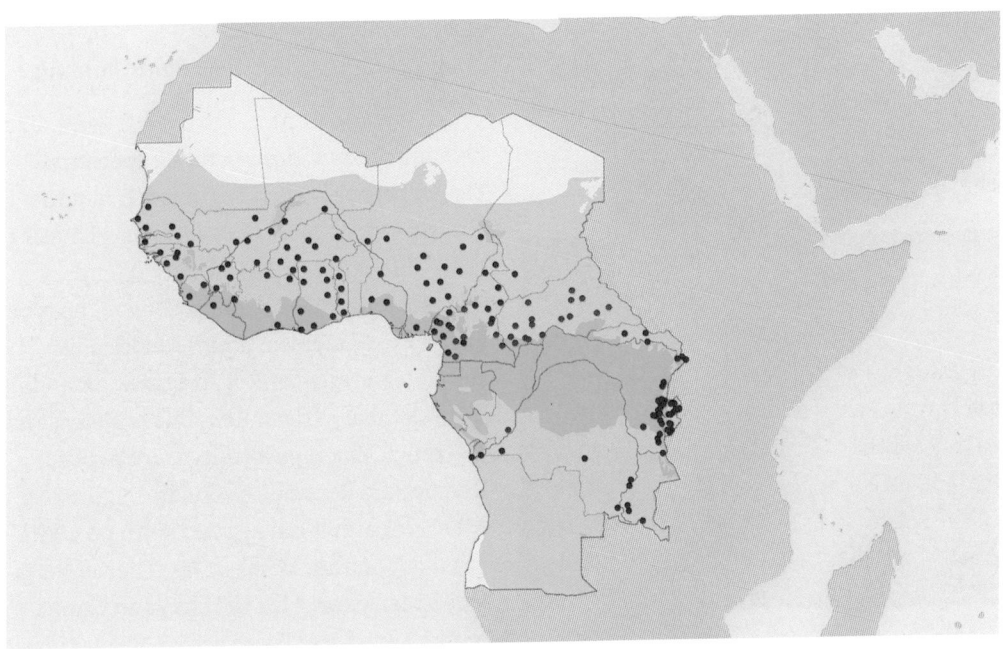

FIGURE 13.72. *Philothamnus hughesi* (red), *Philothamnus irregularis* (blue). By K. Jackson

Irregular Green Bush Snake: *Philothamnus irregularis* (Leach 1819)

The range of *Philothamnus irregularis* extends from Senegal to Chad and Central African Republic. The type locality is Fantee, Ghana.

P. irregularis is distinctive as the only member of the genus in which the buccal epithelium (lining of the inside of the mouth) is black. This is associated with a threat display in which the mouth is opened to display its black lining. The dorsum is green or olive with black interstitial skin. On the nape and the anterior part of the body there is occasionally a series of paired black spots, which may coalesce to form bands. Some dorsal scales may have a white basal spot and/or black edging. The venter is greenish yellow or pale green. There are 18-24 maxillary tooth sockets. The internasals are shorter and narrower than the prefrontals. The length of the loreal is 1.5 times its depth. There is 1 preocular, which is in contact with the frontal. The supraocular is slightly narrower than the frontal. There are 2 postoculars, the superior larger than the inferior. The temporal formula is 1+2. There are 9(4,5,6) upper labials and 8(4) to

FIGURE 13.74. *Philothamnus irregularis* threat display, Ghana. By S. Spawls

11(6) lower labials. The anterior and posterior pairs of chin shields are of roughly equal length. The dorsal scales are arranged in 15 rows. There are 158-186 keeled ventrals (fewer than 177 in males, more than 158 in females) and 93-145 rounded subcaudals (more than 113 in males, fewer than 128 in females). The cloacal scale is divided. The maximum recorded length for this species is 1,146 mm (Broadley and Cock 1975).

Cameroons Bush Snake: *Philothamnus nitidus* (Günther 1863b)

Philothamnus nitidus is a forest species. Two subspecies are recognized: *P. n. nitidus* (Günther 1863b); *P. n. loveridgei* Laurent 1960. The range of *P. n. nitidus* extends from Sierra Leone and Guinea to Congo. The type locality is Demerara, South America (in error). The range of *P. n. loveridgei* extends from Central African Republic to Kenya and Tanzania. The type locality is Itula, Kivu, Democratic Republic of Congo.

The dorsum is dark green, with no white basal dot on the dorsal scales. The venter is pale green except for the chin and throat, which are white. There are 23-33 maxillary tooth sockets. The internasals are longer

FIGURE 13.73. *Philothamnus irregularis*, Guinea. By P. Naskrecki

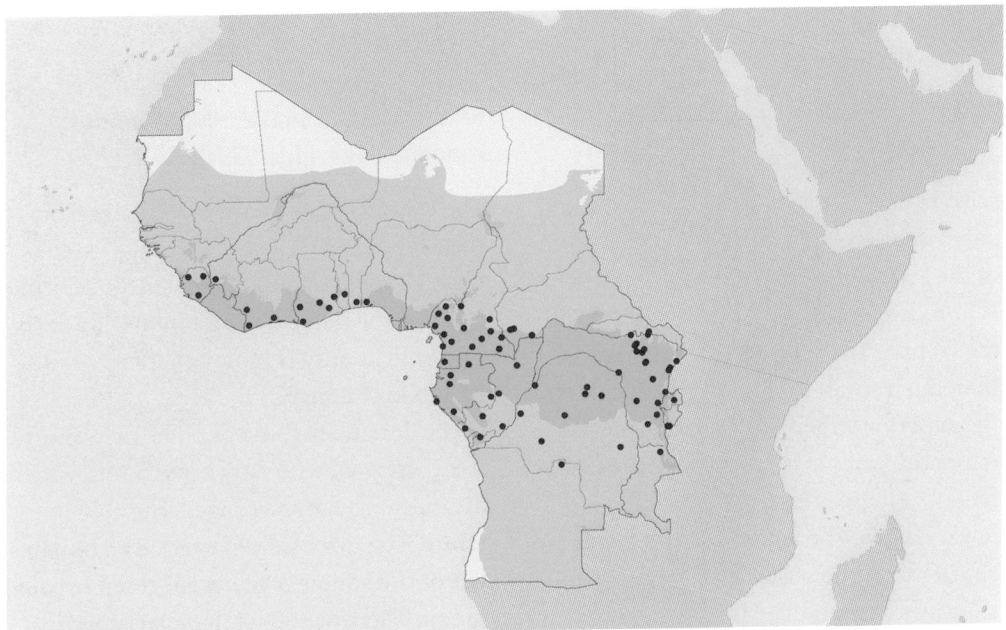

FIGURE 13.75. *Philothamnus nitidus* (blue). By K. Jackson

FIGURE 13.76. *Philothamnus nitidus*, Cameroon. By M. LeBreton

and narrower than the prefrontals. The length of the loreal is twice its depth. There is 1 preocular, which is in contact with the frontal. The supraocular is slightly narrower than the frontal. There are 2 postoculars, the superior larger than the inferior in *P. n.*

nitidus and equal to it in size in *P. n. loveridgei*. The temporal formula is 2+2 in *P. n. nitidus* (1+2 in *P. n. loveridgei*). There are usually 9(4,5,6) upper labials, occasionally 10(5,6,7), and 9(5) lower labials. The anterior chin shields are shorter than the posterior pair. The dorsal scales are arranged in 15 rows. There are 144–176 keeled ventrals (fewer than 175 in males, more than 149 in females) and 126–161 keeled subcaudals (more than 125 in males, fewer than 155 in females). The cloacal scale is divided. The maximum recorded length for this species is 986 mm (Loveridge 1958).

Ornate Bush Snake: *Philothamnus ornatus* Bocage 1872

The range of *Philothamnus ornatus* extends from Cameroon to Angola and Botswana. The type locality is Huilla, Angola.

The dorsum is iridescent green to bronze green, with black interstitial skin, a few scattered black spots on the nape, and a broad, usually yellow-edged brown vertebral stripe extending from the head to the tip of the tail. A few scales may be flecked with white. The body is yellowish cream underneath, and the underside of the tail is greenish. There are 24–31 maxillary tooth sockets. The internasals are as long as and narrower than the prefrontals. The length of the loreal is 1.5 to 2 times its depth. There is 1 preocular, which is not in contact with the frontal. The supraocular is narrower than the frontal. There are 2 postoculars, the superior larger than the inferior. The temporal formula is 1+1. There are 8(3,4,5) or 9(4,5,6) upper labials and usually 9(5) lower labials. The anterior chin shields are shorter than the posterior pair. The dorsal scales are arranged in 15 rows. There are 147–174 keeled ventrals (fewer than 163 in males, more than 152 in females) and 85–104 rounded subcaudals (more than 95 in males, fewer than 102 in females). The cloacal scale is divided. The maximum recorded length for this species is 710 mm (Bocage 1872).

Rwanda Bush Snake: *Philothamnus ruandae* Loveridge 1951b

The Rwanda Bush Snake is a species from the Albertine Rift Valley. Its range extends from Uganda to Rwanda and into bordering areas of the Democratic Republic of Congo. The type locality is Mulungu, Democratic Republic of Congo.

The dorsum is green or olive. In juveniles, it is patterned with dark crossbands, which disappear as the snake ages. The chin is white. The throat is yellowish. The remainder of the venter is yellowish green to pale green. There are 28–34 maxillary tooth sockets. The internasals are shorter and narrower than the prefrontals. The length of the loreal is 1.5 to 2 times its depth. There is 1 preocular, which is not in contact with the frontal. The supraocular is slightly nar-

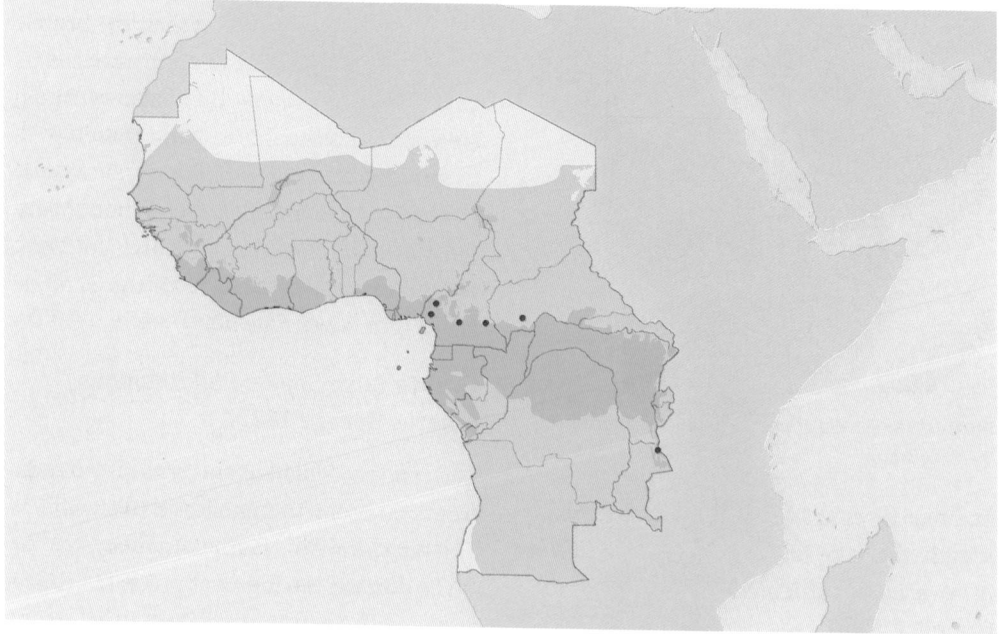

FIGURE 13.77. *Philothamnus ornatus* (red). By K. Jackson

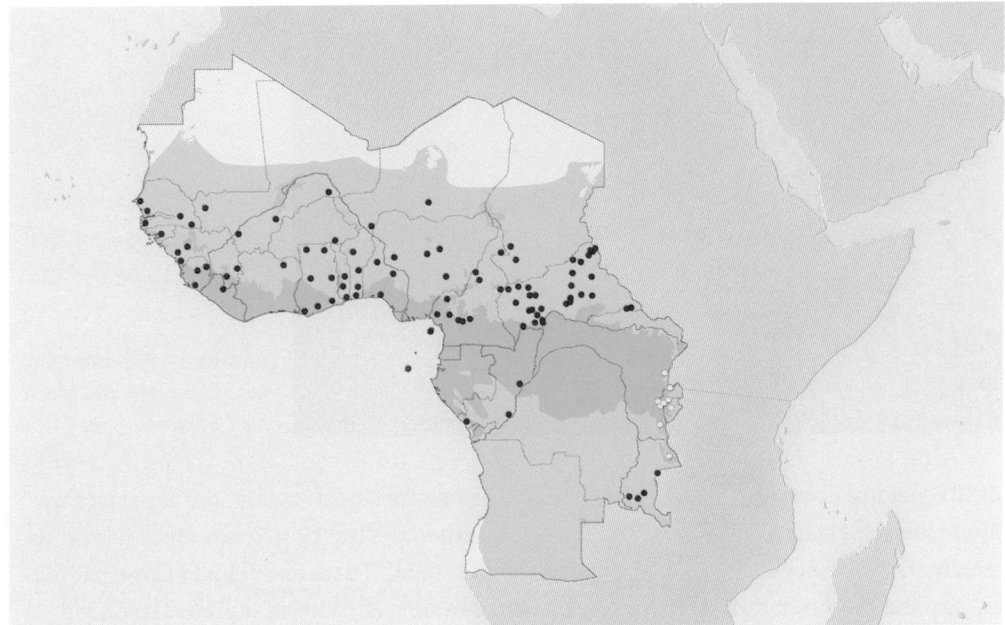

FIGURE 13.78. *Philothamnus ruandae* (yellow), *Philothamnus semivariegatus* (blue), *Philothamnus thomensis* (red). By K. Jackson

rower than the frontal. There are 2 postoculars, the superior larger than the inferior. The temporal formula is 2+2. There are 9(4,5,6) upper labials and 8(4) to 11(6) lower labials. The dorsal scales are arranged in 15 rows. There are 164–181 keeled ventrals (fewer than 173 in males, more than 169 in females) and 84–102 rounded subcaudals (more than 94 in males, fewer than 95 in females). The cloacal scale is single. The maximum recorded length for this species is 962 mm (Loveridge 1958).

Spotted Bush Snake: *Philothamnus semivariegatus* (Smith 1840)

The range of *Philothamnus semivariegatus* extends from Senegal to Ethiopia to South Africa. The type locality is Bushman Flat, Cape Province, South Africa. This species is found in drier habitats (woodland savanna to semi-desert) than other *Philothamnus*,

and its diet consists mostly of lizards rather than frogs.

The dorsum is dark green anteriorly, either uniformly or with a series of brown or black bands on the nape and anterior part of the back. Some dorsal scales usually have a turquoise blue or white streak, with or without black edging. The dorsum is paler green posteriorly, with or without black spots on flanks. The labials, chin, and throat are yellow or white. The body is yellowish green to yellow below. The underside of the tail is pale green or yellow. There are 17–25 maxillary tooth sockets. The internasals are as long as and narrower than the prefrontals. The length of the loreal is 1.5 to 2 times its depth. There is 1 preocular, which is in contact with the frontal. The supraocular is slightly narrower than the frontal. There are 2 postoculars, the superior larger than the inferior. The temporal formula is

FIGURE 13.79. *Philothamnus semivariegatus*, Zambia. By S. Spawls

2+2. There are usually 9(4,5,6), occasionally 8(4,5), upper labials and 9(5) to 10(6) lower labials. The anterior chin shields are shorter than the posterior pair. The dorsal scales are arranged in 15 rows. There are 170–209 keeled ventrals (fewer than 209 in males, more than 171 in females) and 98–166 keeled subcaudals (more than 116 in males, fewer than 162 in females). The cloacal scale is divided. The maximum recorded length for this species is 1,405 mm (Trape and Mané 2000).

São Tomé Bush Snake: *Philothamnus thomensis* Bocage 1882

Philothamnus thomensis is endemic to the island of São Tomé. The type locality is São Tomé Island, Gulf of Guinea.

The dorsum is uniformly olive green, with dark edges to the dorsal scales. The loreal region of the head is blackish. The venter is pale green, the ventral keels sometimes brown. The lips and throat are yellowish. There are 23–27 maxillary tooth sockets. The internasals are as long as and narrower than the prefrontals. The length of the loreal is twice its depth. There is 1 preocular, which is in contact with the frontal. The supraocular is narrower than the frontal.

There are 2 postoculars, the superior larger than the inferior. The temporal formula is 1+1 or 2. There are 9(4,5,6) upper labials and 9–11(6) lower labials. The dorsal scales are arranged in 15 rows. There are 202–215 keeled ventrals, which Loveridge (1958) found showed a reverse of the usual trend in sexual dimorphism, with more than 204 in males and fewer than 210 in females. There are 159–175 keeled subcaudals (more than 169 in males, fewer than 171 in females). The cloacal scale is divided. The maximum recorded length for this species is 970 mm (Bocage 1895b).

Genus *Platyceps* Blyth 1860

Like *Bamanophis*, *Hemorrhois*, *Spalerosophis*, and *Lytorhynchus*, *Platyceps* are Old World Racers, representatives of a Eurasian lineage of colubrines. Racers are slender, fast-moving snakes, active by day in desert, sahel, and dry savanna habitats. *Platyceps* is a genus of 15 species of racers found in desert habitats of northern and east Africa, parts of southern Europe, the Middle East and central Asia. Of these, one species, the Flowered Racer, *Platyceps florulentus*, occurs within our zone.

Flowered Racer, *Platyceps florulentus* (Geoffroy-Saint-Hilaire 1827)

The range of *Platyceps florulentus* extends from Egypt and Somalia in the east to northern Nigeria in the west. The type locality is Egypt. Two subspecies are recognized. Within our zone, *P. florulentus perreti* Schätti 1988 is found in dry savanna and sahel habitats from Nigeria to Cameroon and west as far as Sudan. The type locality for *P. f. perreti* is Soulédé, Mokolo, Cameroon.

The dorsum is light beige with a darker pattern of rectangular lateral blotches. The venter is uniformly ivory. The head is broad with a rounded snout and well-defined neck. The body and tail are long and slender. The eye is small with a round or slightly vertically elliptical pupil. The nasal may be single or divided. The internasals and prefrontals are paired. The loreal is present and longer than deep. There are 2 preoculars, the superior larger than the inferior. The supraocular is narrower than the frontal. There are 2 postoculars, the inferior larger than the superior. There are usually 9(5,6) upper labials. The temporal formula is 2+2 or 2+3. The lower labials range from 9(4) to 11(5). The anterior chin shields are shorter than or equal in length to the posterior pair. The submandibular groove is pronounced. The dorsal scales are smooth, with 2 apical pits, and arranged in 25 straight rows. The vertebral row is not enlarged relative to the other dorsal scale rows. There are 192–231 ventrals (219–231 for *P. f. perreti*). There are 83–108 divided subcaudals (86–95 for *P. f. perreti*). The cloacal scale is divided. The maximum length recorded for this species is 830 mm (Chirio and LeBreton 2007). The maxillary dentition is aglyph, with 16–17 anterior maxillary teeth decreasing in size posteriorly and followed by 2 larger, ungrooved posterior teeth. The hemipenes are unforked, and the sulcus spermaticus is not forked.

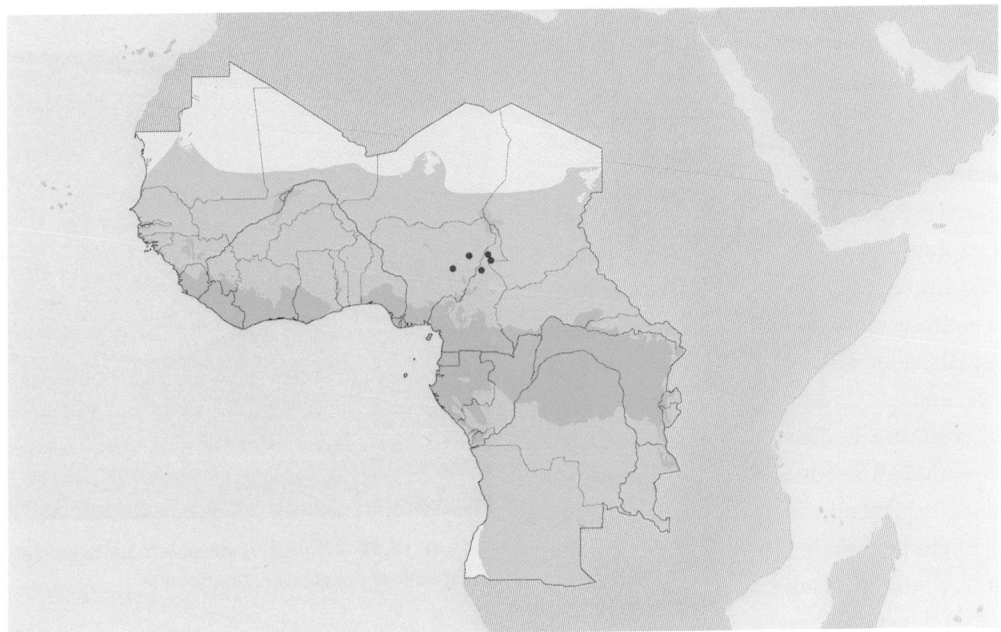

FIGURE 13.80. *Platyceps florulentus perreti* (red). By K. Jackson

Beaked Snakes: Genus *Scaphiophis* Peters 1870

The genus *Scaphiophis* contains two species, one of which occurs within our zone and the other in east Africa. *Scaphiophis* are large, stout-bodied, semi-fossorial snakes. The head is small with the rostral scale enlarged and drawn out into a pointed edge for digging, so that the snout resembles the hooked beak of a bird of prey; hence the common name "Beaked Snakes." The neck is not distinct from the body. The tail is short and slender. The eye is small with a round pupil. The diet consists primarily of rodents caught underground. The maxillary dentition is aglyph, consisting of 12–13 unspecialized maxillary teeth of approximately equal size. Beaked Snakes pose no danger to humans, but when threatened they will sometimes carry out a display involving rearing up and striking (but not biting) with the mouth open to reveal the black buccal epithelium. *Scaphiophis albopunctatus* lays clutches of up to 48 eggs.

The nasal scale is divided. The internasals and prefrontals are paired. A loreal scale is present, sometimes broken up into more than one scale. There are 1–2 preoculars, 2–3 postoculars, 1–2 suboculars, and 1 supraocular, making up 5–7 perioculars and 3 interoculars. There are usually 2 or 3 anterior temporals. There are 5–6 upper labials, none of which is in contact with the eye, and 7–10 lower labials. There is 1 pair of chin shields. The submandibular groove is shallow. The dorsal scales are smooth, with 2 apical pits, and are arranged in 19–31 straight or slightly oblique rows. The vertebral row is not enlarged relative to the other dorsal scale rows. The subcaudals

and the cloacal scale are divided. The hemipenes are unforked with an unforked sulcus spermaticus.

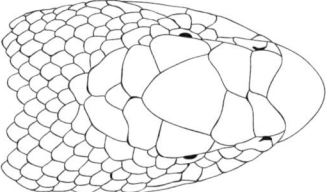

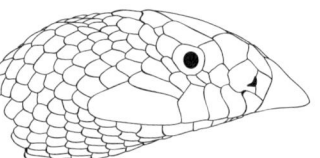

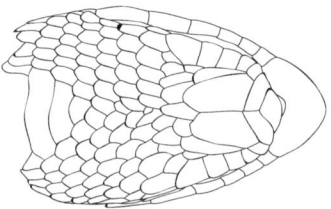

FIGURE 13.81. *Scaphiophis albopunctatus* RMCA 97-021-R-106-TOT 110. By T. Giri

FIGURE 13.82. *Scaphiophis albopunctatus*, Democratic Republic of Congo. By E. Greenbaum

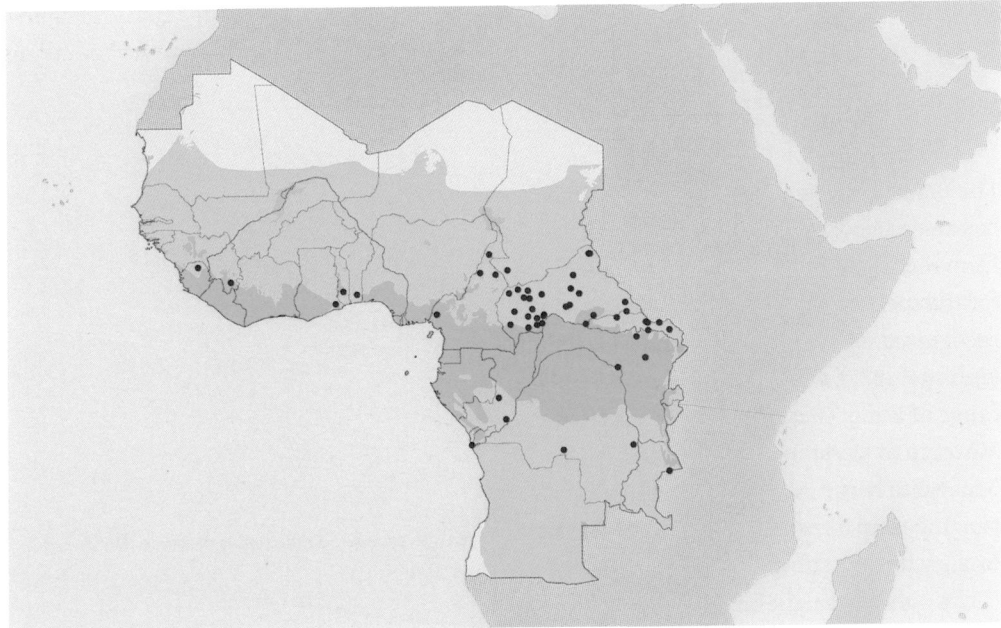

FIGURE 13.83. *Scaphiophis albopunctatus* (blue). By K. Jackson

African Beaked Snake: *Scaphiophis albopunctatus* Peters 1870

The range of *Scaphiophis albopunctatus* extends from Sierra Leone to Kenya and south as far as Zambia and northern Angola. The type locality is Keta, Ghana.

The dorsum is gray, light brown, or pinkish, sometimes with small black-and-white spots. The venter is white. A loreal scale is present and sometimes subdivided horizontally into two or three. There are usually 2 preoculars, 2 postoculars, and 1–2 suboculars. There are 5–7 (usually 6) perioculars altogether. The temporal formula usually 3+4 but sometimes 4+4 or 4+5. There are 5–6 upper labials (none in contact with the eye) and usually 9(3) lower labials. The dorsal scales are smooth, with 2 apical pits, and are arranged in 19–25 straight or slightly oblique rows. There are 170–228 ventrals (fewer than 196 in males, more than 188 in females) and 51–76 divided subcaudals (more than 52 in males, fewer than 73 in females). The numbers of dorsal scale rows and of ventrals increase from west to east across the range of the species. The cloacal scale is divided. The maximum length recorded for this species is 1,512 mm (Broadley 1994).

Genus *Spalerosophis* Jan 1865

Like *Bamanophis*, *Hemorrhois*, *Lytorhynchus*, and *Platyceps*, *Spalerosophis* are Old World Racers, representatives of a Eurasian lineage of colubrines. Racers are slender, fast-moving snakes, active by day in desert, sahel, and dry savanna habitats. *Spalerosophis* is a genus of six species found in desert habitats of northern Africa, the Middle East, and central Asia. Three species occur in Africa, one of which, *Spalerosophis diadema*, is found within our zone.

Diadem Snake: *Spalerosophis diadema* (Schlegel 1837)

The Diadem Snake, *Spalerosophis diadema*, has an enormous distribution, extending from Morocco into Asia. The type locality for the species is India. Two subspecies are recognized, one of which, *S. diadema cliffordi* (Schlegel 1837), occurs within our zone. The range of *S. diadema cliffordi* extends from Morocco to Egypt in the north, and from Senegal to Niger in the south, in habitats ranging from dry savanna to sahel. The type locality for this subspecies is Tripoli, Libya.

The dorsum is light brown, ranging from yellowish to reddish brown, with a darker pattern of narrow rhomboid-shaped spots along the vertebral midline, with smaller ones, outlined in iridescent white, along the flanks. The dorsum of the head is light brown with a marbled pattern of darker brown. The venter is white, sometimes with dark lateral spots. The head is broad with a long, rounded snout and a well-defined neck. The body and tail are long and slender. The eye is small with a round pupil. The nasal is divided. The internasals are paired. The prefrontals are divided up into 2 or 3 pairs of variable size. One pair of prefrontals is in contact with the eye. The loreal is divided into several smaller scales of variable number and shape. There are 2–4 preoculars of similar size. The supraocular is narrower than the frontal. There are 2–5 postoculars of similar size. There are 1–4 suboculars of similar size. There are generally a total of 8–12 periocular scales, all of similar size except for the supraocular, which is larger than the rest. There are usually 11 upper labials, none of which is

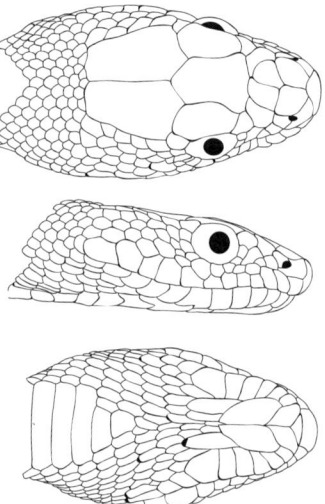

FIGURE 13.84. *Spalerosophis diadema* RMCA 2232. By T. Giri

in contact with the eye. The temporal formula may range from 2+3 to 4+5. There are usually 12(5) or 13(6) lower labials. There is 1 pair of chin shields. The submandibular groove is not pronounced. The dorsal scales are keeled, with 2 apical pits, and arranged in 25–29 straight rows. The vertebral row is not enlarged relative to the other dorsal scale rows. There are 208–248 keeled ventrals and 62–81 divided subcaudals. The cloacal scale is single. The maximum length recorded for this species is 1,800 mm (Villiers 1975). The maxillary dentition is aglyph, with 12–16 anterior maxillary teeth decreasing is size posteriorly and followed after a diastema by 2 ungrooved posterior teeth. The hemipenes are unforked, and the sulcus spermaticus is not forked.

Cat Snakes: Genus *Telescopus* Wagler 1830

A genus of 14 species found in dry habitats throughout Africa and into southeastern Europe and southwestern Asia. Five spe-

FIGURE 13.85. *Spalerosophis diadema*, Egypt. By S. Spawls

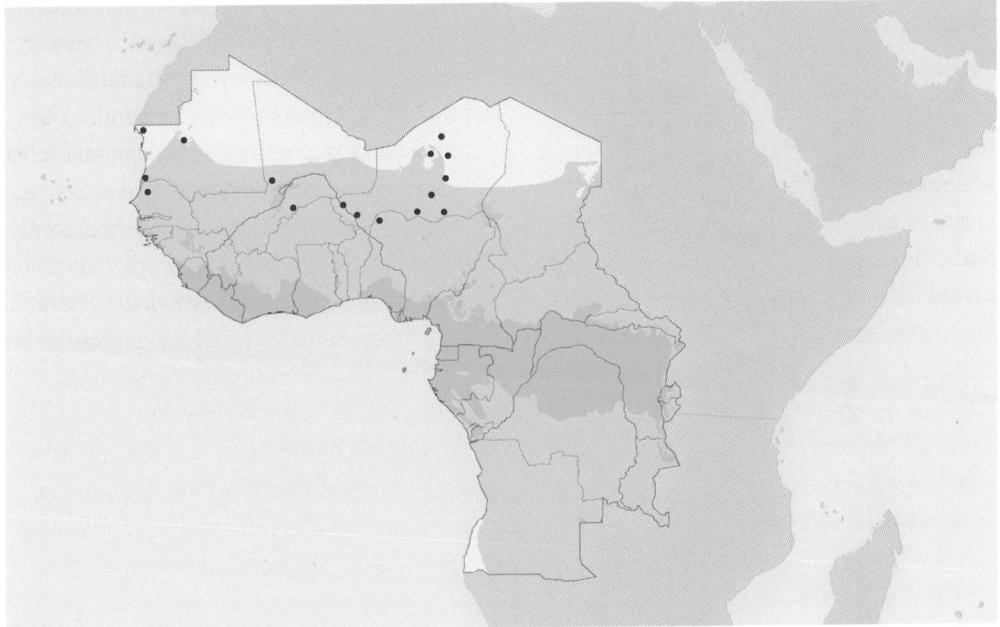

FIGURE 13.86. *Spalerosophis diadema* (blue). By K. Jackson

cies occur within our zone. The common name "Cat Snake" refers to their vertically elliptical pupils, which are similar to those of a cat. *Telescopus* are nocturnal. They are terrestrial or partially arboreal and found in dry habitats such as savanna and sahel.

They use venom, delivered by a grooved posterior fang, as well as constriction to subdue their prey but pose no threat to humans. *Telescopus* prey on lizards, especially arboreal ones such as geckoes and chameleons, as well as other prey such as rodents, birds, bats, and other snakes.

The head is broad and flat, with a well-defined neck. The snout is rounded. The eye is small or medium sized with a vertically elliptical pupil. The body is long and laterally compressed. The tail is short and slender. The rostral is rounded. The nasal is usually divided but sometimes undivided. The internasals and prefrontals are paired. The loreal is present. There is usually 1 preocular, sometimes 2. There are 2 postoculars. There are no suboculars. The frontal is longer than broad and separated from the eye by a supraocular. There are 2 anterior temporals. There are 8–11 upper labials, 2–3 of which are in contact with the eye. There are 9–12 lower labials. There is 1 pair of chin shields. The submandibular groove is pronounced. The dorsal scales are smooth, with 1 apical pit, and are arranged in 17–23 oblique rows. The vertebral row is not enlarged relative to the other dorsal scale rows. The subcaudals are divided. The cloacal scale is divided in all species within our zone except Telescopus finkeldeyi. The maxillary dentition is opisthoglyph, with 5–10 anterior maxillary teeth, decreasing in size posteriorly, followed after a diastema by 2 grooved posterior fangs. The hemipenes are unforked. The sulcus spermaticus is not forked.

Damara Cat Snake: *Telescopus finkeldeyi* Haacke 2013

Telecopus finkeldeyi is a recently described species from the Namib Desert of western and central Namibia. Its range extends into southwestern Angola, within our zone. The type locality is Rössing Uranium mine area, Swakopmund district (2214Db) Namibia.

The Damara Cat Snake differs from other *Telescopus* within our zone in having an undivided cloacal scale and a smaller maximum length. Its color pattern is quite variable. The head ranges from brick red to orange. The dorsum of the body is cream with a variable pattern of 29–45 darker orange to blackish blotches or bands. The nuchal band is usually the darkest and most distinct. The nasal is divided. There is 1 preocular and 2 postoculars. The temporal formula is usually 2+3. There are 9(3,4,5) or occasionally 8(3,4,5) upper labials. There are 12 lower labials. The dorsal scales are

---

Key to Central and Western African Species of the Genus *Telescopus*

1	Cloacal scale single	*T. finkeldeyi*
1'	Cloacal scale divided	2
2(1)	2 upper labials in contact with the eye	3
2'	3 upper labials in contact with the eye	4
3(2)	Head and body, above and below, pale, patterned with irregular dark spots	*T. variegatus*
3'	Head always dark, dorsum of body variable, and venter uniformly white	*T. tripolitanus*
4(2)	19 dorsal scale rows; fewer than 245 ventrals	*T. semiannulatus*
4'	21 dorsal scale rows; more than 240 ventrals	*T. obtusus*

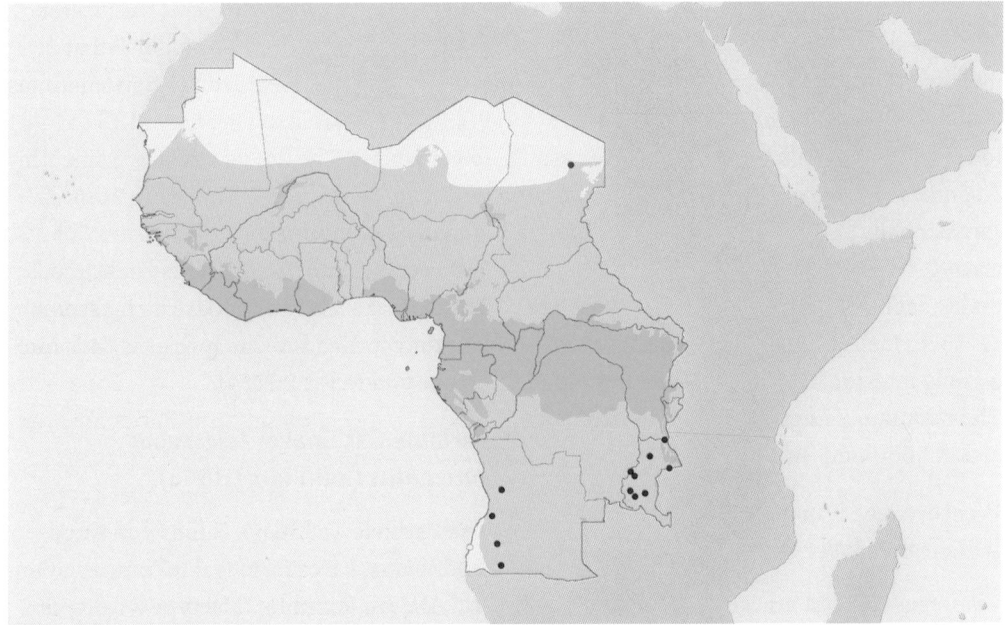

FIGURE 13.87. *Telescopus finkeldeyi* (yellow), *Telescopus obtusus* (red), *Telescopus semiannulatus* (blue).
By K. Jackson

smooth and arranged in 19 oblique rows. There are 191–221 ventrals (fewer than 211 in males, more than 200 in females). There are 48–64 divided subcaudals (more than 54 in males, fewer than 59 in females). The cloacal scale is single. The maximum length recorded for this species is 671 mm (Haacke 2013).

Large-eyed Cat Snake: *Telescopus obtusus* (Reuss 1834)

The range of *Telescopus obtusus* extends from Somalia to Egypt to Sudan, South Sudan, and Chad, without reaching the southern savanna. The type locality is Egypt.

The dorsum is brown to gray with or without a darker pattern. The venter is whitish. The nasal is divided. There is 1 preocular. There are 2 postoculars, the inferior larger than the superior. The temporal formula is usually 2+2. There are usually 10(4,5,6), sometimes 9 or 11(4,5,6), upper labials. There are 11(3) or 12(4) lower labials. The

dorsal scales are smooth, with 1 apical pit, and arranged in 23 (occasionally 21) oblique rows. There are 241–272 ventrals and 70–86 divided subcaudals. The cloacal scale is divided. The maximum length recorded for this species is 1,870 mm (Boulenger 1896).

Tiger Cat Snake: *Telescopus semiannulatus* Smith 1849

The range of *Telescopus semiannulatus* extends from southeastern Democratic Republic of Congo to southern Angola and South Africa. The presence of this species in Rwanda and Burundi is not confirmed. The type locality is South Africa.

The Tiger Cat Snake gets its common name from its coloration. The overall coloration above and below is orange. The head and venter are orange without markings, but the dorsum is patterned with 25–30 black or dark brown blotches or bands. The nasal is divided. There is 1 preocular.

There are 2 postoculars, the inferior larger than the superior. The temporal formula is usually 2+2 or 3, sometimes 3+3. There are 8 or 9(3,4,5) upper labials. There are 11(3) to 12(5) lower labials. The dorsal scales are smooth, with 1 apical pit, and arranged in 19 (occasionally 17 or 21) oblique rows. There are 190–244 ventrals (fewer than 225 in males, more than 214 in females). There are 51–83 divided subcaudals (with slight sexual dimorphism). The cloacal scale is divided. The maximum length recorded for this species is 1,050 mm (Broadley 1983).

Western Cat Snake: *Telescopus tripolitanus* (Werner 1909b)

Telescopus tripolitanus occurs in sahel from Senegal to Niger and north Africa. The type locality is Tripoli, Libya.

The dorsum is brown to gray with a pattern of dark-yellow hexagons. The venter is whitish. The nasal is divided. There is 1 preocular. There are usually 2 postocu-

lars, sometimes 3, the inferior the largest. The temporal formula is usually 2+2 or 3, sometimes 3+3. There are 8(4,5), sometimes 9(4,5), upper labials. There are 11(3) to 14(4) lower labials. The dorsal scales are smooth, with 1 apical pit, and arranged in 21 (occasionally 23) oblique rows. There are 205–230 ventrals and 55–83 divided subcaudals. The cloacal scale is divided. The maximum length recorded for this species is 840 mm (Trape and Mané 2006a).

Variable Cat Snake: *Telescopus variegatus* (Reinhardt 1843)

The Variable Cat Snake is found in woodland savanna from Senegal to southern Central African Republic. The type locality is "Guinea" but presumably meaning "Guinea Coast" rather than the country, Guinea. It is likely that the true type locality is somewhere in Ghana.

The dorsum is brown to gray with darker mottling. The venter is whitish, speckled

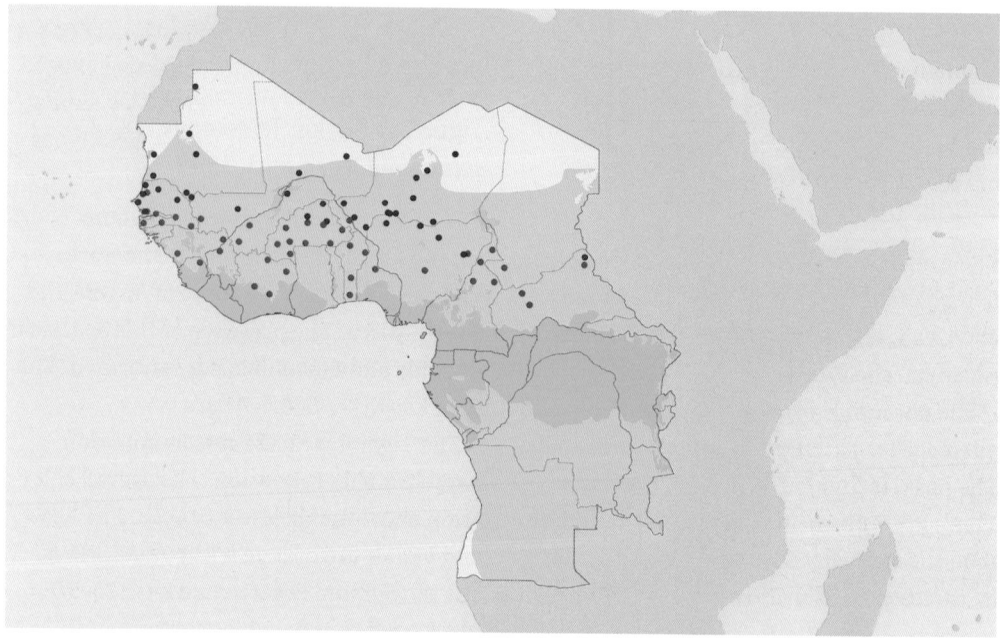

FIGURE 13.88. *Telescopus tripolitanus* (blue), *Telescopus variegatus* (red). By K. Jackson

with brown. The nasal is usually divided, sometimes single. There is 1 preocular. There are 2 postoculars, the inferior larger than the superior. The temporal formula is usually 2+2, sometimes 2+3. There are usually 8(4,5) upper labials. There are usually 9 or 10(4), sometimes 11(4) or (5), lower labials. The dorsal scales are smooth, with 1 apical pit, and arranged in 19 (occasionally 21) oblique rows. There are 198–238 ventrals (fewer than 215 in males, more than 200 in females). There are 53–73 divided subcaudals (more than 60 in males, fewer than 68 in females). The cloacal scale is divided. The maximum length recorded for this species is 900 mm (Villiers 1975).

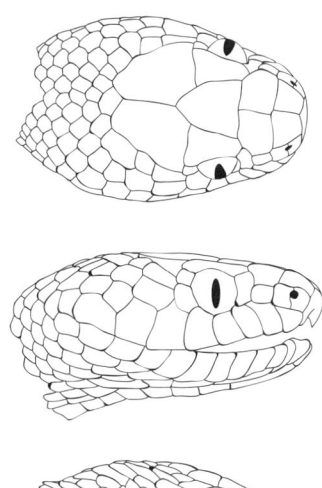
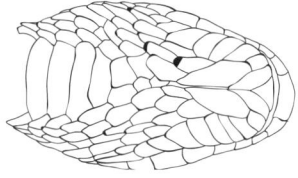

FIGURE 13.90. *Telescopus variegatus* RMCA 19461. By T. Giri

Vine Snakes: Genus *Thelotornis* Smith 1849

Thelotornis is a genus of four species of diurnal arboreal snake found in forests and savannahs throughout sub-Saharan Africa. This genus is one of two genera of African colubrine (the other is *Dispholidus*) whose venom has been known to cause human fatalities. *Thelotornis* is called "Vine Snake" and "Twig Snake" for its arboreal habits and for its habit of staying still and rigid,

FIGURE 13.89. *Telescopus variegatus*, Nigeria. By W. Wüster

so that they resemble a twig or a branch (Henderson and Binder 1980). Vine Snakes feed on amphibians, lizards, birds, snakes, and small mammals. They hunt using a sit-and-wait strategy and by luring potential prey with their brightly colored tongue. *Thelotornis* have a defensive display similar to that of the Boomslang, *Dispholidus typus*, involving puffing up the neck and anterior part of the laterally compressed body. They lay clutches of 4–10 eggs.

The overall shape is characteristic of arboreal snakes, with a slender body; long, tapering tail; and well-defined neck. The head is elongate and narrow with a pointed snout. The eye is small to medium sized, with a horizontally elliptical ("keyhole-shaped") pupil, a feature unique (in Africa) to this genus. The nasal is single. The internasals and prefrontals are paired. The loreal is always present, either in the form of a single loreal scale or divided up into 2 or 3 scales. The temporal formula is 1+2. There is 1 preocular, deeper than long, and 2–4 postoculars. The supraocular scale is broader than the long and narrow frontal, reflecting the overall narrowness of the head. There are generally no suboculars, but sometimes one of the postoculars may be in the position of a subocular. There are 8–9 upper labials, 2 of which are in contact with the eye. There are 8–11 (but occasionally as many as 13) lower labials, of which the first 3–5 are in contact with the anterior sublinguals. There are 2 pairs of chin shields. The submandibular groove is pronounced. The dorsal scales are keeled, without apical pits, and are arranged in 19–21 (occasionally 17) oblique rows. The vertebral row is not enlarged relative to the other dorsal scale rows. There are 146–203 ventrals and 126–175 divided subcaudals. The cloacal scale is divided. The hemipenes are unforked, and the sulcus spermaticus is not forked (Bogert 1940). The maxillary dentition is opisthoglyph, with 11–17 similarly sized anterior teeth followed posteriorly by 3–4 grooved posterior teeth.

The venom of *Thelotornis capensis* contains a prothrombin activator (Rosing et al. 1988; Marsh 1994). The LD50 in mice is on the order of 0.25 mg/kg (Kornalik et al. 1978). Envenomation by *T. c. oatesi* causes defibri-

nation with bleeding at the site of the bite and at other recent wounds (Sadler and Paul 1988) or internal hemorrhaging (Simbotwe 1982). Treatment should be symptomatic. A specific antivenom against the venom of *Dispholidus typus* exists, but it is not effective against the venom of *Thelotornis* (Atkinson et al. 1980). Herpetologist Robert Mertens (1894–1975) died from the bite of a captive *T. capensis*. Like Schmidt with *Dispholidus*, it is likely that Mertens underestimated the potential danger of a bite from *Thelotornis* because it was a colubrid.

Savanna Vine Snake / Twig Snake: *Thelotornis capensis* Smith 1849

Thelotornis capensis is a savanna species. Its distribution extends from southern Angola and southeastern Democratic Republic of Congo to South Africa. The type locality for the species is "Kaffirland and the country toward Port Natal" (meaning Durban). Two subspecies are recognized: *T. c. capensis* Smith 1849; *T. c. oatesi* (Günther 1881). The latter subspecies occurs within our zone. Its type locality is "Matabeleland, Rhodesia" (Zimbabwe). *T. c. capensis* occurs farther south.

The body is silvery gray or brown above with speckled with darker flecks that converge at the sides of the neck. The top of

---

Key to Central and Western African Species of the Genus *Thelotornis*

1	Combined length of the loreal scales is less than three times their depth; upper labials uniformly ivory without black spots .... *T. kirtlandii*
1'	Combined length of loreal scales is greater than three times their depth; upper labials with black spots or dots 2

2(1)	Top of the head uniformly leaf green; usually 8 lower labials .... *T. mossambicanus*
2'	Top of the head with dots; usually 11 lower labials .... *T. capensis*

the head is greenish with speckled pink and a distinct Y-shaped marking. The labials are white with black flecks. The venter is pale gray mottled with darker gray.

The loreal is always divided, usually into 3 smaller scales. The loreals are small and narrow. The combined length of the loreal scales is more than three times their depth. There are usually 3 postoculars, of which the inferior most is the largest and often takes the position of a subocular. There are 8(4,5) or sometimes 9(5,6) upper labials. There are usually 8(4), but sometimes 9(5), lower labials. The anterior chin shields are shorter than the posterior pair. There are

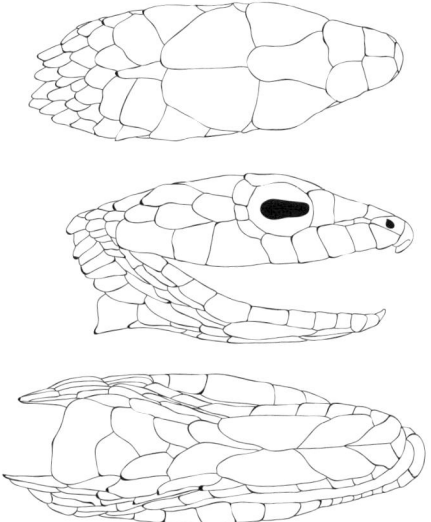

FIGURE 13.91. *Thelotornis capensis* RMCA 78-9-R-9. By T. Giri

FIGURE 13.92. *Thelotornis capensis*, Zambia. By S. Spawls

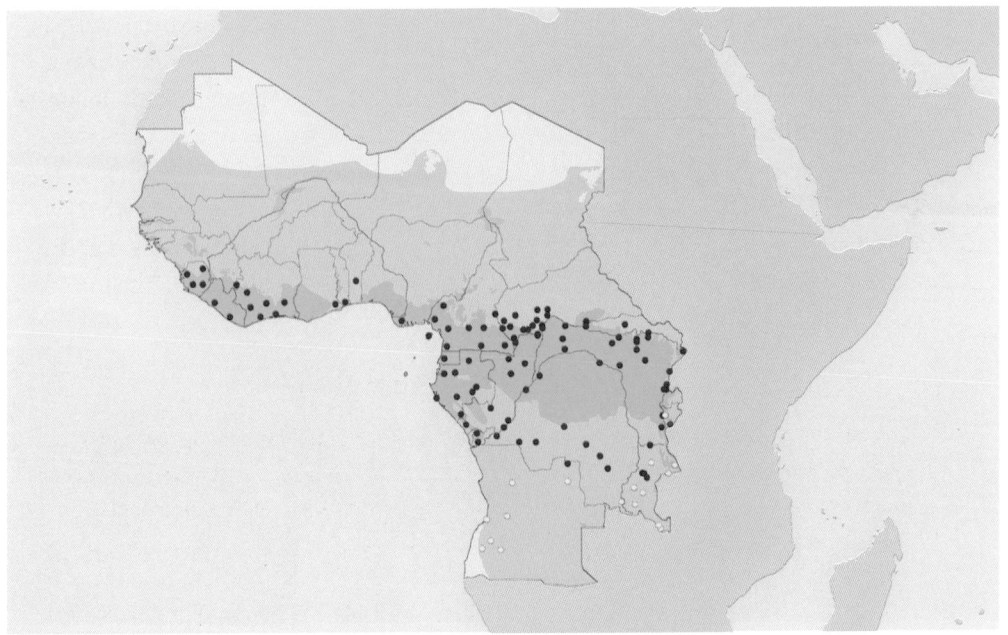

FIGURE 13.93. *Thelotornis capensis* (yellow), *Thelotornis kirtlandii* (blue), *Thelotornis mossambicanus* (red). By K. Jackson

usually 19, sometimes 21, dorsal scale rows. There are 144–177 ventrals (fewer than 178 in males, more than 147 in females) and 148–177 divided subcaudals (more than 132 in males, fewer than 169 in females). The maximum length recorded for this species is 1,400 mm (Spawls et al. 2004).

Forest Vine Snake / Bird Snake: *Thelotornis kirtlandii* (Hallowell 1844)

Thelotornis kirtlandii is a forest species distributed across central and western African forest habitats from Guinea to Uganda. The type locality is Liberia.

The body is mottled gray, green and brown above, usually with alternating light and dark bands. The dorsum of the head is bright green or purplish brown. The upper labials are white, sometimes with dark spots. The venter is pale grayish or pinkish, sometimes patterned with darker streaks or spots. The combined length of the loreals (there may be 1 loreal, or it may be divided into 2) is less than 2–3 times their depth. There are usually 3 postoculars, of which the inferior most is the largest and often takes the position of a subocular. There are 8(4,5) or sometimes 9(5,6) upper labials. There are usually 10(4), but sometimes 9(3) or 11(5), lower labials. The anterior and posterior chin shields are approximately equal in length. There are usually 19, sometimes 21, dorsal scale rows. There are 162–189 ventrals (fewer than 183 in males, more than 161 in females) and 132–172 divided subcaudals (more than 134 in males, fewer than 166 in females). The maximum length recorded for this species is 1,710 mm (Broadley 2001).

FIGURE 13.94. *Thelotornis kirtlandii*, Democratic Republic of Congo. By J. Kielgast

FIGURE 13.95. *Thelotornis kirtlandii*, Ivory Coast. By M.-O. Rödel

Eastern Twig Snake: *Thelotornis mossambicanus* (Bocage 1895a)

Formerly considered a subspecies of *Thelotornis capensis*, *T. mossambicanus* is an east African savanna species with a distribution extending from southern Somalia south along the coast to central Mozambique and west to southern Burundi and the eastern shores of Lake Tanganyika. The type locality is Manica, Mozambique.

The top and sides of the head are green or black. The upper labials and throat are white with black flecks. The dorsum of the body is gray with diagonal whitish bands and brown and reddish spots. The venter of the body is gray with brown striations. The loreal is always divided, usually into 3 smaller scales. The combined loreal scales are more than three times longer than deep. There are usually 3 postoculars, of which the inferiormost is the largest and often takes the position of a subocular. There are 8(4,5) upper labials. There are usually 9–13, usually 11(4) or 11(5), lower labials. The anterior chin shields are shorter than the posterior pair. There are usually 19, occasionally 17 or 21, dorsal scale rows. There are 144–172 ventrals (fewer than 169 in males, more than 144 in females) and 123–167 divided subcaudals (more than 130 in males, fewer than 154 in females). The maximum length recorded for this species is 1,435 mm (Broadley 2001).

Bold-eyed and Dagger-toothed Tree Snakes: Genus *Thrasops* Hallowell 1857

This genus includes six species of diurnal arboreal snakes, five of which occur within our zone. Some authors separate this genus into two: *Thrasops* Hallowell 1857 and *Rhamnophis* Günther 1862. Of the species occurring within our zone, *Thrasops aethiopissa* and *T. batesii* are the two species assigned to *Rhamnophis* by some authors. The basis for the distinctions between these genera were (1) the enlarged vertebral row in *Rhamnophis* but not *Thrasops* and (2) smooth dorsal scales in *Rhamnophis* versus keeled ones in *Thrasops*. There are exceptions to each of these ostensibly diagnostic differences between the two genera, however. The vertebral row is enlarged in both genera though to different degrees. In *Thrasops* is only somewhat enlarged, whereas that of *Rhamnophis* is enlarged to a greater degree, making enlargement of the vertebral row or lack thereof not a clear dichotomous character on which to base the distinction between two genera. As for the character of keeled versus smooth dorsal scales, some *Thrasops* (e.g., females and juveniles of *T. jacksonii*) have smooth or weakly keeled dorsals, making this character problematic also. Because there appear to be no characters that distinguish the two genera from one another, whereas there are other characters, such as dentition, that unite them, it seems reasonable to us to follow authors who have synonymized them (e.g., Leston and Hughes 1968; Trape and Roux-Estève 1995). Other authors (e.g., Broadley and Wallach 2002; Spawls et al. 2004) revive the separation of the two genera while acknowledging their close relatedness as well as the limitations of the characters distinguishing them from one another.

T. aethiopissa inhabiting degraded forest in southeastern Nigeria fed primarily on small birds followed by lizards and occasionally small mammals. They were active year-round, laying eggs twice each year, at the beginning and end of the wet season (Luiselli et al. 2000c). *T. aethiopissa* is rather irascible and presents a defensive display evocative of that of *Thelothornis* or *Dispholidus*. The natural history of other species in the genus is less well documented.

The morphology of *Thrasops* is characteristic of arboreal snakes: The head is short, the neck well defined. The body is long and laterally compressed, and the tail is long and slender. The eye is medium to large, with a round pupil. The maxillary dentition is aglyph, with 17–18 similarly sized anterior maxillary teeth separated by a diastema from 2–3 enlarged but ungrooved ("dagger-like") posterior teeth. The hemipenes are unforked, and the sulcus spermaticus is not forked (Bogert 1940).

The nasal may be single, divided, or semi-divided. The internasals and prefrontals are paired. The loreal is present. There is 1 preocular and 2–3 postoculars. There may be 1 subocular, sometimes considered an inferior postocular. There are 7–9 upper labials, 2 of which are in contact with the eye. There is 1 anterior temporal. There are 2 pairs of chin shields. The submandibular groove is pronounced. There are 8–12 lower labials. The dorsal scales may be smooth or keeled, and in some species this character may be sexually dimorphic or vary with ontogeny. The dorsal scales have 1 apical pit and are arranged in 13–21 oblique rows. The vertebral row may or may not be enlarged relative to the other dorsal scale rows. There are 158–215 ventrals, which may be rounded or keeled. There are 91–159 divided subcaudals. The cloacal scale may be single or divided.

FIGURE 13.96. *Thrasops aethiopissa*, Cameroon, defensive display. By V. Gvoždík

Splendid Dagger-tooth Tree Snake: *Thrasops aethiopissa* (Günther 1862)

The range of *Thrasops aethiopissa* extends from Guinea to the Democratic Republic of Congo. The type locality is West Africa.

The dorsum is blackish or dark green. The dorsal scales are edged with black. The venter is yellowish, fading to pinkish posteriorly. The upper labials are pale with dark edges. The nasal is divided. The loreal is longer than deep. There is 1 preocular, deeper than long. The supraocular is as broad as the frontal. The frontal is slightly longer than broad. There are usually 2 postoculars of similar size, the inferior occupying the position of a subocular. The temporal formula is 1+0. There are usually 8(4,5), occasionally 9(3,4,5), upper labials and 8(4) to 9(5) lower labials. The anterior chin shields are shorter than the posterior pair. The dorsal scales are smooth, with 1 apical pit, and are arranged in 15–19 (usually 17) oblique rows. The vertebral row is distinctly enlarged relative to the other dorsal scale rows. There are 158–179 keeled ventrals and 139–159 subcaudals. The cloacal scale is divided. The

maximum length recorded for this species is 1,500 mm (Loveridge 1944).

One subspecies, *T. aethiopissa ituriensis* (Schmidt 1923), has been proposed on the basis of having 15 dorsal scale rows as opposed to the 17 in the original description of the species (Günther 1862). But subsequent studies (e.g., Laurent 1956; Roux-Estève 1965) indicate greater variability in dorsal scale count for this species than initially thought, which does not justify subspecies status for *T. a. ituriensis*.

Spotted Dagger-tooth Tree Snake: *Thrasops batesii* Boulenger 1908

The range of *Thrasops batesii* extends from Cameroon to the Democratic Republic of Congo. The type localities are Efulen and Akok, Cameroon.

The dorsum is blackish or dark green with pale spots. The venter is yellowish anteriorly with black lateral spots in staggered rows, gradually becoming uniformly dark posteriorly from about halfway along the body. The top of the head is black. The upper labials are pale with dark edges. The nasal is single. The loreal is longer than deep. There is

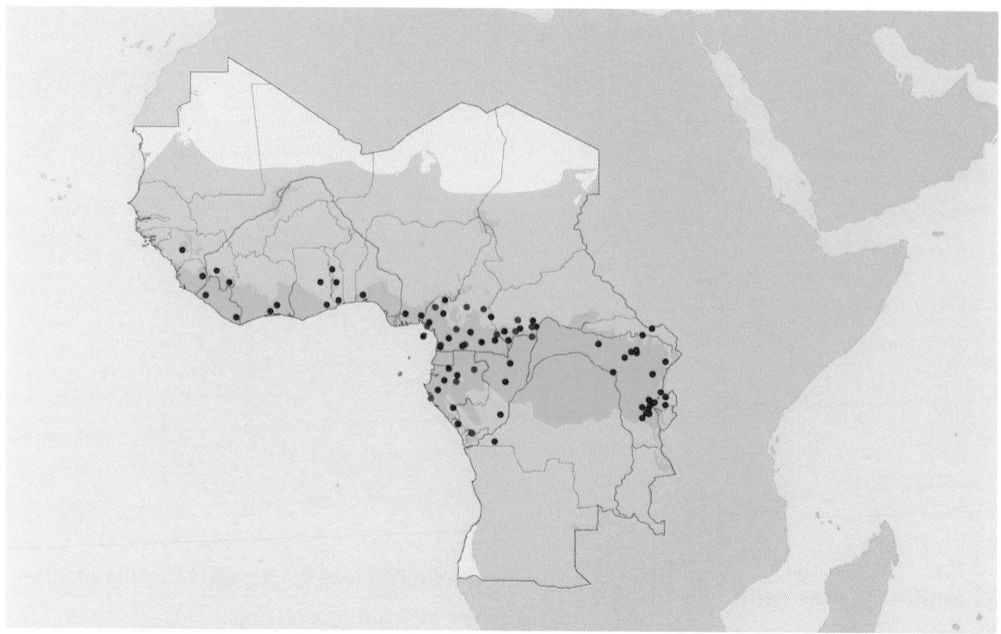

FIGURE 13.97. *Thrasops aethiopissa* (blue), *Thrasops batesii* (red). By K. Jackson

1 preocular, deeper than long. The supraocular is as broad as the frontal. The frontal is slightly longer than broad. There are usually 3 postoculars, the 2 superior ones of equal size, the inferior larger and occupying the position of a subocular. The temporal formula is 1+3 or sometimes 1+0. There are usually 7(4,5), sometimes 6(4,5) or 8(4,5), upper labials. There are usually 8(5) but sometimes 7(4) or 9(6) lower labials. The anterior chin shields are shorter than or equal to the posterior pair. The dorsal scales are smooth, with 1 apical pit, and are arranged in 13 oblique rows. The vertebral row is distinctly enlarged relative to the other dorsal scale rows. There are 163–180 keeled ventrals and 91–116 subcaudals. The cloacal scale is single. The maximum length recorded for this species is 1,800 mm (Boulenger 1908).

Yellow-throated Bold-eyed Tree Snake: *Thrasops flavigularis* (Hallowell 1852)

The range of *Thrasops flavigularis* extends from Nigeria to the Democratic Republic of Congo. The type locality is Gabon.

The dorsum is black, patterned in some individuals with a few yellow spots. The venter is blackish except for the throat, which is yellowish. The nasal is single or semi-divided. The loreal is equal in depth and length. There is 1 preocular, deeper than long, or 2 preoculars, the superior larger than the inferior. The supraocular is as broad as the frontal. The frontal is slightly longer than broad. There are usually 3 postoculars, the 2 superior ones of equal size, the inferior larger and occupying the position of a subocular. The temporal formula is 1+1. There are usually 8(4,5), sometimes 9(4,5), upper labials and 8(3) to 12(5) lower labials. The anterior chin shields are longer than or equal to the posterior pair. The dorsal scales are weakly keeled in adults and smooth in juveniles, with 1 apical pit, and are arranged in 13 or sometimes 15 oblique rows. The vertebral row is not enlarged relative to the other dorsal scale rows. There are 195–215 rounded ventrals

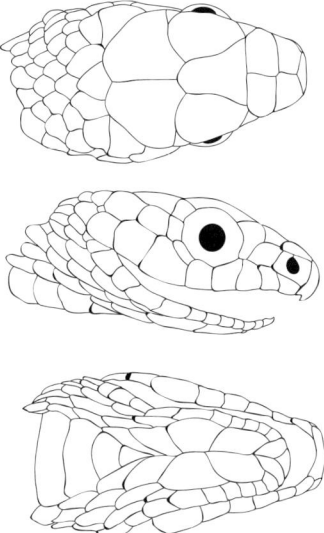

FIGURE 13.98. *Thrasops flavigularis* RMCA RS 1448. By T. Giri

and 128–146 subcaudals. The cloacal scale is divided. The maximum length recorded for this species is 2,400 mm (Stucki-Stirn 1979).

Jackson's Black Tree Snake: *Thrasops jacksonii* Günther 1895

The range of *Thrasops jacksonii* extends from Cameroon to Kenya. The type locality is Kavirondo, Kenya.

The dorsum is uniformly black. The venter is blackish except for the throat, which is paler. Juveniles are black with yellow spots. The nasal is semi-divided. The loreal is deeper than long. There is 1 preocular, deeper than long. The supraocular is as broad as the frontal. The frontal is slightly longer than broad. There are usually 3 postoculars, the 2 superior ones of equal size. The temporal formula is 1+1. There are usually 8(4,5), occasionally 7(4,5) or 9(4,5), upper labials and 10(4) to 12(5) lower labials. The anterior and posterior chin shields are approximately equal in length. The dor-

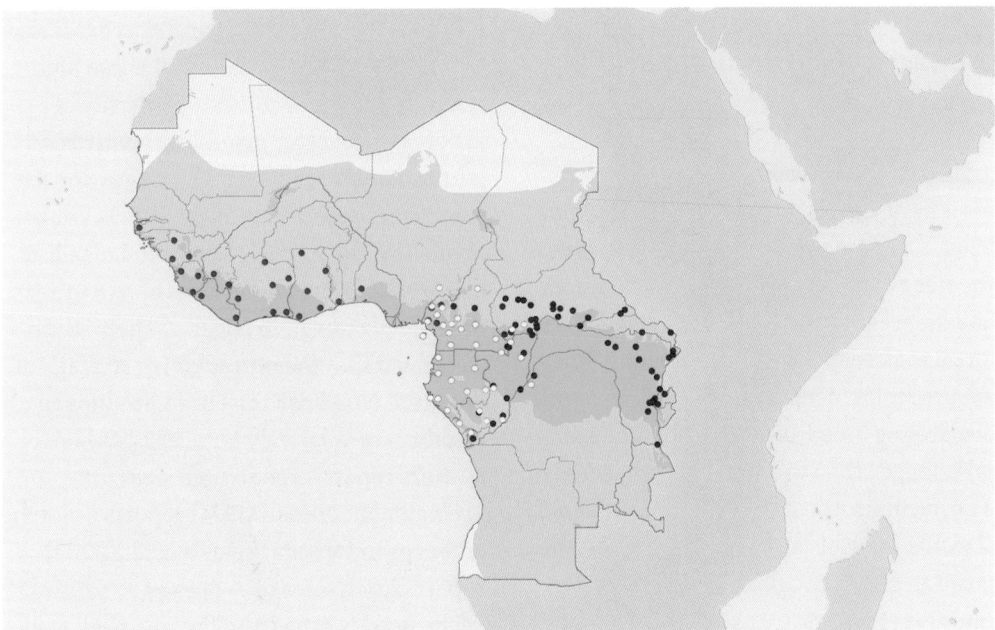

FIGURE 13.99. *Thrasops flavigularis* (yellow), *Thrasops jacksonii* (blue), *Thrasops occidentalis* (red). By K. Jackson

FIGURE 13.100. *Thrasops jacksonii*, Democratic Republic of Congo. By J. Kielgast

sal scales are usually keeled in males but often smooth in females. They each have 1 apical pit, arranged in 17–21 oblique rows. The vertebral row is slightly enlarged relative to the other dorsal scale rows. There are 189–211 ventrals, which are keeled in males, and 130–155 subcaudals. The cloacal scale is divided. The maximum length recorded for this species is 2,160 mm (Schmidt 1923).

Western Black Tree Snake: *Thrasops occidentalis* Parker 1940

The range of *Thrasops occidentalis* extends from Senegal to Cameroon. The type locality is Axim, Ghana.

The dorsum is uniformly black. The venter is dark except for the throat, which is paler, sometimes yellowish. The scales of the sides of the head, especially the upper labials, are pale. The nasal is divided. The loreal is as deep or deeper than long. There is 1 preocular, deeper than long. The supraocular is as broad as the frontal. The frontal is longer than broad. There are usually 3 postoculars, the 2 superior ones of equal size, the inferior larger and occupying the position of a subocular. The temporal formula is 1+1. There are usually 8(4,5) or 9(4,5) upper labials and 8(4) to 10(6) lower

labials. The anterior chin shields are shorter than or equal to the posterior pair. The dorsal scales are keeled, with 1 apical pit, and are usually arranged in 15 oblique rows. The vertebral row is slightly enlarged relative to the other dorsal scale rows. There are 172–187 ventrals and 116–140 subcaudals. The cloacal scale may be single or divided. The maximum length recorded for this species is 2,450 mm (Isemonger 1983).

Broad-headed Tree Snakes: Genus *Toxicodryas* Hallowell 1857

This genus includes two species of large nocturnal arboreal snake, both with large distributions in sub-Saharan Africa. Some authors include *Toxicodryas* with the genus *Boiga* Fitzinger 1826, a genus of 30–plus species of large nocturnal arboreal snakes with an Australasian distribution, a trend started by Schmidt (1923). The main argument against lumping them together is the large geographical discontinuity between the distributions of African *Toxicodryas* and Australasian *Boiga*. The diet of these highly arboreal snakes consists primarily of arboreal lizards for juveniles and birds for adults. Rodents, bats, and bird eggs are also sometimes eaten. Relatively little is known about *Toxicodryas* reproduction. Luiselli et al. (1998b) found clutch sizes of 4–6 in *Toxicodryas blandingii* in Nigeria. Their study suggested that the reproductive strategy of female *T. blandingii* may be to produce small numbers of relatively large eggs. Other authors report larger clutch sizes for *T. blandingii*: Pitman (1974) reports a clutch of 9 eggs in Uganda; Spawls et al. (2004) report clutch sizes of 7–14 eggs.

The head is broad and flat, the neck well defined. The body is laterally compressed,

FIGURE 13.101.
Thrasops occidentalis,
Liberia. By W. Branch

the tail long and slender. The eye is medium sized with a vertical pupil. The maxillary dentition is opisthoglyph, with 10–13 similar-sized anterior teeth followed after a diastema by 2–3 grooved posterior fangs. The rostral is rounded. The nasal is divided. The internasals and prefrontals are paired. The loreal is present. There 1–2 preoculars and 2–3 postoculars. There are no suboculars. The temporal formula is 2+2 or 3. There are 8–9 upper labials, 3 of which are in contact with the eye. There are 10–14 lower labials. There are 2 pairs of chin shields. The dorsal scales are smooth, without apical pits, and arranged in 19-25 oblique rows. The vertebral row is enlarged relative to the other dorsal scale rows. There are 236–289 ventrals and 96–147 divided subcaudals. The cloacal scale may be single or divided. The hemipenes are unforked. The sulcus spermaticus is not forked.

The only reported human cases of envenomation by *Toxicodryas* sp. resulted in mild symptoms, including localized pain and edema (Goodman 1985) and mild neurotoxic symptoms (Amri and Chippaux 2012). The venom of *T. blandingii* contains a neurotoxin (Levinson et al. 1976) that acts like those of elapids by binding to the cholinergic receptors (Broaders et al. 1999). Broaders and Ryan (1997) also observed strong cholinesterase-like activity in the venom of this species.

Blanding's Tree Snake: *Toxicodryas blandingii* (Hallowell 1844)

The range of *Toxicodryas blandingii* extends from Guinea to Kenya and south to Angola and Zambia. The type locality is Liberia.

Blanding's Tree Snake is unusual in having two color phases. In the first of these color phases, seen in large adult males, the

Key to Central and Western African Species of the Genus *Toxicodryas*

1   19 dorsal scale rows midbody .... *T. pulverulenta*
1' 21 to 25 dorsal scale rows midbody .... *T. blandingii*

dorsum is uniformly velvety black or dark blue, and the venter is yellow or cream. In the other color phase, usually seen in females and younger males, individuals are pale gray patterned with dark irregular blotches or bands. Juveniles of both sexes are light brown with a pattern of dark bands. There are usually 2 preoculars, the superior equal to or larger than the inferior. There are usually 2 postoculars, the superior equal to or smaller than the inferior. The frontal is slightly longer than broad. The temporal formula is 2+2, sometimes 2+3. There are usually 9(3,4,5), sometimes 9(4,5,6), occasionally 8 or 10, upper labials. The anterior chin shields are shorter than the posterior pair. The submandibular groove is pronounced. There are from 12(3) to 15(5) lower labials. The dorsal scales are smooth, without apical pits, and arranged in 21–23, occasionally 25, oblique rows. The vertebral row is enlarged relative to the other dorsal scale rows. There are 240–289

ventrals and 115–147 subcaudals, apparently without sexual dimorphism. The cloacal scale is usually divided but sometimes single. The maximum recorded length for this species is 2,740 mm (Rasmussen 1997b).

Orange Tree Snake: *Toxicodryas pulverulenta* (Fischer 1856)

The range of *Toxicodryas pulverulenta* extends from Sierra Leone to the Democratic Republic of Congo to Angola. The type locality is São Tomé Island, probably in error (Schätti and Loumont 1992).

The dorsum ranges from beige to reddish, sometimes with a pattern of dark gray bands. The venter is pink with two dark gray lateral lines. There are usually 2 preoculars, the superior larger than the inferior. There are usually 2 postoculars of equal size. The frontal is equal in length and breadth. The temporal formula is 2+2. There are 8(3,4,5) or 9(4,5,6) upper labials. The anterior chin shields are longer

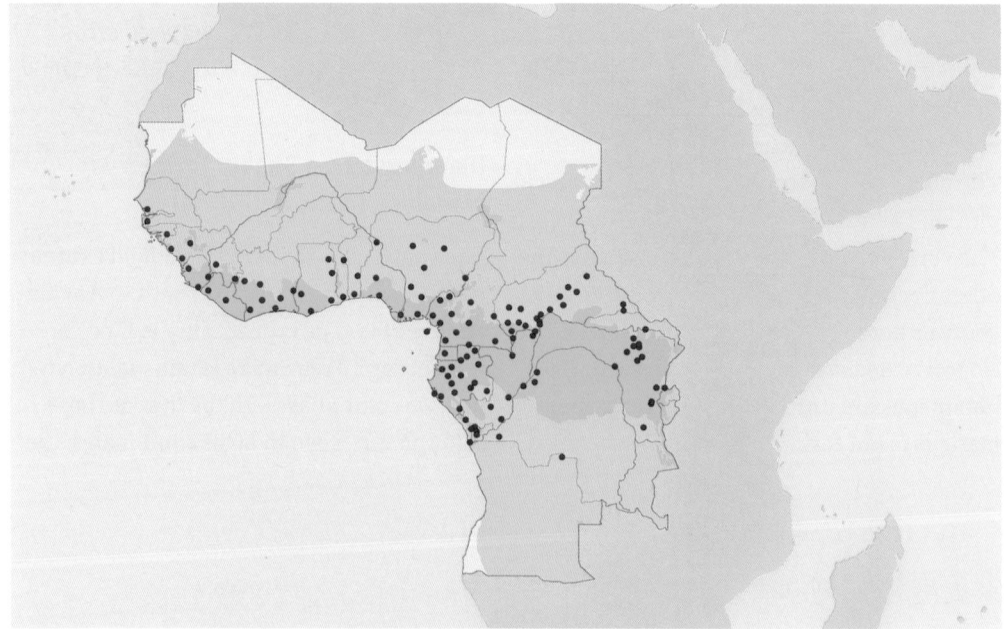

FIGURE 13.102. *Toxicodryas blandingii* (blue). By K. Jackson

than the posterior pair. The submandibular groove is not pronounced. There are from 10(4) to 12(6) lower labials. The dorsal scales are smooth, without apical pits, and arranged in 19 oblique rows. The vertebral row is enlarged relative to the other dorsal scale rows. There are 236–276 ventrals and 96–132 subcaudals. The cloacal scale is single. The maximum recorded length for this species is 1,250 mm (Villiers 1975).

Subfamily Grayiinae

African Water Snakes: Genus *Grayia* Günther 1858

The genus *Grayia* comprises four species, all of which occur within our zone. *Grayia* are large, robust, semiaquatic snakes. Although they superficially resemble natricines,

FIGURE 13.103. *Toxicodryas blandingii*, male, Kenya. By S. Spawls

FIGURE 13.104. *Toxicodryas blandingii* eating a bat, Nigeria. By G. Jesus

FIGURE 13.105. *Toxicodryas blandingii*, female, Democratic Republic of Congo. By J. Kielgast

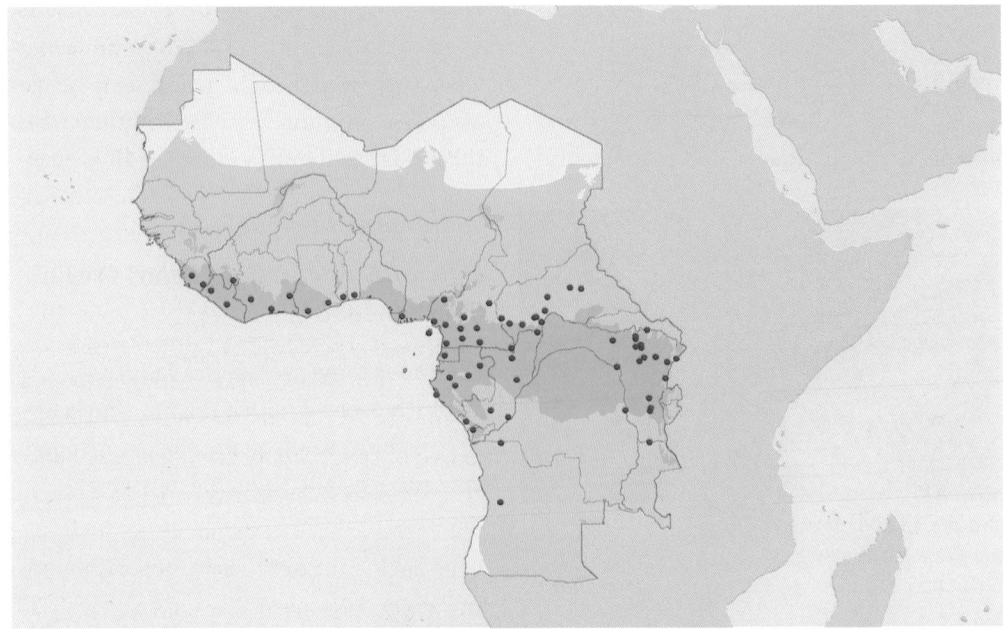

FIGURE 13.106. *Toxicodryas pulverulenta* (red). By K. Jackson

FIGURE 13.107. *Toxicodryas pulveru-lenta*, Democratic Republic of Congo. By J. Kielgast

molecular phylogenies have failed to consistently closely ally them with any larger family (e.g., Vidal and Hedges 2002; Lawson et al. 2005; Vidal et al. 2007). Here, we follow the trend of several recent studies (e.g., Nagy et al. 2005; Pyron et al. 2011, 2013; Zheng and Wiens 2016), assigning them to a separate colubrid subfamily, the Grayiinae.

Of the four species in the genus *Grayia*, two species, *Grayia ornata* and *G. caesar*, have central African ranges, while *G. smithii* and *G. tholloni* have extensive distributions in sub-Saharan Africa. In central Africa, where their ranges overlap, *Grayia* are easily mistaken for the Water Cobra, *Naja annulata*, which they resemble in body size and shape and in general color pattern. *Grayia* are sometimes considered mimics of Water Cobras, but this morphological similarity more likely simply represents convergence to a similar ecological niche.

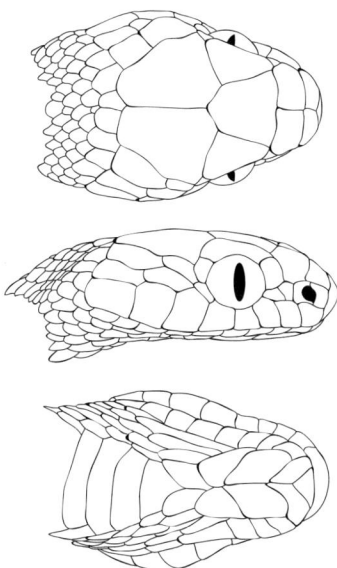

FIGURE 13.108. *Toxicodryas pulverulenta* RMCA 20.271. By T. Giri

In central Africa, both *Grayia* and Water Cobras are frequently caught in gill nets set by fishermen. In swamp forest habitat in the Republic of Congo, *G. ornata* were the snake most frequently captured in gill nets and were much more abundant than *Naja annulata* (Jackson et al. 2007). *G. smithii*, in Nigeria, are ecologically the best studied members of the genus. In Nigeria, *G. smithii* are found in a variety of aquatic habitats, including mangroves, but are strongly associated with rainforest habitats as opposed

to degraded areas near human habitations (Luiselli and Akani 2002). *G. smithii* feed primarily on fishes and to a lesser extent on amphibians (Cansdale 1961; Pauwels et al. 2000). Luiselli (2006c) noted a difference in diet correlated with sex in adults: adult males feed primarily on the aquatic frog, *Silurana tropicalis*, and to a lesser extent on tadpoles and on more terrestrial frogs of the genus *Ptychadena*. Adult females feed primarily on *Ptychadena spp.* and consume significantly more of the highly terrestrial toads (genus *Sclerophrys*, formerly *Bufo*) than do males. Subadults of both sexes feed primarily on tadpoles. *G. smithii* lays eggs during the dry season, and, although the average clutch size is approximately 10 eggs, there is a strong correlation between maternal body length and clutch size (Akani and Luiselli 2001b). Angel et al. (1954) report a female with 23 eggs in the oviduct, each the size of a hen's egg. Jackson (pers. obs.) found 10 eggs, each averaging 2 cm in length, in the oviduct of a *G. ornata* just over a meter in total length.

Grayia have 15–19 rows of straight, smooth dorsal scales, without apical pits. A loreal scale is present. There are 1–2 preoculars and postoculars but never suboculars, so that the eye is always in contact with at

Key to Central and Western African Species of the Genus *Grayia*

1	15 dorsal scale rows midbody	2
1'	17 or 19 dorsal scale rows midbody	3
2(1)	Only 1 upper labial scale in contact with the eye	*G. tholloni*
2'	2 upper labial scales in contact with the eye	*G. caesar*
3(1)	7 upper labials; at least 84 subcaudals	*G. smithii*
3'	8 or 9 upper labials; not more than 88 subcaudals	*G. ornata*

least one of the upper labials. The nasal is divided. The temporal formula is 2+2 or 3. There are always 2 pairs of chin shields, and the submandibular groove is pronounced in all species except *G. smithii*. The cloacal scale may be single or divided. The number of divided subcaudals varies considerably between species. The hemipenes are bilobed, with a divided sulcus spermaticus, with the exception of *G. caesar*, whose hemipenes are unforked. The maxilla is long and the maxillary teeth numerous, homogeneous, and without specialization, even of the posteriormost ones. The hemipenes are unusual in being extremely long. Bogert (1940) describes the hemipenes of *G. ornata*, dissected while retracted inside the base of the tail as extending as far as the seventeenth subcaudal, forking at the twelfth, and with the sulcus dividing at the sixth.

Long-tailed African Water Snake: *Grayia caesar* (Günther 1863b)

The Long-tailed African Water Snake is a central African species with a range extending from Nigeria to Central African Republic. The type locality is the island of Bioko (Equatorial Guinea).

Grayia caesar is blackish brown above with paler bands outlined in black. Females have more of these bands than males. The underside is uniformly pale. The upper labials are yellowish outlined in black. *G. caesar* has 8(4,5) upper labials. The temporal formula is 2+3. There are 10(5) lower labials. There are 15 dorsal scale rows and 123–149 ventrals (fewer than 130 in males, more than 135 in females). The cloacal scale is single in males and divided in females. *G. caesar* earns its common name for its extraordinarily long tail: 142–162 divided subcaudals (more than 150 in males, fewer than 150 in females). The maximum recorded length for this species is 1,148 mm (Roux-Estève 1965).

Ornate African Water Snake: *Grayia ornata* (Bocage 1866b)

The Ornate African Water Snake is a central African species with a range extending from Cameroon to Angola. The type locality is Duque de Bragança, Angola.

The body is dark brown above with darker bands. The underside is pale except becoming dark along the underside of the tail. Grayia ornata has 8(4) or occasionally 9(4) upper labials, but what is distinct about the upper labials of this species is the presence of an accessory scale on the upper lip that is between the fifth and sixth upper labials but does not count as an upper labial itself since it only partially separates the fifth upper labial from the sixth. The temporal formula is 2+3. There are from 10 to 12(4 to 6) lower labials, and the anterior pair of chin shields is longer than the posterior pair. There are 17 or 19 dorsal scale rows and 144–161 ventrals. The cloacal scale is divided, and there are 73–88 divided subcaudals. The maximum length recorded for this species is 1,520 mm (Knoepffler 1966).

Smith's African Water Snake: *Grayia smithii* (Leach 1818)

Smith's African Water Snake has a broad distribution in sub-Saharan Africa, extending from Senegal to Tanzania. The type locality is Boma, Democratic Republic of Congo.

Grayia smithii is dark brown above patterned with even darker bands shaped like chevrons. This pattern gradually becomes less distinct as the snake ages. The underside is pale, sometimes with dark lateral spots. The upper labial scales are pale, edged with black. The underside of the head and

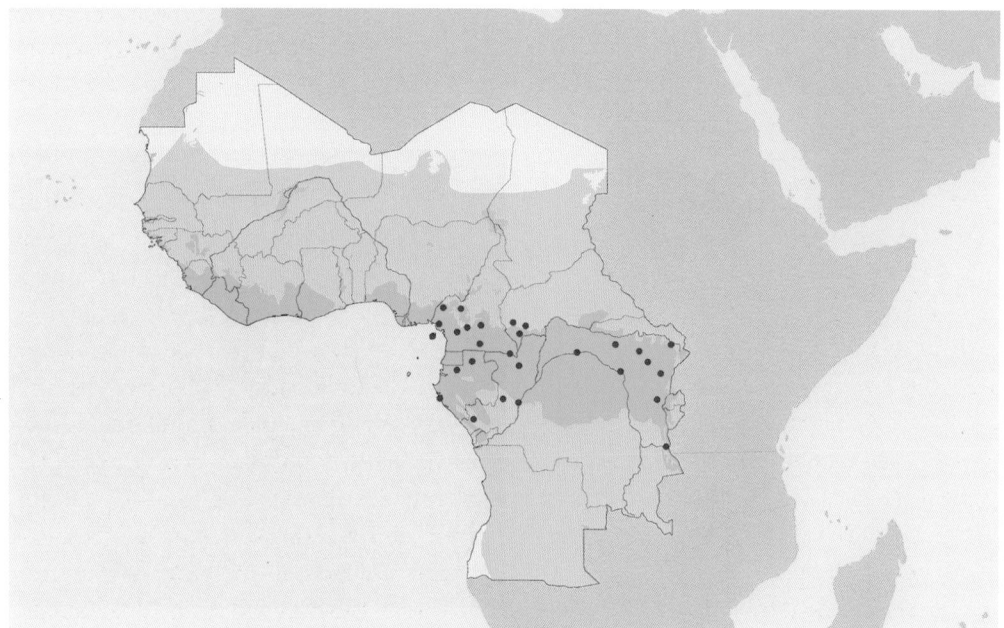

FIGURE 13.109. *Grayia caesar* (red). By K. Jackson

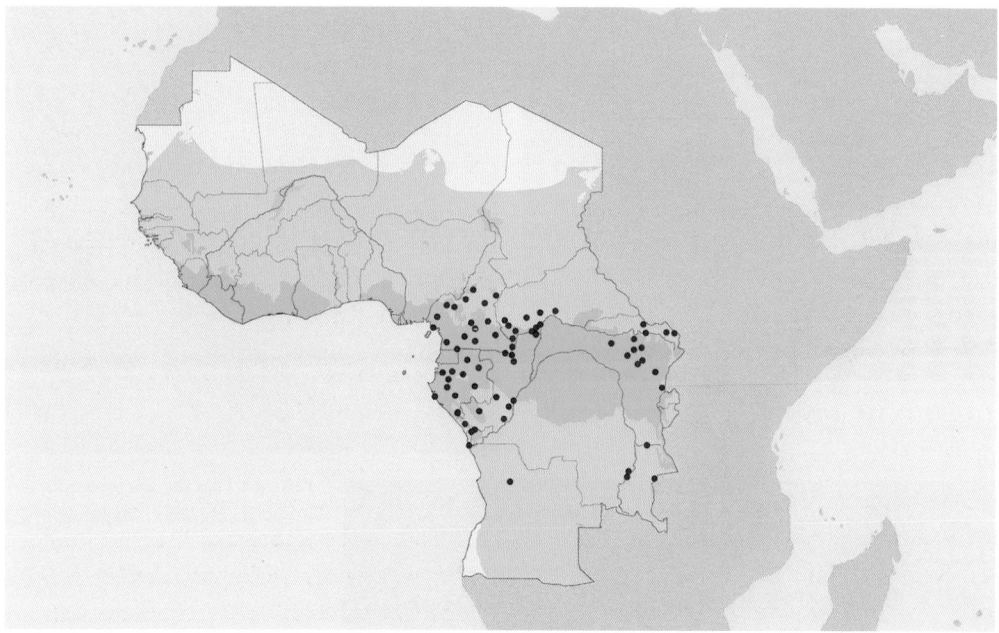

FIGURE 13.110. *Grayia ornata* (blue). By K. Jackson

throat is often bright yellow in life. *G. smithii* has 7(4) upper labials and a temporal formula of 2+3 or occasionally 2+2. There are 11(5) lower labials. There are usually 17 but sometimes 19 dorsal scale rows and 145–168 ventrals (fewer than 161 in males, more than 154 in females). The cloacal scale is divided, and there are 84–106

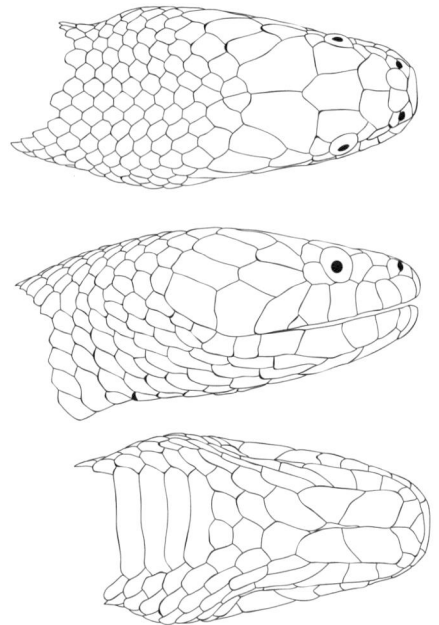

FIGURE 13.111. *Grayia ornata* RMCA
A4-046-R-0004. By T. Giri

FIGURE 13.112. *Grayia ornata*, Republic of Congo.
By K. Jackson

FIGURE 13.113. *Grayia ornata*, Republic of Congo.
By M. Burger

FIGURE 13.114. *Grayia ornata* with
atypical coloration, Republic of Congo.
By M. Burger

FIGURE 13.115. *Grayia ornata* juvenile,
Democratic Republic of Congo. By
E. Greenbaum

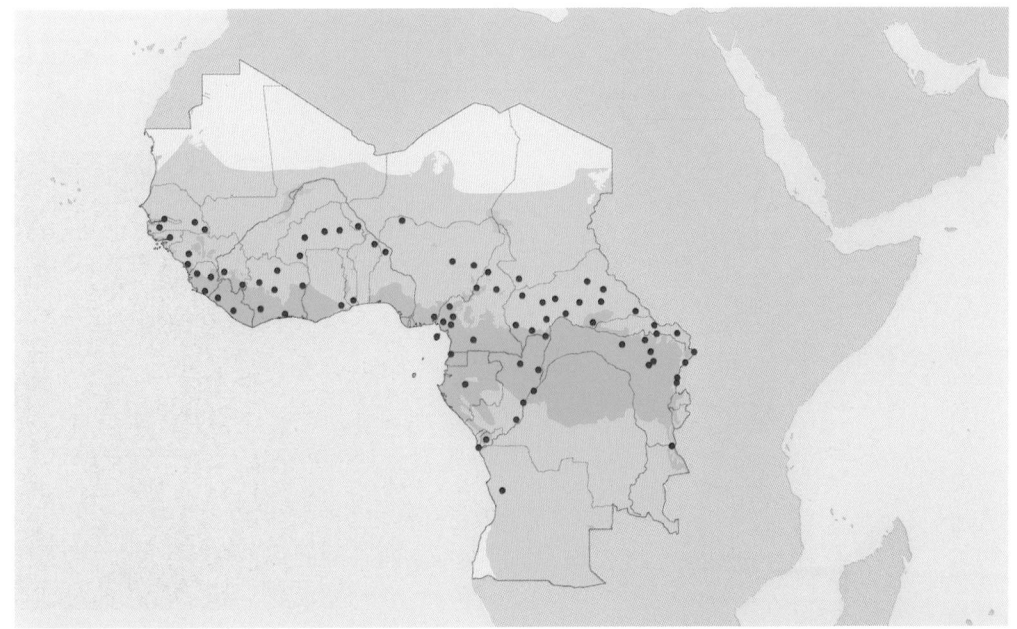

FIGURE 13.116. *Grayia smithii* (blue). By K. Jackson

FIGURE 13.117. *Grayia smithii*, Republic of Congo. By M. Burger

FIGURE 13.118. *Grayia smithii*, Republic of Congo. By K. Jackson

FIGURE 13.119. *Grayia smithii* juvenile, Republic of Congo. By K. Jackson

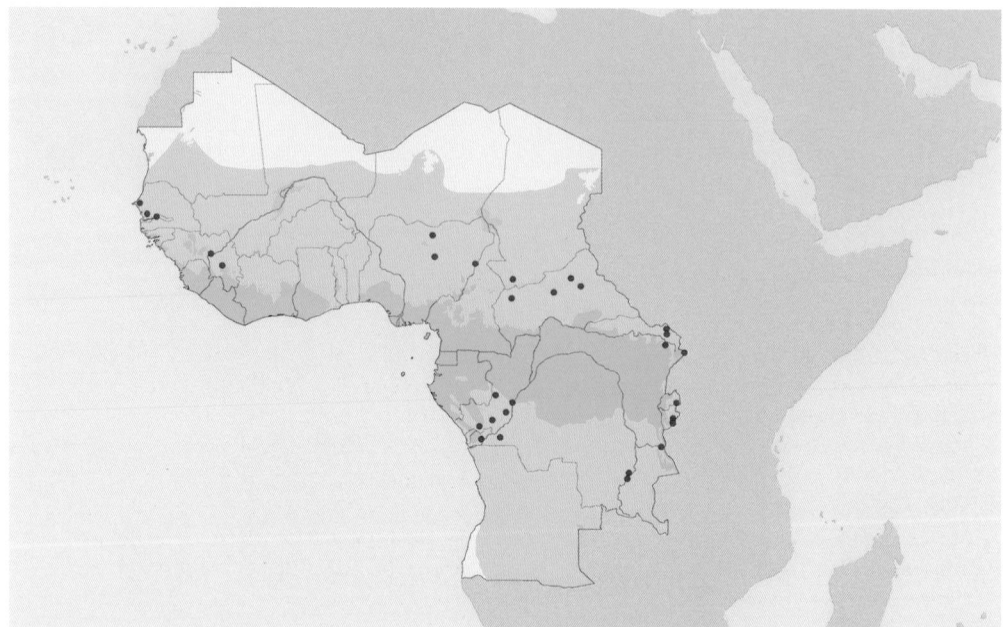

FIGURE 13.120. *Grayia tholloni* (red). By K. Jackson

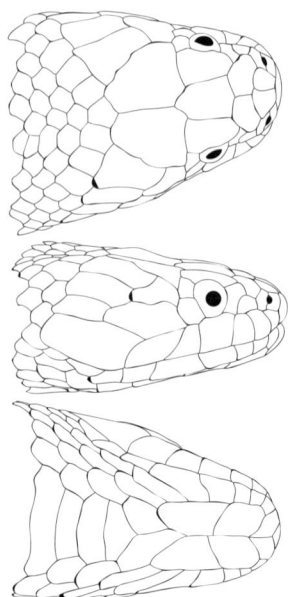

FIGURE 13.121. *Grayia tholloni* RMCA A6-007-R-0007. By T. Giri

divided subcaudals (more than 89 in males, fewer than 100 in females). The maximum recorded length for this species is 2,550 mm (Doucet 1963b).

Thollon's African Water Snake: *Grayia tholloni* Mocquard 1897

Like *Grayia smithii*, *G. tholloni* has a broad distribution in sub-Saharan Africa, extending from Senegal to Kenya to Angola. The type locality is Brazzaville, Republic of Congo.

G. tholloni is blackish brown above with white scales on the sides of the body outlining bands, which disappear with age. The tail is dark and has no pattern. The underside is uniformly pale. The upper labials are pale outlined in black. *G. tholloni* has 7 or 8(4) upper labials. The temporal formula is 2+3, occasionally 2+2. The posterior pair of chin shields is longer than the anterior pair, and there are 10(5) lower labials. There are 15 dorsal scale rows and 130–150 ventrals (fewer than 141 in males, more than 137 in females). The cloacal scale is divided, and there are 110–135 divided subcaudals. The maximum recorded length for this species is 1,200 mm (Villiers 1975).

Bibliography

Adalsteinsson, S. A., W. R. Branch, S. Trape, L. J. Vitt, and S. B. Hedges. 2009. Molecular phylogeny, classification, and biogeography of snakes of the Family Leptotyphlopidae (Reptilia, Squamata). Zootaxa 2244: 1–50.

Akaffou, M. H., J.-P. Chippaux, B. Allali, Z. Coulibaly, and M. Dosso. 2017. Le peuplement ophidien des plantations d'*Hévéa brasiliensis* d'Anguédédou (Sud-est de la Côte d'Ivoire). Bulletin de la Société Française d'Herpétologie 162: 31–38.

Akani, G. C., and L. M. Luiselli. 2001a. Aspects of the natural history of *Natriciteres* (Serpentes, Colubridae) in Nigeria, with special reference to *N. variegata* and *N. fuliginoides*. Herpetological Natural History 7: 162–168.

Akani, G. C., and L. M. Luiselli. 2001b. Ecological studies on a population of the water snake *Grayia smythii* in a rainforest swamp of the Niger Delta, Nigeria. Contributions to Zoology 70(3): http://www.ctoz.nl/vol70/nr03/a02.

Akani, G. C., and L. M. Luiselli. 2009. Aspects of community ecology of amphibians and reptiles at Bonny Island (Nigeria), an area of priority relevance for petrochemical industry. African Journal of Ecology 48: 939–948.

Akani, G. C., L. Luiselli, and E. Politano. 1999. Ecological and conservation considerations on the reptile fauna of the eastern Niger Delta (Nigeria). Herpetozoa 11: 141–153.

Akani, G. C., L. M. Luiselli, F. M. Angelici, C. Corti, and M. A. L. Zuffi. 2001a. The case of rainforest stiletto snakes (genus *Atractaspis*) in southern Nigeria: Evidence of diverging foraging strategies in grossly sympatric snakes with homogeneous body architecture? Ethology, Ecology and Evolution 13: 89–94.

Akani, G. C., D. Capizzi, and L. M. Luiselli. 2001b. *Mehelya crossi* (West African File Snake) diet. Herpetological Review 32(1): 49–50.

Akani, G., L. Luiselli, and F. Angelici. 2002a. Diet of *Thelotornis kirtlandii* (Serpentes: Colubridae: Dispholidini) from southern Nigeria. Herpetological Journal 12: 179–182.

Akani, G. C, L. Luiselli, S. M. Wariboko, L. Ude, and F. M. Angelici. 2002b. Frequency of tail autotomy in the African olive grass snake, *Psammophis* "*phillipsi*" from three habitats in southern Nigeria. African Journal of Herpetology 51: 143–146.

Akani, G. C., E. Eniang, I. Ekpo, F. M. Angelici, and L. M. Luiselli. 2003. Food habits of the snake *Psammophis phillipsi* from the continuous rain- forest region of southern Nigeria (west Africa). Journal of Herpetology 37: 208–211.

Akani, G. C., F. M. Angelici, and L. M. Luiselli. 2005. Ecological data on the Goldie's Tree Cobra, *Pseudohaje goldii* (Elapidae) in southern Nigeria. Amphibia-Reptilia 26: 382–387.

Akani, G. C., N. Ebere, L. Luiselli, and E. Eniang. 2007. Community structure and ecology of snakes in fields of oil palm trees (*Elaeis guineensis*) in the Niger Delta, southern Nigeria. African Journal of Ecology 46: 500–506.

Amri, K., and J. P. Chippaux. 2012. Envenimation bénigne par morsure de *Toxicodryas blandingii* (Hallowell, 1844) en Suisse. Bulletin de la Société Herpétologique de France 142/143: 145–148.

Anderson, J. 1898. Zoology of Egypt. Vol. 1. Reptilia and batrachia. London: Benard Quaritch. 371 pp.

Andersson, L. G. 1901. Some new snakes from Cameroon and South America belonging to the collections of the Royal Museum in Stockholm. Bihang till Kongl. Svenska Vetenskaps-akademiens Handlingar 27: 1–26.

Andersson, L. G. 1916. Notes on the reptiles and batrachians in the Zoological Museum at Gothenburg with an account of some new species. Göteborgs Kungliga Vetenskap och Vitter Hets-Samnalles Hndlingar Sjatte Foljden (Series B, 4) 17(5) [Meddelanden fran Göteborgs Musei Zoologiska Afdelning, No. 9]: 1–41.

Andersson, L. G. 1937. Reptiles and batrachians collected in the Gambia by Gustav Svensson and Birger Rudebeck (Swedish Expedition 1931). Arkiv för Zoologi, A29, 16: 1–28.

Andrews, W. H. 1913. On the effects of the bite of certain opisthoglyphous snakes. South African Journal of Science 9: 269–276.

Angel, F. 1921. Sur les reptiles de la région du Gribingui. Bulletin du Muséum National d'Histoire Naturelle, Paris 27: 141–142.

Angel, F. 1922. Sur une collection de reptiles et de batraciens, recueillis au Soudan francais par le Mission du Dr. Millet-Horsin. Bulletin du Muséum National d'Histoire Naturelle, Paris 28: 39–41.

Angel, F. 1933. Les serpents de l'Afrique Occidentale Francais. Paris: Larose.

Angel, F. 1934. Remarques sur le genre *Oophilositum* Parker (Colubridae, Aglyphe) et description d'une espèce nouvelle. Bulletin de la Société Zoologique de France 59: 417–419.

Angel, F. 1938. Liste des reptiles de Mauritanie recueillis par la mission d'études de la biologie des Acridiens en 1936 et 1937: description d'une sous-espèce nouvelle d'*Eryx muelleri*. Bulletin du Muséum National d'Histoire Naturelle, Paris (série 2) 10: 485–487.

Angel, F., and H. Lhote. 1938. Reptiles et amphibiens du Sahara central et du Soudan. Bulletin du Comité d'Études Historiques et Scientifiques de l'Afrique Occidentale Française 21: 345–384.

Angel, F., J. Guibé, M. Lamotte, and R. Roy. 1954. La réserve naturelle intégrale du Mont Nimba. II. Serpents. Memoirs Institut Française d'Afrique Noire 40: 381–402.

Anthony, J. 1955. Essai sur l'évolution anatomique de l'appareil venimeux des ophidiens. Annales des Sciences Naturelles, Zoologie 11: 7–53.

Atkinson, P. M., B. A. Bradlow, J. A. M. White, H. B. W. Greig, and M. C. Gaillard. 1980. Clinical features of twig snake (*Thelotornis capensis*) envenomation. South African Medical Journal 58: 1007–1011.

Baldé, M. C., D. Dieng, A. P. Inapogui, A. O. Barry, H. Bah, and K. Kondé. 2002. Problématique des envenimations en Guinée. Bulletin de la Société de Pathologie Exotique 95: 157–159.

Baldé, M. C., A. M. B. Camara, H. Bah, A. O. Barry, and S. K. Camara. 2005. Incidence des morsures de serpent: enquête communautaire dans la collectivité rurale de développement (CRD) de Frilguiagbé (République de Guinée). Bulletin de la Société de Pathologie Exotique 98: 283–284.

Barbault, R. 1970. Recherches écologiques dans la savane de Lamto (Côte-d'Ivoire): les traits quantitatifs du peuplement ophidien. La Terre et la Vie 2: 94–107.

Barbault, R. 1971. Les peuplements d'Ophidiens des savanes de Lamto (Côte-d'Ivoire). Annals de l'Université d'Abidjan 4: 133–194.

Barbault, R. 1974. Observations écologiques dans la savane de Lamto (Côte-d'Ivoire): structure trophique de l'herpétocénose. Bulletin d'Ecologie 5: 7–25.

Bennefield, B. L. 1982. Case history of an *Elapsoidea semiannulata boulengeri* bite. Herpetological Association of Africa 27: 15–16.

Bianconi, J. J. 1847. Specimina Zoologica Mosambicana. Memorie della Reale Accademia delle Scienze dell'Istituto di Bologna 1: 171–189

Blackburn, D. G. 1985. Evolutionary origins of viviparity in the Reptilia. II. Serpentes, Amphisbaenia, and Ichthyosauria. Amphibia-Reptilia 5: 259–291.

Blaylock, R. S. 1983. Time of onset of clinical envenomation following snakebite in southern Africa. South African Medical Journal 64: 357–360.

Blyth, E. 1860. Report of curator, zoological department. Journal of the Asiatic Society of Bengal 29(1): 87–115.

Bocage, J. V. B. 1866a. Lista dos reptis das possessoes portuguezas d'Africa occidental que existem no Museu Lisboa. Jornal de Sciencias, Mathematicas, Physicas e Naturaes, Lisboa 1(1): 37–56.

Bocage, J. V. B. 1866b. Reptiles nouveaux ou peu connus recueillis dans les possessions portugaises de l'Afrique occidentale, qui se trouvent au muséum de Lisbonne. Jornal de Sciencias, Mathematicas, Physicas e Naturaes, Lisboa 1(1): 57–78, Plate 1.

Bocage, J. V. B. 1872. Diagnoses de quelques espèces nouvelles de reptiles d'Afrique occidentale. Jornal de Sciencias, Mathematicas, Physicas e Naturaes, Lisboa 4(13): 72–82.

Bocage, J. V. B. 1873. Mélanges herpétologiques. II. Sur quelques reptiles et batraciens nouveaux, rares ou peu connus de l'Afrique occidentale. Jornal de Sciencias, Mathematicas, Physicas e Naturaes, Lisboa 4(15): 209–227.

Bocage, J. V. B. 1879. Subsidios para a fauna das possessoes portuguezas d'Africa occidental. Jornal de Sciencias, Mathematicas, Physicas e Naturaes, Lisboa 7(26): 85–96.

Bocage, J. V. B. 1882. Reptiles rares ou nouveaux d'Angola. Jornal de Sciencias, Mathematicas, Physicas e Naturaes, Lisboa 8(32): 299–304.

Bocage, J. V. B. 1887a. Sur un python nouveau d'Afrique. Jornal de Sciencias, Mathematicas, Physicas e Naturaes, Lisboa 12(46): 87–88.

Bocage, J. V. B. 1887b. Mélanges erpétologiques. II. Reptiles de Dahomey. Jornal de Sciencias, Mathematicas, Physicas e Naturaes, Lisboa 11(44): 192–197.

Bocage, J. V. B. 1887c. Mélanges erpétologiques. IV. Reptiles du dernier voyage de MM. Capello et Ivens à travers l'Afrique. Jornal de Sciencias, Mathematicas, Physicas e Naturaes, Lisboa 11(44): 201–208.

Bocage, J. V. B. 1889. Sur une vipère apparemment nouvelle d'Angola. Jornal de Sciencias, Mathematicas, Physicas e Naturaes, Lisboa (2)2: 127–128.

Bocage, J. V. B. 1890. Sur une pièce nouvelle à ajouter à la faune herpétologique de Saint Thomas et Rolas. Jornal de Sciencias, Mathematicas, Physicas e Naturaes, Lisboa 2: 61–62.

Bocage, J. V. B. 1895a. Herpétologie d'Angola et du Congo. Lisbonne: Imprimerie Nationale, 203 pp., 19 plates.

Bocage, J. V. B. 1895b. Reptiles et batraciens nouveaux ou peu connus de Bioko. Jornal de Sciencias, Mathematicas, Physicas e Naturaes, Lisboa 4: 15–20.

Boettger, O. 1888. Materialen zur Fauna des unteren Congo II Reptilien und Batrachier. Berichte der Senckenbergischen Naturforschenden Gesellschaft in Frankfurt-am-Main 1888: 3–108.

Boettger, O. 1892. Drei neue colubriforme Schlangen. Zoologischer Anzieger 15: 417–420.

Boettger, O. 1893. Übersicht der von Prof. C. Keller anlasslich der Ruspoli'schen Expedition nach den Somalilandern gesammelten Reptilien und Batrachier. Zoologischer Anzieger 16: 113–119 and 129–132.

Bogert, C. M. 1940. Herpetological results of the Vernay Angola Expedition with notes on African reptiles in other collections. I. Snakes, including an arrangement of African Colubridae. Bulletin of the American Museum of Natural History 77: 1–107.

Bogert, C. M. 1943. Dentitional phenomena in cobras and other elapids with notes on adaptative modifications of fangs. Bulletin of the American Museum of Natural History 77: 285–360.

Böhme, W. 1975. Zur Herpetofaunistik Kameruns, mit Beschreibung einens neuen Scinciden. Bonner Zoologische Beiträge, Bonn 26: 2–48.

Böhme, W., and S. De Pury. 2011. A note on the generic allocation of *Coluber moilensis* Reuss 1834 (Serpentes: Psammophiidae). Salamandra 47(2): 120–123.

Boie, F. 1826. General Übersicht der Familien und Gattungen der Ophidia. Isis von Oken, Jena 1826: 981–982.

Boie, F. 1827. Bemerkungen uber Merremis Versuch eines Systems der Amphibiens. 1te Lieferung, Ophidier. Isis von Oken, Jena 20: 508–566.

Boulenger, G. A. 1888. On new or little known South African reptiles. Annals and Magazine of Natural History, London 2(6): 136–141.

Boulenger, G. A. 1890. Description of a new snake of the genus *Glauconia* Gray, obtained by Dr. Emin Pasha on the Victoria Nyanza. Annals and Magazine of Natural History, London 6(6): 91–93.

Boulenger, G. A. 1892. Description of a new snake from Nubia. Annals of Natural History 9: 74–76.

Boulenger, G. A. 1893. Catalogue of the snakes in the British Museum (Natural History). Vol 1. London: Taylor and Francis, 440 pp., 28 plates.

Boulenger, G. A. 1894. Catalogue of the snakes in the British Museum (Natural History). Volume II. Containing the conclusion of the Colubriday aglyphae. London: British Museum (Natural History), 377 pp., 20 plates.

Boulenger, G. A. 1895a. On some new or little-known reptiles obtained by W. H. Crosse Esq. on the Niger. Annals of Natural History 16: 32–34.

Boulenger, G. A. 1895b. Descriptions of two new snakes from Usambara, German east Africa. Annals of Natural History 16: 171–173.

Boulenger, G. A. 1896. Catalogue of the snakes in the British Museum (Natural History). Volume III. Containing the Colubridae (Opisthoglyphae and Proteroglyphae), Amblyycephalidae, and Viperidae. London: British Museum (Natural History), 727 pp., 25 plates.

Boulenger, G. A. 1897a. A list of reptiles and batrachians from the Congo Free State with descriptions of two new snakes. Annals of Natural History 19: 276–281.

Boulenger, G. A. 1897b. Description of a new snake from Sierra Leone. Annals of Natural History 19: 154.

Boulenger, G. A. 1898. Descriptions of two new blind snakes. Annals and Magazine of Natural History, London 1(7): 124.

Boulenger, G. A. 1901. Matériaux pour la faune du Congo: batraciens et reptiles nouveaux. Annales du Musée du Congo (Zoologie), Bruxelles 2: 7–14.

Boulenger, G. A. 1903. Descriptions of new snakes in the collection of the British Museum. Annals of Natural History 12: 350–354.

Boulenger, G. A. 1904a. Descriptions of two new

elapine snakes from the Congo. Annals of Natural History 14: 14–15.

Boulenger, G. A. 1904b. Descriptions of three new snakes. Annals of Natural History 13: 450–452.

Boulenger, G. A. 1905. A list of the batrachians and reptiles collected by Dr. W. J. Ansorge in Angola, with descriptions of new species. Annals of Natural History 16: 105–115.

Boulenger, G. A. 1906a. Report on the reptiles collected by the late L. Fea in west Africa. Annali dell Museo Civico di Storia Naturale di Genova 2: 196–216.

Boulenger, G. A. 1906b. Additions to the herpetology of British east Africa. Proceedings of the Zoological Society of London 1906: 570–573.

Boulenger, G. A. 1908. Description of three new snakes from Africa. Annals of Natural History 2: 93–94.

Boulenger, G. A. 1910. Description of four new African snakes in the British Museum. Annals of Natural History 5: 512–513.

Boulenger, G. A. 1911. Descriptions of three new snakes discovered by Mr. G. L. Bates in south Cameroon. Annals of Natural History 8: 370–371.

Boulenger, G. A. 1913. Description de deux reptiles nouveaux provenant du Katanga. Revue Zoologique Africaine, Tervuren 3: 103–105.

Boulenger, G. A. 1915. A list of the snakes of the Belgian and Portuguese Congo, northern Rhodesia and Angola. Proceedings of the Zoological Society of London 85(2): 193–223.

Boulenger, G. A. 1919a. Descriptions d'ophidien et d'un batracien nouveaux de Congo. Revue Zoologique Africaine, Tervuren 7(2): 186–187.

Boulenger, G. A. 1919b. Batraciens et reptiles recueillis par le Dr. C. Christy au Congo Belge dans les Districts de Stanleyville, Haut-Uélé et Ituri en 1912–1913. Revue Zoologique Africaine, Tervuren 7: 1–29.

Bourgeois, M. 1961. *Atractaspis*—a misfit among the Viperidae? News Bulletin of the Zoological Society of South Africa 3: 29.

Bourgeois, M. 1963a. Note sur *Atractaspis irregularis conradsi* Sternfeld (Viperidae). Structure du crâne et l'appareil de la morusre. Annales de la Société Royale Zoologique de Belgique 93(1): 159–169.

Bourgeois, M. 1963b. Notes sur le crâne de *Miodon* (Colubridae). Annales de la Société Royale Zoologique de Belgique 93(1): 171–178.

Bourgeois, M. 1965. Contribution à la morphologie comparée du crâne des ophidiens de l'Afrique Centrale. Publications de l'Universitè Officielle du Congo à Lubumbashi 18: 293 pp.

Branch, D., and W. R. Branch. 2004. *Bitis arietans*: Puff Adder: arboreal behavior. African Herp News 37: 26–27.

Branch, W. R. 1983. Comments on "Herpetology of Africa: a checklist and bibliography of the orders Amphisbaenia, Sauria and Serpentes, by Kenneth R. G. Welch, 1982." Journal of the Herpetological Association of Africa 29: 22–36.

Branch, W. R. 1986. Hemipenial morphology of African snakes: a taxonomic review. I. Scolecophidia and Boidae. Journal of Herpetology 20: 285–299.

Branch, W. R. 1994. Field guide to the snakes and other reptiles of southern Africa. 2nd ed. Cape Town, South Africa: Struik.

Branch, W. R., and W. D. Håcke. 1980. A fatal attack on a young boy by an African rock python, *Python sebae*. Journal of Herpetology 14: 305–307.

Branch, W. R., G. V. Haagner, and R. Shine. 1995. Is there an ontogenetic shift in mamba diet? Taxonomic confusion and dietary records for Black and Green Mambas (*Dendroaspis*: Elapidae). Herpetological Natural History 3: 171–178.

Brandstätter, F. 1995. Eine revision der gattung *Psammophis* mit berücksichtigung der Schwesterngattungen innerhalf der Tribus Psammophiini (Colubridae, Lycodontinae). Doctoral Dissertation, Universität Saarlandes, Saarbrücken, 480 pp.

Broaders, M., and M. F. Ryan. 1997. Enzymatic properties of the Duvernoy's secretion of Blanding's tree snake (*Boiga blandingi*) and of the mangrove snake (*Boiga dendrophila*). Toxicon 35: 1143–1148.

Broaders, M., C. Faro, and M. F. Ryan. 1999. Partial purification of acetylcholine receptor binding components from the Duvernoy's secretions of Blanding's Tree Snake (*Boiga blandingi*) and the Mangrove Snake (*Boiga dendrophila*). Journal of Natural Toxins 8: 155–166.

Broadley, D. G. 1971a. A revision of the African snake genus *Elapsoidea* Bocage (Elapidae). Occasional Papers of the National Museum of Rhodesia, Salisbury 4: 577–626.

Broadley, D. G. 1971b. A revision of the African snake genera *Amblyodipsas* and *Xenocalamus* (Colubridae). Occasional Papers of the National Museum of Rhodesia, Salisbury 4: 629–697.

Broadley, D. G. 1971c. A review of *Rhamphiophis acutus* (Günther) with the description of a new sub-species from Zambia (Serpentes: Colubridae). Arnoldia Rhodesia 5: 1–8.

Broadley, D. G. 1972. The herpetology of Southern Rhodesia. Part 1: Snakes. Bulletin of the Museum of Comparative Zoology 120(1), 100 pp, 6 Plates.

Broadley, D. G. 1973. Addenda and Corrigenda to the reptiles and amphibians of Zambia. Puku 7: 93–95.

Broadley, D. G. 1974. The Striped Swamp Snake Limnophis bicolor—a new snake for Rhodesia. Journal of the Herpetological Association of Africa 12: 8.

Broadley, D. G. 1975. A review of Psammophis leightoni and Psammophis notostictus in southern Africa (Serpentes: Colubridae). Arnoldia Rhodesia 7: 1–17.

Broadley, D. G. 1979. Predation on reptile eggs by African snakes of the genus Prosymna. Herpetologica 35: 338–341.

Broadley, D. G. 1980. A revision of the African genus Prosymna Gray (Colubridae). Occasional Papers of the National Museum of Rhodesia, Salisbury 6: 481–556.

Broadley, D. G. 1983. FitzSimons snakes of southern Africa. Johannesburg, South Africa: Delta Books, 376 pp.

Broadley, D. G. 1987. Caudal autotomy in African snakes of the genera Natriciteres Loveridge and Psammophis Boie. Journal of the Herpetological Association of Africa 33: 18–19.

Broadley, D. G. 1991a. A review of the Namibian snakes of the genus Lycophidion (Serpentes: Colubridae), with the description of a new endemic species. Annals of the Transvaal Museum 35: 209–215.

Broadley, D. G. 1991b. The herpetofauna of northern Mwinilunga District, northwestern Zambia. Arnoldia Zimbabwe 9(37): 519–538.

Broadley, D. G. 1994. A revision of the African genus Scaphiophis Peters (Serpentes: Colubridae). Herpetological Journal 4: 1–10.

Broadley, D. G. 1997. A review of Hemirhagerrhis viperina (Bocage) (Serpentes: Colubridae), a rupicolous Psammophine snake. Madoqua 19: 161–169.

Broadley, D. G. 1998. A review of the genus Atheris Cope (Serpentes: Viperidae), with the description of a new species from Uganda. Herpetological Journal 8: 117–135.

Broadley, D. G. 2001. A review of the genus Thelotornis A. Smith in eastern Africa, with the description of a new species from the Usambara Mountains (Serpentes: Colubridae: Dispholidini). African Journal of Herpetology 50: 53–70.

Broadley, D. G. 2002. A review of the species of Psammophis Boie found south of latitude 12°S (Serpentes: Psammophiinae). African Journal of Herpetology 51: 83–119.

Broadley, D. G., and E. V. Cock. 1975. Snakes of Rhodesia. Salisbury: Longman Rhodesia, 152 pp.

Broadley, D. G., and B. Hughes. 1993. A review of the genus Lycophidion (Serpentes: Colubridae) in northeastern Africa. Herpetological Journal 3: 8–18.

Broadley, D. G., and V. Wallach. 2002. Review of the Dispholidini, with the description of a new genus and species from Tanzania (Serpentes, Colubridae). Bulletin of the Natural History Museum of London (Zoology) 68: 57–74.

Broadley, D. G., and V. Wallach. 2009. A review of the eastern and southern African Blind-Snakes (Serpentes: Typhlopidae), excluding Letheobia Cope, with the description of two new genera and a new species. Zootaxa 2255: 1–100.

Broadley, D. G., C. T. Doria, and J. Wigge. 2003. Snakes of Zambia: an atlas and field guide. Frankfurt: Edition Chimaira, 280 pp.

Buchholz, R. W., and W. K. H. Peters. 1876. Eine zweite Mittheilung über die von Hrn. Prof. Dr. Buchholz in Westafrika gesammelten Amphibien. Monatsberichte der königlich Akademie der Wissenschaften zu Berlin: 117–123.

Buffrénil, V. D. 1995. Les élevages de reptiles du Bénin, du Ghana et du Togo. Report to the Secretariat of the Convention on International Trade in Endangered Species of Wild Fauna and Flora (CITES), Lausanne, 38 pp.

Bush, S. P., S. M. Green, T. A. Laack, W. K. Hayes, M. D. Cardwell, and D. A. Tanen. 2004. Pressure immobilization delays mortality and increases intracompartmental pressure after artificial intramuscular rattlesnake envenomation in a porcine model. Annals of Emergency Medicine 44: 599–604.

Butler, J. A. 1993. Seasonal reproduction in the African Olive Grass Snake, Psammophis phillipsi (Serpentes: Colubridae). Journal of Herpetology 27: 144–148.

Cadle, J. E. 1984a. Molecular systematics of neotropical xenodontine snakes. I. South American xenodontines. Herpetologica 40: 8–20.

Cadle, J. E. 1984b. Molecular systematics of neotropical xenodontine snakes. II. Central American xenodontines. Herpetologica 40: 21–30.

Cadle, J. E. 1984c. Molecular systematics of neotropical xenodontine snakes. III. Overview of xenodontine phylogeny and the history of New World snakes. Copeia 1984: 641–652.

Cadle, J. E. 1985. The neotropical snake fauna: lineage

components and biogeography. Systematic Zoology 34:1–20.

Cadle, J. E. 1994. The colubrid radiation in Africa (Serpentes: Colubridae): phylogenetic relationships and evolutionary patterns based on immunological data. Zoological Journal of the Linnean Society 110: 103–140.

Calvete, J. J., J. Escolano, and L. Sanz. 2007. Snake venomics of *Bitis* species reveals large intragenus venom toxin composition variation: application to taxonomy of congeneric taxa. Journal of Proteome Research 6: 2732–2745.

Cansdale, G. S. 1961. West African Snakes. London: Longmans.

Chabanaud, P. 1916. Énumération des ophidiens non encore étudiés de l'Afrique Occidentale, appartenant aux collections du muséum avec la description des espèces et des variétés nouvelles. Bulletin du Muséum National d'Histoire Naturelle, Paris 22: 362–382.

Chabanaud, P. 1917. Note complémentaire sur les ophidiens de l'Afrique Occidentale, appartenant aux collections du muséum, avec la description d'une espèce nouvelle. Bulletin du Muséum National d'Histoire Naturelle, Paris 23: 7–14.

Chani, M., H. L'kassimi, A. Abouzahir, M. Nazi, and G. Mion. 2008. A propos de trois observations d'envenimations vipérines graves au Maroc. Annales Françaises d'Anesthésie et de Réanimation 27(4): 330–334.

Chani, M., A. Abouzahir, C. Haimeur, N. D. Kamili, and G. Mion. 2012. Accident vasculaire cérébral ischémique à la suite d'une envenimation vipérine grave au Maroc, traitée par un antivenin inadapté. Annales Françaises d'Anesthésie et de Réanimation 31(1): 82–85.

Chérifi, F., and F. Laraba-Djebari. 2013. Isolated biomolecules of pharmacological interest in hemostasis from *Cerastes cerastes* venom. Journal of Venomous Animals and Toxins Including Tropical Diseases 19(1): 11.

Chérifi, F., A. Namane, and F. Laraba-Djebari. 2014. Isolation, functional characterization and proteomic identification of CC2-PLA$_2$ from *Cerastes cerastes* venom: a basic platelet-aggregation-inhibiting factor. Protein Journal 33(1): 61–74.

Cherlin, V. A. 1990. Taxonomic revision of the snake genus *Echis* (Viperidae). II. An analysis of taxonomy and description of new forms. *In* Borkin, L. J., Reptiles of mountain and arid territories: systematics and distribution. Proceedings of the Zoological Institute, Leningrad, USSR Academy of Science 207: 193–223.

Chiasson, R. B. 1981. The apical pits of *Agkistrodon*. Journal of the Arizona–Nevada Academy of Science 16(3): 69–73.

Chippaux, C., H. L. O'Connor, P. Nosny, J. Plessis, M. Ducloux, and P. Laluque. 1961. Nécroses par morsures de serpent: à propos de douze observations. La Presse Médicale 69: 583–585.

Chippaux, J. P. 1982. Complications locales des morsures de serpents. Médecine Tropicale 42: 177–183.

Chippaux, J. P. 1998a. Snake bites: appraisal of the global situation. Bulletin of the World Health Organization 76: 515–524.

Chippaux, J. P. 1998b. The development and use of immunotherapy in Africa. Toxicon 36: 1503–1506.

Chippaux, J. P. 1999. L'envenimation ophidienne en Afrique: épidémiologie, clinique et traitement. Annales de l'Institut Pasteur / Actualités 10: 161–171.

Chippaux, J. P. 2000. Prevention of snake bites and management of envenomations in Africa. African Newsletter on Occupational Health and Safety 10: 12–15.

Chippaux, J. P. 2002a. Epidémiologie des morsures de serpent au Bénin. Bulletin de la Société de Pathologie Exotique 95: 172–174.

Chippaux, J. P. 2002b. The treatment of snake bites: analysis of requirements and assessment of therapeutic efficacy in tropical Africa. *In* Ménez, A. ed., Perspectives in molecular toxinology. New York: John Wiley & Sons, pp. 457–472.

Chippaux, J. P. 2005. Evaluation de la situation épidémiologique et des capacités de prise en charge des envenimations ophidiennes en Afrique subsaharienne francophone. Bulletin de la Société de Pathologie Exotique 98: 263–268.

Chippaux, J. P. 2006a. Les serpents d'Afrique Occidentale et Centrale. Faune et flore tropicales 35. 3rd ed. Paris: IRD, 311 pp.

Chippaux, J. P. 2006b. Snake venoms and envenomations. Malabar, FL: Krieger, 287 pp.

Chippaux, J. P. 2009. Snakebite in Africa: current situation and urgent needs. *In* MacKessy, S. P., ed., Handbook of venoms and toxins of reptiles. Boca Raton, FL: Taylor & Francis / CRC Press, pp. 435–473.

Chippaux, J. P. 2011. Estimate of the burden of snakebites in sub-Saharan Africa: a meta-analytic approach. Toxicon 57: 586–599.

Chippaux, J. P., and C. Bressy. 1981. L'endémie ophidienne des plantations de Côte d'Ivoire. Bulletin de la Société de Pathologie Exotique 74: 458–467.

Chippaux, J. P., and A. Diallo. 2002. Evaluation de l'incidence des morsures de serpent en zone de sahel sénégalais, l'exemple de Niakhar. Bulletin de la Société de Pathologie Exotique 95: 151–153.

Chippaux, J. P., and M. Goyffon. 1998. Venoms, antivenoms and immunotherapy. Toxicon 36: 823–846.

Chippaux, J. P., B. Courtois, D. Roumet, and R. Eyebi-yi. 1977. Envenimation par morsure de Mamba (Dendroaspis viridis): à propos d'une envenimation à évolution favorable. Médecine Tropicale 37: 545–549.

Chippaux, J. P., G. N-Guessan, F. X. Paris, G. Rolland, and M. Kébé. 1978. Spitting Cobra (Naja nigricollis) bite. Transactions of the Royal Society of Tropical Medicine and Hygiene 72: 106.

Chippaux, J. P., J. Lang, S. Amadi-Eddine, P. Fagot, V. Rage, J. C. Peyrieux, V. Le Mener, and VAO Investigators. 1998. Clinical safety of a polyvalent F(ab')₂ equine antivenom in 223 African snake envenomations: a field trial in Cameroon. Transactions of the Royal Society of Tropical Medicine and Hygiene 92: 657–662.

Chippaux, J. P., J. Lang, S. Amadi-Eddine, P. Fagot, and V. Le Mener. 1999. Short report: treatment of snake envenomations by a new polyvalent antivenom composed of highly purified F(ab')₂: results of a clinical trial in northern Cameroon. American Journal of Tropical Medicine and Hygiene 61: 1017–1018.

Chippaux, J. P., S. Rakotonirina, G. Dzikouk, S. Nkinin, and A. Rakotonirina. 2001. Connaissances actuelles et perspectives de la phytopharmacopée dans le traitement des envenimations ophidiennes. Bulletin de la Société Herpétologique de France 97: 5–17.

Chippaux, J. P., V. Rage-Andrieux, V. Le Mener-Delore, M. Charrondière, P. Sagot, and J. Lang. 2002. Epidémiologie des envenimations ophidiennes au Nord-Cameroun. Bulletin de la Société de Pathologie Exotique 95: 184–187.

Chippaux, J. P., B. Ramos-Cerrillo, and R. P. Stock. 2007a. Study of the efficacy of the black stone on envenomation by snake bite in the murine model. Toxicon 49: 717–720.

Chippaux, J. P., A. Massougbodji, R. P. Stock, and A. Alagón. 2007b. Clinical trial of a F(ab0)2 polyvalent equine antivenom for African snakebites in Benin. American Journal of Tropical Medicine and Hygiene 77: 538–546.

Chippaux, J.-P., R. P. Stock, and A. Massougbodji. 2015a. Antivenom safety and tolerance for the strategy of snake envenomation management. In P. Gopalakrishnakone, H. Inagaki, A. K. Mukherjee, T. R. Rahmy, and C.W. Vogel, eds., Snake venom toxinology. Dordrecht, Netherlands: Springer, pp. 1–16.

Chippaux, J. P., M. C. Baldé, E. Sessinou, M. C. Diallo, M. Y. Boiro, and A. Massougbodji. 2015b. Evaluation d'un nouvel antivenin polyvalent contre les envenimations ophidiennes (Inoserp® Panafricain) dans deux contextes épidémiologiques: le Nord Bénin et la Guinée Maritime. Médecine et Santé Tropicales 25: 56–64.

Chirio, L., and I. Ineich. 1991. Les genres Rhamphiophis Peters, 1854 et Dipsina Jan, 1863 (Serpentes, Colubridae): revue des taxons reconnus et description d'une espèce nouvelle. Bulletin du Muséum National d'Histoire Naturelle, Paris 13: 217–235.

Chirio, L., and I. Ineich. 2006. Biogeography of the reptiles of the Central African Republic. African Journal of Herpetology 55(1): 23–59.

Chirio, L., and M. LeBreton. 2007. Atlas des reptiles du cameroun. Patrimoines Naturels 67. Paris: Muséum National d'Histoire Naturelle, IRD.

Christensen, P. A. 1955. South African snake venoms and antivenoms. Johannesburg: South African Institute for Medical Research, 142 pp.

Colclough, P. 2016. Angolan python care tips and secrets. Reptile Magazine, accessed December 30. http://www.reptilesmagazine.com/Snakes/Snake-Care/Angolan-Python-Care-Tips-and-Secrets/.

Cole, L. R. 1967. The snake Miodon acanthias found with Geotrypetes seraphini (Amphibia: Caeciliidae) as prey. Copeia 1967(4): 862.

Cope, E. D. 1861. Contributions to the ophiology of lower California, Mexico and Central America. Proceedings of the Academy of Natural Sciences of Philadelphia 13(7): 292–306.

Cope, E. D. 1862. Notes upon some reptiles of the Old World. Proceedings of the Academy of Natural Sciences of Philadelphia 14(5): 337–344.

Cope, E. D. 1869. Observations on reptiles of the old world. Proceedings of the Academy of Natural Sciences of Philadelphia 20: 316–323.

Cope, E. D. 1900. The crocodilians, lizards and snakes of North America. Report of the United States National Museum 1898: 153–1270, Plates 1–35.

Corkill, N. L., C. J. Ionides, and C. R. Pitman. 1959. Biting and poisoning by the mole vipers of the genus *Atractaspis*. Transactions of the Royal Society of Tropical Medicine and Hygiene 53: 95–101.

Courtois, B. 1979. A propos de deux spécimens d'*Atractaspis corpulenta* (Hallowell) trouvés en Côte-d'Ivoire. Bulletin de l'Institut Fondamental d'Afrique Noire, Dakar A41: 206–209.

Cundall, D. 2001. Functional morphology. *In* R. A. Siegel, J. T. Collins, and S. S. Novak, eds., Snakes: ecology and evolutionary biology. 2nd ed. Caldwell, NJ: Blackburn Press, pp. 106–142.

Cundall, D., and H. W. Greene. 2000. Feeding in snakes. *In* K. Schwenk, ed., Feeding: form, function and evolution in tetrapod vertebrates. San Diego, CA: Academic Press, pp. 293–333.

Currier, R. B., R. A. Harrison, P. D. Rowley, G. D. Laing, and S. C. Wagstaff. 2010. Intra-specific variation in venom of the African Puff Adder (*Bitis arietans*): differential expression and activity of snake venom metalloproteinases (SVMPs). Toxicon 55: 864–873.

Daudin, F. M. 1803a. Histoire naturelle, générale et particulière des reptiles, Ouvrage faisant suite à l'histoire naturelle générale et particulière composée par Leclerc de Buffon et rédigée de C. S. Sonnini. Tome septième. Paris: F. Dufart, 436 pp.

Daudin, F. M. 1803b. Histoire naturelle, générale et particulière des reptiles, Ouvrage faisant suite à l'histoire naturelle générale et particulière composée par Leclerc de Buffon et rédigée de C. S. Sonnini. Tome cinquième. Paris: F. Dufart, 447 pp.

Deufel, A., and D. Cundall. 2003. Feeding in Atractaspis (Serpentes: Atractaspididae): a study in conflicting functional constraints. Zoology 106: 42–61.

Dollo, L. 1886. Notice sur les reptiles et batraciens recueillis par M. le Capitaine Em. Storms dans la region du Tanganyika. Bulletin du Musée Royal d'Histoire Naturelle de Belgique, Bruxelles 4: 151–160.

Doucet, J. 1963a. Les serpents de la République de Côte d'Ivoire. Serpents venimeux. Acta Tropica 20: 297–340.

Doucet, J. 1963b. Les serpents de la République de Côte d'Ivoire. Généralités et serpents non venimeux. Acta Tropica 20: 201–259.

Doucet, J., and P. Lepesme. 1953. Sur un cas d'envenimation par *Atractaspis*, Vipéridé ouest-africain. Bulletin de l'Institut Fondamental d'Afrique Noire 15: 855–859.

Dowling, H. G. 1951. A proposed method of expressing scale reductions in snakes. Copeia 1951: 131–134.

Dowling, H. G. 1969. Relations of some African colubrid snakes. Copeia 1969: 234–243.

Dowling, H. G., and J. M. Savage. 1960. A guide to the snake hemipenis: a survey of basic structure and systematic characteristics. Zoologica 45: 17–28.

Duméril, A. H. A. 1856. Note sur les reptiles du Gabon. Revue et Magasin de Zoologie 7: 369–375; 417–424; 460–470.

Duméril, A. M. C. 1853. Prodrome de la classification des reptiles ophidiens. Mémoires de l'Académie de Sciences de l'Institut de France, series ii, 23: 399–536, 2 plates.

Duméril, A. M. C., and G. Bibron. 1844. Erpétologie générale ou histoire naturelle complète des reptiles. Tome sixième, comprenant l'histoire générale des ophidiens, la description des genres et des espèces de serpents non venimeux, savoir, la totalité des vermiformes ou des scolécophides, et partie des circuriformes ou azémiophides; en tout vingt-cinq genres et soixante-cinq espèces. Paris: Librairie Encyclopédique de Roret, xii + 609 pp.

Duméril, A. M. C., G. Bibron, and A. H. A. Duméril. 1854a. Erpétologie générale ou histoire naturelle complète des reptiles. Tome septième-Deuxième partie. Comprenant l'histoire des serpents venimeux. Paris: Librairie Encyclopédique de Roret, xii + 781–1536 pp.

Duméril, A. M. C., G. Bibron, and A. H. A. Duméril. 1854b. Erpétologie générale ou histoire naturelle complète des reptiles. Tome septième-Première partie. Comprenant l'histoire des serpents venimeux. Paris: Librairie Encyclopédique de Roret, vii + 1–780.

Dunger, G. T. 1966. A new species of the colubrid genus *Mehelya* from Nigeria. American Museum Novitates 2268: 1–8.

Dunn, E. R. 1928. A tentative key and arrangement of the American genera of Colubridae. Bulletin of the Antivenin Institute of America. 2: 19–24.

Duvernoy, D. M. 1832. Mémoire sur les caractères tirés de l'anatomie pour distinguer les serpents venimeux des serpents non venimeux. Annales des Sciences Naturelles 26: 113–160.

Enwere, G. C., H. A. Obu, and A. Jobarteh. 2000. Snake bites in children in the Gambia. Annals of Tropical Paediatrics 20: 121–124.

Ernst, R., and M. O. Rödel. 2002. A new Atheris species (Serpentes: Viperidae), from Taï National Park, Ivory Coast. Herpetological Journal 12: 55–61.

Fernandez, S., W. Hodgson, J. Chaisakul, R. Kornhauser, N. Konstantakopoulos, A. I. Smith, and

S. Kuruppu. 2014. *In vitro* toxic effects of puff adder (*Bitis arietans*) venom, and their neutralization by antivenom. Toxins 6: 1586–1597.

Fischer, J. G. 1856. Neue Schlangen der Hamburgischen Naturhistorischen Museums. Abhandlungen aus dem Gebiete der Naturwissenschaften herausgegeben von Naturwissenschaftlichen Verein in Hamburg 3: 79–116.

Fischer, J. G. 1888. Herpetologische Mitteilungen. I. Über zwei neue Schlangen und einen neuen Laubfrosch aus Kamerun. Jahrbuch der Hamburgischen Wissenschaftlichen Anstalten, Hamburg 5: 3–10.

Fitzinger, L. J. F. T. 1826. Neue classification der Reptilien nach ihren naturlichen Verwandtschaften. Vienna: J. G. Heubner, 66 pp., 1 plate.

Fitzinger, L. J. F. T. 1843. Systema reptilium. Fasciculus primus, Amblyglossae. Vienna: Braumüller et Seidel, 106 pp.

FitzSimons, V. F. M. 1962. Snakes of southern Africa. London: Macdonald, 423 pp.

FitzSimons, V. F. M. 1974. A field guide to the snakes of southern Africa. London: Collins, 221 pp.

Forskål, P. 1775. Descriptiones Animalium, Avium, Amphibiorum, Piscium, Insectorum, Vermium: quae in Itinere orientali observavit Petrus Forskal . . . post mortem auctoris editit Carsten Niebuhr. Hauniae: Officina Mölleri Typographi apud Heineck et Faber, xxxiv + 164 pp.

Frank, N., and E. Ramus. 1996. A complete guide to scientific and common names of reptiles and amphibians of the world. Pottsville, PA: N. G. Publishing, 377 pp.

Gans, C. 1952. The functional morphology of the egg-eating adaptions in the snake genus *Dasypeltis*. Zoologica 37: 209–243.

Gans, C. 1959. A taxonomic revision of the African snake genus *Dasypeltis* (Reptilia, Serpentes). Annales du Musée Royal du Congo Belge (Série in Octavo, Science Zoologique). Tervuren 74: 1–237.

Gans, C. 1961. Mimicry in procryptically colored snakes of the genus *Dasypeltis*. Evolution 15(1): 72–91.

Gans, C. 1974. Biomechanics: an approach to vertebrate biology. Philadelphia: J. B. Lippincott, 261 pp.

Gans, C., and N. Richmond. 1957. Warning behavior in snakes of the genus *Dasypeltis*. Copeia 1957: 269–274.

Gartner, G. E. A., and H. W. Greene. 2008. Adaptation in the African egg-eating snake: a comparative approach to a classic study in evolutionary functional morphology. Journal of Zoology 275: 368–374.

Geoffroy-Saint-Hilaire, I. 1827. Description des reptiles qui se trouvent en Egypte. *In* Savigny, M. J. C. L., ed., Description de l'Egypte ou recueil des observations et des recherches qui ont été faites en Egypte pendant l'expédition de l'armée française (1798–1801). I. Paris Histoire Naturelle, Imprimerie Impériale, pp. 121–160.

Gervais, P. 1857. Sur quelques ophidiens de l'Algérie. Mémoires de la Section Scientifique, Académie des Sciences et Lettres, Montpellier 3: 511–512.

Gillissen, A., R. D. Theakston, J. Barth, B. May, M. Krieg, and D. A. Warrell. 1994. Neurotoxicity, haemostatic disturbances and haemolytic anaemia after a bite by a Tunisian saw-scaled or carpet viper (*Echis 'pyramidum'*-complex): failure of antivenom treatment. Toxicon 32: 937–944.

Glaw, F., Z. T. Nagy, and M. Vences. 2007. Phylogenetic relationships and classification of the Malagasy pseudoxyrhophiine snake genera *Geodipsas* and *Compsophis* based on morphological and molecular data. Zootaxa 1517: 53–62.

Gmelin, J. F. 1789. Caroli a Linne Systema Naturae per Regna tria Naturae secundum classes, ordines, genera, species cum characteribus, differentiis, synonymis, locis. III. Amphibia et Pisces. Leipzig: Georg. Emanuel Beer, pp. 1038–1516.

Goodman, J. D. 1985. Two record size Blanding's Tree Snakes from Uganda. East African Natural History Society Bulletin 1985: 56–57.

Gower, D., J. Rasmussen, S. Loader, and M. Wilkinson. 2004. The caecilian amphibian *Scolecomorphus kirkii* Boulenger as prey of the burrowing asp *Atractaspis aterrima* Günther: trophic relationships of fossorial vertebrates. African Journal of Ecology 42: 83–87.

Gras, S., G. Plantefève, F. Baud, and J. P. Chippaux. 2012. Snakebite on the hand: lessons from two clinical cases illustrating difficulties of surgical indication. Journal of Venomous Animals and Toxins Including Tropical Diseases 18: 467–477.

Grasset, P. E. 1946. La vipère du Gabon: envenimation par *Bitis Gabonica*: son venin et sérothérapie antiveneuse spécifique. Acta Tropica 3: 97–115.

Gray, B. S. 2011. A study of apical pits using shed snakeskins revisited. Bulletin of the Chicago Herpetological Society 46(10): 125–128.

Gray, J. E. 1842. Monographic synopsis of the vipers, or the family Viperidae. Zoological Miscellany 1: 68–71.

Gray, J. E. 1849. Catalogue of the specimens of snakes

in the collection of the British Museum. London: Edward Newman, pp. i–xv, 1–125.

Gray, J. E. 1858. Description of a new genus of Boidae from Old Calabar and a list of W. African reptiles. Proceedings of the Zoological Society of London 1858: 154–167.

Greene, H. W. 1997. Snakes: the evolution of mystery in nature. Berkeley: University of California Press, pp. 1–365.

Greenbaum, E., F. Portillo, K. Jackson, and C. Kusamba. 2015. A phylogeny of central African Boaedon (Serpentes: Lamprophiidae), with the description of a new cryptic species from the Albertine Rift. African Journal of Herpetology 64(1): 18–38.

Greenham, R. 1978. Spitting Cobra (*Naja mossambica pallida*) bite in a Kenyan child. Transactions of the Royal Society of Tropical Medicine and Hygiene 72(6): 674–675.

Guibé, J. 1952. *Typhlops angeli* (Serpent), espèce nouvelle du Mont Nimba. Bulletin du Muséum National d'Histoire Naturelle, Paris, (2)24(1): 79.

Guillin, M., A. Bezeaud, and D. Menache. 1978. The mechanism of activation of human prothrombin by an activator isolated from *Dispholidus typus* venom. Biochimica et Biophysica Acta 537: 160–168.

Günther, A. 1858. Catalogue of the colubride snakes in the collection of the British Museum. London, 281 pp.

Günther, A. 1859. Description of a new genus of west African snakes and revision of the South American Elaps. Annals of Natural History 4: 161–174.

Günther, A. 1862. On new species of snakes in the collection of the British Museum. Annals of Natural History 9: 124–132.

Günther, A. 1863a. Third account of new species of snakes in the collection of the British Museum. Annals of Natural History 12: 348–365.

Günther, A. 1863b. On some species of tree-snakes (*Ahaetulla*). Annals of Natural History 11: 283–287.

Günther, A. 1864. Report on a collection of reptiles and fishes made by Dr. Kirk in the Zambesi and Nyassa regions. Proceedings of the Zoological Society of London 1864: 303–314.

Günther, A. 1865. Fourth account of new species of snakes in the collection of the British Museum. Annals of Natural History 15: 89–98.

Günther, A. 1866. Fifth account of new species of snakes in the collection of the British Museum. Annals of Natural History 18: 24–29.

Günther, A. 1868. Sixth account of new species of

snakes in the collection of the British Museum. Annals of Natural History 1: 413–429.

Günther, A. 1872. Seventh account of new species of snakes in the collection of the British Museum. Annals and Magazine of Natural History, London 9: 15–37.

Günther, A. 1874. Description of some new or imperfectly known species of reptiles from the Cameroon Mountains. Proceedings of the Zoological Society of London 1874: 442–445.

Günther, A. 1881. Appendix III. Herpetology. *In* Oates, F. Matabele land and the Victoria falls, a naturalist's wanderings in the interior of South Africa. London: C. Kegan Paul, pp. 332–337.

Günther, A. 1888. Contribution to the knowledge of snakes of tropical Africa. Annals of Natural History 1: 322–335.

Günther, A. 1893a. Descriptions of reptiles and fishes collected by Mr. Coode-Hore on Lake Tanganyika. Proceedings of the Zoological Society of London 1893: 628–632.

Günther, A. 1893b. Report on a collection of reptiles and batrachians transmitted by Mr. H. H. Johnston, C. B., from Nyassaland. Proceedings of the Zoological Society of London 1893: 555–558.

Günther, A. 1895. Notices on reptiles and batrachians collected in the eastern half of tropical Africa. Annals of Natural History 15: 523–551.

Haacke, W. D. 1975. Description of a new adder (Viperidae, Reptilia) from southern Africa, with a discussion of related forms. Cimbebasia, Windhoek 4A(5): 115–128.

Haacke, W. D. 2013. Description of a new Tiger Snake (Colubridae, *Telescopus*) from south-western Africa. Zootaxa 3737(3): 280–288.

Haagner, G. V., and R. Smith. 1987. Case history of boomslang (*Dispholodius typus*) envenomation in the eastern Transvaal, South Africa. Herpetological Bulletin 21: 43–45.

Haan, C. C. de. 2003a. Extrabuccal infralabial secretion outlets in *Dromophis*, *Mimophis* and *Psammophis* species (Serpentes, Colubridae, Psammophiini). A probable substitute for "self-rubbing" and cloacal scent gland. Comptes Rendus Biologies 326: 275–286.

Haan, C. C. de. 2003b. Sense-organ-like parietal pits found in Psammophiini (Serpentes, Colubridae) functions, and a cue for a taxonomic account. Comptes Rendus Biologies 326: 287–293.

Hahn, D. E., and V. Wallach. 1998. Comments on the systematics of Old World *Leptotyphlops* (Serpentes:

Leptotyphlopidae), with description of a new species. Hamadryad 23: 50–62.

Håkansson, T., and T. Madsen. 1983. On the distribution of the Black Mamba (*Dendroaspis polylepis*) in west Africa. Journal of Herpetology 17: 186–189.

Hallermann, J., and M. O. Rödel. 1995. A new species of *Leptotyphlops* (Serpentes: Leptotyphlopidae) of the longicaudus-group from west Africa. Stuttgarter Beiträge zur Naturkunde, Stuttgart Series A 532: 1–8.

Hallowell, E. 1842. Description of a new genus of serpents from western Africa. Journal of the Academy of Natural Sciences of Philadelphia 8(2): 336–338.

Hallowell, E. 1844. Description of new species of African reptiles. Proceedings of the Academy of Natural Sciences of Philadelphia 2(6): 169–172.

Hallowell, E. 1852. On a new genus and two new species of African snakes. Proceedings of the Academy of Natural Sciences of Philadelphia 6(6): 203–205.

Hallowell, E. 1854a. Descriptions of new reptiles from Guinea. Proceedings of the Academy of Natural Sciences of Philadelphia 7(5): 193–194.

Hallowell, E. 1854b. Remarks on the geographical distribution of reptiles with descriptions of several species supposed to be new and correction of former paper. Proceedings of the Academy of Natural Sciences of Philadelphia 7(3): 98–105.

Hallowell, E. 1857. Notice of a collection of reptiles from the Gaboon Country, west Africa, recently presented to the Academy of Natural Sciences of Philadelphia, by Dr. Henry A. Ford. Proceedings of the Academy of Natural Sciences of Philadelphia 9(3): 48–72.

Hasson, S. S., R. D. Theakston, and R. A. Harrison. 2003. Cloning of a prothrombin activator-like metalloproteinase from the West African Saw-scaled Viper, *Echis ocellatus*. Toxicon 42: 629–634.

Hatten, B. W., A. Bueso, L. K. French, R. G. Hendrickson, and B. Z. Horowitz. 2013. Envenomation by the Great Lakes Bush Viper (*Atheris nitschei*). Clinical Toxicology 51(2): 114–116.

Hayes, W. K., and S. P. Mackessy. 2010. Sensationalistic journalism and tales of snakebite: are rattlesnakes rapidly evolving more toxic venom? Wilderness and Environmental Medicine 21: 35–45.

Headland, T .N., and H. W. Greene. 2011. Hunter-gatherers and other primates as prey, predators, and competitors of snakes. Proceedings of the National Academy of Sciences 108: 20,865–20,866.

Hedges, S. B. 2011. The type species of the threadsnake genus *Tricheilostoma* Jan revisited (Squamata, Leptotyphlopidae). Zootaxa 3027: 63–64.

Hedges, S. B., A. B. Marion, K. M. Lipp, J. Marin, and N. Vidal. 2014. A taxonomic framework for typhlopid snakes from the Caribbean and other regions (Reptilia, Squamata). Caribbean Herpetology 49: 1–61.

Henderson, R. W., and M. H. Binder. 1980. The ecology and behavior of vine snakes (*Ahaetulla*, *Oxybelis*, *Thelotornis*, *Uromacer*): a review. Milwaukee Public Museum Press 37: 1–38.

Heymans, J. C. 1975. La musculature mandibulaire et le groupe parotidien des Aparallactinae et Atractaspinae (Serpents Colubridae a majorite fouisseurs). Revue Zoologique Africaine 89: 889–905.

Hill, R. E., and S. P. Mackessy. 2000. Characterization of venom (Duvernoy's secretion) from twelve species of colubrid snakes and partial sequence of four venom proteins. Toxicon 2000(38): 1663–1687.

Hoogmoed, M. S., and T. C. S. Avila-Pires. 2011. A case of voluntary tail autotomy in the snake *Dendrophidion dendrophis* (Schlegel, 1837) (Reptilia: Squamata: Colubridae). Boletim do Museu Paraense de História Natural e Etnographia. 6(2): 113–117.

Houghton, P. J., and I. M. Osibogun. 1993. Review article. Flowering plants used against snakebite. Journal of Ethnopharmacology 39: 1–29.

Howes, J. M., M. C. Wilkinson, R. D. Theakston, and G. D. Laing. 2003. The purification and partial characterisation of two novel metalloproteinases from the venom of the west African carpet viper, *Echis ocellatus*. Toxicon 42: 21–27.

Howes, J. M., A. S. Kamiguti, R. D. Theakston, M. C. Wilkinson, and G. D. Laing. 2005. Effects of three novel metalloproteinases from the venom of the West African Saw-scaled Viper, *Echis ocellatus* on blood coagulation and platelet. Biochimica et Biophysica Acta 1724: 194–202.

Hsiang, A. Y., D. J. Field, T. H. Webster, A. D. B. Behlke, M. B. Davis, R. A. Racicot, and J. A Gauthier. 2015. The origin of snakes: revealing the ecology, behavior, and evolutionary history of early snakes using genomics, phenomics, and the fossil record. BMC Evolutionary Biology 15: 87–107.

Hsu, E., J. Davis, and K. Jackson. 2017. Using spreadsheet software to create a multi-access key for central and western African snakes. Herpetological Review 48(4): 747–756.

Hughes, B. 1977. Latitudinal clines and ecogeography of the west African Night Adder, *Causus maculatus*

(Hallowell, 1842), Serpentes, Viperidae. Bulletin de l'Institut Fondamental d'Afrique Noire, Sér. A 39(2): 358–384.

Hughes, B. 1983. African snake faunas. Bonnischer Zoologische Beiträge, Bonn 34(1–3): 311–356.

Hughes, B. 1985. Progress on a taxonomic revision of the African Green Tree Snakes (*Philothamnus spp*). *In* Schuchmann, K.-L., ed., Proceedings of the International Symposium on African Vertebrates: systematics, phylogeny and evolutionary ecology. Berlin: Zoologisches Forschungsinst und Museum Alexander Koenig, pp. 511–530.

Hughes, B. 1999. Critical review of a revision of *Psammophis* by Frank Brandstatter. African Journal of Herpetology 48(1–2): 63–70.

Hughes, B. 2000. The African snake *Bothrophthalmus lineatus* (Peters, 1863). Herpetological Bulletin 74: 28–29.

Hughes, B., and D. H. Barry. 1969. The snakes of Ghana: a checklist and key. Bulletin de l'Institut Fondamental d'Afrique Noire A31: 1004–1041.

Ineich, I. 1998. *Chamaelycus fasciatus* (NCN). Diet. Herpetological Review 29(2): 102.

Ineich, I. 2006. Les élevages de reptiles et de scorpions au Bénin, Togo et Ghana, plus particulierèment la gestion des quotas d'exportation et la définition des codes <<source>> des spécimens exportes. Presented at the Convention sur le Commerce International des Especes de Faune et de Flore Sauvages Menacees d'Extinction, July 18–22, https://www.researchgate.net/profile/Ivan_Ineich/publication/305043720_Les_elevages_de_reptiles_et_de_scorpions_au_Benin_Togo_et_Ghana_plus_particulierement_la_gestion_des_quotas_d%27exportation_et_la_definition_des_codes_source_des_specimens_exportes/links/5834931e08ae004f74c885f5.pdf.

Ineich, I., X. Bonnet, R. Shine, T. Shine, F. Brischoux, M. Lebreton, and L. Chirio. 2006. What, if anything, is a "typical" viper? Biological attributes of basal viperid snakes (genus *Causus* Wagler, 1830). Biological Journal of the Linnean Society 89: 575–588.

Inger, R. F., and H. Marx. 1965. The systematics and evolution of the oriental colubrid snakes of the genus *Calamaria*. Fieldiana: Zoology 49: 1–304.

Ionides, C. J. P. 1954. Nature notes on Bibron's Burrowing Viper (*Atractaspis bibronii rostrata*) in south-east Tanganyika Territory. Africa Wildlife 8: 67–68.

Isemonger, R. M. 1983. Snakes of Africa, southern, central and east. London: Nelson.

Ismail, M., M. S. Al-Ahaidib, N. Abdoon, and M. A. Abd-Elsalam. 2007. Preparation of a novel antivenom against *Atractaspis* and *Walterinnesia* venoms. Toxicon 49: 8–18.

Jackson, K. 2003. The evolution of venom-delivery systems in snakes. Zoological Journal of the Linnean Society 137: 337–354.

Jackson, K., and T. H. Fritts. 1995. Evidence from tooth surface morphology for a posterior maxillary origin of the proteroglyph fang. Amphibia-Reptilia 16: 273–288.

Jackson, K., A. G. Zassi-Boulou, L. B. Mavoungou, and S. Pangou. 2007. Amphibians and reptiles of the Lac Tele Community Reserve, Likouala region, Republic of Congo (Brazzaville). Herpetological Conservation and Biology 2: 75–86.

Jacquet, M. 1896. Sur la présence d'un *Typhlops* en Algérie. Bibliographie Anatomique, Paris & Nancy 4(2): 79–81.

Jakobsen, A. 1997. A review of some east African members of the genus *Elapsoidea* Bocage, with the description of a new species from Somalia and a key for the genus (Reptilia, Serpentes, Elapidae). Steenstrupia 22: 59–82.

Jan, G. 1858. Plan d'une iconographie descriptive des ophidiens, et description sommaire de nouvelles espèces de serpents. Revue et Magasin de Zoologie 9: 438–449 and 514–527.

Jan, G. 1859. Additions et rectifications aux Plan et Prodrome de l'Iconographie descriptive des Ophidiens. Revue et Magasin de Zoologie 11: 503–512.

Jan, G. 1860. Première livraison. Iconographie Générale des Ophidiens Paris, J. B. Baillière et Fils 1(1): Plates 1–6.

Jan, G. 1861. Deuxième livrasion. Iconographie Générale des Ophidiens Paris, J. B. Baillière et Fils 1(2): Plates 1–6.

Jan, G. 1863. Elenco Sistematico Degli Ofidi Descritti e Disegnati per l'Iconografia Generale. Milan: A. Lombardi, 143 pp.

Jan, G. 1864. Iconographie générale des ophidiens. Première famille, les typhlopiens. Paris: J. B. Ballière et fils, 42 pp.

Jan, G. 1865. Prime linee dúna fauna della Persia occidentale. *In* F. de Filippi, Note di un Viaggio in Persia nel 1862. Milan: G. Daelli & C. Editori, pp. 342–357.

Jesus, J., Z. T. Nagy, W. R. Branch, M. Wink, A. Brehm, and D. J. Harris. 2009. Phylogenetic relationships of African Green Snakes (genera *Philothamnus* and *Hapsidophrys*) from São Tomé, Principe and Annobon Islands based on mtDNA sequences, and comments on their colonization and taxonomy. Herpetological Journal 19: 41–48.

Joger, U. 1990. The herpetofauna of the Central African Republic, with description of a new species of *Rhinotyphlops* (Serpentes: Typhlopidae). *In* Peters, G., and R. Hutterer, Vertebrates in the tropics. Bonn: Museum Alexander Koenig, pp. 85–102.

Junqueira-de-Azevedo, I. L., P. F. Campos, A. T. Ching, and S. P. Mackessy. 2016. Colubrid venom composition: an -omics perspective. Toxins (Basel) 8(230): 1–24.

Kardong, K. V. 1980. Evolutionary patterns in advanced snakes. American Zoologist 20: 269–282.

Kelly, C. M. R., N. P. Barker, M. H. Villet, D. G. Broadley, and W. R. Branch. 2008. The snake family Psammophiidae (Reptilia: Serpentes): phylogenetics and species delimitation in the African sand snakes (*Psammophis* Boie, 1825) and allied genera. Molecular Phylogenetics and Evolution 47: 1045–1060.

Kelly, C. M. R., N. P. Barker, M. H. Villet, and D. G. Broadley. 2009. Phylogeny, biogeography and classification of the snake superfamily Elapoidea: a rapid radiation in the Eocene. Cladistics 25: 38–63.

Kelly, C. M. R., W. Branch, D. G. Broadley, N. P. Barker, and M. H. Villet. 2011. Molecular systematics of the African snake family Lamrophiidae Fitzinger, 1843 (Serpentes: Elapoidea), with particular focus on the genera *Lamprophis* Fitzinger 1843 and *Mehelya* Csiki 1903. Molecular Phylogenetics and Evolution 58: 415–426.

Klauber, L. M. 1956. Rattlesnakes: their habits, life histories, and influence on mankind. Vol. 2. Berkeley: University of California Press, 808 pp.

Kley, N. J., and E. L. Brainerd. 1997. Feeding by mandibular raking in a snake. Nature 402: 369–370.

Kloog, Y., I. Ambar, M. Sokolovsky, E. Kochva, Z. Wollberg, and A. Bdolah. 1988. Sarafotoxin, a novel vasoconstrictor peptide: phosphoinositide hydrolysis in rat heart and brain. Science 242: 268–270.

Knoepffler, L. P. 1966. Faune du Gabon (Amphibiens et Reptiles). I. Ophidiens de l'Ogooué-Ivindo et du Woleu-N'tem. Biologia Gabonica 2: 3–23.

Kochva, E., C. C. Viljoen, and D. P. Botes. 1982. A new type of toxin in the venom of snakes of the genus Atractaspis (Atractaspidinae). Toxicon 20(3): 581–592.

Kornalik, F., E. Taborska, and D. Mebs. 1978. Pharmacological and biochemical properties of a venom gland extract from the snake *Thelotornis kirtlandi*. Toxicon 16: 535–542.

Kramer, E., and H. Schnurrenberger. 1963. Systematik, Verbreitung und Ökologie der libyschen Schlangen. Revue Suisse de Zoologie 70(27): 453–568.

Kress, L. F. 1988. The action of snake venom metalloproteinases on plasma proteinase inhibitors. *In* H. Pirkle, F. S. Markland Jr., eds. Hemostasis and animal venoms. New York: Marcel Dekker, pp. 335–348.

Kurnik, D., Y. Haviv, and E. Kochva. 1999. A snake bite by the Burrowing Asp, *Atractaspis engaddensis*. Toxicon 37: 223–227.

Laraba-Djebari, F., M. F. Martin-Eauclaire, G. Mauco, and P. Marchot. 1995. Afaâcytin, an alpha beta-fibrinogenase from *Cerastes cerastes* (horned viper) venom, activates purified factor X and induces serotonin release from human blood platelets. European Journal of Biochemistry 233: 756–765.

Largen, M. J., and S. Spawls. 2010. Amphibians and reptiles of Ethiopia and Eritrea. Frankfurt: Edition Chimaira, 694 pp.

Lataste, F. 1888. Description d'un ophidien diacrantérien nouveau (*Periops Dorri*, n. sp.) originaire du Haut-Sénégal. Le Naturaliste (2)38(10): 227–228.

Laurent, R. F. 1945. Contribution à la connaissance du Genre *Atractaspis* A. Smith. Revue de Zoologie et Botanique Africaine, Tervuren 38: 312–343.

Laurent, R. F. 1947. Notes sur quelques reptiles appartenant à la collection du Musée Royal d'Histoire Naturelle de Belgique. I. Formes Africaines. Bulletin du Musée Royal d'Histoire Naturelle de Belgique 23(16): 1–12.

Laurent, R. F. 1950a. Reptiles nouveaux des Kundelungu. Revue de Zoologie et Botanique Africaine 43: 349–352.

Laurent, R. F. 1950b. Révision du genre *Atractaspis* Smith. Mémoires de l'Institut Royal des Science Naturelle de Belgique 38: 1–49.

Laurent, R. F. 1952. Reptiles et batraciens nouveaux de la Région des Grands Lacs africains. Revue de Zoologie et Botanique Africaine 46: 269–279.

Laurent, R. F. 1954. Subsîdios para estudo da biologgia na lunda, Museo do Dundo. Reptiles et batraciens de la région de Dundo (Angola). (Deuxième note) Publicações Culturais da Companhia de Diamantes de Angola, Dundo 23: 35–84.

Laurent, R. F. 1955. Diagnoses préliminaires de quelques serpents venimeux. Revue de Zoologie et Botanique Africaine 51: 127–139.

Laurent, R. F. 1956. Contribution à l'herpétologie de la région des Grands Lacs de l'Afrique centrale. Annales du Musée Royal du Congo Belge 48: 1–390.

Laurent, R. F. 1958. Notes herpétologiques africaines 11. Revue de Zoologie et Botanique Africaine 58: 115–128.

Laurent, R. F. 1960. Notes complémentaires sur les chéloniens et les ophidiens du Congo oriental. Annales du Musée Royal du Congo Belge 84: 1–86.

Laurent, R. F. 1964. Reptiles et amphibiens de l'Angola (Troisième contribution). Publicações Culturais da Companhia de Diamantes de Angola, Dundo 67: 1–165.

Laurent, R. F. 1968. A re-examination of the snake genus Lycophidion Duméril and Bibron. Bulletin of the Museum of Comparative Zoology 136(12): 461–482.

Laurenti, J. N. 1768. Austriaci viennensis Specimen medicum exhibens Synopsis Reptilium emendatum cum experimentis circa venana et antidota Reptilium Austriacarum. Vienna: Joan Thomae, 214 pp, 5 plates.

Lawson, D. P. 1999. A new species of arboreal viper (Serpentes: Viperidae: Atheris) from Cameroon, Africa. Proceedings of the Biological Society of Washington 112: 793–803.

Lawson, D. P., and P. Ustach. 2000. A redescription of Atheris squamigera (Serpentes: Viperidae) with comments on the validity of Atheris anisolepis. Journal of Herpetology 34: 386–389.

Lawson, D. P., B. P. Noonan, and P. C. Ustach. 2001. Atheris subocularis (Serpentes: Viperidae) revisited: molecular and morphological evidence for the resurrection of an enigmatic taxon. Copeia 2001: 737–744.

Lawson, R., J. B. Slowinski, B. I. Crother, and F. T. Burbrink. 2005. Phylogeny of the Colubroidea (Serpentes): new evidence from mitochondrial and nuclear genes. Molecular Phylogenetics 37: 581–601.

Leach, W. E. 1818. In Tuckey JK, Narrative of an expedition to explore the River Zaire usually called the Congo, in South Africa in 1816 under the direction of Captain, J. K. Tuckey, R. M. To which is added the Journal of Professor Smith C. some general observations on the country and its inhabitants and an Appendix containing the natural History of that part of the Kingdom of Congo through which the Congo flow. London: John Murray, pp. 407–414.

Leach, W. E. 1819. In Bowdich TE, Mission from Cape Coast Castle to Ashantee with statistical account of that kingdom and geographical notices of other parts of the interior of Africa. London: John Murray, pp. 493–496.

Le Berre, M. 1989. Faune du Sahara. I. poissons–amphibiens–reptiles. Paris: Lechevallier & Chabaud, 322 pp.

Le Dantec, P., Y. Hervé, B. Niang, J. P. Chippaux, J. P. Bellefleur, G. Boulesteix, and B. Diatta. 2004. Morsure par vipère Bitis arietans au Sénégal, intérêt de la mesure de pression intracompartimentale. Médecine Tropicale 64(2): 187–191.

Leston, D., and B. Hughes. 1968. The snakes of Tafoo, a forest cocoa–farm locality in Ghana. Bulletin de l'Institut Fondamental d'Afrique Noire A30: 737–770.

Levinson, S. R., M. H. Evans, and F. Groves. 1976. A neurotoxic component of the venom from Blanding's Tree Snake (Boiga blandingi). Toxicon 14: 307–312.

Lichtenstein, M. H. C. 1823. Verzeichniss der Doubletten des zoologischen Museums der Königl. Universität zu Berlin nebst Beschreibung vieler bisher unbekannten Arten von Säugethieren, Vögeln, Amphibien und Fischen. Berlin: T. Trauwein, 118 pp.

Lindholm, W. A. 1905. Beschreibung einer neuer Schlangenart (Dipsadophidium Weileri nov. gen. et. nov. sp.). Jahrbuch des Nassauischen Vereins für Naturkunde, Wiesbaden 58: 183–187.

Linnaeus, C. 1758. Systema naturae per regna tria naturae, Secundum classes, ordines, genera, species cum characteribus differentiis, synonymis, locis. Tomus I. Editio decima, reformata. Stockholm: Laurentii Salvii Holmiae, 824 pp.

Loveridge, A. 1931. A new snake of the genus Typhlops from the Belgian Congo. Copeia 1931(3): 92–93.

Loveridge, A. 1932. New opisthoglyphous snakes of the genera Crotaphopeltis and Trimerorhinus from Angola and Kenya Colony. Proceedings of the Biological Society of Washington 45: 83–86.

Loveridge, A. 1933. Reports on the scientific results of an expedition to the south western highlands of Tanganyika-Territory. VII. Herpetology. Bulletin of the Museum of Comparative Zoology 74: 197–416.

Loveridge, A. 1936. African reptiles and amphibians in Field Museum of Natural History. Zoological Series, Field Museum of Natural History 22: 1–122.

Loveridge, A. 1938. On a collection of reptiles and amphibians from Liberia. Proceedings of the New England Zoological Club 17: 49–74.

Loveridge, A. 1939. Revision of the African snakes of

the genera *Mehelya* and *Gonionotophis*. Bulletin of the Museum of Comparative Zoology 86: 131–162.

Loveridge, A. 1941. Report on the Smithsonian-Firestone Expedition's collection of reptiles and amphibians from Liberia. Proceedings of the United States National Museum 91: 113–140.

Loveridge, A. 1944. Further revisions of African snake genera. Bulletin of the Museum of Comparative Zoology 95: 121–247.

Loveridge, A. 1951a. On reptiles and amphibians from Tanganyika Territory collected by CJP Ionides. Bulletin of the Museum of Comparative Zoology 106: 177–204.

Loveridge, A. 1951b. Synopsis of the African Green Snakes *Philothamnus* and *Chlorophis* with the description of a new form. Bulletin, Institut Royal des Sciences Naturelles de Belgique 27: 1–12.

Loveridge, A. 1953. Zoological results of a fifth expedition to east Africa. III. Reptiles from Nyasaland and Tete. Bulletin of the Museum of Comparative Zoology 110: 141–322.

Loveridge, A. 1958. Revision of five African snake genera. Bulletin of the Museum of Comparative Zoology 119: 1–198.

Loveridge, A. 1959. On a fourth collection of reptiles, mostly taken in Tanganyika Territory by Mr. CJP Ionides. Proceedings of the Zoological Society of London 133(1): 29–44.

Luiselli, L. 2002. Life-history correlates of sub-optimal adaptation to rainforest biota by Spitting Cobras, *Naja nigricollis*, in southern Nigeria: comparative evidences with sympatric Forest Cobras, *Naja melanoleuca*. Revue d'écologie—La Terre et la Vie 57: 123–133.

Luiselli, L. 2003. Do snakes exhibit shifts in feeding ecology associated with the presence or absence of potential competitors? A case study from tropical Africa. Canadian Journal of Zoology 81: 228–236.

Luiselli, L. 2006a. Site occupancy and density of sympatric Gaboon Viper (*Bitis gabonica*) and Nose-horned Viper (*Bitis nasicornis*). Journal of Tropical Ecology 22: 555–564.

Luiselli, L. 2006b. Testing hypotheses on the ecological patterns of rarity using a novel model of study: snake communities worldwide. Web Ecology 6: 44–58.

Luiselli, L. 2006c. Interspecific relationships between two species of sympatric Afrotropical water snake in relation to a seasonally fluctuating food resource. Journal of Tropical Ecology 22: 91–100.

Luiselli, L., and G. C. Akani. 2002. An investigation into to the composition, complexity and functioning of snake communities in the mangroves of south-eastern Nigeria. African Journal of Ecology 40: 220–227.

Luiselli, L., and F. M. Angelici. 2000. Ecological relationships in two Afrotropical cobra species (*Naja melanoleuca* and *Naja nigricollis*). Canadian Journal of Zoology 78: 191–198.

Luiselli, L., G. C. Akani, F. M. Angelici, and I. F. Barieenee. 1998a. Reproductive strategies of sympatric *Bitis gabonica* and *Bitis nasicornis* (Viperidae) in the Niger Delta (Port Harcourt, Nigeria): preliminary data. Amphibia-Reptilia 19: 223–229.

Luiselli, L., G. C. Akani, and I. F. Barieenee. 1998b. Observations on habitat, reproduction and feeding of *Boiga blandingi* (Colubridae) in south-eastern Nigeria. Amphibia-Reptilia 19: 430–436.

Luiselli, L., G. C. Akani, L. D. Otonye, J. S. Ekanem, and D. Capizzi. 1999. Additions to the knowledge of the natural history of *Bothrophthalmus lineatus* (Colubridae) from the Port Harcourt region of Nigeria. Amphibia-Reptilia 20: 318–326.

Luiselli, L., G. C. Akani, and F. M. Angelici. 2000a. Arboreal habits and viper biology in the African rainforest: the ecology of *Atheris squamiger*. Israel Journal of Zoology 46: 273–286.

Luiselli, L., F. M. Angelici, and G. C. Akani. 2000b. Large elapids and arboreality: the ecology of Jameson's green mamba (*Dendroaspis jamesoni*) in an Afrotropical forested region. Contributions to Zoology 69: 147–155.

Luiselli, L., F. M. Angelici, and G. C. Akani. 2000c. Reproductive ecology and diet of the Afro-tropical tree snake *Rhamnophis aethiopissa* (Colubridae). Herpetological Natural History 7: 153–161.

Luiselli, L., F. M. Angelici, and G. C. Akani. 2001a. Food habits of *Python sebae* in suburban and natural habitats. African Journal of Ecology 39: 116–118.

Luiselli, L., G. C. Akani, and F. M. Angelici. 2001b. Diet and foraging behaviour of three little-known African forest snakes: *Meizodon coronatus*, *Dipsadoboa duchesnei* and *Hapsidophrys lineatus*. Folia Zoologica 50(2): 151–158.

Luiselli, L., C. Effah, F. M. Angelici, E. Odegbune, M. A. Inyang, G. C. Akani, and E. Politano. 2002. Female breeding frequency, clutch size and dietary habits of a Nigerian population of Calabar Ground Pythons, *Calabaria reinhardtii*. Herpetological Journal 12: 127–129.

Luiselli, L., G. C. Akani, F. M. Angelici, E. Politano, L. Ude, and S. M. Wariboko. 2003. Diet of the semi-aquatic snake, *Afronatrix anoscopus* (Colubridae) in southern Nigeria. African Journal of Herpetology 52: 123–126.

Luiselli, L., G. C. Akani, F. M. Angelici, E. A. Eniang, L. Ude, and E. Politano. 2004. Local distribution, habitat use, and diet of two supposed "species" of the *Psammophis* "*phillipsii*" complex (Serpentes: Colubridae) sympatric in southern Nigeria. Amphibia-Reptilia 23: 415–423.

Luiselli, L., G. C. Akani, F. M. Angelici, L. Ude, and S. M. Wariboko. 2005. Seasonal variation in habitat use in sympatric Afrotropical semi-aquatic snakes, *Grayia smythii* and *Afronatrix anoscopus* (Colubridae). Amphibia-Reptilia 26(3): 372–376.

Luiselli, L., X. Bonnet, M. Rocco, and G. Amori. 2012. Conservation implications of rapid shifts in the trade of wild African and Asian pythons. Biotropica 44: 569–573.

Madsen, T., and M. Osterkamp. 1982. Notes on the biology of the Fish-eating Snake *Lycodonomorphus bicolor* in Lake Tanganyika. Journal of Herpetology 16: 185–188.

Malukisa, J., M. Collet, S. Bokata, and W. Odio. 2005. Résultats préliminaires d'une enquête herpétologique en plantation de cannes à sucre en République Démocratique du Congo. Bulletin de la Société de Pathologie Exotique 98: 310–311.

Manaças, S. 1955. Saurios e ofidios da Guine Portuguesa. Anais Junta Investigações Ultramar, Lisboa 10: 1–29.

Mané, Y. 1992. Etude systématique et bioécologique des serpents de la région de Dielmo (Sine-Saloum) Sénégal. Dakar: Mémoires à l'université Cheikh-Anta-Diop de Dakar, 87 pp.

Mané, Y. 1999. Une espèce nouvelle du genre *Elapsoidea* (Serpentes, Elapidae) au Sénégal. Bulletin de la Société Herpétologique de France 91: 13–18.

Marrakchi, N., R. Barbouche, S. Guermazi, C. Bon, and M. el Ayeb. 1997a. Procoagulant and platelet-aggregating properties of cerastocytin from *Cerastes cerastes* venom. Toxicon 35: 261–272.

Marrakchi, N., R. Barbouche, C. Bon, and M. el Ayeb. 1997b. Cerastatin, a new potent inhibitor of platelet aggregation from the venom of the Tunisian Viper, *Cerastes cerastes*. Toxicon 35: 125–135.

Marsh, N. A. 1994. Snake venoms affecting the haemostatic mechanism—A consideration of their mechanisms, practical applications and biological significance. Blood Coagulation and Fibrrinolysis 5: 399–410.

Marsh, N., D. Gattullo, P. Pagliaro, and G. Losano. 1997. The Gaboon Viper, *Bitis gabonica*: hemorrhagic, metabolic, cardiovascular and clinical effects of the venom. Life Sciences 61: 763–769.

Marx, H. 1958. Egyptian snakes of the genus *Psammophis*. Fieldiana: Zoology 39: 191–200.

Masterson, G. P. R., B. Maritz, D. Mackay, J. Graham, and G. J. Alexander. 2009. The impacts of past cultivation on the reptiles in a South African grassland. African Journal of Herpetology 58(2): 71–84.

Matschie, P. 1893a. Einige anscheinend neue Reptilien und Amphibien aus West-Afrika. Sitzunsberichte der Gesellschaft Naturforschender Freunde zu Berlin 6: 170–175.

Matschie, P. 1893b. Beiträge zur fauna des Togolandes. Die Reptilien und Amphibien des Togogebietes. Mitteilungen von Forschungsreisenden und Gelehrten aus den Deutschen Schutzgebieten 6(3): 207–215.

McNally, T., G. S. Conway, L. Jackson, R. D. Theakston, N. A. Marsh, D. A. Warrell, L. Young, I. J. Mackie, and S. J. Machin. 1993. Accidental envenoming by a Gaboon Viper (*Bitis gabonica*): the haemostatic disturbances observed and investigation of in vitro haemostatic properties of whole venom. Transactions of the Royal Society of Tropical Medicine and Hygiene 87: 66–70.

Mebs, D., K. Holada, F. Kornalík, J. Simák, H. Vanková, D. Müller, H. Schoenemann, H. Lange, and H. W. Herrmann. 1998. Severe coagulopathy after a bite of a Green Bush Viper (*Atheris squamigera*): case report and biochemical analysis of the venom. Toxicon 36: 1333–1340.

Menzies, J. I. 1966. The snakes of Sierra Leone. Copeia 1966: 169–179.

Merrem, B. 1820. Versuch eines Systems der Amphibien. Marburg: Johann Christian Krieger, 191 pp., 1 plate.

Mertens, R. 1936. Eine neue Natter der Gattung Helicops aus Inner-Afrika. Zoologischer Anzeiger 114: 284–285.

Mertens, R. 1937. Reptilien und Amphibien aus dem südlichen Inner-Afrika. Abhandlungen der Senckenbergischen Naturforschenden Gesellschaft, Frankfurt am Main 435: 1–23.

Mertens, R. 1938. Herpetologische Ergebnisse einer Reise nach Kamerun. Abhandlungen der Senckenbergischen Naturforschenden Gesellschaft, Frankfurt-am-Main 442: 1–52.

Meyer, W. P., A. G. Habib, A. A. Onayade, A. Yakubu, D. C. Smith, A. Nasidi, I. J. Daudu, D. A. Warrell, and R. D. Theakston. 1997. First clinical experiences with a new ovine Fab *Echis ocellatus* snake bite antivenom in Nigeria: randomized comparative trial with Institute Pasteur Serum (Ipser) Africa antivenom. American Journal of Tropical Medicine and Hygiene 56: 291–300.

Mocquard, M. F. 1885. Sur une nouvelle espèce d'*Atractaspis* (*A. leucura*). Bulletin de la Société Philomathique de Paris 10: 14–18.

Mocquard, M. F. 1887. Sur les ophidiens rapportés du Congo par la mission de Brazza. Bulletin de la Société Philomathique de Paris 11: 62–92.

Mocquard, M. F. 1889. Sur une collection de reptiles du Congo. Bulletin de la Société Philomathique de Paris 1: 145–148.

Mocquard, M. F. 1897. Sur une collection de reptiles recueillie par M. Haug, à Lambaréné. Bulletin de la Société Philomathique de Paris 9: 5–20.

Mocquard, M. F. 1902. Sur des reptiles et batraciens de l'Afrique orientale anglaise, du Gabon et de la Guinée française (région de Kouroussa). Bulletin du Muséum National d'Histoire Naturelle, Paris 6: 404–416.

Mocquard, F. 1906. Description de quelques espèces nouvelles de reptiles. Bulletin du Muséum National d'Histoire Naturelle, Paris 7: 464–466.

Mounir, K., A. Belhaj, M. Meziane, S. J. Elalaoui, A. Baite, L. Safi, and M. Atmani. 2009. Une morsure de vipère à corne au niveau du pouce compliquées d'une ischémie aigue d'un membre inférieur. Médecine et Armées 37: 381–384.

Moyer, K., and K. Jackson. 2011. Phylogenetic relationships among the Stiletto Snakes (genus *Atractaspis*) based on external morphology. African Journal of Herpetology 2011: 1–17.

Müller, L. 1911. Zwei neue Schlangen aus den Katanga district, Kongostaat. Zoologischer Anzeiger 38: 357–360.

Myers, C. W. 1982. Blunt-headed Vine Snakes (*Imantodes*) in Panama, including a new species and other revisionary notes. American Museum Novitates 2738: 1–50.

Nagy, Z. T., V. Gvozdik, D. Meirte, M. Collet, and O. S. G. Pauwels. 2014. New data on the morphology and distribution of the enigmatic Schouteden's Sun Snake, *Helophis schoutedeni* (de Witte, 1922) from the Congo Basin. Zootaxa 3755: 96–100.

Nagy, Z. T., U. Joger, M. Wink, F. Glaw, and M. Vences. 2003. Multiple colonization of Madagascar and Socotra by Colubrid Snakes: evidence from nuclear and mitochondrial gene phylogenies. Proceedings of the Royal Society: Biological Sciences 270: 2613–2621.

Nagy, Z. T., R. Lawson, U. Joger, and M. Wink. 2004. Molecular systematics of racers, whipsnakes and relatives (Reptilia: Colubridae) using mitochontrial and nuclear markers. Journal of Zoological Systematics and Evolutionary Research 42: 223–233.

Nagy, Z. T., N. Vidal, M. Vences, W. R. Branch, O. S. G. Pauwels, M. Wink, and U. Joger. 2005. Molecular systematics of African Colubroidea (Squamata: Serpentes). *In* Huber, B. A., B. J. Sinclair, and K. H. Lampe, eds. African biodiversity: molecules, organisms, ecosystems. Bonn: Springer, 443 pp.

Noonan, B. P., and P. T. Chippindale. 2006. Dispersal and vicariance: the complex evolutionary history of boid snakes. Molecular Phylogenetics and Evolution 40: 347–358.

Ota, H., T. Hikida, and J. Barcelo. 1987. On a small collection of lizards and snakes from Cameroon, west Africa. African Study Monographs 8(2): 111–123.

Oukkache, N., M. Lalaoui, and N. Ghalim. 2012. General characterization of venom from the Moroccan snakes *Macrovipera mauritanica* and *Cerastes cerastes*. Journal of Venomous Animals and Toxins Including Tropical Diseases 18: 411–420.

Oyaberu, K. A., and C. J. Shokpeka. 1984. Identification of plantation snakes in Nigeria: an approach to solve occupational hazards. Nigerian Medical Practitioner 7: 151–155.

Parker, H. W. 1931. Some reptiles and amphibians from S.E. Arabia. Annals and Magazine of Natural History, London 10: 514–522.

Parker, H. W. 1936. Dr. Karl Jordan's expedition to south west Africa and Angola: herpetological collections. Novitates Zoologicae, London 40: 115–146.

Parker, H. W. 1940. Undescribed anatomical structures and new species of reptiles and amphibians. Annals and Magazine of Natural History, London 5: 257–274.

Parker, H. W. 1949. The snakes of Somaliland and the Sokotra Islands. Zoologische Verhandelingen, Leiden 6: 1–115.

Pauwels, O. S. G., and A. Ohler. 1999. *Pseudohaje nigra* (Günther 1858). Black Tree Cobra. Diet. African Herpetological News 30: 33–34.

Pauwels, O. S. G., and J. P. Vande Weghe. 2008. Les reptiles du Gabon. Washington, DC: Smithsonian Institution, 272 pp.

Pauwels, O. S. G., G. Lenglet, J.-F. Trape, and A. Dubois. 2000. *Grayia smithii* (Leach, 1818). Smith's

African Water Snake. Diet. African Herpetological News 31: 7–9.

Pauwels, O. S. G., W. R. Branch, and M. Burger. 2004. Reptiles of Loango National Park, Ogooué-Maritime Province, south-western Gabon. Hamadryad 29: 115–127.

Pel, H. S. 1851. Over de jagt aan de Goudkust, volgens eene tienjarige eigene ondervinding. Nederlandsch Tijdschrift voor Jagtkunde 1: 149–173.

Peracca, M. G. 1896. Retili et Anfibi raccolti a Kazungula e sulla strada da Kazungula a Buluwaio dal Rev. Luigi Jalla, Missionario Valdese nell' alto Zambese. Bolletino dei Musei di Zoologia ed Anatomia Comparata della Università di Torino 11 (255): 1–4.

Perret, J. L. 1960. Une nouvelle et remarquable espèces d'Atractaspis (Viperidae) et quelques autres serpents d'Afrique. Revue Suisse de Zoologie 67: 129–139.

Perret, J. L. 1961. Études herpétologiques africaines. III. La faune ophidienne de la région camerounaise. Bulletin de la Société Neuchâteloise des Sciences Naturelle 84: 133–138.

Peters, W. C. H. 1854. Diagnosen neuer Batrachier, welche zusammen mit der früher (24. Juli und 17. August) gegebenen Übersicht der Schlangen und Eidechsen mitgetheilt werden. Berichte über die Bekanntmachunggeeigneten Verhandlungen der Königlich-Preußische Akademie der wissenschaften zu Berlin 1854: 614–628.

Peters, W. C. H. 1857. Über Amblyodipsas, eine neue Schlangengattung aus Mossambique. Monatsberichte der königlich Akademie der Wissenschaften zu Berlin 1856(12): 592–595.

Peters, W. C. H. 1861. Die Beschreibung von zwei neuen Schlangen, Mizodon variegatus aus Westafrika und Bothriopsis quadriscutata. Monatsberichte der königlich Akademie der Wissenschaften zu Berlin 1861: 358–360.

Peters, W. C. H. 1862. Über die von dem so früh in Afrika verstorbenen Freiherrn von Bernim und Dr. Hartman auf ihren Reise durch Aegypten, Nubien und dem Sennâr gesammelten Amphibien. Monatsberichte der königlich Akademie der Wissenschaften zu Berlin 1862: 271–279.

Peters, W. C. H. 1863. Über einige neue oder weniger bekannte Schlangenarten des Zoologischen Museums zu Berlin. Monatsberichte der königlich Akademie der Wissenschaften zu Berlin 1863: 272–289.

Peters, W. C. H. 1867. Über Flederthiere (Pteropus gouldii, Rhinolophus deckenii, Vespertilio lobipes, Vesperugo temminckii) und Amphibien (Hypsilurus godeffroyi, Lygosoma scutatum, Stenostoma narirostre, Onychocephalus unguirostris, Ahaetulla poylepis, Pseudechis scutella). Monatsberichte der königlich Akademie der Wissenschaften zu Berlin 1867 (November): 703–712.

Peters, W. C. H. 1868. Über eine neue Nagergattung, Chiropodomys penicullatus, sowie über einige neue oder weniger bekannte Amphibien und Fische. Monatsberichte der königlich Akademie der Wissenschaften zu Berlin 1868: 448–461.

Peters, W. C. H. 1869. Über neue Gattungen und neue oder weniger bekannte Arten von Amphibien (Eremias, Dicrodon, Euprepes, Lygosoma, Typhlops, Eryx, Rhynchonyx, Elapomorphus, Achalinus, Coronella, Dromicus, Xenopholis, Anoplodipsas, Spilotes, Tropidonotus). Monatsberichte der königlich Akademie der Wissenschaften zu Berlin 1869: 432–447.

Peters, W. C. H. 1870. Über neue Amphibien (Hemidactylus, Urosaura, Tropidolepisma, Geophis, Uriechis, Scaphiophis, Hoplocephalus, Rana, Entomoglossus, Cystignathus, Hylodes, Arthroleptis, Phyllobates, Cophomantis) des Königlish Zoologischen Museums. Monatsberichte der königlich Akademie der Wissenschaften zu Berlin 1870: 641–652.

Peters, W. C. H. 1875. Über die Hrn. Professor Dr. R. Buchholz in Westafrika gesammelten Amphibien. Monatsberichte der königlich Akademie der Wissenschaften zu Berlin 1875: 196–212.

Peters, W. C. H. 1876. Eine zweite Mitthielung über die von Hrn. Professor Dr. R. Buchholz in Westafrica gesammelten Amphibien. Monatsberichte der königlich Akademie der Wissenschaften zu Berlin 1876(2): 117–123.

Peters, W. C. H. 1877. Übersicht der Amphibien aus Chinchoxo (Westafrika), welche von der Afrikanischen Gesellschaft dem Berliner Zoologischen Museum übergeben sind. Monatsberichte der königlich Akademie der Wissenschaften zu Berlin 1877: 611–621.

Peters, W. C. H. 1881. Zwei neue von Herrn Major von Mechow während seiner letzten Expedition nach West-Afrika entdeckte Schlangen und eine Übersicht der von ihm mitgebrachten herpetologischen Sammlung. Sitzungsberichte der Gesellschaft Naturforschender Freunde zu Berlin 1881: 147–150.

Peters, W. C. H. 1882. Zoologie. III. Amphibien. Naturwissenschaftliche Reise nach Mossambique auf Befehl seiner Majestät des Königs Friedrich Wilhem IV. In den Jahren 1842 bis 1848 ausgeführt von Wilhelm C. H. Peters. Berlin: Druck und Verlag von G. Reimer, 191 pp., 33 plates.

Phisalix, M. 1922. Animaux venimeux et Venins. 2 vols. Paris: Masson et Cⁱᵉ.

Piquet, A., C. Toudonou, L. Konetché, B. Sinsin, and J.-P. Chippaux. 2012. Étude préliminaire de la faune ophidienne de la forêt classée de la Lama, Sud Bénin. Bulletin de la Société de Pathologie Exotique 105: 166–170.

Pirkle, H., I. Theodor, D. Miyada, and G. Simmons. 1986. Thrombin-like enzyme from the venom of *Bitis gabonica*. Purification, properties, and coagulant actions. Journal of Biological Chemistry 261: 8830–8835.

Pitman, C. R. S. 1974. A guide to the snakes of Uganda. Rev. ed. London: Wheldon & Wesley.

Pommier, P., and L. de Haro. 2007. Envenomation by Montpellier Snake (*Malpolon monspessulanus*) with cranial nerve disturbances. Toxicon 50: 868–869.

Pook, C. E., U. Joger, N. Stümpel, and W. Wüster. 2009. When continents collide: phylogeny, historical biogeography and systematics of the medically important viper genus *Echis* (Squamata: Serpente: Viperidae). Molecular Phylogenetics and Evolution 53: 792–807.

Pope, C. H. 1958. Fatal bite of captive African rear-fanged snake (Dispholidus). Copeia 1958: 280–282.

Pugh, R. N. H., C. C. M. Bourdillon, R. D. G. Theakston, and H. A. Reid. 1979. Bites by the Carpet Viper in the Niger Valley. Lancet 2: 625–627.

Pugh, R. N. H., R. D. G. Theakston, H. A. Reid, and I. S. Bhar. 1980. Malumfashi Endemic Diseases Research Project. XIII. Epidemiology of human encounters with Spitting Cobra *Naja nigricollis* in the Malumfashi area of northern Nigeria. Annals of Tropical Medicine and Parisitology 74: 523–530.

Pyron, R. A., F. T. Burbrink, G. R. Colli, A. N. M. de Oca, L. J. Vitt, C. A. Kuczynski, and J. J. Wiens. 2011. The phylogeny of advanced snakes (Colubroidea), with discovery of a new subfamily and comparison of support methods for likelihood trees. Molecular Phylogenetics and Evolution 58: 329–342.

Pyron, R. A., F. T. Burbrink, and J. J. Weins. 2013. A phylogeny and revised classification of Squamata, including 4161 species of lizards and snakes. BioMed Central Evolutionary Biology 13: 93.

Pyron, R. A. and V. Wallach. 2014. Systematics of the blindsnakes (Serpentes: Scolecophidia: Typhlopoidea) based on molecular and morphological evidence. Zootaxa 3829(1): 1–81

Rasmussen, A. R., J. C. Murphy, M. Ompi, J. W. Gibbons, and P. Uetz. 2011. Marine reptiles. PLoS One 6(11): e27373.

Rasmussen, J. B. 1986. On the taxonomic status of *Dipsadoboa werneri* (Boulenger), *D. shrevei* (Loveridge), and *Crotaphopeltis hotamboeia kageleri* Uthmöller (Boiginae, Serpentes). Amphibia-Reptilia 7: 51–73.

Rasmussen, J. B. 1989. A taxonomic review of the *Dipsadoboa duchesnei* complex. Bonner Zoologische Beiträge 40: 249–264.

Rasmussen, J. B. 1993. A taxonomic review of the *Dipsadoboa unicolor* complex, including a phylogenetic analysis of the genus (Serpentes, Dipsadidae, Boiginae). Steenstrupia 19: 129–196.

Rasmussen, J. B. 1994. Studies on the taxonomy, phylogeny and zoogeography of the African, rear-fanged Tree-snakes, genus *Dipsadoboa* (Serpentes, Dipsadidae, Boiginae). Copenhagen: Zoologisk Museum, Copenhagen University, 16 pp.

Rasmussen, J. B. 1996. Maxillary tooth number in the African tree-snakes genus *Dipsadoboa*. Journal of Herpetology 30: 297–300.

Rasmussen, J. B. 1997a. On two little known African Water Snakes (*Crotaphopeltis degeni* and *C. barotseensis*). Amphibia-Reptilia 18: 191–206.

Rasmussen, J. B. 1997b. Afrikanske slanger (10). *Boiga blandingii*. Nordisk Herpetologisk Forening 40: 97–103.

Rasmussen, J. B., L. Chirio, and I. Ineich. 2000. The Herald Snakes (*Crotaphopeltis*) of the Central African Republic, including a systematic review of *C. hippocrepis*. Zoosystema 22: 585–600.

Reed, R. N., and G. H. Rodda. 2009. Giant constrictors: biological and management profiles and an establishment risk assessment for nine large species of pythons, anacondas, and the Boa Constrictor. U.S. Geological Survey Open-File Report 2009-1202: 1–302.

Reinhardt, J. T. 1843. Beskrivelse af nogle nye slangearter. Kongelige Danske Videnskabernes Selskabs, Naturvidenskabelige og Mathematiske Afhandlinger, Kjöbenhavn 10: 233–279.

Reinhardt, J. T. 1860. Herpetologiske Meddelelser. II. Beskrivelse af nogle nye til Calamariernes familei henhörende Slanger. Videnskabelige Meddelelser fra den Naturhistoriske Forening i Kjöbenhavn 22: 229–246.

Resetar, A., and H. Marx. 1981. A redescription and generic reallocation of the African colubrid snake *Elapocalamus gracilis* Boulenger with a discussion of the union of the brille and postocular shield. Journal of Herpetology 15: 83–89.

Reuss, A. 1834. Zoologische Miscellen. Reptilien. Ophidier. Museum Senckenbergianum 1: 129–162.

Revault, P. 1994. Serpents, savoirs et santé chez les Mossi. Prise en charge des envenimations par *Echis ocellatus* en Afrique soudano-sahélienne à travers l'exemple du plateau ouagalais. Doctoral Dissertation, Faculté de médecine de Bobigny, Université Paris 13.

Rippey, J. J., E. Rippey, and W. R. Branch. 1976. A survey of snakebite in the Johannesburg area. South African Medical Journal 50: 1872–1876.

Robertson, S. S. D., and G. R. Delpierre. 1969. Studies on African snake venoms. IV. Some enzymatic activities in the venom of the boomslang *Dispholius typus*. Toxicon 7: 189–194.

Roman, B. 1972. Deux sous-espèces de la Vipère *Echis carinatus* (Schneider) dans les territoires de Haute-Volta et du Niger: *Echis carinatus ocellatus* Stemmler-*Echis carinatus leucogaster* n. ssp. Notes et Documents Voltaïques, Ouagadougou 5: 1–15.

Roman, B. 1973. Vipéridés et Elapidés de Haute-Volta. Notes et Documents Voltaïques, Ouagadougou 6: 1–49.

Rosing, J., R. F. A. Zwaal, and G. Tans. 1988. Snake venom prothrombin activators. *In* Pirkle, H., F. S. Markland Jr., eds. Hemostasis and animal venoms. New York: Marcel Dekker, pp. 1–27.

Rossman, D. A., and W. G. Eberle. 1977. Partition of the genus *Natrix*, with preliminary observations on evolutionary trends in natricine snakes. Herpetologica 33: 34–43.

Roux-Estève, R. 1965. Les serpents de La Maboké. Cahiers de La Maboké 3: 51–92.

Roux-Estève, R. 1969. Les serpents de la région de Lamto (Côte-d'Ivoire). Annals de l'Université d'Abidjan 2: 81–140.

Roux-Estève, R. 1974. Révision systématique des Typhlopidae d'Afrique. Reptilia. Serpentes. Mémoires du Muséum National d'Histoire Naturelle, Paris (Série A) 87: 1–313

Roux-Estève, R. 1975. Recherches sur la biogéographie et la phylogénie des Typhlopidae d'Afrique. Bulletin de l'Institut Fondamental d'Afrique Noire 36: 428–50.

Roux-Estève, R. 1979. Une nouvelle espece de Leptotyphlops (Serpentes) du Cameroun: *Leptotyphlops perreti*. Revue Suisse de Zoologie 86(2): 463–466.

Roux-Estève, R., and J. Guibé. 1965a. Étude comparée de *Boaedon fuliginosus* (Boie) et *Boaedon lineatus*

D. et B. (Ophidiens). Bulletin de l'Institut Fondamental d'Afrique Noire A27: 397–409.

Roux-Estève, R., and J. Guibé. 1965b. Contribution à l'étude du genre *Boaedon*. Bulletin du Muséum National d'Histoire Naturelle, Paris 36: 761–774.

Sadler, M., and B. Paul. 1988. Vine snake envenomation. Central African Journal of Medicine 34: 31–33.

Savitzky, A. H. 1980. The role of venom-delivery strategies in snake evolution. Evolution 34: 1194–1204.

Savitzky, A. H. 1983. Coadapted character complexes among snakes: fossoriality, piscivory, and durophagy. American Zoologist 23: 397–409.

Schätti, B. 1988. Systematics and phylogenetic relationships of *Coluber florulentus* Geoffroy 1827 (Reptilia Serpentes). Tropical Zoology 1: 95–116.

Schätti, B., and C. Loumont. 1992. Ein Beitrag zur Herpetofauna von São Tomé (Golf von Guinea). Zoologische Abhandlungen, Staatliches Museum für Tierkunde, Dresden 47: 23–36.

Schätti, B., and J. F. Trape. 2008. *Bamanophis*, a new genus for the west African colubrid *Periops dorri* Lataste, 1888 (Reptilia: Squamata: Colubrinae). Revue Suisse de Zoologie 115: 595–615.

Schlegel, H. 1837. Essai sur la physionomie des serpents. Vol. 2. Partie descriptive. Amsterdam: M. H. Schonekat, 606 pp.

Schlegel, H. 1837–1844. Abbildungen neuer oder unvollständig bekannter Amphibien, nach der Natur oder dem Leben entworfen und mit einem erläuternden Texte begleitet. Düsseldorf: Arnz and Co., xiv + 141 pp.

Schlegel, H. 1848. Over *Elaps jamesonii* Traillust. Bijdragen tot de Dierkunde 1: 5.

Schlegel, H. 1851. Description d'une nouvelle espèce du genre *Eryx*, *Eryx reinhardtii*. Bijjdragen tot de Dierkunde 1: 1–3.

Schlegel, H. 1855. Over eenige nieuwe soorten van vergiftige slangen van de Goudkust. Verslagen en Mededeelingen der Koninklijke Akademie van Wetenschapen Amsterdam (Afdeeling Natuurkunde) 3: 312–317.

Schleich, H. H., W. Kästle, and K. Kabish. 1996. Amphibians and reptiles of north Africa. Koenigstein: Koeltz, 629 pp.

Schmidt, K. P. 1923. Contributions to the herpetology of the Belgian Congo based on the collection of the American Museum Congo Expedition, 1909-1915. II. Snakes, with field notes by Herbert Lang &

James P. Chapin. Bulletin of the American Museum of Natural History 49: 1–146.

Scott Keogh, J., W. R. Branch, and R. Shine. 2000. Feeding ecology, reproduction and sexual dimorphism in the colubrid snake *Crotaphopeltis hotamboeia* in southern Africa. African Journal of Herpetology 49(2): 129–137.

Seifert, S. A., J. White, and B. J. Currie. 2011. Commentary: pressure bandaging for North American snake bite? No! Journal of Medical Toxicology 7: 324–326.

Senter, P. 2000. *Hapsidophrys smaragdinus* (Emerald Snake), *Philothamnus irregularis* (NCN), *Dasypeltis sp.* (Egg-eating Snakes), *Afronatrix anoscopus* (Brown Water Snake), *Grayia smythii* (Smyth's Water Snake), *Natriciteres variegates* (NCN), *Boaedon lineatus* (House Snake), and *Aparallactus niger* (Centipede-Eater). Caudal Prehension. Herpetological Review 31: 246–247.

Shaw, G. 1802. General zoology or systematic natural history. Amphibia. London: G. Kearsley.

Shayer-Wollberg, M., and E. Kochva. 1967. Embryonic development of the venom apparatus in *Causus rhombeatus* (Viperidae, Ophidia). Herpetologica 23: 249–259.

Shine, R. 1983. Reptilian reproductive modes: the oviparity-viviparity continuum. Herpetologica 39: 1–8.

Shine, R. 1985. The evolution of viviparity in reptiles: an ecological analysis. *In* Gans, C., and F. Billet, eds. Biology of the reptilia. Vol. 15. New York: John Wiley, pp. 605–694.

Shine, R., W. R. Branch, P. Harlow, and J. Webb. 1996. Sexual dimorphism, reproductive biology, and food habits of two species of African Filesnakes (*Mehelya*, Colubridae). Journal of Zoology London 240: 327–340.

Shine, R., W. R. Branch, P. Harlow, J. Webb, and T. Shine. 2006a. Biology of Burrowing Asps (Atractaspididae) from southern Africa. Copeia 2006: 103–115.

Shine, R., W. R. Branch, J. K. Webb, P. S. Harlow, and T. Shine. 2006b. Sexual dimorphism, reproductive biology, and dietary habits of Psammophiine snakes (Colubridae) from southern Africa. Copeia 2006(4): 650–664.

Silveira, P. V, and S. A. Nishioka. 1995. Venomous snake bite without clinical envenoming ("dry-bite"). A neglected problem in Brazil. Tropical and Geographical Medicine 47: 82–85.

Simbotwe, M. P. 1982. Survey of snakebite in Zambia. Zambia Museums Journal 6: 88–99.

Sinsin, B., and W. Bergmans. 1999. Rongeurs, ophidiens et relations avec l'environnement agricole au Bénin. Cotonou: Editions du Flamboyant, 200 pp.

Sjöstedt, Y. 1896. *Atractaspis reticulata*, ein neue schlange aus Kamerun. Zoologischer Anzeiger 19: 516–517.

Smith, A. 1829. Contribution to the natural history of South Africa. Zoological Journal, London 4: 433–444.

Smith, A. 1831. Contributions to the natural history of South Africa. 1. South African Quarterly Journal (1)2(5)5: 9–24.

Smith, A. 1838–1849. Illustrations of the zoology of South Africa, consisting chiefly of figures and descriptions of the objects of Natural History collected during an expedition into the interior of South Africa, in the years 1834, 1835, and 1836; fitted out by "The Cape of Good Hope Association for exploring Central Africa," together with a summary of African zoology, and an inquiry into the geographical ranges of species in that quarter of the globe. Reptilia. London: Smith, Elder, Plates I–LXXVIII, 1–28 pp.

Smith, J. B., R. D. Theakston, A. L. Coelho, C. Barja-Fidalgo, J. J. Calvete, and C. Marcinkiewicz. 2002. Characterization of a monomeric disintegrin, ocellatusin, present in the venom of the Nigerian Carpet Viper, *Echis ocellatus*. FEBS Letters 512: 111–115.

Schneemann. M., R. Cathomas, S. T. Laidlaw, A. M. El Nahas, R. D. Theakston, and D. A. Warrell. 2004. Life-threatening envenoming by the Saharan horned viper (*Cerastes cerastes*) causing micro-angiopathic haemolysis, coagulopathy and acute renal failure: clinical cases and review. Quarterly Journal of Medicine 97(11): 717–727.

Spawls, S. 1985. *Dispholidus typus* (Boomslang) Behavior. Herpetological Review 16: 111.

Spawls, S., and W. R. Branch. 1995. The dangerous snakes of Africa. Sanibel Island: Ralph Curtis, 192 pp.

Spawls, S., K. Howell, R. C. Drewes, and J. Ashe. 2004. A field guide to the reptiles of east Africa. London: A & C Black, 544 pp.

Steindachner, F. 1870. Herpetologische Notizen (II). Reptilien gesammelt Während einer Reise in Sengambien. Sitzungsberichte der Kaiserlichen Akademie der Wissenschaften in Wien 62: 326–348.

Stejneger, L. 1893. Description of a new species of blind snake (Typhlopidae) from the Congo Free

State. Proceedings of the United States National Museum 16: 709–710.

Stemmler, O. 1970. Die sandrasselotter aus Westafrika, *Echis carinatus ocellatus* subsp. nov. (Serpentes, Viperidae). Revue Suisse de Zoologie 77: 273–281.

Sternfeld, R. 1908a. Neue und ungenügend bekannte afrikanische shlangen. Sitzungsberichte der Gesellschaft Naturforschender Freunde zu Berlin 1908: 92–95.

Sternfeld, R. 1908b. Die Schlangenfauna Togos. Mit einer Bestimmungstabelle. Bearbetit nach dem Material des Berliner Zoologischen Museums. Mitteilungen aus dem Zoologischen Museum in Berlin 4: 207–236.

Sternfeld, R. 1908c. Die Schlangenfauna von Kamerun mit einer Bestimmungstabelle. Mitteilungen aus dem Zoologischen Museum in Berlin 3: 397–432.

Sternfeld, R. 1910. Neue Schlangen aus Kamerun, Abessynien u. Deutsch-Ostafrika. Mitteilungen aus dem Zoologischen Museum in Berlin 5: 67–70.

Sternfeld, R. 1912. IV. Zoologie II. Lfg. Reptilia. *In* Schubotz., R., ed., Wissenschaftliche Ergebnisse der Deutschen Zentral-Afrika Expedition 1907–1908, unter Führung A. Friedrichs, Herzogs zu Mecklenburg. Leipzig: Klinkhard und Biermann, 197–279.

Sternfeld, R. 1917. Reptilia und Amphibia. *In* Schubotz, H., ed., Wissenschaftliche Ergebnisse der Zweiten Deutschen Zentral-Afrika-Expedition, 1910–1911 unter Führung Adolph Friedrichs, Herzog zu Mecklenburg. Leipzig: Klinkhard und Biermann, [Band] 1, Zoologie, Lieferung 11; S. 407–510.

Streicher, J. W., and J. J. Wiens. 2016. Phylogenomic analyses reveal novel relationships among snake families. Molecular Phylogenetics and Evolution 100: 160–169.

Stucki-Stirn, M. C. 1979. Snake report 721, a comparative study of the herpetological fauna of the former west Cameroon/Africa, with a classification and synopsis of 95 different snakes and description of some new subspecies. Teuffenthal: Herpeto-Verlag, 650 pp.

Sutherland, S. K., A. R. Coulter, and R. D. Harris. 1979. Rationalisation of first-aid measures for elapid snakebite. Lancet 1: 183–186.

Sweeney, R. C. H. 1971. Snakes of Nyasaland, with new added corrigenda and addenda. Amsterdam: Asher, 203 pp.

Taub, A. M. 1966. Ophidian cephalic glands. Journal of Morphology 118: 529–542.

Theakston, R. D. G., and D. A. Warrell. 2000. Crisis in snake antivenom supply for Africa. Lancet 356: 2104.

Thompson, J. C. 1914. Further contributions to the anatomy of the Ophidia. Proceedings of the Zoological Society of London 1914: 379–402.

Thorpe, R. S., and C. J. McCarthy. 1978. A preliminary study, using multivariate analysis of a species complex of African house snakes (*Boaedon fuliginosus*). Journal of Zoology, London 184: 489–506.

Tokar, A. A. 1995. Taxonomic revision of the genus *Gongylophis* Wagler 1930: *G. conicus* (Schneider 1801) and *G. muelleri* Boulenger 1892 (Serpentes Boidae). Tropical Zoology 8: 347–360.

Top, L. J., J. E. Tulleken, J. J. Ligtenberg, J. H. Meertens, T. S. van der Werf, and J. G. Zijlstra. 2006. Serious envenomation after a snakebite by a Western Bush Viper (*Atheris chlorechis*) in the Netherlands: a case report. Netherlands Journal of Medicine 64(5): 153–156.

Tornier, G. 1902. Herpetologisch Neues aus Deutsch-Ost-Afrika. Zoologische Jahrbücher, Abteilung für Systematik, Ôkologie und Geographie der Tiere, Jena 15: 578–590.

Traill, T. S. 1843. Essay on the physiognomy of serpents (Schlegel, H.). Translated by Thos. Stewart Traill. Edinburgh: Maclachlan, Stewart, 256 pp., 1 plate.

Trape, J. F. 1985. Les serpents de la région de Dimonika (Mayombe, République Populaire du Congo). Revue de Zoologie et Botanique Africaine 99: 135–140.

Trape, J. F., and Y. Mané. 2000. Les serpents des environs de Dielmo (Sine-Saloum, Sénégal). Bulletin de la Société Herpétologique de France 95: 19–35

Trape, J. F., and Y. Mané. 2004. Les serpents des environs de Bandafassi (Sénégal oriental). Bulletin de la Société Herpétologique de France 109: 5–34.

Trape, J. F., and Y. Mané. 2005. Une nouvelle espèce du genre *Mehelya* (Serpentes: Colubridae) de Haute-Casamance (Sénégal). Bulletin de la Société Herpétologique de France 115: 23–30.

Trape, J. F., and Y. Mané. 2006a. Guide des serpents d'Afrique occidentale. Savane et désert. [Senegal, Gambia, Mauritania, Mali, Burkina Faso, Niger]. Paris: IRD Editions, 226 pp.

Trape, J. F., and Y. Mané. 2006b. Le genre *Dasypeltis* Wagler (Serpentes: Colubridae) en Afrique de l'Ouest: description de trois espèces et d'une sous-espèce nouvelles. Bulletin de la Société Herpétologique de France 119: 27–56.

Trape, J. F., and R. Roux-Estève. 1990. Note sur une

collection de serpents du Congo avec description d'une espèce nouvelle. Journal of African Zoology 104: 375–383.

Trape, J. F., and R. Roux-Estève. 1995. Les serpents du Congo: liste commentée et clé de détermination. Journal of African Zoology 109: 31–50.

Trape, J. F., Y. Mané, and I. Ineich. 2006. *Atractaspis microlepidota, A. micropholis* et *A. watsoni* en Afrique occidentale et centrale. Bulletin de la Société Herpétologique de France 119: 5–16.

Trape, J. F., L. Chirio, D. G. Broadley, and W. Wüster. 2009. Phylogeography and systematic revision of the Egyptian cobra (Serpentes: Elapidae: *Naja haje*) species complex, with the description of a new species from west Africa. Zootaxa 2236: 1–25.

Trape, S., O. Mediannikov, and J. F. Trape. 2012. When colour patterns reflect phylogeography: new species of *Dasypeltis* (Serpentes: Colubridae: Boigini) from west Africa. Comptes Rendus Biologies 335: 488–501.

Underwood, G., and E. Kochva. 1993. On the affinities of the burrowing asps *Atractaspis* (Serpentes: Atractaspididae). Zoological Journal of the Linnean Society 107: 3–64.

Vaiyapuri, S., R. A. Harrison, A. B. Bicknell, J. M. Gibbins, and G. Hutchinson. 2010. Purification and functional characterisation of rhinocerase, a novel serine protease from the venom of *Bitis gabonica rhinoceros*. PLoS One 5(3): e9687.

Valenta, J., Z. Stach, and M. Svítek. 2010. Acute pancreatitis after viperid snake *Cerastes Cerastes* envenoming: a case report. Prague Medical Report 111: 69–75.

Vidal, N., and S. B. Hedges. 2002. Higher-level relationships of caenophidian snakes inferred from four nuclear and mitochondrial genes. Comptes Rendus Biologies 325: 987–995.

Vidal, N., A. S. Delmas, P. David, C. Cruaud, A. Couloux, and S. B. Hedges. 2007. The phylogeny and classification of caenophidian snakes inferred from seven nuclear protein-coding genes. Comptes Rendus Biologies 330: 182–187.

Vidal, N., W. R. Branch, O. S. G. Pauwels, S. B. Hedges, D. G. Broadley, M. Wink, C. Cruaud, U. Joger, and Z. T. Nagy. 2008. Dissecting the major African snake radiation: a molecular phylogeny of the Lamprophiidae Fitzinger (Serpentes, Caenophidia). Zootaxa 1945: 51–66.

Vidal, N., J. C. Rage, A. Couloux, and S. B. Hedges. 2009. Snakes (Serpentes). *In* Hedges, S. B., and

S. Kumar, eds. The timetree of life. London: Oxford University Press, pp. 390–397.

Vignoli, L., G. H. Segniagbeto, E. A. Eniang, E. Hema, F. Petrozzi, G. C. Akani, and L. Luiselli. 2016. Aspects of natural history in a sand boa, *Eryx muelleri* (Erycidae) from arid savannahs in Burkina Faso, Togo, and Nigeria (west Africa). Journal of Natural History 50: 749–758.

Viljoen, C. C., D. P. Botes, and H. Kruger. 1982. Isolation and amino acid sequence of caudoxin, a presynaptic acting toxic phospholipase A2 from the venom of the Horned Puff Adder (*Bitis caudalis*). Toxicon 20: 715–737.

Villiers, A. 1950. Catalogue VI. La collection de serpents de l'IFAN. Dakar : Institut Française d'Afrique Noire, 155 pp.

Villiers, A. 1951. Mission A. Villiers au Togo et au Dahomey 1950. II. Ophidiens. Etudes Dahoméennes 5: 17–46.

Villiers, A. 1956. Le parc du Niokolo Koba. V. Reptiles. Mémoires de l'Institut Française d'Afrique Noire (Sciences Naturalles), Dakar 48: 150–162.

Villiers, A. 1966. Contribution à la faune du Congo (Brazzaville). Mission A. Villiers et A. Descarpentries. XLII. Reptiles Ophidiens. Bulletin de l'Institut Fondamental d'Afrique Noire A28: 1720–1760.

Villiers, A. 1975. Les serpents de l'ouest Africain. Dakar: Université de Dakar, Institut Français d'Afrique Noire, 195 pp.

Visser, J., and D. S. Chapman. 1982. Snakes and snakebite: venomous snakes and management of snakebite in southern Africa. Cape Town: Purnell, pp. 1–152.

Visser, L. E., S. Kyei-Faried, D. W. Belcher, D. W. Geelhoed, J. Schagen van Leeuwen, and J. van Roosmalen. 2008. Failure of a new antivenom to treat *Echis ocellatus* snake bite in rural Ghana: the importance of quality surveillance. Transactions of the Royal Society of Tropical Medicine and Hygiene 102: 445–450.

Vonk, F. J., J. F. Admiraal, K. Jackson, R. Reshef, M. A. G. de Bakker, K. Vanderschoot, I. van den Berge, M. van Atten, E. Burgerhout, A. Beck, P. J. Mirtschin, E. Kochva, F. Witte, B. G. Fry, A. Woods, and M. K. Richardson. 2008. Evolutionary origin and development of snake fangs. Nature 454: 630–633.

Wagler, J. G. 1830. Natürliches System der Amphibien: mit vorangehender Classification der Säugethiere und Vögel: ein Beitrag zur vergleichenden Zoologie. Munich: In der J.G. Cotta'scchen Buchhandlung, 354 pp.

Wagstaff, S. C., L. Sanz, P. Juárez, R. A. Harrison, and J. J. Calvete. 2009. Combined snake venomics and venom gland transcriptomic analysis of the ocellated carpet viper, *Echis ocellatus*. Journal of Proteomics 71(6): 609–623.

Wallach, V. 2002. *Typhlops etheridgei*, a new species of African blindsnake in the *Typhlops vermicularis* species group from Mauritania (Serpentes: Typhlopidae). Hamadryad 27(1): 108–122.

Wallach, V. 2005. *Letheobia pauwelsi*, a new species of blindsnake from Gabon (Serpentes: Typhlopidae). African Journal of Herpetology 54(1): 85–91.

Wallach, V., and J. Boundy. 2005. *Leptotyphlops greenwelli*: a new wormsnake of the L-bicolor species group from Nigeria (Serpentes: Leptotyphlopidae). Annals of the Carnegie Museum 74(1): 39–44

Wallach, V., W. Wüster, and D. G. Broadley. 2009. In praise of subgenera: taxonomic status of cobras of the genus *Naja* Laurenti (Serpentes: Elapidae). Zootaxa 2236: 26–36.

Warrell, D. A. 2008. Unscrupulous marketing of snake bite antivenoms in Africa and Papua New Guinea: choosing the right product—"What's in a name?" Transactions of the Royal Society of Tropical Medicine and Hygiene 102: 397–399.

Warrell, D. A. 2009. Commissioned article: management of exotic snakebites. Quarterly Journal of Medicine 102: 593–601.

Warrell, D. A., and C. Arnett. 1976. The importance of bites by the Saw Scaled Viper (*Echis carinatus*). Epidemiological studies in Nigeria and a review of world literature. Acta Tropica 33: 307–341.

Warrell, D. A., and L. D. Ormerod. 1976. Snake venom ophtalmia and blindness caused by the Spitting Cobra (*Naja nigricollis*) in Nigeria. American Journal of Tropical Medicine and Hygiene 25: 525–529.

Warrell, D. A., L. D. Ormerod, and N. M. D. Davidson. 1976. Bites by the Night Adder (*Causus maculatus*) and Burrowing Vipers (genus *Atractaspis*) in Nigeria. American Journal of Tropical Medicine and Hygiene 25: 517–24.

Warrell, D. A., N. M. D. Davidson, B. M. Greenwood, L. D. Ormerod, H. M. Pope, B. J. Watkins, and C. R. M. Prentice. 1977. Poisoning by bites of the Saw-scaled or Carpet Viper (*Echis carinatus*) in Nigeria. Quarterly Journal of Medicine 46: 33–62.

Weinstein, S. A., and K. V. Kardong. 1994. Properties of Duvernoy's secretions from opisthoglyphous and aglyphous coulubrid snakes. Toxicon 32: 1161–1185.

Weinstein, S. A., J. J. Schmidt, and L. A. Smith. 1991. Lethal toxins and cross-neutralization of venoms from the African Water Cobras, *Boulengerina annulata annulata* and *Boulengerina christyi*. Toxicon 29: 1315–1327.

Weiser, E., Z. Wollberg, E. Kochva, and S. Y. Lee. 1984. Cardiotoxic effects of the venom of the burrowing asp, *Atractaspis engaddensis* (Atractaspididae, Ophidia). Toxicon 22: 767–774.

Werner, F. 1897. Über Reptilien und Batrachier aus Togoland, Kamerun und Tunis aus dem Kgl. Museum fur Naturkunde in Berlin. Verhandlungen der Zoologisch-Botanischen Gesellschaft in Wein 47: 395–408.

Werner, F. 1899. Über reptilien und batrachier aus Togoland, Kamerun und Deutsch-Neu-Guinea, grosstentheils aus dem Kgl. Museum fur Naturkunde in Berlin. Verhandlungen der Zoologisch–Botanischen Gesellschaft in Wein 49: 132–157.

Werner, F. 1902. Über Westafrikanische Reptilien. Verhandlungen der Zoologisch-Botanischen Gesellschaft in Wein 52: 332–348.

Werner, F. 1907. Ergebnisse der mit Subvention aus der Erbschaft Treitl. Unternommenen zoologischen Forschungsreise Dr. Franz Werner's nach dem ägyptischen Sudan und Nord-Uganda. XII. Reptilien und Amphibien. Sitzungsberichte der Akademie der Wissenschaften in Wien, Mathematisch-Naturwissenschaftliche Klasse 116: 1823–1926.

Werner, F. 1909a. Über neue oder seltene Reptilien des Naturhistorischen Museums in Hamburg. I. Schlangen. Mitteilungen aus dem Naturhistorischen Museum in Hamburg 26: 205–247.

Werner, F. 1909b. Reptilien, Batrachier und Fische von Tripoli und Barka. Zoologische Jahrbücher, Abteilung für Systematik, Ökologie und Geographie der Tiere, Jena 27(6): 595–646.

Werner, F. 1919. Wissenschaftliche Ergebnisse der mit Unterstützung der Kaiserlichen Akademie der Wissenschaften in Wien aus der Erbschaft Treitl von F. Werner unternommenen zoologischen Expedition nach dem Anglo-Aegyptischen Sudan (Kordofan) 1914. IV. Berarbeitung der Fische, Amphibien und Reptilien. Denskschriften der Kaiserlichen Akademie der Wissenschaften in Wien, Mathematische-Naturwissenschaftliche Klasse 96: 437–509.

Werner, F. 1929. Übersicht der Gattungen und Arten der Schlangen aus der Familie Colubridae. III. Teil

(Colubrinae). Zoologische Jahrbücher, Abteilung für Systematik, Ôkologie und Geographie der Tiere, Jena 57: 1–196.

Wildi, S. M., A. Gämperli, G. Beer, and K. Markwalder. 2001. Severe envenoming by a Gaboon Viper (*Bitis gabonica*). Swiss Medical Weekly 131: 54–55.

Witte, G. F. de. 1922. Description d'un ophidien nouveau récolté au Congo par le Dr. Schouteden. Revue de Zoologie et Botanique Africaine 10: 318–319.

Witte, G. F. de. 1930. Un serpent nouveau du Congo Belge (Rhinocalamus rodhaini sp. n.) Revue de Zoologie et Botanique Africaine 19: 1–3.

Witte, G. F. de. 1941. Exploration du Parc National Albert, Mission GF de Witte 1933–1935. Batraciens et Reptiles. Institut des Parcs Nationaux du Congo Belge 1941: 1–261.

Witte, G. F. de. 1952. Amphibiens et reptiles. Exploration Hydrobiologique du Lac Tanganika (1946–1947) 3(3): 1–22.

Witte, G. F. de. 1953. Exploration du Parc National de l'Upemba, Mission G. F. de Witte. Reptiles. Institut des Parcs Nationaux du Congo Belge 6: 1–322.

Witte, G. F. de. 1959. Contribution à la faune herpétologique du Congo Belge. Description de trois serpents nouveaux. Revue de Zoologie et Botanique Africaine 60: 348–351.

Witte, G. F. de. 1962. Genera des serpents du Congo et du Ruanda-Urundi. Annales du Musée Royal de l'Afrique Centrale (Science Zoologique), Tervuren (Série in Octavo) 104: 1–203.

Witte, G. F. de. 1963. The colubrid snake genera Chamaelycus Boulenger and Oophilositum Parker. Copeia 1963: 634–636.

Witte, G. F. de, and R. F. Laurent. 1942. Contribution à la faune herpétologique du Congo Belge. Revue de Zoologie et Botanique Africaine 36: 101–115.

Witte, G. F. de, and R. F. Laurent. 1943. Contribution à la systématique des Boiginae du Congo Belge (Reptilia). Revue de Zoologie et Botanique Africaine 37: 157–189.

Witte, G. F. de, and R. F. Laurent. 1947. Révision d'un groupe de Colubridae Africains, genres Cala-

melaps, Miodon, Aparallactus et formes affines. Mémoires du Musée Royal d'Histoire Naturelle de Belgique 29: 1–134.

Wüster, W., and D. G. Broadley. 2003. A new species of spitting cobra (*Naja*) from north-eastern Africa (Serpentes: Elapidae). Journal of Zoology, London 259: 345–359.

Wüster, W., S. Crookes, I. Ineich, Y. Mané, C. E. Pook, J. F. Trape, and D. G. Broadley. 2007. The phylogeny of cobras inferred from mitochondrial DNA sequences: evolution of venom spitting and the phylogeography of the African spitting cobras (Serpentes: Elapidae: *Naja nigricollis* complex). Molecular Phylogenetics and Evolution 45: 437–453.

Wüster, W., L. Peppin, C. E. Pook, and D. E. Walker. 2008. A nesting of vipers: phylogeny and historical biogeography of the Viperidae (Squamata: Serpentes). Molecular Phylogenetics and Evolution 49: 445–459.

Wynn, A. H., C. J. Cole, and A. L. Gardner. 1987. Apparent triploidy in the unisexual Brahminy Blind Snake, *Ramphotyphlops braminus*. American Museum Novitates 2868: 1–7.

Yaya, G., and A. Danai. 2007. Prise en charge des lésions oculaires dues au crachat de venin d'Elapidae en République Centrafricaine: aspects épidémiologiques et cliniques. Bulletin de la Société de Pathologie Exotique 100: 111–114.

Young, B. A., M. Boetig, and G. Westhoff. 2008. Functional bases of the spatial dispersal of venom during cobra "spitting." Physiological and Biochemical Zoology 82: 80–89.

Zheng, Y., and J. J. Wiens. 2016. Combining phylogenomic and supermatrix approaches, and a time-calibrated phylogeny for squamate reptiles (lizards and snakes) based on 52 genes and 4162 species. Molecular Phylogenetics and Evolution 94(B): 537–547.

Ziegler, T., M. Vences, F. Glaw, and W. Böhme. 1997. Genital morphology and systematics of *Geodipsas* Boulenger 1896 (Reptilia: Serpentes: Colubridae), with description of a new genus. Revue Suisse de Zoologie 104: 95–114.

Index

respiratory, 50, 52, 104, 139; arrest, 50; paralysis, 52, 104; symtoms, 139

Reticulate Centipede-eater. See *Aparallactus lunulatus*

Reticulate Stiletto Snake. See *Atractaspis reticulata*

Rhagerhis, 246, 247

Rhamphiophis, 25, 33, 241, 246, 276, 277, 278; description, 276; diet, 276; extranarial valve, 276; hemipenis, 276; key to species, 276; maxilla, 276; natural history, 276; phylogenic relationship, 33; sexual dimorphism, 276

Rhamphiophis acutus, 276–278, 280; description, 277; distribution, 276, 277; phylogenic relationship, 277–278

Rhamphiophis maradiensis, 246, 276, 278–279; comportment, 278–279; description, 278; distribution, 278; natural history, 278–279

Rhamphiophis oxyrhynchus, 276, 279, 280; description, 279; distribution, 279, 280

Rhamphiophis oxyrhynchus oxyrhynchus, 279

Rhamphiophis oxyrhynchus rostratus, 279

Rhamphiophis rostratus, 246

Rhamphiophis togoensis, 276, 279–280, 281; description, 280; distribution, 279–280, 281

rhinocerase, 81

Rhinoceros Viper. See *Bitis nasicornis*

Rhinoleptus, 57, 58, 59

Rhinoleptus koniagui, 58, 59

Rhombic Egg-eater. See *Dasypeltis scabra*

Rhombic Night Adder. See *Causus rhombeatus*

Rhombic Skaapsteker. See *Psammophylax rhombeatus*

Ringed Water Cobra. See *Naja annulata*

Ringed Wolf Snake. See *Lycophidion semicinctum*

rodent, 48, 49, 60, 61, 63, 64, 69, 73, 81, 95, 142, 190, 198, 239, 247, 264, 358, 362, 374

Rodhain's Purple-glossed Snake. See *Amblyodipsas rodhaini*

Ronier palm, 46

rostral (scale), 8, 9, 17, 18, 19, 20, 22, 23, 24, 25, 26; definition, 8

Rough-scaled Blackbelly Snake. See *Hydraethiops melanogaster*

Rough-scaled Bush Viper. See *Atheris squamigera*

Royal Python. See *Python regius*

rub (rubbing), 95, 98, 242, 243, 244, 246, 250, 270, 276, 308; conspecific rubbing, 250; self-rubbing, 242, 243, 244, 246, 250, 270, 276

Rwanda Bush Snake. See *Philothamnus ruandae*

Sahara Sand Boa. See *Eryx muelleri*

Sahara Sand Viper. See *Cerastes vipera*

Sahel Egg-eater. See *Dasypeltis sahelensis*

Sand Boa. See *Eryx*

Sand Racer. See *Psammophis*

Sand Snake. See *Psammophis*

Sand Snake. See *Psammophis leopardinus*

Sand Viper. See *Cerastes*

Sanzinia, 233

Sanziniinae, 29, 53, 60, 61; phylogenic relationship, 29, 53, 61

São Tomé Bush Snake. See *Philothamnus thomensis*

sarafotoxin, 139

Savanna Lesser File Snake. See *Gonionotophis grantii*

Savanna Vine Snake. See *Thelotornis capensis*

Saw-scaled Viper. See *Echis*

Scaphiophis, 23, 33, 358; comportment, 358; description, 358; diet, 358; hemipenis, 358; maxilla, 358; natural history, 358; phylogenic relationship, 33

Scaphiophis albopunctatus, 358, 359; description, 359; distribution, 359; natural history, 258

Schlegel's Giant Blind Snake. See *Afrotyphlops schlegelii*

Schokari Sand Snake. See *Psammophis schokari*

Schouteden's Mud Snake. See *Helophis schoutedeni*

scincid, 198

Sclerophrys, 88, 379

Scolecophidia, 2, 17, 21, 28, 29, 53; key to families, 54; phylogenic relationship, 28, 29, 53

scorpion, 98, 164

Scutophis, 246

Sea Krait. See *Laticauda*

Semiornate Smooth Snake. See *Meizodon semiornatus*

Semiornate Snake. See *Meizodon semiornatus*

Senegal Garter Snake. See *Elapsoidea trapei*

Serine protease, 51, 81, 96

sexual dimorphism, 95, 157, 167, 237, 242, 243, 249, 250, 254, 256, 270, 296, 305, 309, 318, 321, 323, 326, 347, 356, 364, 376, 381

Shea tree, 46

Shield-nose Snake. See *Aspidelaps*

shock, 81, 99; hypovolemic, 99

Short-snouted Tree Snake. See *Dipsadoboa brevirostris*

Shovel-snout. See *Prosymninae*

Shovel-snout Snake. See *Prosymna*

Shreve's Tree Snake. See *Dipsadoboa shrevei*

shrew, 60, 73, 190, 331

Sibynophiinae, 29, 72, 105, 137, 189, 241, 289, 302; phylogenic relationship, 29, 72, 105, 137, 189, 241, 289, 302

Side–stabbing Snake. See *Atractaspis*

Silurana tropicalis, 288, 379

Skaapsteker. See *Psammophylax*; *Psammophylax rhombeatus*

Thirteen-scaled Bush Snake. See *Philothamnus carinatus*

Thollon's African Water Snake. See *Grayia tholloni*

Thrasops, 24, 30, 33, 45, 369–371; description, 370; hemipenis, 370; key to species, 371; maxilla, 370; natural history, 370; phylogenic relationship, 33

Thrasops aethiopissa, 369, 370, 371, 372; comportment, 370; description, 371; diet, 370; distribution, 371, 372; natural history, 370

Thrasops aethiopissa ituriensis, 371

Thrasops batesii, 369, 371–372; description, 371–372; distribution, 371, 372

Thrasops flavigularis, 371, 372–373; description, 372–373; distribution, 372, 373

Thrasops jacksonii, 369, 371, 373–374; description, 373–374; distribution, 373; sexual dimorphism, 373–374

Thrasops occidentalis, 371, 373, 374, 375; description, 374; distribution, 373, 374

Thread Snake. *See* Leptotyphlopidae

thrombin, 81

thrombin-like enzyme, 96, 329

thrombosis, 96

Tiger Cat Snake. See *Telescopus semiannulatus*

toad, 72, 88, 107, 304, 320, 326, 379

Togo Beaked Snake. See *Rhamphiophis togoensis*

tongue, 8, 365

tortoise, 133

Toxicodryas, 24, 30, 33, 45, 51, 374–375; description, 374–375; diet, 374; envenomation, 51, 375; hemipenis, 375; key to species, 375; natural history, 374; phylogenic relationship, 33, 374; venom, 51

Toxicodryas blandingii, 374, 375–376, 377; description, 375–376; distribution, 375, 376; natural history, 374; venom, 375; sexual dimorphism, 376

Toxicodryas pulverulenta, 375, 376–377, 378, 379; description, 376–377; distribution, 376, 378

toxin, 50, 51, 107, 139; cytotoxin, 119; dendrotoxin, 107; muscarinic toxin, 107; neurotoxin, 50, 82, 112, 375; neurotoxin-β, 82; three-finger toxin, 51; sarafotoxin, 139

Trachylepis, 73, 336

Tree Cobra. See *Pseudohaje*

Tree Snake. See *Dipsadoboa*

Tricheilostoma, 57, 58–59

Tricheilostoma bicolor, 58

Tricheilostoma broadleyi, 58

Tricheilostoma greenwelli, 58

Tricheilostoma sundewalli, 58

Trilepida, 58

Tropidophiidae, 29, 53, 61

True Cobra. See *Naja*

True Sea Snake. *See* Hydrophiinae

turtle, 296

Twig Snake. See *Thelotornis; Thelotornis capensis*

Two-headed Snake. See *Chilorhinophis; Chilorhinophis gerardi*

Two-striped Night Adder. See *Causus bilineatus*

Typhlopidae, 17, 18, 20, 21, 28, 29, 53–56, 61, 233; diet, 54; key to genera, 54; phylogenic relationship, 28, 29, 53, 61

Typhlops, 55, 155, 157, 159, 164, 172

Underwood's Tree Snake. See *Dipsadoboa underwoodi*

Ungaliophiinae, 29, 53, 60, 61; phylogenic relationship, 29, 53, 61

upper labial or "supralabial" (scale), 1, 9, 10, 12, 13, 17, 20, 21, 22, 23, 26, 27; definition, 9

Uraeus, 118, 130, 131, 132

Uropeltidae, 29, 53, 61; phylogenic relationship, 29, 53, 61

Varanus niloticus, 69

Variable Cat Snake. See *Telescopus variegatus*

Variable Stiletto Snake. See *Atractaspis irregularis*

Variegated Marsh Snake. See *Natriciteres variegata*

venom, 35, 49, 50, 51, 52, 72, 73, 81, 89, 90, 91, 92, 93, 94, 96, 98, 99, 104, 105, 107, 112, 118, 119, 137, 139, 140, 142, 143, 145, 146, 147, 148, 150, 152, 154, 160, 163, 166, 172, 185, 329, 330, 361, 365, 366, 375; composition, 50, 51, 73, 81, 96, 98, 107, 112, 119, 139, 329, 366, 375; delivery, 35, 104, 119, 137, 140, 166, 172, 185, 361; gland, 36, 72, 89, 90, 91, 92, 93, 94, 99, 104, 119, 137, 139, 142, 143, 145, 146, 147, 148, 150, 152, 154

ventral (scale), 1–3, 4, 7, 8, 13, 14, 18, 21, 22, 23, 24, 26; definition, 1–2

vertebra (bone), 30, 31, 34, 189, 233, 238, 241, 249, 280, 302, 308, 318

vertebral row, 6, 7, 8, 23; definition, 7

Vine Snake. See *Thelotornis*

viperid. *See* Viperidae

Viperidae, 16, 17, 19, 28, 29, 30, 31, 36, 50, 52, 72–103, 105, 136, 137, 138, 139, 241, 289, 302; envenomation, 50, 52; maxilla, 136, 138; natural history, 72–73; phylogenic relationship, 28, 29, 30, 72, 105, 137, 189, 241, 289, 302; skull, 138

Waprin-like proteins, 51

Water Cobra. See *Boulengerina; Naja annulata; Naja christyi*

Water Snake. See *Lycodonomorphus*